Disaster Risk Reduction

Methods, Approaches and Practices

Series Editor

Rajib Shaw, Keio University, Shonan Fujisawa Campus, Fujisawa, Japan

About the Series

Disaster risk reduction is a process that leads to the safety of communities and nations. After the 2005 World Conference on Disaster Reduction, held in Kobe, Japan, the Hyogo Framework for Action (HFA) was adopted as a framework for risk reduction. The academic research and higher education in disaster risk reduction has made, and continues to make, a gradual shift from pure basic research to applied, implementation-oriented research. More emphasis is being given to multi-stakeholder collaboration and multi-disciplinary research. Emerging university networks in Asia, Europe, Africa, and the Americas have urged process-oriented research in the disaster risk reduction field. With this in mind, this new series will promote the output of action research on disaster risk reduction, which will be useful for a wide range of stakeholders including academicians, professionals, practitioners, and students and researchers in related fields. The series will focus on emerging needs in the risk reduction field, starting from climate change adaptation, urban ecosystem, coastal risk reduction, education for sustainable development, community-based practices, risk communication, and human security, among other areas. Through academic review, this series will encourage young researchers and practitioners to analyze field practices and link them to theory and policies with logic, data, and evidence. In this way, the series will emphasize evidence-based risk reduction methods, approaches, and practices.

More information about this series at http://www.springer.com/series/11575

Juan M. Pulhin · Makoto Inoue · Rajib Shaw
Editors

Climate Change, Disaster Risks, and Human Security

Asian Experience and Perspectives

Springer

Editors
Juan M. Pulhin
Department of Social Forestry and Forest Governance
College of Forestry & Natural Resources (CFNR)
University of the Philippines Los Baños
Laguna, Philippines

Makoto Inoue
Faculty of Human Sciences
Waseda University
Saitama, Japan

Rajib Shaw
Graduate School of Media and Governance
Keio University
Kanagawa, Japan

ISSN 2196-4106 ISSN 2196-4114 (electronic)
Disaster Risk Reduction
ISBN 978-981-15-8851-8 ISBN 978-981-15-8852-5 (eBook)
https://doi.org/10.1007/978-981-15-8852-5

This Springer imprint is published by the registered company Springer Nature Singapore Pte Ltd.
The registered company address is: 152 Beach Road, #21-01/04 Gateway East, Singapore 189721, Singapore

Foreword

Regarded as the world's most vulnerable region to natural hazards, Asia has been the focus of various studies on the impacts of climate change and disaster risks that have continuously threatened human security. And yet, in the face of adversity, it has accumulated a wealth of valuable experience and innovative responses that can serve as foundation for other countries and regions to pursue a climate-resilient and human-centered development. Given the increasing frequency and intensity of climate-induced and other forms of disasters, the need to delve deeper into the linkages among climate change, disaster risk and human security has never been more urgent.

This book draws attention to Asia's experience with climate change and disaster risks and how they threaten different elements of human security, highlighting recent disasters that have greatly affected the economy and the environment in the region. Each chapter is written by esteemed experts and budding scholars whose local and regional expertise and diverse outlooks underline the importance of the human security lens to fully understand climate change and disaster risks. Case studies illustrate the primacy of emphasizing resilience in Asia, while also encouraging other countries and regions to heighten their awareness and preparedness for worst-case scenarios. The recommendations aim to guide policies and programs that reduce vulnerability and promote people's right to a free and dignified life.

I would like to commend the editors for their rich insights on this timely and important topic. This pioneering book will be a valuable reference for academics, researchers, and students who seek new perspectives and a more extensive understanding of the intricate relationship between climate change, disaster risks, and human security. It provides a sound empirical basis for policy and decision-makers as well as development practitioners to strengthen the nexus between science, policy, and practice towards improving our response to the mounting challenge of human insecurity in the age of climate change and increasing disasters in Asia and around the world.

Ambassador Knut Vollebaek
Chair, United Nations Advisory Board on Human Security

Preface

The genesis of this book dates back in the early 2010s during our preparation of the Intergovernmental Panel on Climate Change Fifth Assessment Report (IPCC-AR5) where I had the opportunity to serve as one of the Coordinating Lead Authors (CLAs) of the human security chapter of the Working Group II contribution to the report. As we systematically review the then available growing volume of materials on climate change and human security, it became very apparent to me, coming from the Asia region myself, that there is poverty of literature in the Asian context that explores the complex linkages between climate change and human security. Even much less explored is the interactions between climate change and disaster risks and how these impacts on or threaten human security despite the strong historical evidence showing that Asia is the most vulnerable region of the world in terms of climate-induced and other forms of disasters. My recent experience on building the capacity of the Philippine local government units on climate change and disaster risk planning also adds impetus to investigate further the theme of this book.

My two-month visiting professorship at the University of Tokyo in 2016 enabled me to develop further the concept of the book with four major thematic areas in collaboration with my two co-editors. It likewise allowed us to invite recognized experts and budding scientists with interest on the subject matter to contribute to this volume. The unique collaboration of seasoned and emerging scientists adds richness to the content and analysis of the different chapters even as it contributes to the capacity building effort of young scientists in Asia. Such capacity development efforts should receive top priority if we are to pursue climate-resilient and human-secured Asian development pathways in the context of worsening climate and intensifying disasters—a key strategy that is mentioned in the concluding chapter of this book.

The major push toward the completion of the book happened during my 20-day stint as Visiting Scholar at the Waseda University in 2018 where I worked closely with my host professor and co-editor, Dr. Makoto Inoue in reviewing and providing critical comments to most of the draft chapters. The book also benefitted from the comments of some anonymous reviewers who shared their valuable time and expertise to refine the contents of the different chapters. Insights from the discussions of the ongoing preparation of the IPCC Sixth Assessment Report also enriched the introductory and concluding chapters with my co-editor, Prof. Rajib Shaw and myself serving as CLA and Lead Author (LA), respectively, for the Asia regional chapter.

The now more than five-month lockdown in some parts of the Philippines brought about by the COVID-19 pandemic has contributed to the delay in the publication of this book. The pandemic has nonetheless affirmed the value of looking at climate change and disaster risks from the human security perspective considering the multiple-stressors that Asia and the rest of the world is currently facing that undermines human's capacity to achieve "freedom from fear and want" and to live with dignity. Drawing from the Asian experience and perspectives, we hope that this book will serve as valuable materials for academics interested on this subject matter but also to policy and decision makers and development practitioners who want to contribute towards advancing climate-resilient, human-secured and sustainable development pathways not only in Asia but also for the rest of the world.

We want to thank the following organizations and individuals who contributed in one way or another towards the publication of this book:

- The University of the Philippines Los Baños for providing me the flexibility and freedom to devote a significant portion of my official time in leading the publication of this book;
- The University of Tokyo and Waseda University for hosting me as Visiting Professor and Visiting Scholar in 2016 and 2018, respectively, that enabled me to work with my two co-editors, to conceptualize and review the chapters of the book. Further, Waseda University provided financial assistance to fund the language editing of the book;
- My colleagues at the human security chapter of the IPCC-AR5 for the intensive assessment of literature and tireless intellectual discussions on climate change and human security that enthused me to pursue this book project;
- All the chapter authors of the book for sharing their research findings and knowledge on the subject as well as for their patience in the series of revisions to address the different comments made by the reviewers and the Springer staff;
- Ms. Teresita Rola and Ms. Perlyn Pulhin-Yoshida for the valuable comments, proof-reading and editing the draft chapters that contribute to the improvement of the book quality;
- Ms. Millicent Joyce Q. Pangilinan for all the hard work of helping secure the copyright permissions, formatting of all the chapters, and facilitating compliance to all the other manuscript submission checklist of the Springer Nature; and
- Mr. Dinesh Natarajan of Springer Nature for his effective guidance and facilitation of the publication process.

LosBaños, Philippines Juan M. Pulhin
August 2020

Contents

Editors and Contributors

About the Editors

Dr. Juan M. Pulhin is full Professor and former Dean of the College of Forestry and Natural Resources, University of the Philippines Los Baños, Philippines. He was Lead Author of the 2007 Fourth Assessment Report of the Intergovernmental Panel on Climate Change (IPCC), Coordinating Lead Author of the Human Security Chapter of the 2014 IPCC Fifth Assessment Report and currently Lead Author of the Asia chapter of the IPCC Sixth Assessment Report. He has been a four-time a visiting professor at The University of Tokyo in 2007, 2010, 2016 and 2017 and a visiting scholar at Waseda University, Tokyo in 2018. He has written and edited 11 books and published more than 100 technical papers and book chapters on natural resources policy and governance, and on the human dimension of climate change and disaster risks. He is a fellow, invited expert, and council member of several international organizations and a member of the National Academy of Science and Technology (NAST), Philippines.

Dr. Makoto Inoue is Professor at the Faculty of Human Sciences, Waseda University, Japan and former Professor at The University of Tokyo. He serves as Dean of Graduate School of Human Sciences, Waseda University since September 2020. His specialty is environmental sociology, forest governance, and Southeast Asian studies. He has written and edited/co-edited 34 books, and published more than 140 original articles in peer-reviewed international and national journals on the social aspects of forestry and the environment. He serves as President of The Japanese Association for Environmental Sociology (June 2019–present), Vice President of Japanese Society for Current Anthropology (January 2020–present), Vice President of Society for Environmental Economics and Policy Studies (April 2018–March 2020), Councilor of The International Association for the Study of the Commons (January 2019-present), a board member of various academic societies in Japan, and an editorial board member of several international journals.

Dr. Rajib Shaw is Professor at the Keio University in Japan and former Executive Director of the Integrated Research on Disaster Risk (IRDR) programme. He is also Senior Fellow of Institute of Global Environmental Strategies (IGES) Japan, and Chairperson of SEEDS Asia, and CWS Japan, two Japanese NGOs. Previously, he was Professor at the Graduate School of Global Environmental Studies, Kyoto University, Japan. His expertise includes community-based disaster risk management, climate change adaptation, urban risk management, and disaster and environmental education. He is the editor of a book series on disaster risk reduction, published by Springer, Editor-in-Chief of the journal Progress in Disaster Science. He severs as Chair of United Nations Science Technology Advisory Group for disaster risk reduction, and is Co-chair of the UNDRR's Asia-Pacific Science Technology Academic Advisory Group (ASTAAG). Prof. Shaw has published more than 45 books and more than 300 academic papers and book chapters.

List of Contributors

Ma Louiella Rose O. Catudio obtained her bachelor's degree in Forestry from the University of the Philippines Los Baños (UPLB). She is currently a graduate student of MS Forestry specializing in silviculture and forest influences in the same university, and currently serves as a graduate assistant for a forestry project on the assessment of the Community-based Forest Management (CBFM) program in the Philippines. Prior to this, she was involved in the studies of water resource management and community and program development of the Department of Environment and Natural Resources (DENR). Her researches delve into the study of social forestry, forest resource management, agroforestry, and climate change.

Mirza Fahmi is a lecturer at Civil Engineering Department of Politeknik Negeri Lhokseumawe, Indonesia. He obtained his master's degree from Universitas Syiah Kuala in 2017, majoring in Water Resource Management. Recently, he has been developing his research interests on tsunami mitigation, spatial analysis, and coastal zone management. Besides his academic activities, he is also actively involved in promoting coastal ecology and conservation through community-based activities.

Kumiko Fujita is a researcher at the International Consortium on Landslides. She has more than 20 years of experience in education, research, and development for natural disaster risk reduction. Her expertise is in applying Japanese technologies of mitigating and preventing natural disasters, such as landslides, debris flow, erosion, and flood to Asian countries. She had coordinated training courses for foreign engineers provided by the Japan International Cooperation Agency (JICA). She also coordinated science programs at the United Nations Educational Scientific and Cultural Organization (UNESCO) in Indonesia. Recently, she works as a researcher for a project titled "Development of Early Warning Technology of Rain-Induced Rapid and

Long-Travelling Landslides". She has a doctorate degree in Global Environmental Studies from Kyoto University.

Ana Veronica S. Gabriel started her career in research at the OML Center through projects on climate change and agroforestry, as well as loss and damage linked with climate change adaptation and disaster risk reduction. For the latter, she had the opportunity to showcase the findings at various international and national scientific meetings. She also served as Associate Managing Editor of the Climate, Disaster and Development Journal before becoming Project Officer at the Forest Foundation Philippines, a non-profit organization that provides grants to organizations that empower the people to protect the forests.

Dr. Dixon T. Gevaña is a faculty member of the Department of Social Forestry and Forest Governance, College of Forestry and Natural Resources, University of the Philippines Los Baños (UPLB). He teaches courses related to forest sustainability and conducts researches on coastal forest management and climate change mitigation. Most of his publications delved on the role of mangroves in addressing climate change. He obtained both his undergraduate (BS Forestry) and master's (MS Community Development) degrees from University of the Philippines, and doctorate degree in Forest Environmental Science from Seoul National University, South Korea.

Dr. Oscar A. Gómez (Ph.D., Tohoku University) is Assistant Professor at the College of Asia Pacific Studies, Ritsumeikan Asia Pacific University. His main interest is in global governance and human security practice, with special emphasis on humanitarian crises. He was Research Fellow at Japan International Cooperation Agency (JICA) Research Institute for almost 5 years and, before that, Japan Society for the Promotion of Science Postdoctoral Fellow at the Graduate School of Global Studies, Doshisha University. He co-wrote background papers for UNDP's Global Human Development Reports 2014, 2016, and 2020. He is a member of the Scientific Committee of the Humanitarian Encyclopedia project. Recent publications include co-edited volumes on the humanitarian-development nexus (Routledge) and human security norms in East Asia (Palgrave).

Dr. Brendan Howe (Ph.D., TCD) is Professor of International Relations at the Graduate School of International Studies, Ewha Womans University. He has held fellowships at the Freie Universität Berlin, University of Sydney, East-West Center, Georgetown University, UNIMAS, and Beijing Foreign Studies University. His research focuses on governance, human security, and post-conflict development in Asia. He has authored/edited more than 90 publications including Regional Cooperation for Peace and Development: Japan and South Korea in Southeast Asia (Routledge, 2018); National Security, State-Centricity, and Governance in East Asia (Palgrave, 2017); Peacekeeping and the Asia-Pacific (Brill, 2016); Democratic Governance in Northeast Asia (Palgrave, 2015); Post-Conflict Development in East Asia (Ashgate, 2014); and The Protection and Promotion of Human Security in East Asia (Palgrave, 2013).

Mikio Ishiwatari is Senior Advisor on Disaster Management and Water Resources Management at Japan International Cooperation Agency, and Visiting Professor at the Graduate School of Frontier Sciences, The University of Tokyo. He has been engaged in projects and research works on disaster risk reduction (DRR), climate change adaptation, and water. He led the formulation of the Japanese assistance policies of climate change adaptation and community-based disaster management. He worked at the World Bank as Senior Disaster Risk Management Specialist, and produced the "Learning from Megadisaster: Lessons from the Great East Japan Earthquake." He worked at various positions at the Ministry of Land, Infrastructure, and Transport, Japan for 17 years.

Takako Izumi is Associate Professor at the International Research Institute of Disaster Science (IRIDeS), Tohoku University, Japan. She also serves as Program Director of the Multi Hazards Program under the Association of Pacific Rim Universities (APRU), which comprises 48 universities and academic institutes in the Pacific Rim. Her research interests include international and regional frameworks/strategies for disaster risk reduction, international humanitarian assistance, and DRR initiatives at the local and community levels. In May 2015, she has been appointed as a member of the UNISDR Asian Academia, Scientific and Research Advisory Group (ASTAAG) to support the effort of governments and stakeholders in the implementation of the Sendai Framework for DRR. She holds a Ph.D. in Global Environmental Study from Kyoto University, Japan.

Dr. Alfredo Mahar Francisco A. Lagmay is Academician of the National Academy of Science and Technology (NAST) and Professor at the National Institute of Geological Sciences, University of the Philippines (UP). He is Executive Director of the UP Resilience Institute and Director of the UP NOAH Center. He received his bachelor's and master's degrees from the University of the Philippines and holds a Ph.D. in Earth Sciences from the University of Cambridge. In 2013, he was awarded the Outstanding Professional of the Year Award in the field of Geology. Two years after, the European Geosciences Union (EGU) recognized his outstanding achievements in interdisciplinary natural-hazard research and natural-disaster engagement in the Philippines with a Plinius Medal.

Dr. Rodel D. Lasco has more than 38 years of experience in natural resources and environmental research, conservation, education, and development at the national and international level. His work has focused on issues related to natural resources conservation, climate change, and land degradation. Since 2000, he is an author of several reports by the IPCC, the 2007 co-winner of the Nobel Peace Prize. He is also a member of the National Academy of Science and Technology (NAST) in the Philippines. He is the Philippines Coordinator of the World Agroforestry Centre (ICRAF). Concurrently, he is Executive Director of the OML Center, a private foundation whose mission is to promote research on climate adaptation and disaster risk reduction.

Jose Ma. Luis Montesclaros is Associate Research Fellow with the Centre for Non-Traditional Security Studies (NTS Centre), S. Rajaratnam School of International Studies (RSIS), Nanyang Technological University, Singapore (NTU). He conducts policy analysis with dynamic models of food security, economics, agriculture, and climate change in ASEAN, and has provided inputs to Singapore's Inter-Ministry Committee on Food Security (IMCFS). As consultant, he co-authored the World Bank's "2014 ASEAN Equitable Development Monitor Report". Holding a master's degree in Public Policy from the Lee Kuan Yew School of Public Policy (LKYSPP), National University of Singapore (ASEAN Scholar), and a BS Economics degree from the University of the Philippines Diliman, he was among LKYSPP's "Leaders of Tomorrow", 2014 St. Gallen Leadership Awards (Switzerland).

Dr. Agus Nugroho is a lecturer at the Agribusiness Department, Universitas Syiah Kuala in Banda Aceh, Indonesia. He is also Coordinator for Socio Economic of Disaster Research Cluster at Tsunami and Disaster Mitigation Research Center (TDMRC) at Universitas Syiah Kuala. He obtained his doctoral degree from Kyoto University, Japan in May 2016, majoring in Natural Resource Economics. His research interests are on social and economic aspects of disaster, natural resource risk management, and disaster risk financing and insurance (DRFI). He was also involved as a panelist in developing of DFRI Strategies in Indonesia established by the Ministry of Finance, Republic of Indonesia.

Yosuke Okita is pursuing his Ph.D. degree at the Graduate School of Media and Governance, Keio University in Japan while working for Japan International Cooperation Agency (JICA)'s Indonesia office in Jakarta. He has more than 10 years of experience working in international disaster and emergency management, including some emergency missions as a member of the Japan Disaster Relief team and the United Nations Disaster Assessment and Coordination (UNDAC) team. From 2014 to 2017, he worked at the Office for the Coordination of Humanitarian Affairs (OCHA) in Geneva as Asian regional focal point for International Search and Rescue Advisory Group (INSARAG) and UNDAC. He received an M.A. in International Relations and a Master of Diplomacy from The Australian National University.

Millicent Joyce Q. Pangilinan is currently a researcher for a forestry project at the University of the Philippines Los Baños. She has 10 years of experience in research and project management in both the private and development sectors. She started her career in Japan doing economic data and market analysis intended for mergers and acquisitions projects in Asia and in the USA. After 4 years, she transitioned into the development field, where she was engaged in research projects on agriculture, climate change, and forestry in Southeast Asia and the Philippines. She earned her master's degree in International Public Policy (Economics track) at Osaka University, Japan and bachelor's double degree in Economics and Finance at De La Salle University Manila, Philippines.

Gulsan Ara Parvin is a researcher of Laboratory of Rural Planning and Sustainable Rural Development, Graduate School of Global Environmental Studies, Kyoto University, Japan. She was Associate Professor of Urban and Regional Planning Department of Bangladesh University of Engineering and Technology. She earned her Ph.D. from the Urban Engineering Department of the University of Tokyo. Her expertise is in disaster risk reduction and community resilience. She has also conducted different researches related to women empowerment and microfinance in Bangladesh. She has published about 50 research papers in different journals and books. She has research and working experiences in Bangladesh, Japan, Malaysia, Thailand, and USA.

Dr. Rose Jane J. Peras is Professor and current Chair of the Department of Social Forestry and Forest Governance, College of Forestry and Natural Resources, University of the Philippines Los Baños. She has more than 20 years of experience in research and development and 15 years in forestry education and public service. Her research involvements in forestry, community-based forest management, integrated watershed management plan development, human dimension of climate change, disaster risk reduction/climate change adaptation, and forest governance led her to author/co-author 29 technical publications. She earned her Bachelor of Science in Forestry and Master of Science in Forestry from the University of the Philippines Los Baños and Ph.D. in Forest Governance from The University of Tokyo.

Dr. Ross Michael Pink teaches Sustainability, Politics, and International Law at Kwantlen Polytechnic University in Canada. Dr. Pink is the author of two recent book, *Water Rights in Southeast Asia and India* and *The Climate Change Crisis: Solutions and Adaption for a Planet in Peril*, both published by Palgrave Macmillan. Dr. Pink is also the Co-Founder of Global Water Rights (www.globalwaterrights.com).

Dr. S. V. R. K. Prabhakar works on mainstreaming climate change adaptation and disaster risk reduction into development, adaptation metrics, adaptive policies, risk insurance, regional adaptation planning for transboundary climate risks, loss and damage, vulnerability assessment, and training needs assessment in South and Southeast Asian countries. He has been Coordinating Lead Author for the UNEP Geo-6 Asia report, Lead Author for the HIMAP Report, and Contributing Author to the Fifth and Sixth Assessment Reports of the IPCC Working Group II. Prabhakar obtained Ph.D. in Agronomy from IARI, India. He possesses more than 17 years of experience in participatory research and development with international and national research and development organizations including IGES, Kyoto University, UNDP, NIDM, CIMMYT-RWC, IARI, and ICRISAT.

Perlyn Pulhin-Yoshida is a freelance environmental science researcher-writer and editor. She holds a master's degree in Environmental Science and a bachelor's degree in Development Communication. She has more than 18 years of work experience in the areas of environmental research, science-policy initiatives, knowledge management, and communications and development in international and national research organizations including OML Center, ICRAF, and APN. She has also served as Managing Editor of the Climate, Disaster and Development Journal. Her research and publication interests include climate change, disaster risk management, resilience, and loss and damage. She has co-authored a number of scholarly publications and presented in local and international conferences. She recently moved to Japan and is currently working as a full-time English teacher.

Mr. Mark Anthony M. Ramirez is Executive Director of the Resources, Environment and Economics Center for Studies (REECS) based in the Philippines. He has more than 20 years of experience in the field of watershed management, community forestry, natural resources management, forest land use planning, climate change and DRR and has co-written articles and publications in these fields. A licensed forester, he has a master's degree in Social Development from the Ateneo de Manila University.

Clarissa D. Ruzol is a student of MRes/PhD Anthropology at the London School of Economics and Political Science. They worked as Assistant Professor of Anthropology at the Department of Social Forestry and Forest Governance, University of the Philippines Los Baños from 2017 to 2020. They were a graduate fellow of the Asia Research Institute in the National University of Singapore in 2015 and had received grants from the University of Montana, the Southeast Asian Regional Center for Graduate Study and Research in Agriculture, the Swedish Institute, and the UNICEF in the Kyrgyz Republic to present their works in conferences. They have done research on topics such as cultural risk theory, social network analysis, local knowledge, environmental policy, and social impact assessment. They also write about structural violence, power, scientific knowledge, and infrastructure development in the Philippines.

Dr. Lorena L. Sabino is Assistant Professor with the Department of Social Forestry and Forest Governance, College of Forestry and Natural Resources, University of the Philippines Los Baños (UPLB). She obtained her doctorate degree in Environmental Science with minors in Community Development and Social Forestry at UPLB. In her dissertation, she used a human security perspective in assessing how the vulnerability to climate variability and change is linked to household livelihood security. She is currently involved in climate change and disaster risk-related projects, sustainable mangrove rehabilitation, vulnerability, and human security research projects.

Joy Santiago is a licensed environmental planner and a lecturer specializing in geographic information system (GIS), vulnerability and risk assessment, and land use planning. She was part of the Nationwide Operational Assessment of Hazards which efficiently mapped hydrometeorological hazards of the entire Philippines. She was awarded the 2016 outstanding student poster and PICO awards by the European Geosciences Union. She was also part of several projects which focuses on hazard, climate and disaster risk assessment, formulation of risk-sensitive comprehensive local development plans, and capacity building in collaboration with the Asian Development Bank, national and local government agencies, and non-government organizations. She also presented in numerous local and international conferences and co-authored internationally peer-reviewed scientific publications.

Dr. Linda Anne Stevenson is Head of Communication and Scientific Affairs at the Asia-Pacific Network for Global Change Research (APN), which is an intergovernmental organization with 22 member governments. With 17 years of experience in APN, she oversees over 70 national and regional projects annually across a broad range of disciplines, and works closely with researchers and decision-makers in Asia and the Pacific. She has significant experience in scientific capacity building that includes developing capacity of early-career researchers; plans dialogues that cross the science-policy; and engages with UNFCCC, IPCC, and IPBES. She has over eight years of lecturing experience in Japanese universities on "Global Environmental Systems." She holds a doctorate degree in Applied Chemistry from the University of Strathclyde, UK.

Dr. Syamsidik is a lecturer at the Civil Engineering Department, Universitas Syiah Kuala in Banda Aceh, Indonesia. He is also Coordinator for Tsunami Research Cluster at Tsunami and Disaster Mitigation Research Center (TDMRC) at Universitas Syiah Kuala. He obtained his doctoral degree from Toyohashi University of Technology in Japan in March 2009, majoring in Coastal Engineering. Since then, he has been developing his research interests on tsunami mitigation, tsunami numerical simulations, disaster preparedness, and sustainable disaster recovery. He was a visiting scholar at the International Research Institute of Disaster Science (IRIDeS) in Tohoku University, Japan in 2018 and in 2019. He also served as a visiting scholar at the University of Pittsburgh and at Georgia Technology Institute in the USA in 2017 and 2018, respectively.

Paul S. Teng is Adjunct Senior Fellow (Food Security) at the NTS Centre, RSIS, NTU, and concurrently Managing Director of the National Institute of Education (NIE) International Pte Ltd. He has over thirty years of experience in developing countries from positions at the World Fish Centre, the International Rice Research Institute, and Monsanto Company. Paul has conducted research on new agri-technologies, science-based entrepreneurship, food security and sustainable development, with recent focus on urban food security and agtech. Professor Teng has been recognized with an Honorary Doctor of Science by Murdoch University, Australia; the Eriksson Prize in Plant Pathology and as a Fellow of the American Phytopathological Society; and The World Academy of Sciences. He has published over 250 technical papers.

Abbreviations and Acronyms

AASA	Association of Academies of Science in Asia
AAWS	Adopt-A-Wildlife Species
ADB	Asian Development Bank
ADESA	Average Dietary Energy Supply Adequacy
ADRC	Asian Disaster Reduction Center
ADRRN	Asian Disaster Reduction and Response Network
ADSDPP	Ancestral Domains Sustainable Development and Protection Plan
AMCDRR	Asian Ministerial Conference on Disaster Risk Reduction
APN	Asia-Pacific Network for Global Change Research
AR5	IPCC 5th Assessment Report
ASEAN	Association of Southeast Asian Nations
AUSAID	Australian Agency for International Development
AWG-SF	ASEAN Working Group on Social Forestry
AXO	Abandoned Explosive Ordnance
BAPPENAS	Indonesian Development and Planning Agency
BAU	Business-As-Usual
BCCRF	Bangladesh Climate Change Resilience Fund
BCCSAP	Bangladesh Climate Change Strategy and Action Plan
BFSC	Bangladesh Food Security Cluster
BHC	Barangay Health Centers
BIOHAZ	Panel on Biological Hazards
BMGP	Bacon Manito Geothermal Project
BMKG	Geophysics, Climate, and Meteorological Agency
BNPB	Badan Nasional Penanggulangan Bencana/National Disaster Management Agency (Indonesia)
BPBA	Aceh Disaster Management Agency (Indonesia)
BREACH	Building Resilient and Economically Adept Communities and Households
BRR Aceh Nias	Tsunami Rehabilitation and Reconstruction Agecy for Aceh and Nias
BSR	The Business of a Better World

BWDB	Bangladesh Water Development Board
CADT	Certificate of Ancestral Domain Title
CAIT	Climate Analysis Indicators Tool
CARE	Cooperative for Assistance and Relief Everywhere
CAREERS	College Admission Review and Readiness
CBA	Community-based Adaptation
CBACC	Community-based Adaptation to Climate Change
CBDRM	Community-based Disaster Risk Management
CBDRR	Community-based Disaster Risk Reduction
CBFM	Community-based Forest Management
CBMM	Community-based Mangrove Management
CBNRM	Community-based Natural Resource Management
CC	Climate Change
CCA	Climate Change Adaptation
CCA-DRR	Climate Change Adaptation-Disaster Risk Reduction
CCC	Climate Change Commission
CCJD	Center for Community Journalism and Development
CCTF	Climate Change Trust Fund
CCVA	Climate Change Vulnerability Analysis
CDA	Cooperative Development Authority
CDO	Cagayan de Oro
CEO	Chief Executive Officer
CEP	Coastal Embankment Project
CERP	Coastal Embankment Rehabilitation Program
CFM	Community Forest Management
CFRI	Climate-Fragility Risk Index
CFRs	Climate-Fragility Risks
CFUGs	Community Forest User Groups
CFW	Cash-For-Work
CHED	Commission on Higher Education
CHS	Commission on Human Security
CIA	Central Intelligence Agency
CIMMYT-RWC	International Maize and Wheat Improvement Center-Rice-Wheat Consortium
CLUP	Comprehensive Land Use Plan
CO_2	Carbon Dioxide
COP	Conference of Parties
COVID-19	Coronavirus Disease
CRED	Center for Research on the Epidemiology of Disasters
CRI	Climate Risk Index
CSCAND	Collective Strengthening of Community Awareness for Natural Disasters
CSO	Civil Society Organization
CSR	Corporate Social Responsibility
CTV	Canterbury Television

CWS	Christian World Service
DAO	Department of Environment and Natural Resources Administrative Order (Philippines)
DCED	Donor Committee for Enterprise Development
DENR	Department of Environment and Natural Resources (Philippines)
DepEd	Department of Education (Philippines)
DER	Disaster and Emergency Response
DFID	Department for International Development (United Kingdom)
DHA	Department of Humanitarian Affairs
DJF	December, January, February
DMB	Disaster Management Bureau (Bangladesh)
DMC	Disaster Management Center (Sri Lanka)
DOH	Department of Health (Philippines)
DOST	Department of Science and Technology (Philippines)
DPNeT	Disaster Preparedness Network (Nepal)
DPRK	Democratic People's Republic of Korea
DREAM	Disaster Risk and Exposure Assessment for Mitigation
DRM	Disaster Risk Management
DRR	Disaster Risk Reduction
DRRM	Disaster Risk Reduction Management
DRVRM	Disaster Risk and Vulnerability Reduction and Management
DS	Divisional Secretary
DTM	Digital Terrain Model
EAP	East Asia and the Pacific
ECA	Ecologically Critical Area
EDC	Energy Development Corporation
EFSA	European Food Safety Authority
EIA	Environmental Impact Assessment
EM-DAT	Emergency Events Database
EMR	Eastern Mediterranean Region
ENSO	El Niño/Southern Oscillation
EOC	Emergency Operations Center
EP-NDRCC	Expert Panel of National Disaster Reduction Committee of China
ERW	Explosive Remnants of War
ESCAP	Economic and Social Commission for Asia and the Pacific
ESIA	Environmental and Social Impact Assessments
EU	European Union
EWS	Early Warning System
FAO	Food and Agriculture Organization of the United Nations
FAP	Flood Action Plan
FCD/I	Flood Control, Drainage and/or Irrigation
FCPF	Forest Carbon Partnership Facility
FEMA	Federal Emergency Management Agency

FFWC	Flood Forecasting and Warning Centre
FLO-2D GDS PRO	FLO-2D Grid Developer System Professional
FMB	Forestry Management Bureau (Philippines)
G7	Group of 7 Countries
GAM	Free Aceh Movement
GBM River Basin	Ganges, Brahmaputra and Meghna Basins
GDH	Growing Degree Hour
GDP	Gross Domestic Product
GED	General Economics Division
GEF	Global Environment Facility
GEJE	Great East Japan Earthquake
GFDRR	Global Facility for Disaster Reduction and Recovery
GGGI	Global Green Growth Institute
GHG	Greenhouse Gases
GICHD	Geneva International Center for Humanitarian Demining
GLOF	Glacial Lake Outburst Flood
GM	Genetically modified
GNDR	Global Network of Civil Society Organization for Disaster Reduction
GoB	Government of the People's Republic of Bangladesh
GRiSP	Global Rice Science Partnership
HDR	Human Development Report
HELEn	Health, Education, Livelihood, and Environment
HFA	Hyogo Framework for Action
HIF	Humanitarian Innovation Fund
HLURB	Housing and Land Use Regulatory Board
HOA	Hotel Owners Association
IARI	Indian Agricultural Research Institute
ICC	International Criminal Court
ICCCAD	International Centre for Climate Change and Development
ICRAF	World Agroforestry Centre
ICRISAT	International Crops Research Institute for the Semi-Arid Tropics
ICT	Information and Communication Technology
IDMC	Internal Displacement Monitoring Center
IDR	Indonesian Rupiah
IEC	Information, Education and Communication
IEDM	International Environment and Disaster Management
IFAD	International Fund for Agricultural Development
IFPRI	International Food Policy Research Institute
IFRC-RCS	International Federation of Red Cross and Red Crescent Societies
IFSAR	Interferometric Synthetic Aperture Radar
IGES	Institute for Global Environmental Strategies
IGO	Intergovernmental Organizations

IIED	International Institute for Environment and Development
ILO	International Labor Organization
InaTEWS	Indian Ocean Tsunami Early Warning System
INGO	International Non-Government Organization
INSARAG	International Search and Rescue Advisory Group
IOT	Internet of Things
IPCC	Intergovernmental Panel on Climate Change
IRP	Income Recovery Program
IRRI	International Rice Research Institute
ISAAA	International Service for the Acquisition of Agri-biotech Applications
IT	Information Technology
ITC	International Trade Centre
IUCN	International Union for Conservation of Nature
IWRM	Integrated Water Resource Management
JFM	Joint Forest Management
JICA	Japan International Cooperation Agency
JJA	June, July, August
JMA	Japan Meteorological Agency
JST	Japan Science and Technology Agency
KEITECH	Kananga-EDC Institute of Technology
KINU	Korea Institute for National Unification
Lao PDR	Lao People's Democratic Republic
LAPA	Local Adaptation Plans for Action
LCCAP	Local Climate Change Action Plan
LCG	Local Consultative Group
LDRRMP	Local Disaster Risk Reduction and Management Plan
LED	Light-Emitting Diode
LGU	Local Government Unit
LiDAR	Light Detection and Ranging
LTTE	Liberation Tigers of Tamil Eelam Movement
MAM	March, April, May
MCDEM	Ministry of Civil Defense and Emergency Management (New Zealand)
MDB	Multilateral Development Bank
MDGs	Millennium Development Goals
MGB	Mines and Geosciences Bureau (Philippines)
MICIC	Migrants in Countries in Crisis Initiative
MoDMR	Ministry of Disaster Management and Relief (Bangladesh)
MOE	Ministry of Environment (Japan)
MoEF	Ministry of Environment and Forests (Bangladesh)
MoFA	Ministry of Foreign Affairs (Japan)
MoFDM	Ministry of Food and Disaster Management (Bangladesh)
MoHA	Ministry of Home Affairs and Disaster Preparedness Network (Nepal)

MSME	Micro, Small and Medium Enterprise
MSWRR	Ministry of Social Welfare, Relief and Resettlement (Myanmar)
MtC	Million Tonnes Carbon
NAMRIA	National Mapping and Resource Information Authority (Philippines)
NAP	National Action Plan
NAPA	National Adaptation Programme of Action
NAST	National Academy of Science and Technology
NCCAMP	National Climate Change Adaptation Master Plan
NCCAP	National Climate Change Action Plan
NCCR	Swiss National Centre of Competence in Research
NCCSP	Nepal Climate Change Support Programme
NCDM	National Council for Disaster Management
NCEI	National Centers for Environmental Information
NDCC	National Disaster Coordinating Council (Philippines)
ND-GAIN	Notre Dame Global Adaptation Index
NDRC-MCA	National Disaster Reduction Center, Ministry of Civil Affairs (China)
NDRRM	National Disaster Risk Reduction and Management
NDRRMC	National Disaster Risk Reduction & Management Council (Philippines)
NDRRMP	National Disaster Risk Reduction and Management Plan
NEDA	National Economic and Development Agency (Philippines)
NESDP	National Socio-Economic Development Plan
NGA	National Government Agency
NGO	Non-Government Organizations
NGP	National Greening Program
NIDM	National Institute of Disaster Management, India
NLD	National League for Democracy
NOAA	National Oceanic and Atmospheric Administration (USA)
NOAH	Nationwide Operational Assessment of Hazards (Philippines)
NPDM	Bangladesh Climate Change Strategy and Disaster Management
NPR	Nepalese Rupee
NPTE	National Panel of Technical Experts
NRA	National Regulatory Authority for UXO/Mine Action Sector in Lao PDR
NRDC	National Resources Defense Council
NSMNP	Northern Sierra Madre National Park (Philippines)
NTDTV	New Tang Dynasty Television
NTFP	Non-Timber Forest Product
NTS Centre	Centre for Non-Traditional Security Studies
NTU	Nanyang Technological University
NWMP	National Water Management Plan

NWP	National Water Plan
OC	Philippine Disaster Resilience Foundation Operations Center
OCHA	Office for the Coordination of Humanitarian Affairs
ODA	Official Development Assistance
ODI	Overseas Development Institute
OECD	Organisation for Economic Cooperation and Development
OHCHR	Office of the High Commissioner for Human Rights
OML Center	Oscar M. Lopez Center
OSOCC	On-Site Operations Coordination Center
PAGASA	Philippine Atmospheric, Geophysical, and Astronomical Services Administration
PCA	Prudential Corporation Asia
PDRF	Philippine Disaster Recovery Foundation
PEER	Partnership Enhanced Engagement in Research
PES	Payments for Environmental Services
PHIVOLCS	Philippine Institute of Volcanology and Seismology
PHP	Philippine Peso
PICs	Pacific Island Countries
PM	Particular Matter
PNA	Peacebuilding Need and Impact Assessment
PPP	Public-Private Partnership
PRC	Pew Research Center
PRC	Philippine Red Cross
PSF	People Survival Fund
PV	Photovoltaic
PWC	Pricewaterhouse Coopers
PWD	Persons-With-Disability
R&D	Research and Development
R2P	Responsibility to Protect
RADA	Reconstruction and Development Agency (Sri Lanka)
RAPID	Resilience and Preparedness toward Inclusive Development Program
RCP	Representative Concentration Pathway
RDC	Reception/Departure Center
READY	Hazards Mapping and Assessment for Effective Community-based Disaster Risk Management
REALU	Reducing Emission from Agriculture and Land Use
REBUILD	Resilience Capacity Building for Cities and Municipalities to Reduce Disaster Risks from Climate Change and Natural Hazards
RECOFTC	The Center for People and Forest
REDD	Reducing Emissions from Deforestation and Forest Degradation
RIDF	Rainfall Intensity Duration Frequency
RIMES	Regional Integrated multi-hazard Early-warning System

RMB	Chinese renminbi (yuan)
ROK	Republic of Korea
RSIS	S. Rajaratnam School of International Studies
SA	South Asia
SAMP	Special Area Management Plans
SAR	Search and Rescue
SARS-CoV	Severe Acute Respiratory Syndrome
SDGs	Sustainable Development Goals
SEAR	Southeast Asian Region
SEPA	State Environmental Protection Administration
SFDRR	Sendai Framework for Disaster Risk Reduction
SitRep	Situational Reports
SL	Sustainable Livelihood
SLCP	Sloping Lands Conversion Programme
SLF	Sustainable Livelihood Framework
SLR	Sea Level Rise
SMEs	Small and Medium Enterprises
SNA	Support to National REDD+ Action: Global Programme Framework 2011–2015
SON	September, October, November
SST	Sea Surface Temperature
SUC	State University and Colleges
TAFLOL	Task Force for Logistic and Law and Order
TAFREN-CNO	Task Force to Rebuild the Nation-Center for National Operation
TESDA	Technical Education and Skills Development Authority (Philippines)
ToR	Terms of Reference
TRIGRS	Transient Raifall Infiltration and Grid-based Regional Slope-Stability Model
UK	United Kingdom
UN	United Nations
UNCED	United Nations Conference on Environment and Development
UNDAC	United Nations Disaster Assessment and Coordination
UNDP	United Nations Development Programme
UNEP	United Nations Environment Programme
UNESCAP	United Nations Economic and Social Commission for Asia and the Pacific
UNESCWA	United Nations Economic and Social Commission for West Asia
UNFCCC	United Nations Framework Convention on Climate Change
UNGA	United Nations General Assembly
UNHCR	United Nations High Commissioner for Refugees
UNICEF	United Nations International Children's Emergency Fund
UNISDR	United Nations International Strategy for Disaster Reduction

UNRISD	United Nations Research Institute for Social Development
UP	University of the Philippines
UPLB	University of the Philippines Los Baños
US	United States
USAID	United States Agency for International Development
USD	United States Dollars
USDA	United States Department of Agriculture
UXO	Unexploded Ordnance
VDC	Village Development Committee
VMR	Vegetative Material Reproduction
W.I.S.H.	Water, Infrastructure, Sanitation and Health
WARPO	Water Resources Planning Organization
WFP	World Food Programme
WHO	World Health Organization
WPR	Western Pacific Region
WTO	World Tourism Organization
YEU	YAKKUM Emergency Unit
YISEDA	Young Innovators for Sustainable Environmental Development Association

Chapter 1
Climate Change and Disaster Risks in an Unsecured World

Juan M. Pulhin, Makoto Inoue, Rajib Shaw, Millicent Joyce Q. Pangilinan, and Ma Louiella Rose O. Catudio

Abstract For the past century, climate change impacts and disaster risks have resulted not only in millions of deaths but also in a long-term compromise on human security at various scales, from local to global. Theoretical and empirical bases both reveal how the intertwining relationship of climate change and disaster risk compromises the freedom of people for a decent and dignified life. Particularly imperiled are the livelihoods, health, food, safety, and even the culture of the vulnerable population. Asia, in particular, has been the world's most vulnerable region in terms of quantity and magnitude of impacts from various forms of disasters, climate-induced or otherwise. Amid the escalating threats, the region continues to adapt and innovate building on the lessons it gained from its vast experiences in pursuit of resiliency and security. This chapter elaborates on the interrelationship of climate change, disaster risk, and human security gleaned through the experiences of Asia while also providing a grounded analysis of the key concepts to bridge the major gap in understanding their linkages, hence, embodying the unifying theme of the book.

Keywords Climate change · Disaster risk · Human security · Asia · Resilience

---

J. M. Pulhin (✉)
Department of Social Forestry and Forest Governance, College of Forestry and Natural Resources (CFNR), University of the Philippines Los Baños, Laguna, Philippines
e-mail: jmpulhin@up.edu.ph

M. Inoue
Faculty of Human Sciences, Waseda University, Saitama, Japan

R. Shaw
Graduate School of Media and Governance, Keio University, Kanagawa, Japan

M. J. Q. Pangilinan · M. L. R. O. Catudio
University of the Philippines Los Baños, Laguna, Philippines

J. M. Pulhin et al. (eds.), *Climate Change, Disaster Risks, and Human Security*, Disaster Risk Reduction, https://doi.org/10.1007/978-981-15-8852-5_1

1.1 Introduction

Since the establishment of the Intergovernmental Panel on Climate Change (IPCC) in 1988, the discussion on climate change and its related issues has been given utmost priority. To date, the IPCC's published assessment reports have been the scientific basis for governments to develop policies that respond to the threats of climate change.

The importance of climate change gained more traction when the IPCC was co-awarded the 2007 Nobel Peace Prize with former United States Vice President Al Gore. Through the joint award for the work on climate change research and education, the Norwegian Nobel Committee aspired "to contribute to a sharper focus on the processes and decisions that appear to be necessary to protect the world's future climate, and thereby to reduce the threat to the security of mankind" (Norwegian Nobel Committee 2007). The link between climate change and human security was further highlighted, given the awardees' argument that conflict may arise due to environmental issues. Since then, there has been a heightened interest from the academe, international organizations, and funding agencies to conduct interdisciplinary and cross-sectoral research on climate change. Discussions and researches on the topic have also been more inclusive in recent years. From looking at climate change in the prospect of natural and geographical sciences, researchers have started exploring the issue through the lens of various fields of study such as economics, political science, sociology, anthropology, and international relations, among others.

To ensure a common understanding of key concepts, the most commonly used terms in this book have been adopted from various sources and summarized in Box 1.

Box 1. Definition of terms

Extreme events "comprise a facet of climate variability under stable or changing climate conditions. They are defined as the occurrence of a value of a weather or climate variable above (or below) a threshold value near the upper (or lower) ends ('tails') of the range of observed values of the variable." (Lavell et al. 2012: 30)

Disasters are "severe alterations in the normal functioning of a community or a society due to hazardous physical events interacting with vulnerable social conditions, leading to widespread adverse human, material, economic, or environmental effects that require immediate emergency response to satisfy critical human needs and that may require external support for recovery." (Lavell et al. 2012: 31)

Disaster risk is "the likelihood over a specified time period of severe alterations in the normal functioning of a community or a society due to hazardous physical events interacting with vulnerable social conditions, leading to widespread adverse human, material, economic, or environmental effects that require immediate emergency response to satisfy critical human needs and that may require external support for recovery." (Lavell et al. 2012: 32)

Exposure refers to "the presence (location) of people, livelihoods, environmental services and resources, infrastructure, or economic, social, or cultural assets in places that could be adversely affected by physical events and which, thereby, are subject to potential future harm, loss, or damage." (Lavell et al. 2012: 32)

Vulnerability is "the propensity or predisposition to be adversely affected" (Lavell et al. 2012: 32). It "includes the characteristics of a person or group and their situation that influences their capacity to anticipate, cope with, resist, and recover from the adverse effects of physical events" (O' Brien et al. 2008: 14) with reference to Wisner et al. (2004).

Climate change is "an alteration in the state of the climate that can be identified by changes in the mean and/or the variability of its properties, and that persists for an extended period, typically decades or longer." (Lavell et al. 2012: 29). It is also the change attributed "directly or indirectly to human activity that alters the composition of the global atmosphere and which is in addition to natural climate variability observed over comparable time periods" (United Nations Framework Convention on Climate Change [UNFCC] 1992) or the change identified by "changes in the mean and/or the variability of its properties, and that persists for an extended period, typically decades or longer." (Hegerl et al. 2007: 667).

Disaster risk management is defined as "the processes for designing, implementing, and evaluating strategies, policies, and measures to improve the understanding of disaster risk, foster disaster risk reduction and transfer, and promote continuous improvement in disaster preparedness, response, and recovery practices, with the explicit purpose of increasing human security, well-being, quality of life, and sustainable development." (Lavell et al. 2012: 34).

Human security "in the context of climate change is a condition that exists when the vital core of human lives is protected, and when people have the freedom and capacity to live with dignity. The vital core of human lives includes the universal and culturally specific, material and non-material elements necessary for people to act on behalf of their interests." (Adger et al. 2014)

Resilience is "the ability of a system and its component parts to anticipate, absorb, accommodate, or recover from the effects of a potentially hazardous event in a timely and efficient manner, including through ensuring the preservation, restoration, or improvement of its essential basic structures and functions." (Lavell et al. 2012: 35)

The interrelationship of concepts is further presented in Fig. 1.1. As explained by Lavell et al. (2012), weather and climate events, such as typhoons, floods, droughts, and heat waves, are natural occurrences that may turn into a disaster due to vulnerable conditions and exposure. Blaikie et al. (1994) and Wisner et al. (2004) explained about the social causation of disasters, and argued that socioeconomic (e.g., job location, class, gender, ethnicity, age group, disability, immigration status) and political factors (e.g., power relations, political system at national and international scales) also contribute to disasters. For example, earthquake preparedness is affected by socioeconomic factors, particularly highlighting that people with low-income status (Najafi et al. 2015; Shapira et al. 2018) and lower levels of education (Muttarak and Pothisiri 2013; Paul and Bhuiyan 2009) tend to be more vulnerable. Countries with outdated infrastructure due to lack of resources, weak enforcement of proper building codes, and poor governance proved to have more exposure to the impacts of disasters (Halvorson and Hamilton 2007; Marks 2015). Floods and droughts have greater impact on poor households, especially in terms of property damages and the financial losses they experience as they rely on the environment for their livelihood (De Silva and Kawasaki 2018; Parvin et al. 2016). Weak governance has also been seen to aggravate disasters (Howe and Bang 2017), especially due to political (Woodward et al. 1998), as well as incoherent policies and underdeveloped decentralization strategies (Marks and Lebel 2016; Ng 2016).

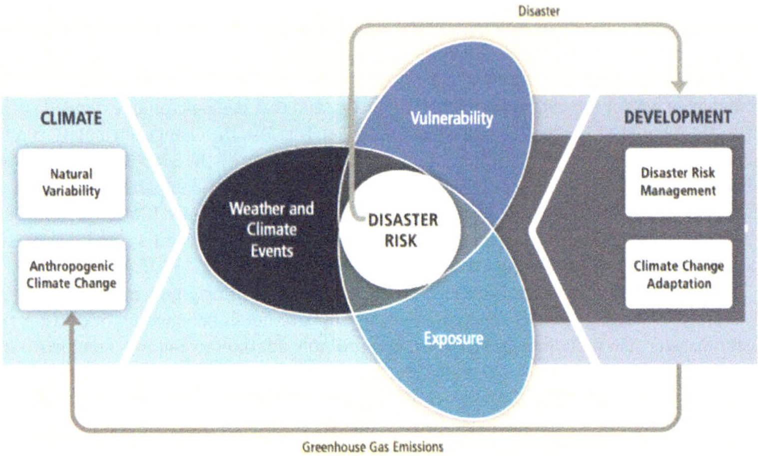

Fig. 1.1 Climate change and its interrelationship with disaster risk and development *Source* Lavell et al. (2012:31)

Experts have recognized that climate change is a "threat multiplier," especially with the impacts it has exhibited on agriculture and forestry, health, politics, and other facets of the society. Adger et al. (2014) support the argument of Scheffran et al. (2012) that climate systems and human security along with natural resources and societal stability are interconnected. The research reiterated that interconnection makes it difficult to determine specific interactions between climate change and human security; hence, concluding that a "good evidence on the chain of causality," and not a highly confident statement, supports the argument. In recent history, climate change coupled with disaster risks have been observed to threaten human security, thereby affecting livelihoods, health, food, safety, and even culture, including other factors that hamper people from living a decent and dignified life.

1.2 Climate Change and Disaster Risks in Asia

Oliver-Smith (1994) and Warren (2016) as cited in Aguilar et al. (2016) stated that the root causes of disasters involve a historical process and manifest many years before the actual event as hazards called disaster risks, which are usually in the form of physical and social conditions (Hewitt 1983; Lewis 1999; Bankoff 2001; Wisner et al. 2004 as cited in Lavell et al. 2012). Fortunately, since 1999, world disaster data is collated through the Emergency Events Database (EM-DAT), a database maintained by the Center for Research on the Epidemiology of Disasters (CRED) housed at the School of Public Health of the Université Catholique de Louvain in Brussels, Belgium. The EM-DAT has been one of the most important tools for researchers

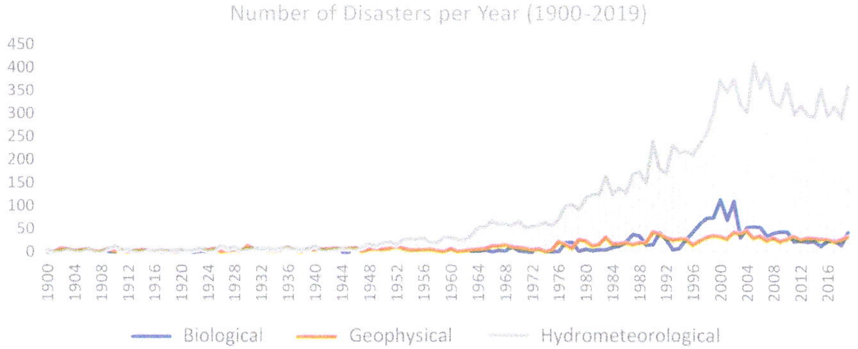

Fig. 1.2 Number of disasters per year from 1900 to 2019 *Source* EM-DAT database, accessed June 2020

and policymakers as it presents historical and disaster figures that serve as scientific basis in disaster management planning, vulnerability assessments, and humanitarian activities.

World disasters are attributable to the three kinds of natural hazards: (1) biological, such as plagues and disease epidemics; (2) geophysical, such as earthquakes, landslides, tsunamis, and volcanic activity; and (3) hydrometeorological, such as food, drought, cyclones, and typhoons. Data shows that disasters have been increasing per year since 1900, ultimately recording a total of 440 cases in 2019. It can be noted that hydrometeorological hazards trigger the most number of disasters every year (Fig. 1.2).

For almost 120 years, Asia has experienced more disasters than any other continent in the world, with a total of 6,213 disasters, accounting for 40% of the world's 15,401 total disasters (Fig. 1.3). Its regions located in the Pacific Ocean and Ring of Fire are regarded as one of the most disaster-prone areas that have recorded major losses and fatalities. Yet again, hydrometeorological events have constantly brought the most number of disasters in all regions.

In 2020, Germanwatch released its 15th edition of the Climate Risk Index (CRI) report which aims to assist in evidence-based policymaking and global discourse through a historical analysis of the climate change impacts in the last 20 years. Eckstein et al. (2020: 3) define the CRI as an indicator of "the level of exposure and vulnerability to extreme events, which countries should understand as warnings in order to be prepared for more frequent and/or more severe events in the future," which only focuses on extreme weather events. Countries are ranked by their CRI scores, which was computed using a weighted total score of the following loss indicators: (1) number of deaths the highest-ranked country, with a weight of 1/6; (2) number of deaths per 100,000 inhabitants, with a weight= 1/3; (3) sum of losses in USD in purchasing power parity (PPP), with a weight=1/6; and (4) losses per unit of gross domestic product (GDP), with a weight of 1/3. A higher CRI rank indicates that the country is more vulnerable to impacts of weather-related loss events. To have a wider perspective of how countries have been impacted in the last 20 years, the

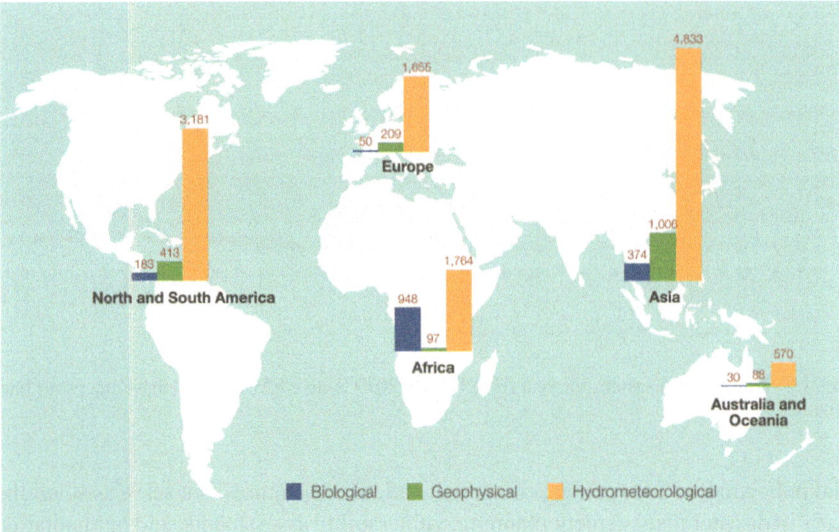

Fig. 1.3 Number of disasters per continent (1900–2019) *Source* EM-DAT database, accessed June 2020

CRI report also computed a long-term CRI by getting the average of CRI scores from 1999 to 2018 which is further elaborated in Fig. 1.4. According to Eckstein et al. (2020), countries in the top 10 rank high in the list either because of extreme disasters experienced or due to the high frequency of severe weather events affecting them each year. The results further prove how Asia has been seriously impacted by extreme events considering that six out of ten countries in the list are from the region. Specifically, these countries are located in the subregions of South Asia and Southeast Asia.

To further scrutinize the impact of these extreme events to Asia, Fig. 1.5 illustrates the total number of fatalities per region for the period 1900–2019. Most fatalities occurred in Asia with deaths reaching a total of 25,996,996, which accounts for 80% of the total world fatalities. Moreover, Fig. 1.6 delves deeper into the data at the country level, and presents the top 10 countries that have experienced the most fatalities since 1900. Data is further broken down to the specific extreme event that contributed to the fatalities. The top five extreme events that accounted for the most fatalities are as follows: (1) drought, 9.6 million; (2) impact, 6.8 million; (3) epidemic, 6.5 million; (4) earthquake, 1.6 million; and storm, 1.3 million. Among all Asian countries, China recorded the highest number of fatalities amounting to 12,521,960 deaths, which is mostly attributable to the 1931 China floods, followed by India with 9,106,069 deaths, and Bangladesh with 2,990,892 deaths.

While all aforementioned information explains that Asia has indeed experienced a greater share of disasters and impacts compared to the rest of the world, it is also important to look at socioeconomic indicators that may have increased the region's vulnerability and exposure. The indicators were obtained from the databases of the

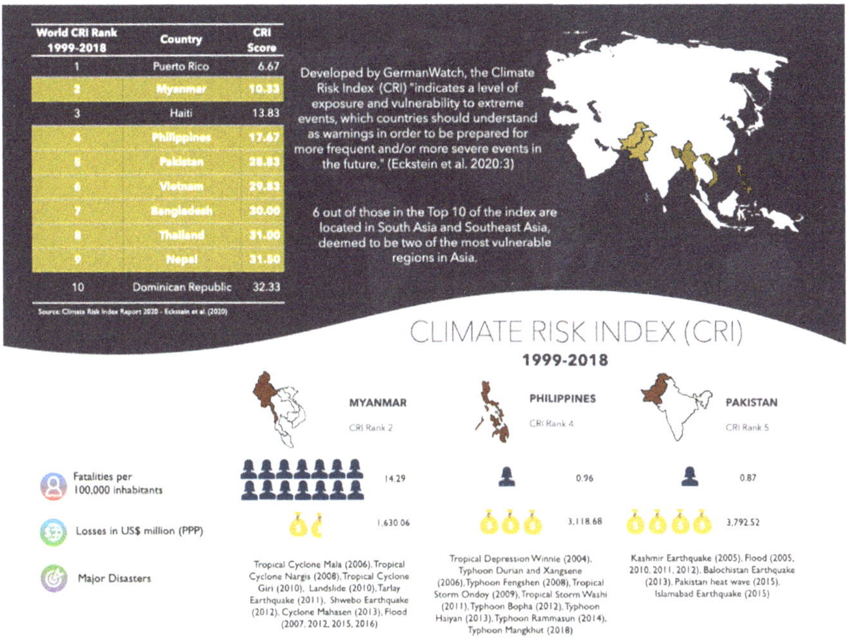

Fig. 1.4 Long-term Climate Risk Index from 1999 to 2018 *Source* Eckstein et al. (2020)

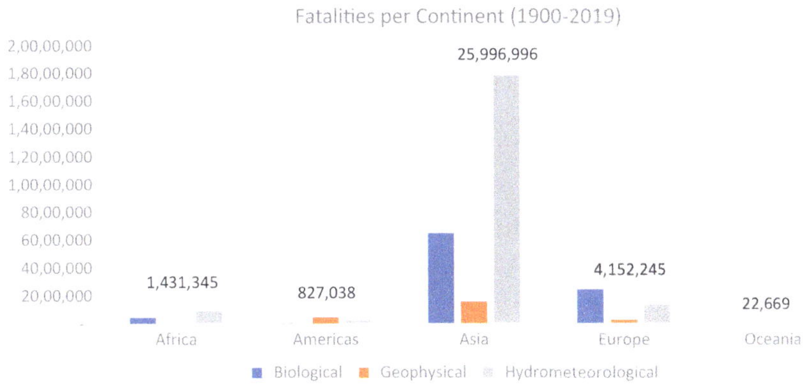

Fig. 1.5 Number of fatalities per continent from 1900 to 2019 *Source* EM-DAT database, accessed June 2020

United Nations Global Sustainable Development Goals (SDG) and World Bank, and were chosen to represent a category of human security as discussed in the Human Development Report (United Nations Development (UNDP) 1994). Table 1.1 shows Asian data as compared to those from the rest of the world for a clearer perspective of the region's current conditions. In 2018, around 89 million Asians lived below the

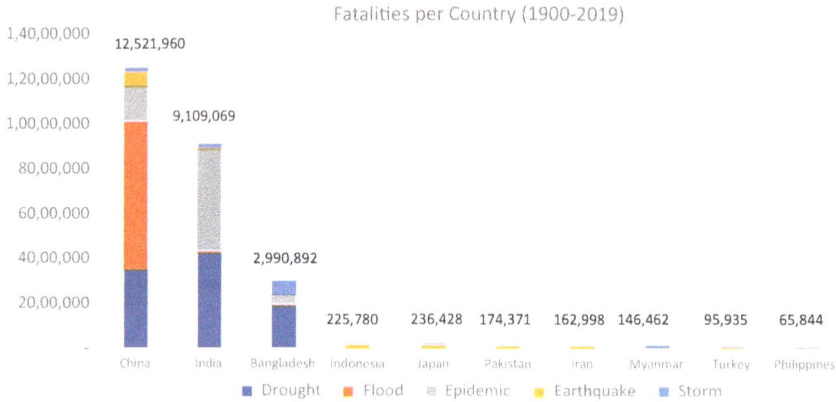

Fig. 1.6 Top 10 Asian countries with the most fatalities from 1900 to 2019 *Source* EM-DAT database, accessed June 2020

international poverty line of USD 1.9 per day, representing 2% of the total regional population. For the other indicators, Asia lagged behind the rest of the world. For the food security aspect, 11.14% of the Asian population are undernourished, compared to the global level of 10.80%. For the environmental security aspect, 93.47% of the Asian population, or 4,168,857,585 people are exposed to PM 2.5 air pollution, compared to the global level of 91.30%. Finally, compared to the rest of the world, more Asians are displaced due to disasters (0.28% vs. 0.25%), and due to conflict and violence (0.38% vs 0.31%). Therefore, the combination of the frequency of natural occurrences and relatively poor socioeconomic figures in various human security aspects has undoubtedly put Asia at a disadvantage, making it more vulnerable to climate change and disaster risks.

1.3 Human Security in the Context of Climate Change and Disaster Risks: A Framework of Analysis

Narratives on the unprecedented climate change impacts and natural disasters in Asia have portrayed the intertwining relationship of climate change, disaster risks, and human security across time. Statistics have proved the vulnerability of Asia not only because of its exposure to these hazards but also due to the underlying sociopolitical and socioeconomic conditions (Redclift et al. 2013). Widespread poverty, undernourishment, unsafe drinking water services, worsening air pollution, high level of internal displacement due to conflict and violence, are few of the manifestations of persistent issues in Asia. These factors, coupled with the worsening threats from climate change and disaster risks, do not only increase the vulnerability of the people but also decrease both their coping and adaptive capacities.

Table 1.1 Asia at a glance: selected human security indicators and statistics for 2017 and 2018

Human security categories	Indicator	Year	Number of People	% share in total Asia population	World percentage
Economic security	Population in poverty (international, below $1.9/day)	2018	89,331,056	2	10
Food security	Prevalence of undernourishment	2017	497,701,600	11.14	10.80
Health security	People using at least basic drinking water services	2017	4,168,857,585	93.34	89.60
Environmental security	PM2.5 air pollution, population exposed to levels exceeding WHO guideline value	2017	4,174,955,909	93.47	91.30
Personal security	Internally displaced persons, new displacement associated with disasters	2018	12,701,108	0.28	0.25
Community security	Internally displaced persons, total displaced by conflict and violence	2018	17,050,000	0.38	0.31

Source United Nations Global SDG Database (https://unstats.un.org/sdgs/indicators/database/), accessed August 2020; World Development Indicators (https://databank.worldbank.org), accessed August 2020

From a humanist perspective, discussions on these threats have drawn attention to the necessity to prioritize the welfare of human beings regardless of gender, race, religion, ethnicity, citizenship, or other distinguishing characteristics (Gasper 2015). It is the dignity and sanctity of human life that has provided enough reason to guarantee the fullest possible development and rights for every human being. Since the initiation of the discussion on human security, its concept has been evolving to articulate the most effective means for its operationalization. From the threats of war among nations, the configuration of threats for human security has greatly changed in this post-Cold War period. Although some threats are still considered as ancient and persistent while others unprecedented, this contemporary world has much to offer in terms of addressing these threats (Alkire 2003). Extensive researches, technological and analytical advances, political changes, strong international cooperation and collaboration are some of the significant changes that have been utilized to widen the understanding and efforts to ensure human security. Yet, the complexities and multidimensionality of human security continue to challenge various responsible institutions, especially that environmental threats have escalated beyond their perceived manageable impacts because of climate change and natural hazards (Alkire 2003; Gasper 2013; Leaning and Arie 2000).

Consequently, the vagueness of human security, encompassing everything from physical security to psychological well-being, has required it to be viewed in various lenses to magnify the specific elements that are being undermined (Alkire 2003). In terms of climate change and natural disasters, Fig. 1.7 provides the analytical structure of how the concepts of climate change and disaster risk undermine human security. The figure shows the different systems on which climate change manifests its alterations with emphasis on the physical systems that overlap with the hazards from the hydrometeorological subgroup of natural disasters. Aside from the hydrometeorological subgroup, natural disaster persists threatening the vulnerable population through the biological and geophysical hazards that similarly took thousands of lives in the past century. Further, specific elements of human security are provided to narrow down the scope by which human security can be examined in terms of the freedom and capacity to live with dignity. Likewise, the enumerated specific elements of human security embody the vital core of human lives consistent with the definition of human security stated by Adger et al (2014).

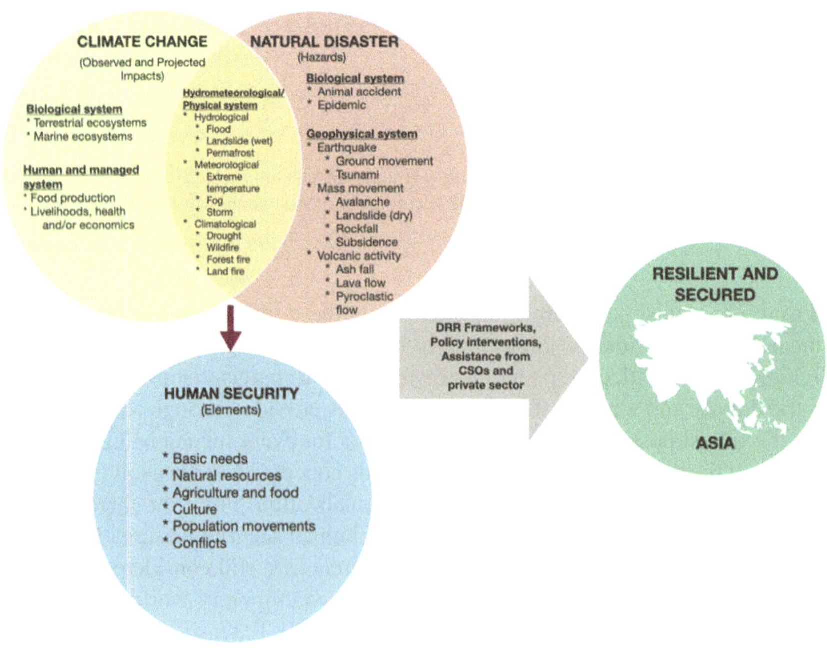

Fig. 1.7 Analytical framework of climate change, disaster risk, and human security employed in the book

Climate change, as defined previously, presents the means of where it has been manifesting, the observed and projected impacts. Observed impacts have utilized the palaeo-climatic evidence to prove that toward the end of the nineteenth century, global (combined land and ocean) average warming of 0.45±0.15 °C had been recorded beyond the natural climate variability. Such fluctuations have been attributed to the increasing greenhouse gases (GHG) brought by anthropogenic activities like the burning of fossil fuels, deforestation, and urbanization (IPCC et al. 1990; Steffen et al. 2011). Analysis of the observed global average surface temperatures for more than the past half-century had shaped the climate models used for the projection of future climate conditions (Buis 2020). Unfortunately, the increased accuracy of these climate models revealed the continuous rise of global mean temperature by 2.6–4.8 °C for the next decades (Hayhoe et al 2017). Thus, suggesting greater impacts on the biological, human, and physical systems that directly undermine human security.

For the biological systems, climate change has significantly altered the biogeochemical cycles of the Earth, hence, causing modifications to different ecosystems including their functioning (Steffen et al. 2011). Both observed and projected impacts of climate change reveal the increasing threat to the environment resulting in higher extinction risks and an abrupt and irreversible global and regional-scale change in the composition, structure, and function of the ecosystem (Field et al. 2014). Similarly, alterations in the biogeochemical cycles and the biosphere have induced changes in the physical system through the increasing frequency, intensity, spatial extent, duration, and timing of weather and climate extremes, which lead to unprecedented extremes (Seneviratne et al. 2012). Coupling the impacts of climate change in the biological and physical systems provides the bigger picture of how climate change undermines specific elements of human security through the changes in the human and managed systems. Clearly, the occurrence of weather and climate events hinders the provision of ecosystem goods and services (Osborn 2019) which directly imperil the food production, livelihoods, socioeconomic condition, and health of the affected population.

Interestingly, the physical system of climate change overlaps with the hydrometeorological subgroup of natural disasters. With this, the enumerated events in the physical system of climate change can also be viewed as natural hazards. Yet, the pervasive impacts of climate change consider it as a threat multiplier for disaster, security, and conflict (Huntjens and Nachbar 2015). Climate change directly enhances these natural hazards by increasing their intensity and frequency while also exacerbating the existing environmental degradation and mismanagement, demographic changes, rapid and unplanned urbanization, and scarcity of livelihood options for the poor (Cardona et al. 2012). Disaster risk differs from climate change because of its widened scope that covers beyond the occurrence of the natural hazards. In the context of disaster risk, natural hazard itself does not produce disaster and is not the sole concern. Other factors such as the vulnerability of the affected population, their exposure to these natural hazards and their coping capacity are also essential considerations.

For instance, the intersection of climate change and disaster risk and how they undermine specific elements of human security are vividly manifested through circumstances such as the sea-level rise and ocean warming that have been affecting the frequency and magnitude of typhoons and storm surges. Hence, resulting in frequent wiping out of lives, shelters, properties, infrastructures, transport systems, and livelihoods (Laffoley and Baxter 2016), thereby increasing the vulnerability of the exposed population, in this case, the coastal community. Typically, these events will be followed by widespread flooding and possible landslides that worsen the situation because of the expansion of the affected population coupled with other societal, political, and economic concerns. Drought, on the other hand, has intensified and extended, thereby, affecting millions of lives and aggravating the problems on food insecurity, water pollution and scarcity, crop damage and failure, and poverty (UNESCAP 2020). Likewise, these examples could also ignite conflict within the affected population and sometimes including the concerned states due to resource scarcity and competition (Huntjens and Nachbar 2015). On the other hand, projected climate change impacts, increased exposure, and vulnerability could also lead to population displacement as an act of adaptation particularly if the threats greatly compromise their livelihood sources. Cultural dimensions of the community could also be threatened since most of their cultural values and practices are embedded in their livelihoods and territories that are being undermined by climate change and human security (Adger et al. 2014).

Apart from the hydrometeorological hazards, other sources of risks could also come from biological and geological hazards. Biological hazards have organic origins or conveyed by biological vectors like pathogenic microorganisms, toxins, and other biotic substances. These hazards pose risks to animals, livestock, and plants that could result in outbreaks, epidemics, and/or pandemics. Among the prominent disasters from biological hazards that have taken countless lives are the Ebola Virus Disease outbreak in West Africa, Zika virus infection in the Americas and the Pacific, Yellow fever in multiple countries in Africa, Severe Acute Respiratory Syndrome (SARS-CoV) in Asia (UNISDR 2018) including the current Coronavirus disease (COVID-19) that has been afflicting the whole world. On the contrary, geological hazards differ from the previous hazards because it originates from earth's internal processes such as earthquakes, volcanic activity, and emissions and other related geophysical processes like mass movements and landslides. Oftentimes, these processes are enhanced by the hydrometeorological hazards, for instance, tsunamis. Although it is triggered by undersea earthquakes and other geologic events, it is also considered as a coastal water-related hazard since it takes part in the oceanic process (UNISDR 2009). In the same way, exposure of the vulnerable population to these hazards could directly curtail the vital core of human security.

Defining the intertwining relationship of climate change and disaster risk and how they undermine human security is necessary to articulate the pervasive threats that thwart the operationalization of human security. Clarity in this aspect opens up the broader opportunity on how this contemporary world could tailor its advances to ensure human security in its simplest form. Hence, sparing the vulnerable population from receiving a response mechanism that misfits the existing localized threats brought by climate change and natural hazards. Recent initiatives, particularly the international cooperation among nations, unify them through intergovernmental agreements and strategies that have been evolving since the 1970s to provide solutions to these pressing concerns. For instance, Hyogo Framework for Action (HFA) 2005–2015 has emphasized the necessity of building resilient nations and communities to disasters in order to reduce the loss of lives including social, economic, and environmental assets.

Afterward, the Sendai Framework for Disaster Risk Reduction has been proposed as the successor of HFA. The framework is a 15-year, voluntary, and non-binding agreement that involves the State and other stakeholders including the local government unit (LGU) and private sectors. The State has the primary role to reduce disaster risk while other stakeholders share the responsibility to attain substantial reduction of disaster risk and losses of lives, livelihoods, and health as well as the economic, physical, social, cultural, and environmental assets of persons, business, communities, and countries (Aitsi-Selmi et al. 2015). Implementation of the said framework including the national climate change adaptation and mitigation strategies entails political intervention in order to harmonize them with the national and local plans to fulfill the end goal of having a resilient and secured Asia.

In essence, the concept of human security pertains to the provision of freedom from fear and freedom from want (United Nations Development [UNDP] 1994) for every human being amid threats of climate change and disaster risk. However, the current concept still remains to have difficulty in its operationalization. Perceptions of people adversely affected by these threats have their own ideas of threats/risks and security based on the reality that they are living and surrounded with. Local perspectives on human security may still be consistent with the proposed concepts and theories yet, it still outweighs the gravity of the localized threats in the form of poverty, food and water scarcity, localized climate change effects on crops, regular occurrences of flooding and landslides, awareness of the presence of lawless elements and many others (Atienza 2015). Thus, viewing human security on a local basis provides the closest possible means of providing appropriate solutions. For this reason, this book gleans toward the rich experience of Asia to unravel both the stories of loss and the glimmers of hope that this region unceasingly holds. Consequently, the lessons that they have acquired from their vast experience has greatly shaped the region in their pursuit of resiliency and sustainability. Asia, then, is a picture of a progressive region in terms of ensuring human security amid the increasing threat from climate change and disaster risks.

1.4 Organization of the Book

This book revolves on themes carefully selected by the editors highlighting the Asian context, elements, specific cases, and responses associated with the interlinked nature of climate change, disaster risk, and human security. The four major themes are as follows: (1) understanding the context of climate change, disaster risk, and human security in Asia, (2) breaking down the elements of human security threatened by climate change and disaster risks, (3) drawing lessons from cases in the Asian context on how climate change and disaster risks undermined human security, and (4) recognizing current Asian initiatives that address challenges posed by climate change and disaster risk on human security. Drawing from their vast multidisciplinary and diverse experiences in the academe, research and policymaking, and field observations, each chapter author brings attention to the growing interest of climate change and disaster risk, and their relationship to human security. Further, the authors attempt to substantiate lessons learned and best practices using actual cases in Asia that have addressed various pressing environmental and social issues. Figure.1.8 is a map of Asian country cases presented and discussed in the book.

The first theme gives emphasis to ensuring that readers have a common understanding of the current situation of climate change, disaster risk, and human security in Asia. It lays down the foundation of these issues by looking at both historical and current perspectives to determine how they have changed over the years.

The second theme breaks down the elements of human security to explain how they are affected by climate change and disaster risks. These elements are scrutinized under environmental, socioeconomic, and political lenses to provide more substantial details on how they are threatened at present.

The third theme exhibits actual cases that have changed the Asian landscape in terms of its view on human security after being subjected to issues related to climate change and disaster risks. With Asia being considered as one of the most vulnerable

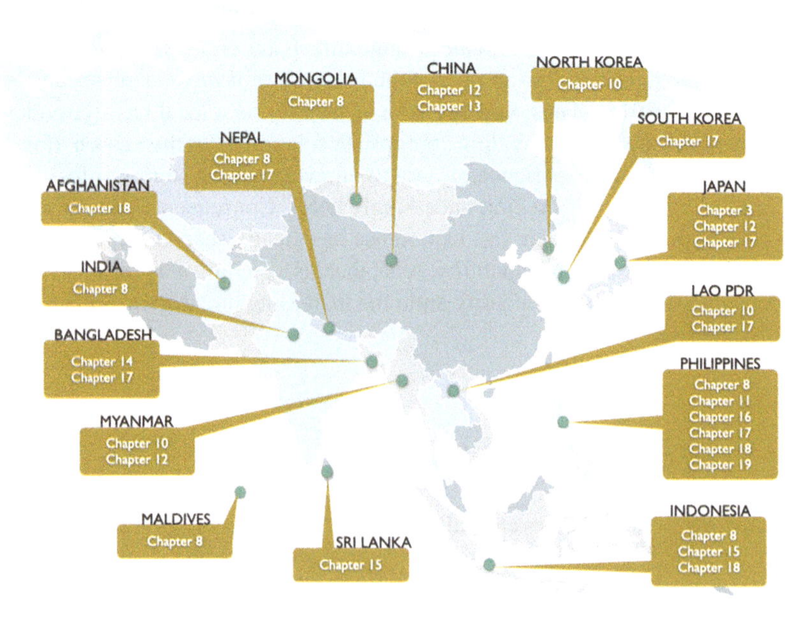

Fig. 1.8 Country cases discussed in the book

regions in the world, this theme captures actual lessons learned that can be helpful for other countries to ensure their preparedness and lessen, if not prevent, the impact of climate change and disasters.

The fourth and last theme highlights notable initiatives by Asian countries to address the dangers brought about by climate change and disaster risks on human security. It underscores best practices that can be emulated by different countries and incorporated in all levels of their policy and decision-making.

References

Adger W N, Pulhin J M, Barnett J, Dabelko G, Hovelsrud G, Levy M L, Oswald Spring U, & Vogel C (2014) Human security. In FieldC B, Barros V R, Dokken D J, Mach K J, Mastrandrea M D, Bilir T E, Chatterjee M, Ebi K L, Estrada Y.O, Genova R C, Girma B, Kissel E S, Levy A N, MacCracken S, Mastrandrea P R, & White L L (Eds) Climate change 2014—Impacts, adaptation and vulnerability: Part A: Global and sectoral aspects: Working group II contribution to the IPCC Fifth Assessment Report: Volume 1: Global and Sectoral Aspects (pp 755–791). Cambridge University Press

Aitsi-Selmi A, Egawa S, Sasaki H et al (2015) The sendai framework for disaster risk reduction: Renewing the global commitment to people's resilience, health, and well-being. Int J Disaster Risk Sci 6: 164–176 (2015). https://doi.org/10.1007/s13753-015-0050-

Alkire S (2003) A conceptural framework for human security. CIRSE working paper 2. Queen Elizabeth House, Oxford

Atienza MEL (2015) People's views about human security in five Philippine municipalities. Disaster Prevention and Management 24(4):448–467. https://doi.org/10.1108/DPM-12-2014-0277

Bankoff G (2001) Rendering the world unsafe: 'Vulnerability' as western discourse. Disasters 25(1):19–35. https://doi.org/10.1111/1467-7717.00159

Blaikie P, Cannon T, Davis I, & Wisner B (1994) At risk: Natural hazards, people vulnerability and disasters 1st edition. https://doi.org/10.4324/9780203428764

Buis A (2020) Study confirms climate models are getting future warming projections right. News Articles, NASA Global Climate Change

Cardona OD, Van Aalst MK, Birkmann J, Fordham M, Mc Gregor G, Rosa P, Pulwarty RS, Schipper ELF, Sinh BT, Décamps H, Keim M, Davis I, Ebi KL, Lavell A, Mechler R, Murray V, Pelling M, Pohl J, Smith AO, Thomalla F (2012) Determinants of risk: Exposure and vulnerability. Managing the Risks of Extreme Events and Disasters to Advance Climate Change Adaptation: Special Report of the Intergovernmental Panel on Climate Change 9781107025:65–108. https://doi.org/10.1017/CBO9781139177245.005

De Silva MMGT, Kawasaki A (2018) Socioeconomic vulnerability to disaster risk: A case study of flood and drought impact in a Rural Sri Lankan community. Ecological Economics 152(June):131–140. https://doi.org/10.1016/j.ecolecon.2018.05.010

Eckstein D, Künzel V, Schäfer L, & Winges M (2020) *Global climate rate index 2020 Who suffers most from extreme weather events ? Weather-related loss events in 2018 and 1999 to 2018*

Field C, Barros V, Maach K, Mastrandrea M (2014) Technical summary. In: Part A (ed) Impacts, Adaptation, and Vulnerability. Global and Sectoral, Aspects. contribution of working group II to the Fifth assessment report of the Intergovernmental Panel on Climate Change, pp 35–94

Gasper D (2013) Concepts of human security. Human Insecurity in a Global World, December 2011, 15–39

Gasper D (2015) Human security: From definitions to Investigating a Discourse. Routledge Handbook of Human Security, 10872. https://doi.org/10.4324/9781315885926.ch2

Halvorson SJ, Hamilton JP (2007) Vulnerability and the erosion of seismic culture in mountainous Central Asia. Mountain Research and Development 27(4):322–330. https://doi.org/10.1659/mrd.0900

Hayhoe K, Edmonds J, Kopp RE, LeGrande AN, Sanderson BM, Wehner MF, Wuebbles J (2017) Climate models, scenarios, and projections. Climate Science Special Report: Fourth National Climate Assessment I:133–160. https://doi.org/10.7930/J0WH2N54.U.S

Hegerl G, Zwiers F, Braconnot P, Gilett N, Luo Y J A, Marengo Orsini J, Nicholls N, Penner J, & Stott P (2007) Understanding and attributing climate change. In Solomon S, Qin v, Manning M, Chen Z, Marquis M, Averyt K, Tignor M, & Miller H (Eds) Climate change 2007: The physical science basis. Contribution of working group i to the fourth assessment report of the intergovernmental panel on climate change (pp 664–745). Cambridge University Press

Hewitt K (1983) The idea of calamity in a technocratic age. In Hewitt K (Ed) Interpretations of calamity from the viewpoint of human ecology (pp 3–32). Allen and Unwin Inc. https://doi.org/10.4324/9780429329579-1

Howe B, Bang G (2017) Nargis and Haiyan: The politics of natural disaster management in Myanmar and the Philippines. Asian Studies Review 41(1):58–78. https://doi.org/10.1080/10357823.2016.1265484

Huntjens P, & Nachbar K (2015) Climate change as a threat multiplier for human disaster and conflict policy and governance recommendations for advancing climate security climate change as a threat multiplier for current and future conflict policy and governance recommendations for Adva. May. http://www.thehagueinstituteforglobaljustice.org/working-paper-9

Intergovernmental Panel on Climate Change., Houghton, J. T., Intergovernmental Panel on Climate Change., & World Meteorological Organization (1990) IPCC first assessment report. WMO, Geneva

Laffoley D, & Baxter J M (2016) Explaining ocean warming : (Issue September).

Lavell A, Oppenheimer M, Diop C, Hess J, Lempert R, Li J, Muir-Wood R, Myeong S, Moser S, Takeuchi K, Cardona O-D, Hallegatte S, Lemos M, Little C, Lotsch A, & Weber E (2012) Climate change: New dimensions in disaster risk, exposure, vulnerability, and resilience. In Christopher B Field, Q. Dahe, T. F. Stocker, & V. Barros (Eds.), Managing the risks of extreme events and disasters to advance climate change adaptation: Special report of the intergovernmental panel on climate change (pp 25–64). Cambridge University Press. https://www.cambridge.org/core/books/managing-the-risks-of-extreme-events-and-disasters-to-advance-climate-change-adaptation/climate-change-new-dimensions-in-disaster-risk-exposure-vulnerability-and-resilience/BB084965B525470AE958E68BAFEC133F

Leaning J, & Arie S (2000) Human security: A framework for assessment in conflict and transition. December, 78. http://www.certi.org/publications/policy/humansecurity-4.PDF

Lewis J (1999) *Development in disaster-prone places*. IT Publications (Practical Action). https://doi.org/10.3362/9781780442013

Marks D (2015) The urban political ecology of the 2011 floods in Bangkok: The creation of uneven vulnerabilities. Pacific Affairs 88:623–651. https://doi.org/10.5509/2015883623

Marks D, & Lebel L (2016) Disaster governance and the scalar politics of incomplete decentralization: Fragmented and contested responses to the 2011 floods in Central Thailand. Habitat International, 52: 57–66. https://doi.org/https://doi.org/10.1016/j.habitatint.2015.08.024.

Muttarak R, & Pothisiri W (2013) The role of education on disaster preparedness: Case study of 2012 Indian Ocean earthquakes on Thailand's Andaman coast. Ecology and Society 18(4): 51. https://doi.org/10.5751/ES-06101-180451

Najafi M, Ardalan A, Akbarisari A, Noorbala A A, & Jabbari H (2015) Demographic Determinants of Disaster Preparedness Behaviors Amongst Tehran Inhabitants, Iran. PLoS Currents, 7, ecurrents.dis.976b0ab9c9d9941cbbae3775a6c5fbe6. https://doi.org/10.1371/currents.dis.976b0ab9c9d9941cbbae3775a6c5fbe6

Ng S (2016) Governance beyond the government: Responding to a reactionary flood governance regime in Ayutthaya, Thailand. Habitat International 52: 11–19. https://doi.org/10.1016/j.habitatint.2015.08.029

Norwegian Nobel Committee (2007, October 12) The nobel peace prize for 2007. https://www.nobelprize.org/prizes/peace/2007/press-release/

O' Brien K, Leichenko R, & Vogel C (2008) dister risk reduction, climate change adaptation and human security. In eport pre—pared for the Royal Norwegian Ministry of Foreign Affairs by the Global Environmental Change and Human Security (GECHS) Project, GECHS Report 2008:3. https://doi.org/ISSN: 1504-5749

Oliver-Smith A (1994) Peru's five hundred-year earthquake: Vulnerability in historical context. In A. Varley (Ed.), Disasters, development, and environment (pp 3–48). J. Wiley

Osborn D (2019) The breadth of climate change impacts on biological systems. Emerging Topics in Life Sciences 3(2): 107–113. https://doi.org/10.1042/ETLS20180114

Parvin G A, Shimi A C, Shaw R, & Biswas C (2016) Flood in a changing climate: The impact on livelihood and how the rural poor cope in Bangladesh. Climate 4(4). https://doi.org/10.3390/cli4040060

Paul B, Bhuiyan R (2009) Urban earthquake hazard: Perceived seismic risk and preparedness in Dhaka City, Bangladesh. Disasters 34:337–359. https://doi.org/10.1111/j.1467-7717.2009.01132.x

Redclift MR, Grasso M, Redclift MR (2013) *Handbook on climate change and human security*. Edward Elgar Publishing, Cheltenham

Scheffran J, Brzoska M, Kominek J, Link PM, Schilling J (2012) Climate change and violent conflict. Science 336(6083):869–871. https://doi.org/10.1126/science.1221339

Seneviratne SI, Nicholls N, Easterling D, Goodess CM, Kanae S, Kossin J, Luo Y, Marengo J, McInnes K, Rahimi M, Reichstein M, Sorteberg A, Vera C, Zhang X (2012) Changes in climate extremes and their impacts on the natural physical environment. In: Field CB, Barros V, Stocker TF, Qin D, Dokken DJ, Ebi KL, Mastrandrea MD, Mach KJ, Plattner G-K, Allen SK, Tignor M, Midgley PM (eds) Managing the risks of extreme events and disasters to advance climate change adaptation. A special report of working groups I and II of the Intergovernmental Panel on Climate Change (IPCC) (pp 109–230). Cambridge University Press, Cambridge, UK, and New York, NY, USA

Shapira S, Aharonson-Daniel L, Bar-Dayan Y (2018) Anticipated behavioral response patterns to an earthquake: The role of personal and household characteristics, risk perception, previous experience and preparedness. International Journal of Disaster Risk Reduction 31(April):1–8. https://doi.org/10.1016/j.ijdrr.2018.04.001

Steffen W, Grinevald J, Crutzen P, Mcneill J (2011) The Anthropocene : Conceptual and historical perspectives. https://doi.org/10.1098/rsta.2010.0327

UNESCAP. (2020). Ready for the Dry Years. https://www.unescap.org/sites/default/files/publications/ReadyfortheDryYears.pdf

UNISDR (2009) UNISDR Terminology on Disaster Risk Reduction. https://www.unisdr.org/files/7817_UNISDRTerminologyEnglish.pdf

UNISDR (2018) UNISDR Annual Report 2017. https://www.unisdr.org/files/58158_unisdr2017annualreport.pdf

United Nations Development (UNDP) (1994) World Development Report 1994. In Human Development Report 1994: New Dimensions of Human Security. https://doi.org/10.1596/978-0-1952-0992-1

United Nations Framework Convention on Climate Change (UNFCC). (1992).

Warren J (2016) Typhoons and the Inequalities of Philippine Society and History. In *Philippine Studies*: Historical and Ethnographic Viewpoints (Vol. 64). https://doi.org/10.1353/phs.2016.0036

Wisner B, Blaikie P, Cannon T, Davis I (2004) *At Risk: Natural Hazards*, 2nd edn. Routledge, London, UK

Woodward A, Hales S, & Weinstein P (1998) Climate change and human health in the Asia Pacific region: Climate Research 11(1) 31–38. http://www.jstor.org/stable/24865974

Dr. Juan M. Pulhin is full Professor and former Dean of the College of Forestry and Natural Resources, University of the Philippines Los Baños, Philippines. He was Lead Author of the 2007 Fourth Assessment Report of the Intergovernmental Panel on Climate Change (IPCC), Coordinating Lead Author of the Human Security Chapter of the 2014 IPCC Fifth Assessment Report and currently Lead Author of the Asia chapter of the IPCC Sixth Assessment Report. He has been a four-time a visiting professor at The University of Tokyo in 2007, 2010, 2016 and 2017 and a visiting scholar at Waseda University, Tokyo in 2018. He has written and edited 11 books and published more than 100 technical papers and book chapters on natural resources policy and governance, and on the human dimension of climate change and disaster risks. He is a fellow, invited expert, and council member of several international organizations and a member of the National Academy of Science and Technology (NAST), Philippines.

Dr. Makoto Inoue is Professor at the Faculty of Human Sciences, Waseda University, Japan and former Professor at The University of Tokyo. He serves as Dean of Graduate School of Human Sciences, Waseda University since September 2020. His specialty is environmental sociology, forest governance, and Southeast Asian studies. He has written and edited/co-edited 34 books, and published more than 140 original articles in peer-reviewed international and national journals on the social aspects of forestry and the environment. He serves as President of The Japanese Association for Environmental Sociology (June 2019–present), Vice President of Japanese Society for Current Anthropology (January 2020–present), Vice President of Society for Environmental Economics and Policy Studies (April 2018–March 2020), Councilor of The International Association for the Study of the Commons (January 2019-present), a board member of various academic societies in Japan, and an editorial board member of several international journals.

Dr. Rajib Shaw is Professor at the Keio University in Japan and former Executive Director of the Integrated Research on Disaster Risk (IRDR) programme. He is also Senior Fellow of Institute of Global Environmental Strategies (IGES) Japan, and Chairperson of SEEDS Asia, and CWS Japan, two Japanese NGOs. Previously, he was Professor at the Graduate School of Global Environmental Studies, Kyoto University, Japan. His expertise includes community-based disaster risk management, climate change adaptation, urban risk management, and disaster and environmental education. He is the editor of a book series on disaster risk reduction, published by Springer, Editor-in-Chief of the journal Progress in Disaster Science. He severs as Chair of United Nations Science Technology Advisory Group for disaster risk reduction, and is Co-chair of the UNDRR's Asia-Pacific Science Technology Academic Advisory Group (ASTAAG). Prof. Shaw has published more than 45 books and more than 300 academic papers and book chapters.

Millicent Joyce Q. Pangilinan is currently a researcher for a forestry project at the University of the Philippines Los Baños. She has 10 years of experience in research and project management in both the private and development sectors. She started her career in Japan doing economic data and market analysis intended for mergers and acquisitions projects in Asia and in the USA. After 4 years, she transitioned into the development field, where she was engaged in research projects on agriculture, climate change, and forestry in Southeast Asia and the Philippines. She earned her master's degree in International Public Policy (Economics track) at Osaka University, Japan and bachelor's double degree in Economics and Finance at De La Salle University Manila, Philippines.

Ma Louiella Rose O. Catudio obtained her bachelor's degree in Forestry from the University of the Philippines Los Baños (UPLB). She is currently a graduate student of MS Forestry specializing in silviculture and forest influences in the same university, and currently serves as a graduate assistant for a forestry project on the assessment of the Community-based Forest Management (CBFM) program in the Philippines. Prior to this, she was involved in the studies of water resource management and community and program development of the Department of Environment and Natural Resources (DENR). Her researches delve into the study of social forestry, forest resource management, agroforestry, and climate change.

Chapter 2
Disaster Risk Reduction, Climate Change Adaptation, and Human Security: A Historical Perspective Under the Hyogo Framework and Beyond

Rajib Shaw, Juan M. Pulhin, and Makoto Inoue

Abstract Disasters affect economic and physical development gains, disrupt the social fabric, and pose challenges at different scales. Occurrence of disasters, number of reported disaster events, loss and damage due to disasters, population affected–all these show increasing trends. This significantly affects human security and enhances development. Asia is an epicenter of urbanization, although majority of the population still live in rural areas and heavily depend on agriculture. Moreover, barring a few, Asian countries still face the myriad challenges of poverty, lack of basic services and health facilities, poor infrastructure, weak governance and inability to deliver, and so on. Climate change is further threatening development aspirations through a series of extreme and slow-onset hydrometeorological events, while geohazards occur in the region regularly. The challenges posed by disasters multiply due to existing inefficiencies and weaknesses of developing countries. The underlying risk factors cannot be reduced unless such weaknesses are targeted. This chapter reviews the progress and challenges in disaster risk reduction (DRR), climate change adaptation (CCA), and related human security issues during an earlier global framework (the Hyogo Framework for Action) and suggests specific actions for the next global framework (the Sendai Framework for Disaster Risk Reduction). Recommendations include (1) reducing underlying risk factors for poverty reduction and sustainable development, (2) enhancing the economic viability of risk reduction measures, (3) institutionalizing community-based disaster risk reduction, (4) addressing risk reduction in recovery, and (5) enhancing DRR education to help reduce the underlying risk factors.

R. Shaw (✉)
Graduate School of Media and Governance, Keio University, Kanagawa, Japan
e-mail: shaw@sfc.keio.ac.jp

J. M. Pulhin
Department of Social Forestry and Forest Governance, College of Forestry and Natural Resources (CFNR), University of the Philippines Los Baños, Laguna, Philippines

M. Inoue
Faculty of Human Sciences, Waseda University, Saitama, Japan

J. M. Pulhin et al. (eds.), *Climate Change, Disaster Risks, and Human Security*, Disaster Risk Reduction, https://doi.org/10.1007/978-981-15-8852-5_2

Keywords Underlying risk factors · Disaster vulnerability · Climate change adaptation · Human security · Sustainable development goals

2.1 Introduction

Disaster risk reduction (DRR) was first talked about among members of the international community during the International Decade of Natural Disaster Reduction: 1990–1999. The field turned 25 years old in 2015, when the third World Conference on Disaster Reduction was held in Sendai, Japan in 2015. While tremendous progress has been made in several fronts, which helped reduce disaster-related mortality and losses, there remains a wide gap in research and action toward reducing exposure and underlying risk factors (UN 2015a).

Climate change adaptation (CCA) achieved prominence in the Nairobi Conference of Parties (COP) in 2005, and the Nairobi Work Program was initiated with the United Nations Framework Convention on Climate Change (UNFCCC). It got a boost in 2007 in the Bali COP as the Bali Action Plan strongly recommended the establishment and disbursement of an adaptation fund (Shaw and Tran 2012). The Paris agreement of 2015 also clearly mentioned CCA and its relation to damage and loss (UN 2015b).

In contrast, the concept of human security is rather new, becoming prominent only in 2000, when then Japanese Prime Minister Kenzo Obuchi floated this idea during the UN Millennium Summit and started the Human Security Commission, co-chaired by Madam Sadako Ogata and Nobel Laureate Amartya Sen. In the famous document "*Human Security Now*," Ogata and Sen (2003) emphasized the importance of understanding human security in the context of freedom from want and freedom from fear. Sen (1998) argued then that development needs to be measured as the true sense of degree of freedom, where there needs to be a balance between people-centered security and development versus state-centered security. Human security is to protect people from different types of "downward risks," which originate from equity, dignity, human rights, and freedom. Different types of development needs that involve hunger, poverty, water, population, health, and the environment must be considered. At that time, it was linked to the eight Millennium Development Goals (MDGs), which, in the current context, were changed to Sustainable Development Goals (SDGs) (UN 2015c).

Ogata and Sen (2003) pointed out that, "At the center of sustainable development is the delicate balance between human security and the environment." They also mentioned that, "Governments and other stakeholders are increasingly aware of the relationship between ecological stability and human security. The emphasis is more on environment management. However, there has been little concrete action at the local level to ensure the participation of affected communities and people in such a management scheme. Critical to this is the need to explicitly plan for improved environment management and sustainable development to disaster prevention and preparedness."

This chapter attempts to highlight the ways of reducing exposure and underlying risk factors, which are linked to enhancing human security. It analyzes the progress in DRR with respect to the underlying risk factors mentioned in an earlier global framework (the Hyogo Framework for Action [HFA]) and tries to link it to CCA. The HFA was the first global framework for DRR and was implemented between 2005 and 2015. This was followed by the Sendai Framework for Disaster Risk Reduction (SFDRR: 2015–2030). Finally, the chapter provides some directions for future relevant DRR approaches within the context of the new global disaster framework.

2.2 Reflections on the Progress in Reducing Underlying Risk Factors

In the HFA, Priority for Action Area 4 dealt with "Reducing the underlying risk factors." Six core indicators are used to determine progress. In general, self-assessment scores among the five HFA priorities revealed that Priority 1 (related to governance) obtained the highest average score (3.11) and Priority 4 received the lowest (2.92). This essentially denotes that Priority for Action area 4 is under-achieved. It was further observed that no country has essentially made "comprehensive achievement" across all core indicators under this priority area as only 13% of the countries under review in 2013 report either comprehensive or substantial achievement across all indicators (UN ISDR 2013a). Moreover, Priority for Action 4 has been recognized as the most challenging. (HFA 2013). Therefore, its significance remains highly contextual and needs further comprehensive action.

2.2.1 Major Achievements in Priority Areas

Three specific areas are looked at here, all related to human security: (1) environmental aspects, (2) DRR and CCA linkage, and (3) link between DRR and infrastructure projects.

2.2.1.1 Improvement in Environmental Legislation and Availability of Environmental Impact Assessment

Environmental safeguard measures are a prerequisite for DRR and such consideration exists from effective natural resource management, judicious use of environmental resources, conservation, and sustainable development. The core indicators broadly deal with environmental aspects related to DRR. More precisely, core indicator six describes the urgency of putting procedures in place to assess the disaster risk impacts of major development projects, especially impacts, which directly denotes the use of environmental impact assessment (EIA) of development projects. Importantly, much progress in pursuing this Priority for Action has been achieved through the development and adoption of protected area legislations, addressing DRR in climate change adaptation programs and projects, availability and application of EIA, as well as incorporation of DRR into national and sectoral public investment systems (UN ISDR AP 2013), and availability of procedures for DRR incorporation into major development and infrastructure projects. For example, Bangladesh has introduced the Wild Conservation and Preservation Act and Forest Transit Root 2011, developed ecologically critical area (ECA) management guidelines, and introduced EIA Guidelines for five sectors (Bangladesh Country Report National progress report on the implementation of the Hyogo Framework for Action, 2011–2013). Similarly, in India, a system of disaster resilience audit on self-certification basis was applied to all centrally sponsored initiatives from the project's planning stage (UN ISDR AP 2013). In Sri Lanka, the government developed special area management plans (SAMP) incorporating DRR concerns for six specific coastal locations. On the other hand, a national working group on DRR mainstreaming has been established in Pakistan to conduct impact assessments on mega-projects, such as construction of dams, highways, and irrigation facilities, at an early stage of project development (UN ISDR AP 2013). Therefore, it can be seen that significant progress in formulating legislations, guidelines, and environmental plans has been made by majority of the countries. Moreover, environmental impact assessment has been made a prerequisite for development projects in most of them. This is especially relevant to safeguarding the lives and livelihood of the poor and vulnerable, thereby enhancing human security.

2.2.1.2 Integration of Disaster Risk Reduction (DRR) in Climate Change Adaptation (CCA) Schemes

The progress and achievement of the different HFA priorities have been documented through a review of national progress reports (UN ISDR 2013a). It showed that legislative arrangements and bodies have been established to directly address DRR and/or CCA. Of the 36 countries that have done self-assessment related to HFA progress, 32 (89%) reported establishing new DRR/DRM/CCA bodies specifically to deal with climate and disaster risks. Different governments have undertaken measures to mainstream DRR and CCA into their economic and social sectors through appropriate policy, legislation, and funding mechanisms. Countries commonly known for their attention to capacity building in DRR/DRM/CCA include Japan, the Republic of Korea, China, Iran, and Australia. Bangladesh and India have also made substantial improvements in capacity building to deal with DRR along with CCA. Progress in incorporating DRR in CCA has been also reported in the Maldives, Bhutan, Pakistan, and the Philippines. In the Pacific, the Disaster Risk Reduction and Disaster Management Framework for Action 2005–2015, emphasized the development of a Joint National Action Plan for DRM and CCA. Pacific countries are working toward integrating regional platforms on DRM and CCA in their Roadmap for Integrated DRM and CCA. In fact, these two components are so tightly linked to development initiatives that it is no wonder that most of the Pacific Island countries (PICs) are already engaged in the process of DRR/DRM and CCA integration into their respective budgetary and planning processes. Furthermore, most PICs are in the middle of establishing multisectoral platforms for DRR, with special attention to CCA issues.

2.2.1.3 Incorporation of DRR in Development and Infrastructure Projects

Incorporating DRR in development projects aims to stimulate the adoption of risk-sensitive strategies in all sectors of development planning. In the 2005–2015 period, national governments have realized the significance of mainstreaming DRR into policies, plans, and infrastructure development. In the past, most infrastructure built in hazard-prone areas have negatively affected the pace of the country's economic growth. Realizing the significance and nature of linkages between disasters and infrastructure development, federal governments have successfully initiated various strategies for devising and enforcing targeted approaches to minimize vulnerabilities and mitigate the adverse consequences of disasters in development programs. The broad mission is the achievement of sustainable socioeconomic and physio-ecological development goals through reducing exposure and underlying risk factors. Such DRR policies and plans have guided key stakeholders involved in decision-making and implementation of development plans. As mentioned, in India, DRR is now an integral part of all new development projects under the close monitoring of its finance ministry. This has been made possible by the federal governments' commitment in incorporating DRR into policymaking and planning. Similarly, Bangladesh has expedited DRR mainstreaming into all sectors through risk-sensitive development planning and implementation and reinforcing climate—and disaster-resilient development. Governments are now paying serious attention to establishing mechanisms for capacity enhancement and operationalizing plans in their respective functional areas. In Pakistan, a national working group has been established to mainstream DRR in mega-projects at an early stage of project planning and development. The federal governments are now closely collaborating with regional and local planning authorities in mainstreaming DRR in all infrastructure development planning and programs.

2.2.2 *Major Gaps and Challenges*

Based on the official report of the United Nations International Strategy for Disaster Reduction (UNISDR) and the country self-assessment reports, the weakest areas of progress under the Priority for Action Area 4 can be identified as follows: (1) little or no consideration of DRR/DRM in national and sectoral public investment in most development sectors and non-recognition of DRR as a cross-cutting issue within different sectoral investments; (2) absence of land-use policies and zoning systems, resulting in inappropriate and unsafe development; (3) insufficient institutional capacity in economic and productive sectors; (4) absence of risk financing and risk transfer mechanisms; and (5) disregard for community-based adaptation and DRR.

2.2.2.1 Little or No Consideration of DRR in National and Sectoral Public Investment

There is a strong interconnection between DRR and sectors such as infrastructure, construction, irrigation, agriculture, education, health, and livelihood. However, how much investment is actually being spent on DRR in each of these sectors has not generally been clear. Strong coordination and integration are necessary to achieve the desired DRR outcomes as expressed in the HFA. Concerted efforts in integrated planning and implementation, supported by specific by-laws and concrete enforcement mechanisms, are required to translate sectoral public investment in DRR into action and intended impacts. Credible evidence gathered through quality research should be utilized to make the business case of sectoral DRR investments in local communities. Demonstrating the positive return on investment for actions taken to reduce underlying risk factors may improve public commitment to such initiatives (UN ISDR 2013a). Reducing vulnerability and exposure as significant drivers of risk is a challenging task for national and local governments. However, with the sectoral approach, it is hoped that actions to be taken will be more focused and direct, and therefore, more effective (UN ISDR AP 2013).

2.2.2.2 Enforcement of Land-Use Policies and Zoning System

Land-use policy and zoning system is one of the key strategies in reducing exposure and underlying risk factors. The ministers attending the 5th Asian Ministerial Conference on Disaster Risk Reduction (AMCDRR) in Yogyakarta called on DRR stakeholders to build and sustain capacities and legal mandates of national and local governments and the private sector to integrate DRR in land-use policies and plans. Several countries have devised and adopted legislation mechanisms to regulate building codes, land-use zoning, and urban development, but, generally, they have significantly poor capacity to operationalize and enforce such laws and regulations. For example, in Indonesia, the major challenge, in addition to overlapping regulations, is poor enforcement of the zoning system. Efforts have been made to implement these regulatory and policy instruments in true spirit, but, so far, the process is not successful. Physical development and encroachments onto the active floodplain have increased human exposure to flood risk. However, flood risk assessment and mapping have not been carried out to designate risk-sensitive zoning and enforcement. In their respective country reports, the governments of assessment and Pakistan admit not having regulations for risk-sensitive land-use zoning and for private development in the floodplain, coastal zones, and hilly areas. Similarly, in Turkey, construction is undertaken without taking into consideration effective building codes. In Afghanistan, lack of financial resources is the major barrier in implementing building codes and guidelines for earthquake-resistant designs. The countries recognize the need for concerted efforts in integrated planning and implementation, supported by specific by-laws and mechanisms for enforcement. Dissemination of hazard risk information in vulnerable areas would help further strengthen the enforcement procedure.

2.2.2.3 Insufficient Institutional Capacity in Economic and Productive Sectors

In general, countries lack institutional capacity in the economic and productive sectors. It is therefore essential that governments pay special attention to enhancing institutional capacity for hazard mitigation, preparedness, response, and recovery. Although countries have already prepared national-, state-, and district-level disaster management plans, these have not been effectively implemented primarily because of weak institutional capabilities and lack of financial resources. In the case of Afghanistan, climate change brought increased frequency of droughts, reducing agricultural production. The country's weak institutional capacity poses a major challenge. Another challenge is the low DRR knowledge and understanding at the institutional level and low technical expertise. Financial constraints and lack of appropriate budget allocation to implement HFA at local and regional levels constitute a great barrier. This gap can be addressed by involving line ministries at every stage in the process, from formulating policies to their execution. The immediate prerequisite is to assess technological, institutional, and policy options to rationalize the role of organizations to further strengthen institutional capacities in the economic and productive sectors. The needs of marginalized groups, including children, and gender inequality must be properly addressed in all phases of disaster risk management.

2.2.2.4 Disregard for Community-Based Climate Change Adaptation and Disaster Risk Reduction

Since the implementation of HFA, there has been a growing concern among governments to consider CCA vis-a-vis DRR. In the past three assessment cycles, the institutional approach has gained momentum and several federal governments have adopted multiple schemes for CCA and DDR. The institutionalization of CCA and DRR has certainly been a welcome change; however, it is extremely important to involve the communities at risk in a more proactive manner. The existing approach undertaken by a majority of the countries has been, by and large, a top-down hierarchical approach. To "reduce the underlying risk factors," several social and economic schemes were put in place to enhance the resilience of the vulnerable population. However, in spite of the existence of such policies, ground-level implementation has been poor. Therefore, it behooves governments to involve the communities into the mainstream CCA and DRR process. For example, Bangladesh has recently started a pilot project, the Community-Based Adaptation to Climate Change (CBACC) project, through afforestation in the coastal vulnerable areas of four districts. Importantly, the community-based DRR approach is now strongly emerging, but this requires institutionalization and legal sanctions, which have been lacking in most of the countries. In the wake of such need, more community-based approaches should be used at both policy and implementation levels.

2.2.2.5 Absence of Risk Financing and Risk Transfer Mechanisms

Many governments report the lack of capacity of the domestic insurance sector as a significant barrier to progress in risk financing and risk transfer (UN ISDR 2013a). Insurance pricing mechanisms neither reflect risk levels nor provide adequate incentive for risk-sensitive business investment, particularly in countries with low insurance penetration rates but rapidly growing markets (UN ISDR 2013a). In China, for example, only 3% of the property is insured against earthquakes and only 5% against typhoons and floods (UN ISDR 2013a). Some countries, such as Vietnam, report a general lack of insurance culture. Crop insurance, for example, is available to farmers, but it is rarely availed of (UN ISDR 2013a). It has been shown that businesses that have invested in risk management may economically outperform their competitors (UN ISDR 2013a). Similarly, recent case studies in the agricultural sectors of selected countries in the Asia-Pacific region showed that "risk insurance has a potential to provide a cost-effective means of covering financial shocks to the insured" (Prabhakar et al. 2017). The same study, however, noted that a comprehensive assessment framework is needed to differentiate various forms of insurance products and establish their effectiveness. Although governments report significant progress in investing more to address risks, the required shift to anticipate risks in non-government public and private investments remains a challenge.

2.3 Discussion and Key Recommendations

2.3.1 Reduce the Underlying Risk Factors for Poverty Reduction and Sustainable Development

It is recognized worldwide that efforts related to DRR must be systematically integrated into policies, plans, and programs for sustainable development and poverty reduction (UN ISDR 2007). Of the three strategic goals of HFA, the first one is crucial as it aims to achieve a more effective integration of disaster risk considerations into sustainable development policies, plans, and programs at all levels (UN ISDR 2007). However, a systematic action to address disaster risk in the context of sustainable development is a big challenge, as predicted by scholars way back in

1994 in drafting the Yokohama Strategy. Although sustainable development, poverty reduction, and DRR are considered mutually supportive objectives after two decades of implementing the Yokohama Strategy, a systematic and long-term integration of sustainable development and poverty reduction programs with disaster management is rare (Yodmani 2001; UN ISDR AP 2013). The first Global Assessment Report (UN 2009) on Disaster Risk Reduction, Risk, and Poverty in a Changing Climate has shown how different risk drivers, including disasters, enhance poverty. UN (2009) emphasized that these underlying risk drivers are crucial to achieve not only HFA but also sustainable development and CCA.

There is no doubt that disasters have a disproportionate impact on the poor. They suffer most and they are more affected (UN 2009; UN ISDR 2007; UN ISDR AP 2013; Parvin and Shaw 2013; Johnson 2006; Yodmani 2001). In the last two decades, both poverty reduction and disaster management approaches have a paradigm shift. The first approach has shifted from income poverty to human poverty and the second has shifted from top-down relief and response to intersectoral risk management. There are, therefore, wider rooms for the poor in the present DRR approach. At the same time, donors are pushing to incorporate sustainability issues and environmental considerations into poverty reduction strategies (Strange and Bayley 2008). But still, it is a long way to go to transform a poverty reduction strategy into a sustainable development strategy. On the other hand, national progress reports related to HFA show rising awareness and efforts to foster resilience and introduce DRR concepts to poor and vulnerable communities. However, all these attempts are disconnected, inadequate, and unable to fully capture all underlying risk factors associated with poverty and sustainable development (HFA-Asia Pacific 2011–2013).

Holistic and comprehensive approaches addressing DRR, poverty reduction, and sustainable development are needed. Developing mutually supportive, innovative, multidimensional, and intersectoral policies, plans, and programs for risk reduction, poverty alleviation, and sustainable development is of utmost importance. Governments, non-government organizations (NGOs), civil societies, the private sector, and local communities will have to work together to bring CCA, DRR, poverty alleviation, and sustainable development on a single platform.

2.3.2 Enhance Economic Viability of Risk Reduction Measures

The number of reported disasters as well as economic losses associated with disaster events is increasing across the globe, and the Asia-Pacific region is no exception. A global assessment report on DRR (UN ISDR 2013b) stated that "Over the coming years, trillions of dollars of new business investment are set to pour into hazard-exposed regions. How the private sector—accounting for 70–85% of total investment—decides to place its funds will largely determine how much disaster risk is accumulated and how underlying risk drivers are addressed." The report also estimates average potential losses from future disaster risk reduction (DRR) and cyclonic winds alone at USD 189 billion per year.

Despite these alarming estimates of loss and damage, economic viability of risk reduction measures remained an underrated issue. The last decade witnessed enormous international effort toward bringing a significant change in the regime of risk reduction, which is now changed from reactive to proactive, albeit at a varying degree. While this change has galvanized governments to prepare disaster management plans, improve response-preparedness, enhance early warning systems, build capacity of government functionaries, etc., it is still far from reaching out to the private sector, especially small and medium enterprises (SMEs). Large corporations, especially those that experienced major catastrophes in developed countries such as the Kobe earthquake and Hurricane Katrina, have been working on business continuity planning to ensure that their businesses have the least disruption in case of a disaster. However, due to resource and capacity constraints, SMEs are unable to realize the benefit of investing in risk reduction measures.

It has been often quoted in DRR-related documents that an investment of a dollar during the pre-disaster phase can save about USD 7 in the recovery/rebuilding phase. However, in developing countries of the Asia-Pacific region, governments find it difficult to even cater to the developmental aspirations of society. Education, health, livelihood and employment, infrastructure, provision of basic services, etc. receive a major share of national budgets. Mainstreaming DRR calls for dovetailing risk reduction measures in development programs, however, barring a few countries, mainstreaming remains an unfinished agenda. Thus, higher education institutions, local business groups, and NGOs need to work together on several fronts:

- Carry out a number of localized loss and damage studies in specific sectors of the economy. These studies should be able to spell out cost-benefit advantage in investing in DRR measures.
- Investigate innovative, practical, and implementable tools to reduce losses and damages, especially for SMEs. Collaborative and cooperative risk-sharing and risk transfer mechanisms may supplement innovative tools to effectively distribute disaster-related losses.
- Raise awareness through targeted training programs, focused group discussions, engaging local media and schools for effectively conveying the benefits of investing in DRR measures.
- Organize need-based capacity development programs for local businesses, financial institutions, and informal sectors of the economy to inform and engage in finding optimum solutions to reduce risk via available tools.
- Implement simple risk reduction measures requiring minimal investment such as acting on early warning, raising plinth of houses and shops in flood-prone areas, building seismic-resistant tenements for livelihoods and residence, forest and mangrove rehabilitation, etc., are some of the low-hanging fruits offering no-regret solutions to sensitively invest in DRR.

2.3.3 Institutionalize Community-Based Disaster Risk Reduction

The importance of community-based disaster risk reduction (CBDRR) has already been documented and viewed as one of the key factors in achieving the goals of HFA (Shaw 2012). Past disasters such as the Great Hanshin Earthquake in 1995 have introduced important CBDRR concepts like "self-help" and "mutual-help," which advocate enabling the communities to become effective first responders to disasters before public help is available. The East Japan Earthquake and Tsunami in 2011 revealed that, even with substantial investments in DRR infrastructure and technology, it is ultimately the coping capacity of communities that determines the fate of their survival, especially for mega-disasters. Recognizing the need to continuously strengthen CBDRR, various governments, international donors, and community-based organizations in the Asia-Pacific region have developed and implemented projects with different issues targeted. On the other hand, the effectiveness and sustainability of CBDRR efforts have repetitively been challenged because there is still a strong tendency for CBDRR to be project- or pilot-based, thus making continuity of activities difficult for local stakeholders when the project period is finished due to lack of funding and guidance. In this situation, CBDRR may be conducted only on the short-term or on an ad hoc basis, making evaluation problematic because measurable outcomes from CBDRR activities commonly take longer to see.

Effective and sustainable implementation of CBDRR lies in placing efforts to institutionalize good practices conducted at the community level. First, good practices should be compiled and studied at the central and/or regional government levels and made part of DRR policy frameworks and action planning. Second, provision for establishing a general account budget, specifically for CBDRR at the local level, is essential to ensure continuous implementation of activities. Regular monitoring and evaluation of activities through community participation is also important for local governments and communities to understand their strengths and weaknesses as well as to enable them to adjust to new DRR needs and challenges. Building partnerships of various stakeholders from different sectors is of utmost importance, but it is not common to see specific cases of these partnerships being institutionalized. Finally, as CBDRR efforts also contribute to building sustainable communities, these must be integrated into community development plans and also be incorporated into community events, such as traditional festivals and other social functions of the locality.

The primary actor for institutionalizing CBDRR is the local government, but it is important that other community stakeholders participate in the process. In some countries, where ODA plays an important role, a system can be devised to institutionalize community consent over investment decisions for development that may pose risks for community members.

2.3.4 Address Risk Reduction in Recovery

Recovery from disasters can be considered a development opportunity, which is often used or misused (Shaw 2013). In the aftermath of a disaster, there is always pressure to quickly restore support systems and livelihood and repair damages. In most of the cases, this undermines the quality of relief, reconstruction, and rehabilitation work. Time pressure and other constraints such as difficulties in communication and transport in the post-disaster environment make it difficult to restore lives and livelihood with enhanced resilience. However, recovery is a balance between speed and quality. Speed is higher when it is done in a centralized way, by a single agency. But when it comes to cooperation and collaboration among different stakeholders, departments, and agencies, the process becomes slow.

There is a general consensus to address risk reduction or resilience building through the recovery process. The recovery program instituted after the 2001 earthquake in Gujarat, India, was used effectively to enhance building safety in the region and to address development concerns such as earthquake, health, education, etc. A similar trend was also observed after the 2004 tsunami in Aceh, where land-use planning and coastal regulations were improved. The post-2010 Pakistan flood also introduced strong land-use regulations. In post-2011 East Japan earthquake and tsunami, the roles of schools and vital infrastructure in the coastal areas were revisited. This resulted in specific changes in the education system, with more focus on governance issues and disaster education and raising community awareness.

Shaw (2013) has introduced the governance, education, and technology framework in post-disaster recovery, which leads to resilience building of local communities. It suggests specific measures for regulatory and institutional changes (as they relate to governance), education of and raising awareness among different stakeholders, and dissemination of technology-related information, which can be linked to longer term issues and thereby give way to effective DRR practices.

2.3.5 Enhance DRR Education to Help Reduce Underlying Risk Factors

The goal of developing disaster-resilient communities heavily depends on the success of DRR education (Petal 2008 as cited in Shaw et al. 2011a). The continuous implementation of formal and non-formal DRR education, linked to community-based DRR, promises the development of a culture of safety that can make societies less vulnerable and more resilient to the impacts of disasters in the future. In some countries, DRR education has been integrated at the primary, secondary, and tertiary levels as well as in the training of teachers. DRR education and training modules should also be prepared for cities, municipalities, and provinces to enhance the capacity of local government officials and specialized DRR personnel to develop, manage, and administer non-formal DRR education programs and information and education

(IEC) efforts for the citizenry. Given the number of local government units across countries, this is expected to be a massive undertaking for providers of DRR education at all levels. To help reduce underlying risk factors, DRR education should include the impacts of natural hazards as well as the effects of changing social, economic, and environmental conditions on the risk exposure and vulnerability of citizens. Learners should be aware of the nature of every disaster and their impacts on health, education, agriculture, livelihood, and infrastructure.

Mechanisms of support can help build an enabling environment for DRR education to be successful. First, strong leadership at the education ministry is needed in pursuing nationwide school-based initiatives to tackle DRR. Second, school support networks are necessary. Schools lack the resources to implement DRR education initiatives on their own. Thus, partnerships and sponsors are needed to sustain such projects. Private sector companies, NGOs, and civil society organizations can partner with public schools to provide support, including facilities, infrastructure, textbooks, computers, science laboratory equipment, and teaching and skill development. To ensure a broad and diverse base of resources and inputs, school officials should reach out to local emergency agencies, government committees, service organizations, parent–teacher associations, and businesses. Primary and secondary schools should also develop partnerships with tertiary schools, especially for support in teacher training. Third, the potential of using social media should be explored as the Internet and mobile phones are becoming very popular across all age groups. Social media can enable educators to reach a wider audience in a cost-effective manner. And, finally, a cross-sectoral approach should be used. DRR is not an issue confined to one specific sector, so addressing it effectively requires inputs from a wide variety of stakeholders, some of whom may have a better understanding of what is needed at the local level. Budgetary constraints remain one of the challenges in implementing DRR education strategies so this issue should be addressed by providers and supporters of DRR education.

The other important part of education is higher education. In recent years, several universities in the Asia-Pacific region have started masteral programs on DRR (Shaw et al. 2011a, b). This is important to develop DRR as an academic discipline. With 24 years of experience, risk reduction can be recognized as an academic discipline through its incorporation in the university curriculum, generating young professionals who can tackle the subject matter with academic rigor.

2.4 Final Remarks

The new Sendai Framework for Disaster Risk Reduction (SFDRR) (UN 2015a) has a strong focus on CCA, sustainable development, and human security. The two important features of the SFDRR are the synergies among the global frameworks and initiatives and the focus on local implementation. In a 25-page document, SFDRR mentions sustainable development 20 times and climate change, 15 times. Also, it uses "local" 48 times in the context of government, community, knowledge, priority, and DRR strategies. Thus, in the last several years, DRR and CCA have come closer to each other, with human security and sustainable development goals providing the overall umbrella for these frameworks. Institutional barriers are evident globally, regionally, nationally, and locally, and different institutions and organizations are in charge of these separately. However, synergies and interconnectivity are gradually increasing.

The new urban agenda is also another important global framework, especially in the context of increasing trends of urbanization. Thus, it is important to reduce the risks faced by human settlements, in particular those in urban areas, through risk-sensitive management. It is key to reducing exposure and underlying risk factors and achieving community resilience. This is especially relevant to Asia, which is the epicenter of urbanization. Population concentration is not only limited to large cities; medium and small cities are also registering unprecedented growth. The scale and volume of development in these cities are challenging the entire urban growth paradigm, including urban systems, urban governance, and urban planning. Economic growth is also observed in most countries of Asia, but, this has resulted in a widening gap among different income groups and has given rise to inequality. If not urgently dealt with, this trend may negatively impact the fragile fabric of community cohesion and thus hinder community resilience.

The other focus is to enhance government-science and technology collaboration to improve evidence-based and informed decision-making to address underlying risk factors. Incorporating DRR in higher education schemes should facilitate this process by generating young professionals who can transform decisions into actions. The Asian region brings a stark contrast in this regard as a significant proportion of its population is still struggling without access to basic amenities and infrastructure to be able to have decent quality of life. Governments in the developing countries of Asia are striving to strike a balance between achieving development priorities and reducing underlying risk factors. Both technical and professional capacities, especially at the local level, do not match the increasing demand for risk reduction professionals. Academic and technical organizations, including universities, national and regional research organizations, expert NGOs, and others will be helpful in bridging this gap. Initially this collaboration should be strengthened at the national level but should gradually percolate to the local government level as well.

References

Johnson JD (2006) Natural disaster and vulnerability. OECD Development Center Policy Brief no. 29. OECD Press, Paris

Ogata S, Sen A (2003) Human security now. Human Security Commission, New York

Parvin GA, Shaw R (2013) Role of microfinance institutions (MFIs) in coastal community's disaster risk reduction, response and recovery: a case study of Hatiya. Bangladesh Disasters J 37(1):165–184

Petal M (2008) Concept note: formal and informal education for disaster risk reduction. http://www.riskred.org/activities/ddredislamabad.pdf

Prabhakar SVRK, Solomon DS, Abu-Bakar A, Cummins J, Pereira JJ, Pulhin JM (eds) (2017) Case studies in insurance effectiveness: some insights into costs and benefits. Southeast Asia Disaster Prevention Research Institute, Bangi, Malaysia, p 123
Sen A (1998) Development as freedom. Oxford Press, London
Shaw R, Tran P (2012) Environment and disaster linkages. Emerald Publisher, UK, p 354
Shaw R, Takeuchi Y, Gwee Q, Shiwaku K (2011a) Disaster education: an introduction. In: Shaw R, Shiwaku K, Takeuchi Y (eds) Disaster education. Emerald Group Publishing, Bingley, UK, pp 1–22
Shaw R, Mallick F, Takeuchi Y (2011b) Essentials of higher education in disaster risk reduction: prospects and challenges. Disaster education. Emerald Publisher, UK, pp 95–113
Shaw R (2012) Community-based disaster risk reduction. Emerald Publisher, UK, p 402
Shaw R (2013) Disaster recovery: used or mis-used development opportunities. Springer, Tokyo
Strange T, Bayley A (2008) Sustainable development: linking economy, society, environment. Organization for Economic Cooperation and Development, France
UN (United Nations) (2009) Global assessment report on disaster risk reduction, risk and poverty in a changing climate. UN, Geneva
UN (United Nations) (2015a) Sendai framework for disaster risk reduction 2015–2030. UN ISDR document. UN, Sendai, Japan, p 25
UN (United Nations) (2015b) Transforming our world: the 2030 agenda for sustainable development. UN General Assembly Resolution. United Nations, New York, USA, p 35
UN (United Nations) (2015c) Adoption of the Paris Agreement. UNFCCC COP 21st Session. United Nations, France, Paris, p 32
UN ISDR (United Nations International Strategy for Disaster Reduction) (2007) Hyogo framework for action, 2005–2015: building the resilience of nations and communities to disasters. UNISDR, Geneva
UN ISDR AP (United Nations International Strategy for Disaster Reduction–Asia and Pacific) (2013) The Hyogo framework for action in Asia and the Pacific: regional synthesis report 2011–2013. http://www.unisdr.org/we/inform/publications/32851
UN ISDR (United Nations International Strategy for Disaster Reduction) (2013a) Implementation of the Hyogo framework for action: summary of reports 2007–2013. http://www.unisdr.org/we/inform/publications/32916
UN ISDR (United Nations International Strategy for Disaster Reduction) (2013b) Hyogo framework for action in Asia and the Pacific. UNISDR, Geneva
Yodmani S (2001) Disaster risk management and vulnerability reduction-Protecting the poor. Paper presented at Social Protection Workshop 6: Protecting Communities—Social Funds and Disaster Management under the Asia and Pacific Forum on Poverty, 5–9 Feb 2001, Asian Development Bank, Manila, Philippines

Dr. Rajib Shaw is Professor at the Keio University in Japan and former Executive Director of the Integrated Research on Disaster Risk (IRDR) programme. He is also Senior Fellow of Institute of Global Environmental Strategies (IGES) Japan, and Chairperson of SEEDS Asia, and CWS Japan, two Japanese NGOs. Previously, he was Professor at the Graduate School of Global Environmental Studies, Kyoto University, Japan. His expertise includes community-based disaster risk management, climate change adaptation, urban risk management, and disaster and environmental education. He is the editor of a book series on disaster risk reduction, published by Springer, Editor-in-Chief of the journal Progress in Disaster Science. He severs as Chair of United Nations Science Technology Advisory Group for disaster risk reduction, and is Co-chair of the UNDRR's Asia-Pacific Science Technology Academic Advisory Group (ASTAAG). Prof. Shaw has published more than 45 books and more than 300 academic papers and book chapters.

Dr. Juan M. Pulhin is full Professor and former Dean of the College of Forestry and Natural Resources, University of the Philippines Los Baños, Philippines. He was Lead Author of the 2007 Fourth Assessment Report of the Intergovernmental Panel on Climate Change (IPCC), Coordinating Lead Author of the Human Security Chapter of the 2014 IPCC Fifth Assessment Report and currently Lead Author of the Asia chapter of the IPCC Sixth Assessment Report. He has been a four-time a visiting professor at The University of Tokyo in 2007, 2010, 2016 and 2017 and a visiting scholar at Waseda University, Tokyo in 2018. He has written and edited 11 books and published more than 100 technical papers and book chapters on natural resources policy and governance, and on the human dimension of climate change and disaster risks. He is a fellow, invited expert, and council member of several international organizations and a member of the National Academy of Science and Technology (NAST), Philippines.

Dr. Makoto Inoue is Professor at the Faculty of Human Sciences, Waseda University, Japan and former Professor at The University of Tokyo. He serves as Dean of Graduate School of Human Sciences, Waseda University since September 2020. His specialty is environmental sociology, forest governance, and Southeast Asian studies. He has written and edited/co-edited 34 books, and published more than 140 original articles in peer-reviewed international and national journals on the social aspects of forestry and the environment. He serves as President of The Japanese Association for Environmental Sociology (June 2019–present), Vice President of Japanese Society for Current Anthropology (January 2020–present), Vice President of Society for Environmental Economics and Policy Studies (April 2018–March 2020), Councilor of The International Association for the Study of the Commons (January 2019-present), a board member of various academic societies in Japan, and an editorial board member of several international journals.

Chapter 3
State Fragility and Human Security in Asia in the Context of Climate and Disaster Risks

S. V. R. K. Prabhakar and Rajib Shaw

Abstract Climate-fragility risks are on the rise globally and developmental paths taken by countries could be attributed to this rise. To understand the link between development and climate-fragility risks, this chapter presents the findings of the survey conducted by the authors with development practitioners and experts from Asia. The purpose of the survey was to elicit responses regarding the perceptions of stakeholders engaged in disaster risk reduction and climate change adaptation on climate-fragility risks, their understanding of the subject, and their opinion on how to address them. To explain differences in survey findings, the chapter also presents a Climate-Fragility Risk Index (CFRI) that provides a systematic analysis and presentation of various climate-fragility indicators. The survey indicated diverse views on what constitutes climate fragilities with clear differences between developed countries and developing countries. Majority of respondents opined that the climate-fragility risks are additional to disaster risks while others said that such risks emerge as a result of interaction between climate change and developmental pressure. Most respondents from Japan had a low understanding of climate-fragility risks (CFRs) at the policy level, while majority of responses from developing countries in Asia revealed a very low understanding on the subject at the policy level. Climate change as a major driver affects livelihoods, worsens disaster impacts, and leads to transboundary conflicts. Major challenges such as dysfunctional institutions and developmental deficit, including migration and internal displacement, are directly bearing upon the fragility risks of the developing countries. For developed countries such as Japan, demographic stress such as aging population is a significant factor to be addressed. As a way forward, this chapter advices the countries to formulate and implement well-coordinated and evidence-based policies in the increasingly integrating world with better investments in understanding drivers and pressures behind CFRs and better information sharing for well-informed decision-making.

S. V. R. K. Prabhakar (✉)
Institute for Global Environmental Strategies (IGES), Kanagawa, Japan
e-mail: prabhakar@iges.or.jp

R. Shaw
Keio University, Kanagawa, Japan

J. M. Pulhin et al. (eds.), *Climate Change, Disaster Risks, and Human Security*, Disaster Risk Reduction, https://doi.org/10.1007/978-981-15-8852-5_3

Keywords State fragility · Human security · Asia · Disaster risk reduction · Climate-fragility risk index

3.1 Introduction

Climate change is one of the key global security challenges of the twenty-first century. Its impacts are a "threat multiplier" that will increase state fragility, fuel social unrest, and potentially result in violent conflict. Existing state fragility is simultaneously hampering efforts at adaptation, particularly among vulnerable populations. This threatens to lock many societies into "fragility traps."

The foreign ministers of the Group of Seven Countries (G7) representing Canada, France, Germany, Italy, Japan, and the USA discussed and welcomed the independent report *A New Climate for Peace: Taking Action on Climate and Fragility Risks* that has championed the concept of climate-fragility risks, which was commissioned by G7 members in April 2015 (Ruttinger et al. 2015). The report identified seven compound climate-fragility risks that pose serious threats to the stability of states and societies in the decades ahead (these are further elaborated in the following section): (a) local resource competition, (b) livelihood insecurity and migration, (c) extreme weather events and disasters, (d) volatile food prices, (e) transboundary water management, (f) sea level rise and coastal degradation, and (g) unintended effects of climate policies. The report emphasized the need for developing integrated policy agenda and brought to the fore the importance of addressing the compounding of risks at various scales and policy domains.

This paper identifies the specific climate-fragility risks and their implications on the developmental discourse both in developed countries, as represented by responses from Japan, and developing countries, as represented by responses from Afghanistan, Bangladesh, China, India, Indonesia, Malaysia, Nepal, Pakistan, the Philippines, Sri Lank, Thailand, and Vietnam, and the nature of renewed collaboration that is necessary between these countries to address this issue in a new climate regime. It also provides a deeper understanding of the nature of climate-fragility risks for developed countries such as Japan face and identifies ways forward, keeping in view the developing and developed countries.

3.2 Climate-Fragility Risks are Significant

The concept of climate-fragility risks looks at climate change as the ultimate "threat multiplier" and says that it will aggravate fragile situations and may even contribute to social upheaval and even violent conflict. Risks are compounded when the impacts of climate change interact with other problems that states are already facing. Even though individual risks may appear to be simple at first sight, the combination of these risks and their interaction with climate change could ultimately overburden even the strongest countries. Consequently, such compounding may prevent countries from adapting successfully to climate change, unable to address security issues, thus trapping countries in a vicious cycle.

Globally, climate-fragility risks are growing, albeit more rapidly in some regions than in other regions. Several climate-fragility risks have come to fore that are important to be addressed in order to achieve the goal of sustainable development. The most important climate-fragility factors include the (a) competition for resources, (b) livelihood insecurity and migration, (c) extreme weather events and disasters, (d) volatile food prices, (e) transboundary resource management, (f) sea level rise, and (g) unintended effects of climate policies (based on Ruttinger et al. 2015).

1. Competition for resources: While competition for resources is already increasing globally, the trends of declining access to natural resources, particularly water and arable land, increasing demand of growing populations, and rapid economic development will further spur competition over essential resources, increasing tension, and contributing to local conflicts in the years to come. The impact of this growing resource competition in G7 countries could be significant and can have severe consequences as it interacts with geopolitical issues.
2. Livelihood insecurity and migration: Livelihoods of people dependent on natural resources will be endangered by climate change. Impacts in terms of dried-up water sources and threats to jobs connected to climate-sensitive economic sectors are very much relevant for the developed countries as well. Increased pressure on agriculture could spur migration to already heavily urbanized regions, leading to high unemployment. While migration could be a way of coping with climate change shocks, unplanned migration could put undue pressure on the resources and deprive regions of valuable human resources, pushing them further down the line of deprivation.
3. Extreme weather events and disasters: G7 countries are already vulnerable to several extreme weather events and climate change is expected to alter the course of extreme events to detrimental impacts. This means that existing insurance mechanisms and other methods to cope with the impacts of disasters may fall insufficient and government assistance may be called more often. Increasing disaster intensity in the already vulnerable and poorer regions could lead to enhanced cross-border migration, putting pressure on some of the G7 countries. When combined with the internal security situation, the pressure of immigrant influx would further increase the demand for concerted action among G7 countries.
4. Volatile food prices: G7 countries have relatively well-placed food pricing and public support systems. However, the increasing reliance of G7 countries on food imports from developing regions means greater exposure to climate change impacts in countries with increasing food price volatility.
5. Transboundary resource management: Transboundary resources such as river systems play an important role in the well-being of millions of people due to their enormity and climate change could pose a serious threat to these water resources in combination with the unplanned development happening in their

catchments. As a result, there is renewed interest in managing the transboundary water resources through intergovernmental mechanisms whose functioning in turn depends on the relationship between countries. Increasing pressure on governance structures, especially when water management is eclipsed by political considerations or affected by power asymmetries, needs special attention for an effective management of transboundary water resources.

6. Sea level rise and coastal degradation: Rising sea levels already threaten the economic and physical viability of low-lying areas and no country will be entirely free from this threat. High-value investments have been made along the coastal areas in G7 countries and these investments are at high risk of loss and damage. Implications of changing coastline on maritime boundaries need to be considered to avoid conflicts over territorial and sea lanes and ocean resources.
7. Unintended effects of climate policies: Well-intentioned policies could impair economic development, undermine political stability, and exacerbate insecurity if these policies are made only keeping the short time horizons in view and if they do not consider the climate change impacts in the long run. G7 countries are in general rated well in various governance indicators, including regulatory quality, governance effectiveness, and rule of law. However, unforeseen effects often arise due to the lack of cross-sectoral coordination and conflict-sensitive planning, which are common issues even in G7 countries. There is a need to discuss the implications of climate-fragility risks on various major policy areas in G7 countries to ensure effective implementation in the years to come.

While the report was based in the context of the G7 countries, it also identifies the connections between G7 countries with the rest of the world and how these connections could influence the climate fragility of G7 countries. Based on a thorough assessment of existing policies on climate change adaptation, development cooperation and humanitarian aid, and peacebuilding, the report recommends that the G7 countries take concrete action, both as individual members and jointly, to tackle climate-fragility risks and increase resilience. Even though the report and the G7 foreign ministers took an explicit geopolitical perspective on climate change, the compound risks identified need to be addressed on a priority basis, including addressing issues such as energy security, climate mitigation and adaptation, and disaster risk reduction (DRR).

Further to the release of the report, recognizing the urgency of addressing the climate-fragility risks, the G7 Foreign Ministers Meeting in Hiroshima, disaster risk reduction, in April 2016 stressed that fragility issues be addressed through promoting resilience measures among the member states. The joint communique of G7 Meeting stated the following:

- "We reiterate that climate change poses a serious threat to **global security** and economic prosperity and shared the view that foreign policy must contribute to addressing this challenge effectively."
- "We will work to prioritize prevention of **climate fragility risks** by aligning our efforts toward the common goal of increasing resilience and reducing fragility in the face of global climate change, including taking steps to integrate climate-fragility considerations across our national governments."

A strong nexus between development and disasters is emerging as an important issue as developing regions are highly exposed to natural disasters and the developing and poor countries are disproportionately impacted by these natural disasters. As a result of natural disasters and other reasons such as state failure to provide gainful livelihood options and development services and conflict situations in part, internal displacement is on the rise. Globally, natural disasters play a major role in internal displacement than conflicts and internal security issues. The internal displacement in South and East Asia is characterized by natural disasters, whereas that in west Asia could be due to conflict situations. Despite the importance of internal displacement, countries in these regions have not done well in addressing these issues proactively and strategically.

Connected to this context is also the issue of price instability of food and other resources in the region. Several countries were severely impacted by the 2008 and 2011 inflation of food prices with more disproportionate impact on the urban poor, including casual and unskilled laborers, petty traders, laborers, and peri-urban agriculturists. Instances of 10–15% decline in food consumption (15–20% increase in food expenditure) in 50–70% of poor households from 2007 to 2008 (WFP 2009), food riots, and poverty impacts were reported. Despite this, governments have not been well prepared to deal with this situation, leading to significant impact on food security, poverty, and livelihood. Reports also linked the peak food prices with that of the Arab spring (Guardian 2011).

Transboundary rivers play an important role in the well-being of regions as big river basins harbor millions of people. Despite the importance of these river resources and the density of population and possible impact on their well-being, countries often have failed to reach an agreement on how best to use these water resources, leaving conflict especially in South and East Asian countries. Most countries in Asia have not signed or ratified any international treaty on the use of transboundary water resources. The UN Watercourses Convention has not been ratified by countries in Asia (since its inception in 1997, only 40+ countries either have ratified/signed/approved the convention text and the convention has not yet come into force due to the delay). The UN Convention on the Protection and Use of Transboundary Watercourses and International Lakes brought out in 1992 has been ratified only by 44 countries, none from Asia that have significant transboundary rivers. Regional processes such as the South Asian Association for Regional Cooperation are far from being successful in addressing these issues. The water and energy nexus is in particular an important issue that has led to conflicts among central Asian countries. In 2012, a major dispute developed over water allocation and management among the Central Asian countries of Tajikistan, Kyrgyzstan, and Uzbekistan (Janusz-Pawletta and Gubaidullina 2015). In April 2012, Tajikistan protested that Uzbekistan stopped the natural gas supply because of a disagreement over a hydropower dam.

Other fragility risks include the negative impact of policies on resources such as overexploitation of groundwater as can be seen in western Bangladesh, saltwater intrusion-related impact on drinking water quality, disputes along the local river (Indian rivers Kaveri and Krishna shared between the states of Tamil Nadu,

Karnataka, and Andhra Pradesh have been under dispute for several years), and canal courses on the use of these waters, which have not received much policy attention.

Keeping this context in view, an online survey was carried out to understand the perception of stakeholders engaged in development, disaster risk reduction, and climate change adaptation (discussed in subsequent sections of this chapter).

3.3 Climate-Fragility in the Context of Japan

In the case of Japan, the impact of climate-fragility needs to be seen in both internal and external contexts.

3.3.1 Internal Factors of Climate-Fragility

The internal climate-fragility risk is mostly related to disaster events in Japan. One of the key issues for Japan is its demographic context, especially high aged population and high rate of depopulation in the rural areas. The small- and medium-sized towns and villages in the rural areas can be considered as highly climate fragile. As an example, in the year 2011, the city of Nachikatsuura in Wakayama Prefecture was affected strongly by Typhoon No. 12 (Chiba and Prabhakar 2017). Due to severe landslides and mudflows, several areas became uninhabitable, and a new zonation map was issued based on technical studies. Some areas were denoted vulnerable, and people had to move out of these areas. This is a significant loss of a small and already depopulated and aged community, where one-third of the population had to move out of the neighborhood. This created pressure on the social cohesion of the community, and the regular social events and traditional seasonal festivals were on the verge of disappearance. Thus, climate fragility here not only triggered migration but also impacted strongly on local culture and community cohesion.

Although not a climate event, the East Japan Earthquake and tsunami and the nuclear meltdown also prompted a major in-country, out-prefecture migration, which is the maximum in number after World War II. This caused not only economic or social pressure on the surrounding cities and prefectures, but the uncertainty of the reconstruction and recovery of the affected areas became a major issue. A climatic event of similar magnitude could pose similar threats to the Japanese society in the future.

The changing demography of Japan has severe implications due to heat and cold waves and drastic seasonal changes. The 2007 heat wave has taken the largest number of casualties in Japan. It seriously impacted the health of the aged population, and heat strokes have been seen common and increasing during harsh summers in central and western Japan. An expansion of the distribution area of the Asian tiger mosquito transmitting dengue fever and other diseases is also observed in recent years. A new invasion of mosquitoes transmitting Japanese encephalitis from Southeast Asia is also reported (IEDM 2013).

The interactions between quality of agricultural products and climatic events are coming to the fore in Japan in terms of heat waves affecting the quality of rice. These impacts are especially important since the focus of Japan's agriculture is to enhance the export of high-quality rice and rice products, targeting the increasingly wealthy sections of the Asian markets such as China. A survey done by the International Environment and Disaster Management (IEDM 2013) shows that rice farmers saw increasing temperature and prolonged hot days as the most serious threat to rice cultivation. Introduction of heat-tolerant rice varieties is found to be one of the most urgent needs of farmers. The heat waves mainly affect the high-quality rice production, affecting food prices in due course.

Hirota et al. (2006) have mentioned that warming temperatures will have a substantial effect on Japan's agriculture industries. The increase in temperature will lead to an expansion of rice cultivation in Hokkaido, the northernmost island of Japan, which will be beneficial to the government policy of increasing rice-self-sufficiency and high-quality rice exports, while fruit quality would be highly affected (Case and Tidwell 2009). The impact of climate change on fruit quality has also been confirmed by a study of IEDM (2014).

An analysis of apple and orange production in Aomori and Wakayama prefectures shows increasing impacts on fruit quality in terms of texture and size, which also affects the taste. Shifting of temperature grid in Japan will possibly enable the larger fruit and juice makers to relocate their business to different parts of Japan. However, the small fruit farmers will have difficulties in adapting to the changing climate and in maintaining their income sources. The WWF study by Case and Tidwell (2009) also suggested that climate change would affect fisheries production in Japan, with impact on both freshwater and saltwater fish. This may have an impact on the food habit and Japanese cuisine.

3.3.2 External Factors of Climate Fragility

The external climate-fragility issues are related to Japan's dependency on food imports. Although Japan's self-sufficiency rate for rice, eggs, whale meat, and mandarin oranges exceeds 90%, the rate for essential ingredients for Japanese cuisine, including soybeans, is a mere 5%, and just 13% for daily necessities like cooking oil. A recent report by the Nikkei Asian Review (2016) shows that in spite of a strong Japanese yen, imported food is becoming costlier. This is because of a change of import scenario. While Japan used to get most of its shrimps from Thailand and bananas from Philippines, the impacts of drought and diseases of the shrimps and bananas have affected quality production in these two countries. Climate change issues are one of the key factors for the drought and the disease. This has resulted in the shift of the import point to India for shrimp and Ecuador for banana. Thus, the cost incurred to import the same commodity is becoming high, with a larger food mile per kilogram. Moreover, the shrimp demand in some countries like China is increasing due to the higher number of wealthy people. And the Japanese exporters are increasingly facing challenges and price competition. This complex phenomenon contributes to the increase in food prices in Japan, which, in turn is related to climate-fragility issues.

Other complex and external climate fragilities have a deep impact on Japan's fishing industries, especially in international waters. Nishimura (2016) has emphasized that, due to climate change, and associated ocean temperature, currents, and nutrient changes, the fish movement is becoming different. Japanese fishermen need to travel long distances to catch fish, which increases fuel cost and eventually fish prices. This issue becomes more complex due to increasing demand for fish in neighboring countries and regions like China, Korea, and Taiwan, where more fishermen and fishing boats are competing with Japanese fishing boats. This, in turn, makes climate-fragility risk more regional and the solutions more complex.

3.4 Methodology

3.4.1 Stakeholder Perspectives on Climate-Fragility Risks

An online survey was conducted by the authors, in collaboration with Adelphi, Germany, to elicit responses from experts and practitioners engaged in disaster risk reduction, climate change adaptation, and related fields, including peace and security, to understand the currents state of knowledge, important climate-fragility risks, and policy issues to address climate-fragility risks. The questionnaire consisted of both multiple-choice and open-ended questions divided into three sections. The first section consisted of questions to assess the awareness of the respondents on climate-fragility risks and implications of climate change on fragility state of the country the respondents represented. The second section consisted of questions related to current state of actions to address climate-fragility risks and to identify a way forward in terms of specific policy suggestions. The last section comprised questions related to the background of the respondents in terms of expertise, experience, and demographic details.

A total of 179 responses were obtained from developing (represented largely by India, Philippines, Bangladesh, Vietnam, Thailand, Indonesia, China, Malaysia, Nepal, Pakistan, and Sri Lanka) and developed countries (represented by Japan, Australia, United Kingdom, Germany, and USA) to understand and compare responses between two economic groups. All the responses were analyzed and presented as a percentage of total responses. Most respondents represented universities, NGOs, governmental bodies, and think tanks. Most have expertise in CCA, DRR, environment, and sustainable development and have worked at community, national, and international levels.

3.4.2 Comparing Countries on Climate-Fragility Risks

Comparing climate-fragility risks in developing and developed countries would help us to understand where countries stand in terms of specific fragility risks and if the developmental status of a country has any influence on the nature of fragility risks. For cross-comparison purposes, indicators related to specific fragility risks were identified based on the inputs from the online survey. The indicators considered for building the Climate-Fragility Risk Index (CFRI) are presented in Table 3.1 along with the rationale and limitations. For building the CFRI, the following countries were considered based on the availability of data representing developing and developed countries: Australia, Bangladesh, Cambodia, China, India, Indonesia, Japan, Korea, Lao PDR, Malaysia, Myanmar, Pakistan, Philippines, Sri Lanka, Thailand, and Vietnam.

These indicators were transformed using linear normalization with saturation levels. Saturation levels were either adjusted to full (i.e., wherever the published saturation levels are not available) or adjusted to the published maximum and minimum

Table 3.1 Indicators used in developing the climate-fragility risk index (CFRI)

Indicator	Proxy indicator	Rationale and limitations	Source of data
Local competition for water	Baseline water stress	The higher the stress, the more the competition. However, water stress may not always lead to local competition, depending on the local governance and social systems which are addressed by the governance indicator	WRI (2016)
Extreme events	Climate risk index	Climate risk index is the most comprehensive risk index covering climatic hazards and has been regularly produced for most countries	Germanwatch (2016)
Migration and internal displacement	Percentage of population affected by migration and internal displacement	The data provided by the Internal Displacement Monitoring Center (IDMC) give a clear picture of the number of people affected by internal displacement and migration. These numbers were converted into % of population	IDMC (2015)
Food price volatility		Food price volatility was calculated as standard deviation of principal food crop prices in the past decade in local currency	FAOSTAT (2016)
Sea level rise (SLR)	Percentage of population affected by SLR	% of population affected by SLR reflects social and economic impacts better than a mere change in SLR	Climate Central (2015)

(continued)

Table 3.1 (continued)

Indicator	Proxy indicator	Rationale and limitations	Source of data
Unintended effects of policies	World Bank regulatory quality indicator	There are no verifiable measures for unintended effects of policies yet; however, the World Bank regulatory quality indicator provides a close assessment of policy effectiveness, assuming least unintended effects of policies are expected with higher regulatory quality	World Bank (2016)

levels (as in the case of Climate Risk Index). All the transformed indicator values were added without weightages (i.e., all indicators have equal weightage in the final Climate-Fragility Index value). This is to avoid ambiguity in applying weightages, which may vary from one expert to another and from one context to another. The results were shown as a heat map using the web-based Choropleth Tool Carto (Carto 2018).

3.5 Results and Discussion

3.5.1 Stakeholder Perceptions of Climate-Fragility Risks

The online survey was conducted to understand the current state of knowledge and policy issues for addressing climate-fragility risks in developing and developed countries. One hundred three respondents from 17 developing countries (including India, Philippines, Bangladesh, Vietnam, Thailand, Indonesia, China, Malaysia, Nepal, Sri Lanka, and Pakistan) and 76 respondents from five developed countries (Australia, Germany, Japan, UK, and USA) were included. These respondents came mainly from universities, NGOs, governmental bodies, and think tanks. Most have expertise in climate change adaptation, disaster risk reduction, environment, and sustainable development and have worked at community and national levels (Asia) or at national and international levels (Japan). Most were in the age group of 30–40 (56%, rest of Asia) and 50–60 (28%, Japan); most respondents were male (68% rest of Asia, 73% Japan).

The survey indicated a high urgency for addressing climate change-related threats as majority of the respondents regarded these as either imminent or near-future threats, while others considered them already a threat (Fig. 3.1a). However, a marginally higher number of developed country respondents opined it as already or imminent threat and a relatively higher number from developing countries looked at it as a relatively long-term threat compared with respondents from the developed countries. There is a wide variation in the awareness about climate-fragility risks. While 25% of developing country respondents indicated climate-fragility risks as those emerging out of interaction of climate change with developmental pressures, 33% reported it as being an added aspect to current disaster risks. Relatively fewer respondents thought climate-fragility risks can undermine the peace and stability of countries, even though the number is marginally high among the developing countries. Understanding of climate-fragility risks at the policy level varied significantly among the developing and developed country respondents. More respondents from developed countries reported a better understanding among policymakers than their developing country counterparts who reported significantly lower awareness levels. Overall, these differences indicate the greater need for capacity building among policymakers and experts on issues related to climate-fragility risks.

It was apparent that climate change has a significant influence on fragility risks, even though these implications could differ between developing and developed countries (Fig. 3.1d). For example, climate change could aggravate fragility risks through impacting livelihoods in developing countries; in developed countries, it could impact by affecting natural disasters. Developing countries could also experience fragile conditions through aggravated resource scarcity and impacts on state sovereignty. As a result, the nature of climate-fragility risks may vary between developing and developed countries (Fig. 3.2a). Developing countries could face more migration and internal displacement issues as a result of the severe impact on development and dysfunctional institutions in these countries. On the other hand, developed countries may face issues related to demographic stress such as an aging population that could have a disproportionate impact on their economies. In the case of Japan, depopulation of rural areas has been considered one of the important fragility risks after aging population.

The survey revealed that policy domains differ in their potential to address climate-fragility risks (Fig. 3.2b). In developing countries, agriculture- and water-related policies will have more potential to address the fragility risks than measures related to internal security or foreign policy. In contrast, the potential to address fragility risks is rather equally distributed among policies such as climate change adaptation, disaster risk reduction, agriculture and water, and developmental policies in developed countries. Respondents reported the least potential for security-related policies among developed countries compared with that in developing countries. Compared with this potential, respondents rated current achievements in these policy domains relatively well. They opined that these policy domains are achieving what they could achieve at the given awareness and capacity levels. However, they thought that foreign policy, developmental policies, and peacebuilding measures can do better than the current level. As a result, respondents have rated developmental policies higher than the rest of the policies in both developing and developed countries (Fig. 3.2c), while the difference between these countries was higher in the area of migration and internal displacement where developed countries are reportedly lacking in effectiveness. To address climate-fragility risks in general and current policy ineffectiveness in specific terms, there is a need to invest in sharing better information, developing integrated programs, and investing in research and development to understand the drivers behind climate-fragility risks in developing countries (Fig. 3.2d). In developed countries,

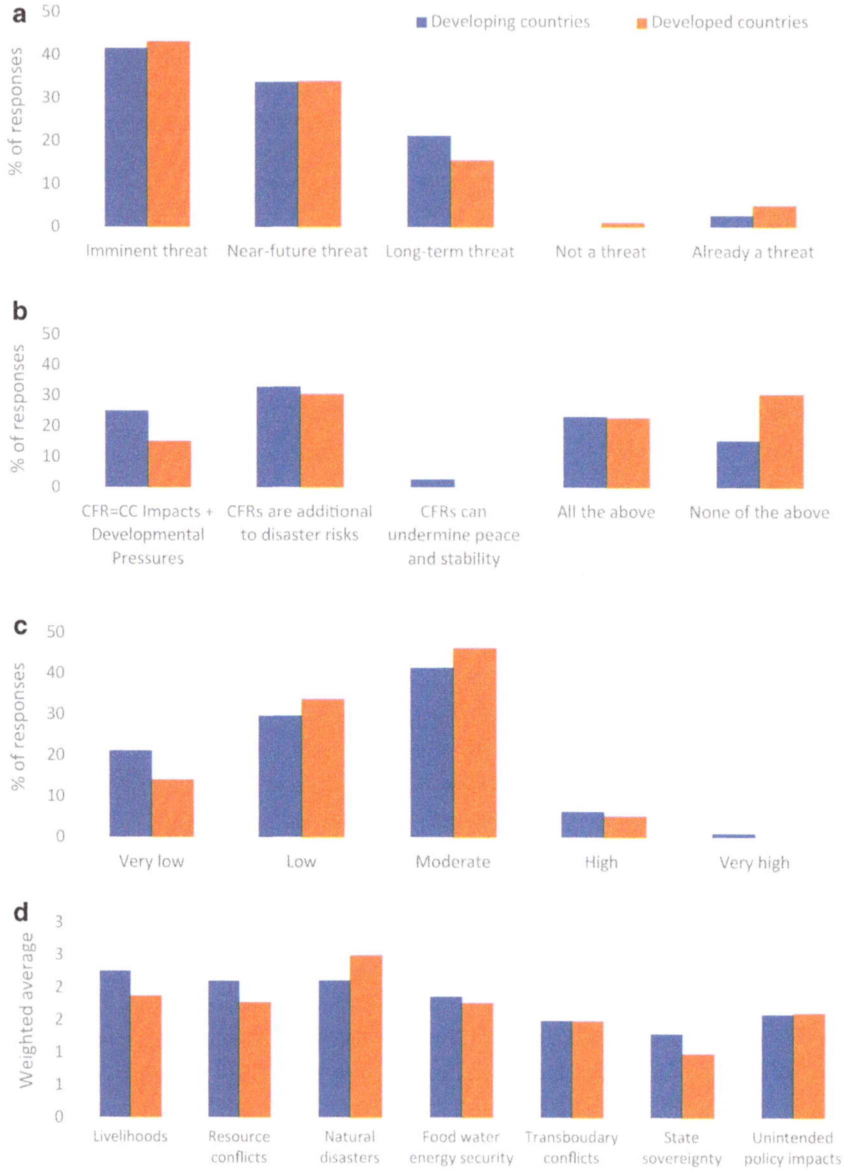

Fig. 3.1 Respondents' understanding of climate-fragility risks. **a** Respondents' perception regarding the urgency of climate change as a threat. **b** Respondents' understanding of climate-fragility risks. **c** Respondents' understanding of climate-fragility risks at the policy level. **d** Major climate change impacts that have fragility implications

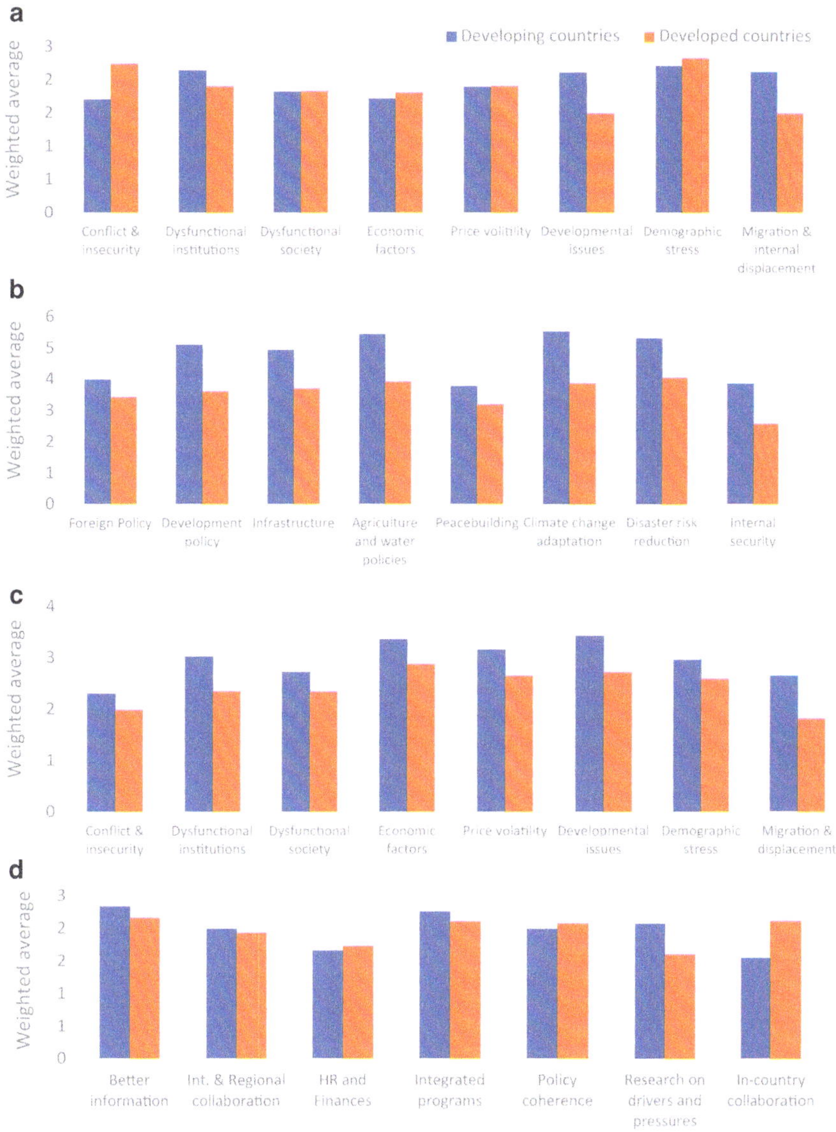

Fig. 3.2 Means and measures of addressing climate-fragility risks. **a** Major climate-fragility risks. **b** Expected potential of different policy domains to address climate-fragility risks. **c** Effectiveness of current policies in addressing climate-fragility risks. **d** Priority interventions for addressing climate-fragility risks

efforts could focus on in-country collaboration and policy coherence, issues equally relevant to developing countries.

3.5.2 Fragility Risks in Japan

A total of 63 survey responses were received from Japan. The respondents have the following characteristics: 71% were males and 26% were in the 50–60 age group; 26% and 24% of the respondents had specialization in environmental management, and disaster risk reduction, respectively, with 60% having more than seven years' experience in their field of specialization. Most have spent their career either at the international (26%) or national level (25%). None have expressed not having any knowledge in the field of climate-fragility risk, while 29% came to know about the subject by reading reports, followed by those attending expert workshops (21%).

Climate change is an imminent threat to threat (48% of responses) or it is already going on through the impacts of climate change (3%). As a result, the country faces challenges such as increased intensity of natural weather-related disasters that affect its food security and energy security. These have impacts on the fragility of the country in the areas of urbanization and depopulation of rural areas and food, water, and energy price volatility (Fig. 3.3). Most respondents (89%) thought that climate change influences the fragile status of the country and it does so by impacting through natural disasters, undermining livelihoods, and creating resource conflicts in the country.

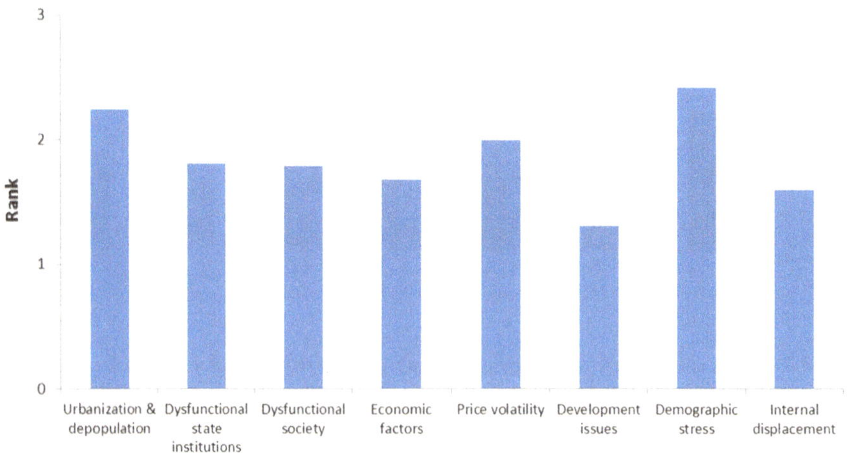

Fig. 3.3 Top factors contributing to fragility risks in Japan

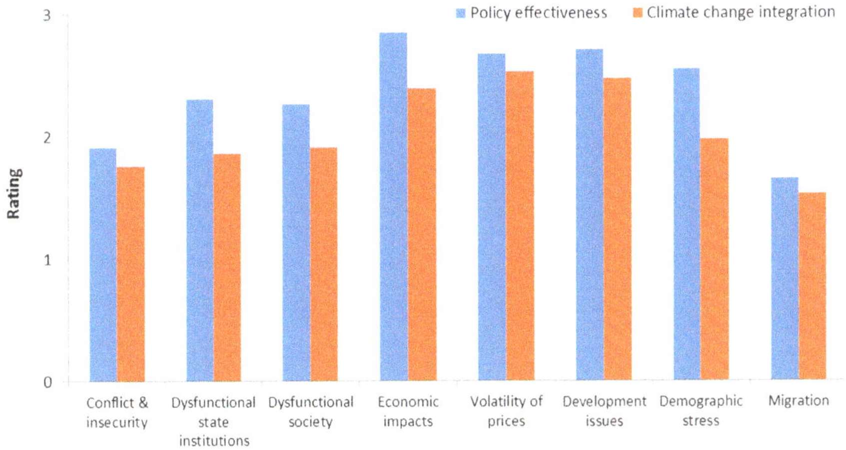

Fig. 3.4 Effectiveness of policies in specific fragility risk areas and the extent of integration of climate change aspects into these policies

Overall, most respondents were either low (38%) or moderately (43%) satisfied in the way national policies are addressing the climate-fragility issues in the country. In terms of interventions in specific policy areas, most have received similar ratings in terms of the extent to which they are addressing the climate-fragility issues (Fig. 3.4). However, areas such as the economy, developmental initiatives, and price volatility get relatively better rating than internal displacement and conflict and security areas. Respondents opined that price volatility, the economy, and developmental areas fare well in terms of integrating climate change concerns than in rest of the policy areas.

In terms of specific policy areas with high potential to address fragility risks, disaster risk reduction and agriculture and water areas, followed by climate change adaptation were rated high by the respondents. Compared with this potential, policy areas such as climate change adaptation, internal security, and disaster risk reduction are doing much better than their perceived potential, whereas foreign policy and developmental policies can be improved (Fig. 3.5). There is a need to address this policy gap through investing in research and development and human resource development and fostering regional and international collaboration. Thirty percent of the respondents thought that these measures should be implemented at the national level, while 27% wanted them at all levels.

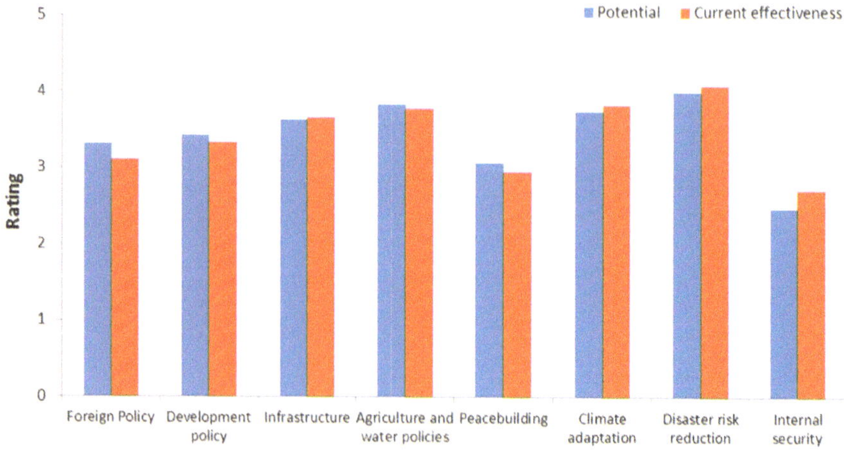

Fig. 3.5 Potential of different policy domains to address climate-fragility risks and their current performance compared with the perceived potential

3.5.3 Comparing Countries in Terms of Climate-Fragility Risks

The climate-fragility risk index (CFRI) that compared major countries in Asia and Oceania has clearly shown that countries differed in climate fragility and nature of issues, hence the need for country-specific solutions (Fig. 3.6). The average CFRI for developing countries comprising Bangladesh, Cambodia China, India, Indonesia, Lao PDR, Malaysia, Myanmar, Pakistan, Sri Lanka, Thailand, and Vietnam stood at 0.76. It was 0.61 for the developed countries comprising Australia, Japan, South Korea, the USA, and United Kingdom. The differences between countries were largely due to variations in exposure to sea level rise (Vietnam and Thailand are highly vulnerable) and food price volatility (Australia scored highest in this regard). Diversity was much less in indicators of internal displacement and regulatory quality of country governance systems. The analysis indicated close association between CFRI and per capita GDP of countries. The exponential relationship between these two seems to imply a critical threshold level, suggesting a critical level of per capita income below which countries tend to have higher climate-fragility risks. This linkage between developmental status of countries and climate risks has already been widely recognized (Hallegatte 2013).

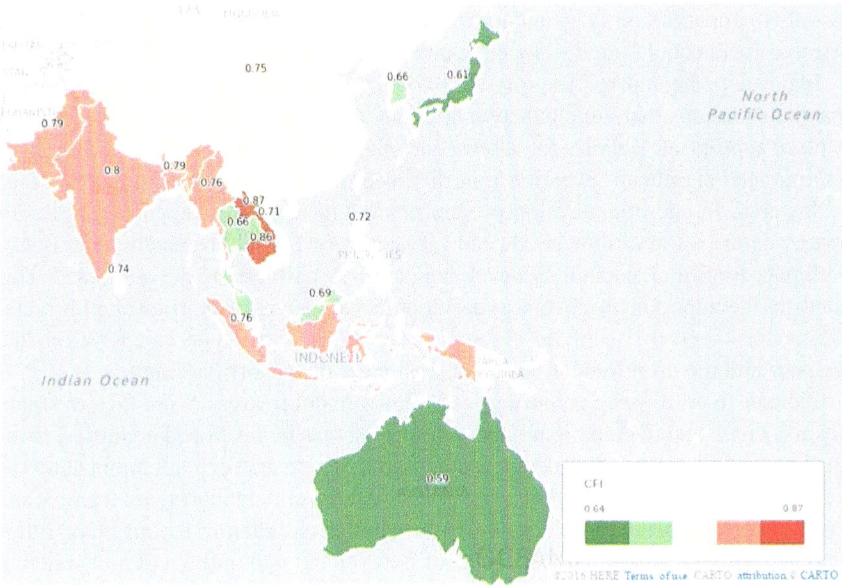

Fig. 3.6 Climate fragility of selected countries in Asia and Oceania (*Source* Authors)

3.6 Conclusions

We are already experiencing the intense signs of climate-fragility risks in several areas as discussed in this paper. Overall, climate change could worsen the frequency and intensity of natural disasters and it could impact economic conditions such as food prices and exacerbate current social and security problems, which include internal conflicts and increases in internally displaced populations.

The survey responses have clearly shown a divide between developing and developed countries, reflecting how developmental and related differences influence climate-fragility risks. For example, fewer respondents from developing countries felt that climate change is an immediate threat, while more people from Japan regarded it to be so. These differences in opinions could influence the nature and urgency with which countries could pursue their climate risk reduction policies. Differences in the level of understanding climate-fragility risks could influence the degree of stress given by these countries in implementing necessary measures to mitigate these risks. Responses from both developing and developed countries emphasize the need for addressing these issues across levels with relatively greater focus on the implementation of policies on the ground, leading to tangible outcomes. Both developing and developed countries are facing significant challenges from the climate-fragility paradigm. Developing countries are more vulnerable because of underlying socioeconomic factors, weak institutions that deal with conflicts, and developmental deficit that prevents them from meeting the basic needs of the people. Climate change is thought to impact fragility by undermining livelihoods in developing countries, while extreme events could put developed countries at greater risk.

In developed countries, fragility risks could be magnified through specific demographic situations that are characteristic of an aging society. There is a need to put in place appropriate policies for addressing emerging demographic issues. Japan has accumulated significant experience in this regard, thus enabling it to take the lead in this area. In addition, developed countries are increasingly dependent on developing countries for various goods and services, which could be easily undermined if climate-fragility risks in their developing country partners are not addressed. This confirms the importance of overseas development assistance to help developing countries address security issues, development deficits, and other economic issues so that the poor and the disenfranchised can at least meet their daily basic needs.

Policies in developing countries need improvement in areas of conflict, dysfunctional society, and dysfunctional institutions, whereas in developed countries, focus must be given to addressing demographic issues. There are regional implications as well. Increasing internal conflict and its implications for neighboring prefectures and countries means there is an urgent need to strengthen mutual trust among them. Information exchange through key channels of government and introduction of coherent policies, for example, become necessary. In this respect, a third party or a multilateral mechanism could play an important role, though the existing ones are not working very well. Regional mechanisms such as the South Asian Association for Regional Cooperation, the Association of Southeast Asian Nations and the like should take a lead in providing an enabling environment for strengthening that trust.

There are complementarities between developed and developing countries in how they can mutually address the issue of climate fragility. This is in terms of coordinated policy development between them due to increasing dependency on each other and the implications of CFRs in one country or another. They can also share integrated risk assessment and design policies that can deliver benefits for both, as in the case of immigration policies. For this to happen, there is a need for countries in the region to collaborate and cooperate at various levels, including through agreements, treaties, and platforms, and to develop coordinated policies that take into account cause-and-effect relations of factors that are geographically placed in different countries but are very much connected in their pathways.

Japan's internal climate-fragility risk arises from the increasing number of disaster events, its intensity, and frequency, Japan's demographic character (like aged population and depopulation), and impacts of heat and cold waves on certain vulnerable groups, as well as primary industries like agriculture, fisheries, etc. The external climate-fragility risk is more complex for Japan due to its higher dependency on food imports, related costs, and also on foreign policy issues, especially in East and Southeast Asia.

In terms of policy options, internal and external climate-fragility risks need to be addressed in a holistic way, considering the interdependency of water, agriculture, and fishery sectors, as well as linkage to foreign policy. While disaster risk reduction and climate change adaptation approaches are more internal to Japan's own policies, they must be analyzed in a more synergistic way to link it to different specific sectors such as food, agriculture, fisheries, and health. National adaptation planning covers the climate change impacts on all the abovementioned sectors. Natural disasters and coastal areas have specific provisions in the national adaptation plan. Thus, linking different sectoral policies is the key point.

The other issue is to bring policy focus to the local level and apply it in the local governance context. As seen in the examples, the local governments are the first responders in case of disasters. Most of these local government units have their disaster preparedness plans. The integration of adaptation planning in the local disaster preparedness plan is very important. Fujii (2016) pointed out that the Ministry of Environment (MOE) of Japan has formulated local adaptation plans in nine prefectures and two cities since 2015. It will be important to replicate those in other smaller cities and towns where disaster- and adaptation-related problems are quite distinct from those of the larger cities. In addition, it is important that these plans integrate the climate-fragility risks identified in this chapter.

Finally, linking sectoral policies, disaster risk reduction policies, and climate change adaptation policies together with foreign policy is the most important issue, which is the core of climate-fragility risk in this interconnected world. This can be done through continuous dialogue among stakeholders engaged in these processes and through integration with the overall development plan of the Government of Japan, preferably under the umbrella of the Prime Minister's Office.

Acknowledgements The authors are grateful that part of this research was supported by the Environment Research and Technology Development Fund (2-1801) of the Environmental Restoration and Conservation Agency of Japan and IGES Strategic Research Fund 2018. Collaboration with Adelphi, Germany, is likewise acknowledged.

References

Carto (2018) Available at www.carto.com. Accessed 9 March 2018

Case M, Tidwell A (2009) Nippon changes: climate impacts threatening Japan: today and tomorrow. WWF, Japan, p 13

Central Climate (2015) Mapping choices: carbon, climate, and rising seas our global legacy. Climate Central, USA

Chiba Y, Prabhakar SVRK (2017) Priority practices for addressing non-economic loss and damage caused by typhoons in Japan: a case study of Nachikatsuura town. IGES Research Report May 2017. Hayama, Japan: Institute for Global Environmental Strategies, 38

FAOSTAT (2016) FAO, Rome, Italy. faostat.fao.org. Accessed 9 December 2016. Available at http://www.fao.org/faostat/en/. Accessed 8 October 2018

Fujii S (2016) National adaptation planning in Japan, in Joint Adelphi-IGES Expert Workshop on Climate-fragility risks in Japan: implications for Asia and the globe, June 2016, Tokyo. Institute for Global Environmental Strategies, Japan

Germanwatch (2016) Global Climate Risk Index 2017. Who suffers most from extreme weather events? Weather-related loss events in 2015 and 1996 to 2015. Germany: Germanwatch

Guardian (2011) Are food prices approaching a violent tipping point? Available at https://www.theguardian.com/environment/damian-carrington-blog/2011/aug/25/food-price-arab-middle-east-protests. Accessed 26 June 2019

Hallegatte S (2013) An exploration of link between development, economic growth and natural risk. World Development Report 2014. USA: World Bank

Hirota T, Iwata Y, Hayashi M, Suzuki S, Hamasaki T, Sameshima R, Takayabu I (2006) Decreasing soil-frost depth and its relation to climate change in Tokachi, Hokkaido, Japan. J Meteorol Soc Jpn 84:821–833. http://www.jstage.jst.go.jp/article/jmsj/84/4/821/_pdf. Accessed 8 November 2016

IDMC (Internal Displacement Monitoring Center) (2015) Global report on internal displacement. Switzerland: IDMC

IEDM (2013) Climate change and rice production: implication and adaptation option for Japan. Kyoto University IEDM Seminar series. Japan: Kyoto University

IEDM (2014) Climate change impacts on local juice company's production in Japan. Kyoto University IEDM Seminar series. Japan: Kyoto University

Janusz-Pawletta B, Gubaidullina M (2015) Transboundary water management in central Asia legal framework to strengthen interstate cooperation and increase regional security. *L'eau en Asie centrale*, 25. Available at http://journals.openedition.org/asiecentrale/3180. Accessed on 26 June 2019

Nikkei Asian Review (2016) Supply demand pressure driving up Japan's food import cost. http://asia.nikkei.com/Markets/Commodities/Supply-demand-pressures-driving-up-Japan-s-food-import-costs. Accessed 20 December 2016

Nishimura M (2016) Keynote address in Joint Adelphi-IGES Expert Workshop on Climate-fragility risks in Japan: implications for Asia and the globe, June 2016, Tokyo. Institute for Global Environmental Strategies, Japan

Ruttinger L, Smith D, Stang G et al (2015) A new climate for peace: taking action on climate and fragility risks. Adelphi, International Alert, Woodrow Wilson International Center for Scholars, European Union Institute for Security Studies

WFP (2009) World Hunger Series: Hunger and Markets. Rome, Italy: World Food Programme. Available at https://www.wfp.org/publications/world-hunger-series-hunger-and-markets. Accessed 26 June 2019

World Bank (2016) Worldwide governance indicators. World DataBank. USA: World Bank

WRI (Water Resources Institute) (2016) Aquaduct. Aqueduct country and river basin rankings. USA: WRI

Dr. S.V.R.K. Prabhakar works on mainstreaming climate change adaptation and disaster risk reduction into development, adaptation metrics, adaptive policies, risk insurance, regional adaptation planning for transboundary climate risks, loss and damage, vulnerability assessment, and training needs assessment in South and Southeast Asian countries. He has been Coordinating Lead Author for the UNEP Geo-6 Asia report, Lead Author for the HIMAP Report, and Contributing Author to the Fifth and Sixth Assessment Reports of the IPCC Working Group II. Prabhakar obtained Ph.D. in Agronomy from IARI, India. He possesses more than 17 years of experience in participatory research and development with international and national research and development organizations including IGES, Kyoto University, UNDP, NIDM, CIMMYT-RWC, IARI, and ICRISAT.

Dr. Rajib Shaw is Professor at the Keio University in Japan and former Executive Director of the Integrated Research on Disaster Risk (IRDR) programme. He is also Senior Fellow of Institute of Global Environmental Strategies (IGES) Japan, and Chairperson of SEEDS Asia, and CWS Japan, two Japanese NGOs. Previously, he was Professor at the Graduate School of Global Environmental Studies, Kyoto University, Japan. His expertise includes community-based disaster risk management, climate change adaptation, urban risk management, and disaster and environmental education. He is the editor of a book series on disaster risk reduction, published by Springer, Editor-in-Chief of the journal Progress in Disaster Science. He severs as Chair of United Nations Science Technology Advisory Group for disaster risk reduction, and is Co-chair of the UNDRR's Asia-Pacific Science Technology Academic Advisory Group (ASTAAG). Prof. Shaw has published more than 45 books and more than 300 academic papers and book chapters.

Chapter 4
Policies and Institutions Shaping Human Security in an Era of Changing Climate and Increasing Disasters

S. V. R. K. Prabhakar

Abstract Policies are the channels through which governments influence society and make course corrections for intended consequences. Institutions have the power to influence policy processes and outcomes and can determine the way governments engage with society. Policy formulation and implementation thus far have largely been based on populist assumptions and preferences with simple developmental outcomes in a simpler world. However, the advent of human security principles and efforts to apply these principles to policymaking has proved that the current policy and institutional systems are inadequate to fully address the human security needs of societies and countries. Climate change is one factor that has contributed to even more complexity to the ever-changing topology of issues that policies are expected to address today and in the future. Climate change has necessitated a need to revisit the current institutional and policy mechanisms so as to uphold human security. Keeping this need in view, this chapter presents a conceptual background on the role of institutions and policies in shaping human security in Asia. There is a need for conceptual clarity on what constitutes human security for these principles to start making a significant impact on policies and institutions as practical application of such principles is yet to emerge significantly. The available measures of human security consider a limited set of policy and institutional indicators and are incremental advancements over measures such as the human development index. Hence, they are yet to fully capture the multiple dimensions of human security in a practical manner.

Keywords Human security · Climate change · Policies · Institutions · Human development

---

S. V. R. K. Prabhakar (✉)
Institute for Global Environmental Strategies (IGES), Kanagawa, Japan
e-mail: prabhakar@iges.or.jp

J. M. Pulhin et al. (eds.), *Climate Change, Disaster Risks, and Human Security*, Disaster Risk Reduction, https://doi.org/10.1007/978-981-15-8852-5_4

4.1 Introduction

Policies and institutions are important agents shaping human security through the ways they respond to threats of climate change and disasters. These may also negatively influence human security through policies that undermine them since there is no full understanding of the underlying causal factors that linking the interventions to human security. Hence, it is very important to understand the ways and means through which policies and institutions may influence human security.

Countries in Asia have a long history of formulating policies to address their developmental aspirations but they are relatively new entrants into the area of making policies for addressing disaster risks and threats posed by climate change. Further, the concepts of human security, including climate security, are new to governments in the region. They are yet to install institutional mechanisms to incorporate human security-related issues into their policy processes. However, current human security issues in Asia could be understood as a product of a long policy and institutional history that the region has, spanning decades if not centuries, in traditional areas of development. Assessing these policy and institutional contexts as a long-term product of current human security condition is daunting. It is because of the difficulty in establishing a cause-and-effect relationship between historical policy experiences and current human security outcomes. Such a task is also daunting for the reason that the thought of human security is more recent in nature both in terms of its understanding and its use in the policy domain. Hence, a major part of the past policy and institutional context that may have shaped today's human security in Asia was largely developed without viewing policies and institutional interventions from the lens of human security. However, an ex-post analysis of policies and institutions would provide interesting insights into the factors that may have played an important role in bringing about those policies and institutional contexts, leading to today's human security conditions and those in the near future.

A range of policies and institutions can mold human security as we understand it today. While several policies and institutional processes can have some relevance to human security, it is important to prioritize them to keep the scope of the paper succinct and focused. It is also important to understand how policies and institutions shape human security, especially knowing that policies in the past were not explicitly designed with human security considerations. Actual policy interventions related to human security in Asia will be dealt in detail in the other chapters. This chapter in particular provides a conceptual background and an understanding of how policies and institutions shape human security, how policies and institutions are related to each other, and how these would lead to human security outcomes. The chapter first elaborates on how policies and institutions affect human development in general, deriving from the wealth of knowledge available in this field. Subsequently, the chapter dwells on the literature, touching upon human security as an outcome of policies and institutions and trying to identify specific factors that would contribute to this outcome.

4.2 Role of Policies and Institutions in Human Development

In this section, an effort has been made to describe the role of policies and institutions in human development as a background before discussing their role in achieving human security. Human security and human development are highly related, if not the same (Gomez and Gasper 2016; United Nations 2009), and concepts of human development have a much longer history than human security. While human development impinges upon growth and equity, human security stresses the 'downturn with security'—i.e., about sustainability of gains made as a result of human development efforts. Policies and institutions have been at the forefront of shaping human development in the past century and hence are highly relevant to human security. It is thus important to understand the role played by policies and institutions in human development itself as a first step of appreciating their role in human security due to the synergy between human development and human security.

4.2.1 State and Development

The state influences development through a set of policies, programs, and laws. There are several definitions of what 'policy' is (Torjman 2005). However, for the purpose of this chapter and from the viewpoint of public administration, policy can be defined as 'a purposive course of action followed by an actor or a set of actors in dealing with a problem or a matter of concern' (Anderson 1984). There is a body of literature on why governments enact policies (Considine 2005; Gerston 2010; Ingram and Smith 1993; Kay 2006; Torjman 2005). Most of these opinions converge to state that collective action would enable society to consume public goods and that a combination of several market failures would affect the way public goods are produced, distributed, and consumed (Weimer and Vining 1992). Hence, the origin of the role of government in enacting policies is to enable equitable use of public resources (public goods).

Understanding the notion of development has been an important part of modern human history and several scholars have made different attempts to give a shape to the thought of development—i.e., what is development, how can development be achieved, and what are the roles of different stakeholders in development. The discussion on development dates back to the advent of welfare economics pioneered by Adam Smith, who said that welfare increases with increase in production (Smith 1776). Scholars such as Jeremy Bentham, who pioneered social welfare, stated that welfare is equal to the utilities derived by individuals in society. These ideas ushered in societal welfare and welfare economics and have further led to discussions on the role of the state and private markets in creating wealth and welfare.

Subsequently, a significant consensus emerged that welfare cannot be solely created by markets due to market imperfections and failures that result in income inequalities and unequal resource distribution and access. It may even be able to help address the market imperfections in creating wealth and this is where public policies come into play (Wallis and Dollery 1999). As a result, governments and the public policies they put forth have gained importance in ushering development in the right direction. The discussion on policies and their role in development can go beyond the scope of this chapter. However, it is important to note that governments have not always figured out the best way to form policies and to implement them. Questions such as who decides policies, who benefits from policies, and what mechanisms are optimum to ensure that those negatively affected by the policies are identified and addressed in subsequent iterations of policy formulation and implementation processes have not been well addressed. Consequently, government policies aimed at societal development have not reached even the fullest scale of maturity, even in some of the most developed countries. It could also be said that only a handful of countries are able to do well in certain policy domains.

4.2.2 *Development Effectiveness and Policies*

According to the World Bank (2017), only high-income OECD countries show higher levels of government effectiveness and regulatory quality, followed by high-income non-OECD countries (Fig. 4.1). The data trends indicate a direct relation between income status of a country and its policy effectiveness expressed through government effectiveness and regulatory quality. Such relational trends between the development state of a country and its policy effectiveness have long been well understood (Azulai et al. 2014). One contributory factor to such trends is lack of proper state capacity to formulate policies that will result in certain stated policy outcomes (Azulai et al. 2014).

The important question is how and what policies have led to development. There are a number of countries, each with a different story to tell, and hence, it is often difficult to generalize how certain countries are able to achieve development. Only broad generalizations are possible. There are references to how different economists viewed the association between policy effectiveness and the development state of a country. For example, liberal economists believe that countries with strong markets will have better policy governance abilities, whereas heterodox economists believe in growth-enhancing governance focusing on the capacity to actively address market failures (Khan 2007). It has also been argued that the current policy environment tends to focus more on previous means of development and tends to ignore the latter—i.e., being able to address market failures in enabling equitable distribution of assets and ensuring political stability in the changing socioeconomic circumstances (Khan 2007).

The basic difference between these views seems to lie on the fact that governance capacities that the states need for ensuring economic growth are different from those needed for good governance. It has been argued that state capacities for good governance are those related to maintaining efficient markets and being able

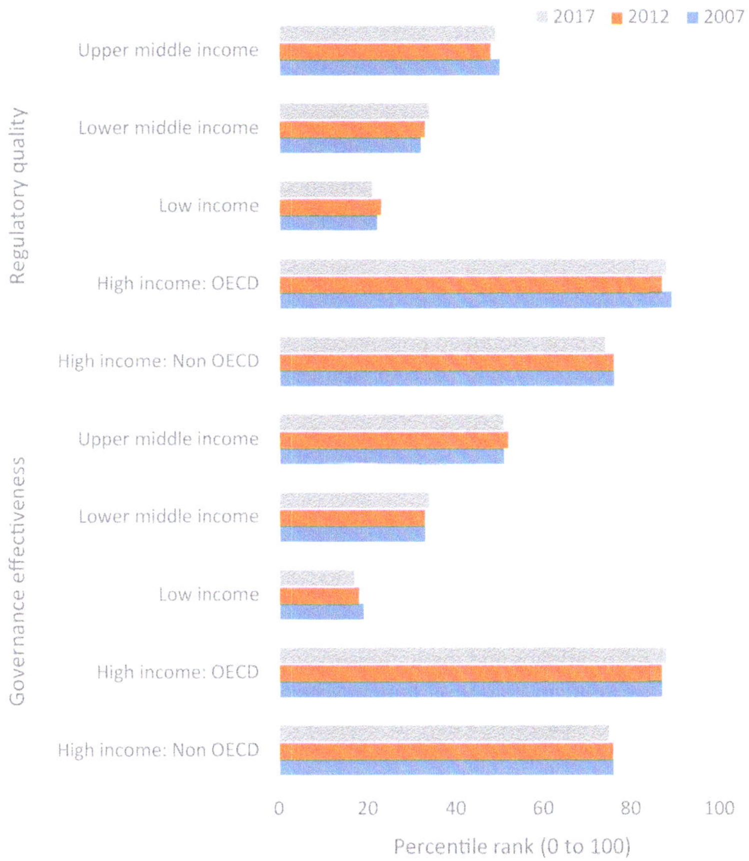

Fig. 4.1 Policy effectiveness among developed and developing countries, expressed in terms of government effectiveness and regulatory quality (The World Bank 2017)

to focus on providing public goods and services to avoid rent-seeking and government failure (Khan 2007). On the other hand, institutional economists believe that capacities needed for economic development are those that involve increased and sustained investments and policies that help acquire and learn new technologies. In the case of countries with low development, especially in Africa, it has been argued that the focus on governance reforms is misguided and that the emphasis should be on investment in critical infrastructure and health (Sachs et al. 2004) since there was no relationship found between country performance and quality of governance after accounting for differences in development level.

Notwithstanding these different opinions, the World Bank approach to development, especially in low developed countries, has been through promoting capacity development toward capable and accountable states and calling for engaged societies that demand change with civil society and private sector providing check and balance on the executive while recognizing the fact that approaches could differ from country to country (Kaunda 2005). In this context, the role of aid cannot be overlooked, where development effectiveness of aid has been viewed in terms of ownership of the recipient country, transparency and accountability, result-driven allocation, and partnerships that are inclusive in nature (OECD 2011). Several development partners, including the European Union, have affirmed their support to these principles of effectiveness (European Commission 2017).

While the indicators of policy and institutional effectiveness for achieving human security will be dealt later in the chapter, it is in particular relevant to say here that development effectiveness must be considered by the states in their formulation and implementation of development policies and programs inasmuch as elements of accountability, transparency, and participation have long been recognized for their critical role in development in general and development decision-making in particular (Harvard Kennedy School 2016; Atwood 2012; Results for Development 2017; Carstens 2005).

4.3 Human Security and Policies

To understand what policies and institutions may have shaped human security, it is important to look at the definition of human security (what it is) and understand its various dimensions (what affects human security). In the context of climate change, human security has been defined 'as a condition that exists when the vital core of human lives is protected and when people have the freedom and capacity to live with dignity' (Adger et al. 2014). Adger et al. (2014) suggested that human security is influenced by markets, the state, and civil society where policies and institutions play a major role as conduits of influence that these agents have. They have further elaborated on the concept of human security in terms of elements such as livelihoods, culture, political stability, indigenous people, and migration. Adger et al. (2014) also considered food security and public health and well-being as important dimensions of human security. They likewise regarded as core dimensions culture, indigenous people and traditional knowledge, migration, conflict, and transboundary resources.

Outside the context of climate change, efforts were made to define human security as a means of bringing out from the traditional security paradigm of defining it within the ambit of military security and securing a territory from external aggression into the scope of the common man by emphasizing those aspects that they face on a daily basis such as job security, income security, energy security, and food security (UNDP 1994). By putting the common man at the center of human security, it was hoped that a paradigm shift in the development discourse would emerge.

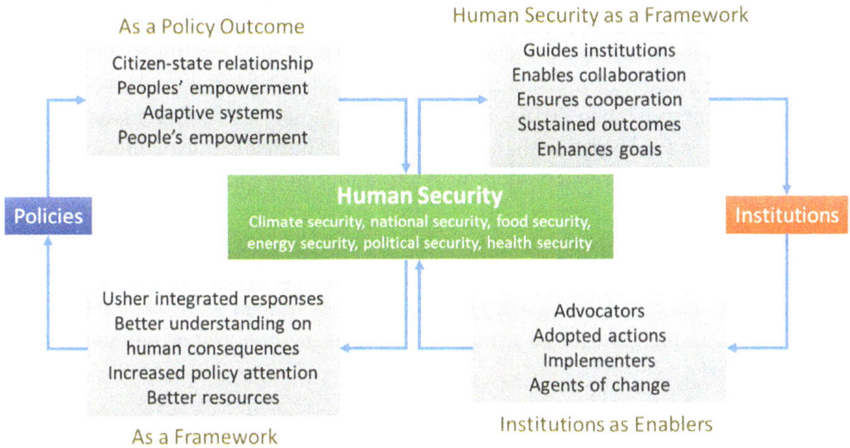

Fig. 4.2 Human security as a policy framework and as an outcome of policies and institutional actions (*Source* Author)

Since human security can cover such a wide array of subjects, assessing the policy and institutional factors shaping human security along these lines would have to be done with careful consideration to avoid vagueness and to have focused discussion. To do so, this chapter proposes that human security be viewed both as a policy framework in itself and as an outcome of policies (Fig. 4.2). However, the discussion presents evidences that human security is increasingly playing a role as a policy framework, while there is also strong evidence for it being viewed as a policy outcome with more and more civil society organizations designing programs around human security and with increasingly more governments embedding human security into their development policies, either domestic or overseas.

4.3.1 Human Security as a Policy Framework

Human security itself has been viewed as a policy framework if not a policy in itself because doing so brings about various advantages. Human security helps obtain integrated responses, capitalizing on the capacities of various actors that engage in human development with coherence in the allocation of responsibilities and resources (United Nations 2014a). Human security puts people ahead of state and territory; as a result, many countries have started integrating human security into their foreign policy and agenda. For example, the Government of Japan has included human security principles into its overseas development agenda and has considered human security as an international principle for cooperation in a significant manner to enhance the relevance of assistance it provides to developing countries (Ministry of Foreign Affairs of Japan 2014). As a result, the Japanese ODA received international recognition for operationalizing the conceptual human security into something practical and outcome-oriented, thereby changing the perception about human security (Fukuda-Parr and Messineo 2012). Other governments, such as the Canadian government, have viewed human security as a means of putting in place an integrated action agenda encompassing public safety, conflict prevention, protection, peace and governance, and accountability. They believe that only human security could provide an all-encompassing conceptual framework needed for its foreign policy and trade (Akiyama 2004).

4.3.2 *Human Security as an Outcome of Policies*

In the current conditions of unsustainable development where several sections of the populace are deprived of human security, human security gains more prominence as a much-needed policy outcome. The human security index, developed as a trinity of environmental, economic, and social components, revealed several countries such as West Asia and Africa doing poorly compared with many other countries (Hastings 2010). Such studies and many others indicate a range of factors affecting human security, which include discriminatory policy and regulatory environment, social contexts, and impacts of natural disasters and climate change. There is a need to implement policies that can address these factors, leading to human security as an outcome. In this process, governments are important stakeholders in human security since the policies they implement affect the survival, livelihood, and dignity of their populations (United Nations 2014a).

Different approaches have been hypothesized, advocated, or implemented to ensure human security outcomes of policies. It is evident that no single approach can do this, but that rather a collection of approaches is necessary, depending on location specificity (Table 4.1). Policies that can promote protection and empowerment have, in particular, been stressed in the context of human security. It has been argued that such a dual framework can accommodate the existence of dual systems such as the top-down nature of governments that ensures protection and the bottom-up nature of civil society organizations that ensures empowerment (United Nations 2014a).

Table 4.1 Policy outcomes for ensuring human security

Policy outcome for ensuring human security	Supporting arguments	Reference
Climate security	Climate change undermines human security through impacts on livelihoods, compromising culture, migration, and state capacity to act	(Adger et al. 2014)
Human development	The need for protection from diseases, unemployment, crime, and hunger overweighs that of traditional national security fears emerging from war and other international conflicts	(UNDP 1994)
Human development	Natural disasters threaten human survival and affect the economic and social foundations of people	(Yamada 2015)
Protection and empowerment	Can accommodate actors engaged in top-down and bottom-up interventions	(United Nations 2014a; JICA 2010)
Adaptive systems	Well accommodate the uncertainty brought about by climate change	(IISD 2011)
People empowerment	Enables and enhances state-citizen partnership in security policymaking	(Wall 2014)
Enhanced policy capacity	Brings to the front the human vulnerability component ahead of the state-centric traditional security notions	(Fukuda-Parr and Messineo 2012)
Freedom from fear	Special consideration for the most vulnerable and shielding local communities from local and global threats	(JICA 2010)
Energy security	Energy security is a necessary condition for human security where energy is considered a public good (deep energy security)	(Karlsson-Vinkhuyzen and Jollands 2015; United Nations 2009)
Food security	Full access to adequate food as a fundamental human right	(FAO 2016; United Nations 2009)
Environmental security	Human security is an outcome of the impact of environment on people and their well-being	(Elliott 2015; United Nations 2009)

(continued)

Table 4.1 (continued)

Policy outcome for ensuring human security	Supporting arguments	Reference
Health security	Health threats put extreme pressure on people and communities with implications on their overall security	(Chestrad International 2013; Caballero-Anthony 2002)
State security	Individuals and communities and their well-being are affected by threats posed against the state from outside and inside	(United Nations 2009)
Economic security	Economic fluctuations of short, medium, and long term affect the lives of individuals and communities, social priorities, and balance of power	(Fontanel and Corvaisier-Drouart 2014)

4.3.3 Human Security and Disaster Risk Reduction Policies

Disasters are an important threat to human security as they affect human survival by undermining the economic and social foundations of people (Yamada 2015). disaster risk reduction is an important policy outcome for ensuring human security in several parts of Asia, one of the regions that is most prone to natural hazards, and where issues such as internal conflict and traditional security are not major threats compared with other parts of the world. Countries in Asia have already started strengthening their disaster risk reduction efforts by moving from response-oriented approaches to ways with greater focus on prevention and mitigation. However, such prevention and mitigation measures are still largely technical and infrastructure-oriented, leaving the human element to response and relief. This is even so when several countries promote community-based disaster risk management plans that are heavily preparedness-oriented rather than focused on relief and response.

Part of the limitation lies in the fact that prevention and mitigation demand considerable human, financial, and technical resources that are often lacking at the grassroots level where community-based disaster risk management plans are being prepared. Social approaches such as sound economic policies that buffer income variations, risk diversification by varying livelihood investments, risk pooling, and savings could have a profound impact on human security of individuals and communities (Holzmann et al. 2003; Rasheed 2015), but these are largely missing from the mainstream risk reduction strategies. Only recently are risk insurance and community-based microfinance and micro-insurance programs being piloted, and these initiatives take time to get fully mainstreamed into disaster risk reduction policies (Prabhakar 2017).

Human mobility is an important issue that affects the security of several countries, communities, and individuals, both positively and negatively. Identifying positive mobility from negative ones is not easy because of the complex risk and economic and environmental outcomes human mobility can have (Guadagno 2016a). The Sendai Framework for Disaster Risk Reduction 2015–2030 in particular makes progress by including human mobility in it, which is considered an important development within the global policy dialogue. More importantly, it recognizes the negative aspects of mobility, i.e., producing risks, as well as its positive aspects of mobility, i.e., resilience outcomes (Guadagno 2016b). The translation of these global policy developments into national and subnational outcomes is an important but challenging task ahead.

Promoting social risk management strategies needs two considerations. First, the solution to the conflict between individual risk vs collective risk has to be addressed for policy solutions to make sense to policymakers and the masses; governments are to deal with collective risk in their policy measures, while it is the individual prerogative to address much of the individual risk (OECD 2014). Second, systemic risk, the component of risk outside the hands of the individuals and that lies in factors beyond their influence, needs to be resolved for individual risk management to make an impact. The current policy environment is not well aligned to start making sense of these two issues, which can provide an important breakthrough to ensure human security.

4.3.4 Human Security and Climate Change Policies

One important and emerging policy area that is highly relevant to this book is the interaction of climate change and human security. Climate security denotes the threats posed by climate change to various aspects of human security. Similar to many other policy outcomes previously discussed and listed in Table 4.1, climate security too can be considered an outcome of several policies embedded in traditional sectors and hence has similarities to human security. However, the securitization of climate change, which implies 'climatization of security,' has happened very recently and has been part of a gradual change that climate change risk management underwent in the past 30 years (Oels 2013). According to Oels (2013), climate change as a security issue is a form of risk management through 'contingency' as opposed to 'science-based' risk management in the fields of climate change adaptation and mitigation—i.e., the response to climate change since 2007 has been to prepare and manage the inevitable impacts of unmitigated climate change as a result of failure to take up ambitious climate change mitigation actions.

The climate security concepts have helped expand the scope of discussion on climate change from traditional fields of food and energy security to national security, some ways opposite to shifts happening from traditional security to human security. The securitization of climate change is not far from criticism. Despite acknowledging that climate security could inform political responses to climate change, such advances may not do much to the cause of climate change and it may be unlikely to inform an effective response to climate change (McDonald 2013). Others revealed that there has been a certain level of overemphasis on the deterministic linkages between climate change and national security but concluded that the linkages are far more complex and require thorough analysis across disciplines (Gemenne et al. 2014).

In the context of climate change, the effectiveness and adaptiveness of policies and systems have in particular received special attention to address certain challenges that climate change has brought. Though the concept of adaptive policies is not new, the use of this term in the context of climate change can be traced to an International Institute for Sustainable Development (IISD) project entitled 'Designing Policies in a World of Uncertainty, Change and Surprise' (IISD 2011). The basic notion of a policy being dynamic has strong roots in a branch of policy science called 'policy dynamics' (Baumgartner and Jones 2002). This branch studies feedback connections between conditions and actors that are responsible for the development of policy over a time period. Few similarities and contrasts can be drawn between the concept of adaptive policies and policy dynamics. The similarity between the two is that both deal with how a policy evolves over a period of time and how they deal with the dynamic pressures that exist within the domain where a policy is made to operate.

The concept of adaptive policies states that policies have to deal with both known and unknown conditions operating within the sphere of influence that they have and these may lead to unknown and unintended consequences and probably may not be as effective as they are designed to be (IISD and TERI 2006). This, in the science of policy dynamics, is considered positive and negative feedback processes that induce equilibrium and stability in the system (Baumgartner and Jones 2002). Both the concepts of adaptive policies and policy dynamics deal with institutions that are involved in designing and implementing policies and how (i.e., processes through which) policies are made. Hence, it can be concluded that a good understanding of policy dynamics can help the climate change adaptation community well. The understanding from policy dynamics and adaptive policies suggest a broad range of conditions within which policies are expected to work in designing and implementing policy solutions for better climate change adaptation since such solutions are able to deal with uncertainties that are inherent to the problem.

The question is how adaptive policies contribute to human security in the changing climate. For instance, climate change is ridden with uncertainties in terms of future projections on nature and degree of impacts (Manning et al. 2004), hindering credible and proactive actions, including policy interventions to mitigate the negative impacts. However, uncertainties should not be the reason for inaction and that principles of adaptive management and adaptive policies would help in handling the greater part of uncertainty (IISD and TERI 2006). The concept of adaptive systems and adaptive management and policies hinges upon the fact that they help develop alternative hypotheses, help identify gaps in knowledge, and help set priorities (Peterson et al. 1997). Because of this basic nature of adaptive systems, they contribute to an enabling environment where human security can thrive well.

4.3.5 Policy Effectiveness for Human Security

For human security to be a policy outcome, it is important to note that there will often be a very limited number of policies that are expected to address the whole gamut of issues that societies face. Hence, it is essential to design policies based on a robust understanding of issues and underlying factors and outcomes of policies being implemented, which is often difficult to discern. This is particularly the case in areas where policies are being formulated following ideological expectations rather than assured evidence-based outcomes. Evidence-based policymaking is increasingly gaining attention, but it has not yet become a norm as a result of gaps at the end of the policymakers and policy advocates, including researchers who lack a clear understanding of how policymaking works and how policies work on the ground.

Werthes et al. (2011) opined that, for human security to be a policy outcome, it needs to be assessed in the policy spheres of fiduciary duty, responsibility to prevent, responsibility to react, and responsibility to rebuild. They concluded that the current policy development can largely be stated as strong in the sphere of responsibility to act and does poorly in the spheres of prevention and rebuilding. This conclusion could be largely valid as the human security literature emerged from the field of humanitarian crisis. This observation is especially important in the context of disaster risk reduction, where reaction to natural hazards has evolved stronger than prevention and mitigation. There is ample literature on policymaking (Cairney 2016; Parkurst 2017) but a slowly growing but limited evidence of policies making human security impacts is just starting to show.

In addition to being a policy outcome, human security has also been increasingly integrated into programming by various NGOs and civil society organizations that started changing the conceptual outlook of human security to more practical perspectives. Such developments are proving that the traditional state-centric security notions have often ignored the vulnerability of people who are subjected to various forms of conflicts and suppression (Fukuda-Parr and Messineo 2012). For example, international organizations such as the International Federation of Red Cross and Red Crescent Societies have been able to address human security issues in their humanitarian value and assistance programs where the culture of non-violence and sustainable development were promoted in diverse capacity-building programs with implications to good governance (Niskala 2008). Even more substantial interventions could be found in the case of the Japan International Cooperation Agency, which is a bilateral assistance agency of Japan. It has successfully integrated human security perspectives into its country assistance programs, promoting freedom from fear, considering the socially vulnerable, protecting and empowering the poor, and addressing global risks by working with national governments, local governments, and civil society organizations (JICA 2010).

From the foregoing discussion, it is evident that any outcomes of policies that are listed in Table 4.1, including disaster risk reduction, environmental security, climate security, food security, energy security, and national security, can have human security implications in ways that are explained in this section. From this discussion, it became clear that human security needs an integrated policy framework rather than a collection of policies working in isolation. The outcomes that constitute human security could range from one as specific as protection from certain diseases and social conflicts to overarching security outcomes such as food, energy, and economic security. Some of these outcomes, such as food security, do not themselves result from one policy intervention but from several policies and they often have to target various geographic, social, and economic conditions in which they need to operate. Because of this dynamic nature and practical focus, the United Nations in particular sees human security as an enabling intervention for addressing issues concerning governments and people.

Policies cannot by themselves produce appropriate outcomes and they need institutions for translating policies into human security outcomes. The next section looks into the ways in which institutions shape human security.

4.4 Human Security and Institutions

Policies and institutions are two inseparable components of human security inasmuch as institutions are vehicles that deliver policies to people and enable people to engage with the policy process (Fig. 4.2). As discussed earlier, human security aims at integrated approaches that are necessarily multi-stakeholder processes and hence, human security approaches can have complex implications for institutions. Institutions, whether of state or civil society, have a paramount role in ensuring human security; they can use human security either as their own policy framework or to enable designing and implementing policies that have broader human security outcomes (Fig. 4.2).

As a framework, human security principles help institutions foster collaboration across levels and sectors and ensure cooperation on matters of importance to human security. Institutions can adopt such principles as a guiding framework for refining their institutional agenda and activities, making their activities more relevant to their stakeholders and contributing to the overall goal of sustainable development. Institutions, in turn, can make a significant contribution to human security mainly as enablers of human security outcomes of policies. Here, they are advocating human security through their action, adopting new actions necessary to promote human security, implementing them within the human security domain, and helping human security to evolve from a principle and a frameworks to something that is actionable, leading to tangible outcomes that enhance people's well-being.

4.4.1 Formal vs. Informal Institutions

Though the literature focuses on the role of formal institutions in human security, including economic, legislative, and constitutional institutions, this section covers a limited scope of formal institutions and focuses more on informal institutions such as civil society organizations and those that work with formal institutions from outside the governance system. Such focus on informal institutions is necessary since formal institutions have long been at the center of human rights and post-conflict situations (Mehta 2006), which is the subject matter of the traditional security paradigm. The shift of policy focus from traditional security to human security means a greater need for emphasis on informal institutions. Such shift also highlights the need to move from a largely reactive nature of formal institutions in the traditional security era to a proactive role of informal institutions in the human security era. The focus on informal institutions should not entirely be construed as a necessity arising out of state and policy failure, there may be an element of it, but rather as a need to complement the state and its policies for more effective outcomes.

4.4.2 Social and Solidarity Economy and Human Security

Social and solidarity economy (SSE) refers to a collection of all the organizations, including cooperatives, associations, social enterprises, etc., that work for the good of the community through the pursuit of social and environmental objectives following the principles of cooperation, solidarity, ethics, and democratic self-management (ILO 2016; UNRISD 2014). They promote the culture of collaboration and communities of cooperation as against the competition enforced and perpetuated by the dominant economy (SolidaritNYC 2016); the stress here is more on the principles, values, and practices. Hence, SSE provides a basis, a set of principles, values, and practices, for the way governments should engage with civil society in meeting the social, environmental, and economic objectives of the whole (i.e., civil society and government). The need for SSE has emerged over the years with the increasing realization of the failure of the current development system in balancing the social, economic, and environmental objectives of development. In a way, all informal institutions can be referred under SSE. Though some scholars treat SSE as an alternative to the mainstream economy and development, in essence, its advocates do not see it as an alternative but as a way to correct and strengthen the existing economic and development approach.

SSE approaches have been used with some success in various fields of development that are relevant to human security. For example, Kumbamu (2017) has demonstrated how the SSE approach has a positive impact on India's agriculture sector: a case study in southern India showed farmers successfully 'reclaiming' control over the local agri-food system toward sustainability. The principles of SSE have also been

Table 4.2 Commonalities between principles of SSE and practices reported in disaster and climate risk reduction disaster risk reduction in Asia

Principle of SSE	Examples from risk reduction	References
Collaboration and cooperation	Collaboration and cooperation have been found important for formulating effective disaster risk reduction policies and outcomes in Thailand, Bangladesh, and Myanmar	(Raungratanaamporn et al. 2014; ActionAid 2015; Habib et al. 2012)
Solidarity	Solidarity between communities and local organizations was found to be an important component in sustaining community-based risk reduction initiatives in Malaysia, the Philippines, and Nepal	(Izumi and Shaw 2012; MercyCorps 2009)
Collective management of shared resources (e.g., collective farming)	Management of collective resources such as water, forests, and agricultural land has been found to enhance adaptive capacities of communities in Nepal and improve rice yields in Indonesia with significant implications to climate change objectives	(Gentle et al. 2013; ADB 2017)

successfully employed to demonstrate their relevance to the UN sustainable development goals in the local context as in the case of Seoul addressing vulnerability, exclusion, and unsustainable development (UNRISD 2017b) and of India addressing gender issues (UNRISD 2017a), climate change (Global Green Growth Institute 2011), social justice (UNESCWA 2014), and disaster risk reduction (Ofreneo and Hega 2016). The community-based approaches being promoted under climate change adaptation and disaster risk reduction in Asia (Table 4.2) have strong compatibility with some of the principles and values of SSE. Hence SSE approaches can promote risk reduction as they are able to promote collaboration, solidarity, and collective management of shared resources.

4.4.3 Private Institutions and Human Security

The power and hence the role of business and financial institutions (collectively private institutions) have substantially risen in this increasingly globalized world. The role of private institutions in human security thus deserves scrutiny. On one hand, private institutions are able to bring new opportunities for capacity-building and empowerment of communities through their corporate social responsibilities and innovative solutions. On the other hand, these institutions have also been known to have created insecurity as they bring pressures of globalization to the local level (Ghai 1997). It has been argued that these private institutions have started undermining the power and reach of the state, and hence there is much to be clarified in terms of the net result of the role played by private institutions in human security.

4.4.4 Institutional Failure and Human Insecurity

Institutional failure can stem from various sources. While the earlier discussion on institutions depict the functional view of institutions, there is an elitist view of institutions that suggests that institutions can also serve the interest of the elite or the more powerful sections of society. It is the absence of representative institutions that fail to help govern in the common interest (Barker 2013). There is also what is called institutional bias, wherein institutions tend to discriminate against certain sections of society or groups of individuals or certain values or certain government policies. Institutional bias has been widely reported in the social justice system, labor markets, education, law, and health care (Henry 2016). The source of institutional bias is reported to arise from an understanding of institutions about the people they are representing, mostly originating from the model of representative democracy (i.e., governments and policies formed by those representing the people rather than by the people themselves as in the case of direct democracy) (Ragan 2013), how access to power and exercise of power manifests in implementing the institutional agenda, and ideological commitments of those governing the institutions since institutions are governed by individual actors who may, in fact, have their own political agenda (Rothstein and Teorell 2005).

The limited capacity of institutions to implement policies and programs is an important aspect to be considered in any discourse that is related to institutional failure. Countries lacking capable institutions have been recognized as an important reason behind the failure of development efforts (Graham 2002). Governments have often used institutions as a decoy to manage opposition to policies and views (Paksoy 2006). Termed 'decoy institutions,' these are deliberately formed and organized in such a way that they lack any valid capacity to implement the stated goals and objectives but are meant to act as a front to avoid actual formulation and implementation of policies (Dimitrov 2020).

Human insecurity (Brauch 2011) generally and disasters specifically (Güiza et al. 2016) have been viewed as an outcome of policy and institutional failure. There is evidence showing that policies merely do not produce outcomes, but it is the institutions acting as preconditions that help translate policies into successes so much so that a major category of policy success could be stated as institutional success (Hudson 2015). For a policy to be successful, its compatibility with constitutionally protected institutional rights is as important as the constitutionality of the policy itself. Hence, it has been argued that policies would have to be checked for their institutional vulnerabilities even before they are enacted and implemented. However, institutional vulnerability of risk reduction policies enacted and implemented in most of the countries in Asia has not been thoroughly studied and represents an important research gap.

Analysis of institutional vulnerability is sparse in Asia and the available literature suggests that institutional vulnerabilities have not been well considered in various disaster risk assessments and that institutional vulnerability can play a major role in how effectively policies are translated into success (Lassa 2010). Turning from institutional vulnerability to institutional success requires identification of determinants of institutional success. Some of the determinants include technical and organizational governance represented in terms of qualified human resources (Gandhi and Herath 2007; Agrawal et al. 2009), institutional quality expressed in terms of legitimacy (Alonso and Garcimartín 2009), extent of integration of human security principles into institutional practices (Muggah and Krause 2008), how adaptive institutions are to changing societal needs (IISD 2011), social efficiency of institutions by weighing social costs against benefits (Siba 2008), coordination and communication among institutions (Wang et al. 2016), ability to represent different perspectives, and support of participatory principles (Agrawal et al. 2009).

4.5 Policy and Institutional Indicators of Human Security

It is evident that institutions and policies are major determinants of human security and the extent these two contribute to human security could vary, depending on the socioeconomic developmental context. The nature of contribution of these two on human security outcomes could be evaluated through indicators. While human security has been seen as a principle, as a policy framework, and as a policy outcome, identifying policy and institutional indicators will help one to visualize clearly whether human security is mainstreamed into policies and institutions. Identifying the indicators of human security also means viewing it as a practical concept and as something that can help identify specific tenets underpinning the stated outcome that poses a limitation. Indicators are also needed for the reason that, although the normative nature of human security is largely agreed upon (Fukuda-Parr and Messineo 2012; Owen 2008), its analytical rigor has not been clear, posing the challenge to identify specific indicators. Part of the problem in identifying specific indicators is the wide scope of the areas that can be related to and which affect the security of individuals and communities, making it conceptually and analytically impossible to capture in a practical manner. However, Owen (2008) opined that human security is still worth measuring because it will help clarify its ambiguous nature, turning it into something more practical and usable, and thereby gain an understanding of the causality factors affecting human security.

Notwithstanding these limitations, efforts have been made to measure human security or its different facets. It can be stated that these efforts are evolving and are yet to provide full and practical insights into the operational aspects of human security (Table 4.3). Keeping in view the focus of this chapter, this section identifies

Table 4.3 Some policy and institutional indicators for measuring human security

Indicator	Reference
Political stability	(Hastings 2010)
Unintended impacts of climate change policies	(Adelphi 2017)
Investments in security-related fields, including foreign assistance, humanitarian assistance, disaster risk reduction	(Finel and Bartolf 2009)
Public services, human rights, rule of law	(The Fund for Peace 2017)
Democracy and political freedom (percent of adults able to participate in elections)	(Owen 2008)
Public expenditure on defense vs education, human freedom index, degree of democratization	(Lonergan et al. 1999)
Political freedom measured as societal freedom	(Eldering 2010)

those indicators that are related to policy and institutional aspects and does not cover the entire scope of human security, which could be vast. Due to limited efforts to quantify human security, indicators from other relevant security fields that have close relationship with human security are also referred to in this section. The literature indicates an essential shift in understanding and evolution from the traditional notion of security to the broader concept of human security (Owen 2008).

The human security index, one among the first efforts to quantify human security, defined human security as a triad of environmental, economic, and social aspects interacting with each other (Hastings 2010). It identified financial economic governance, environmental protection policies, and governance as important areas of the economic, environmental, and social domains of the human security index. The climate security index, which is an initiative under the American Security Project, identified funding for security-relevant fields, including renewable energy, humanitarian emergency, and conflict as important indicators of the country's climate security (Finel and Bartolf 2009). Due to its significant role, the foreign assistance allocated to human security fields was also considered an important aspect of national security. The fragile states index developed by The Fund for Peace, though not entirely a human security index but with significant relevance, includes several political and policy-relevant indicators such as public services, openness of government relationship with its people, human rights, and rule of law (The Fund for Peace 2017). The openness indicators and human rights indicators in particular also look at the public confidence on state institutions and their legitimacy. The human insecurity index (Werthes et al. 2011) was developed as a composite of five dimensions comprising economic, environmental, food, health, personal and community, and political dimensions. Unlike other indices discussed earlier, this index was developed as the opposite of security and deducting the human insecurity index value from 100 provides an equivalent picture of human security though the authors did not provide any such interpretation. The main difference between human security index and human insecurity index, as described by Werthes et al. (2011), is that the former just improves the human development index in terms of equitability rather

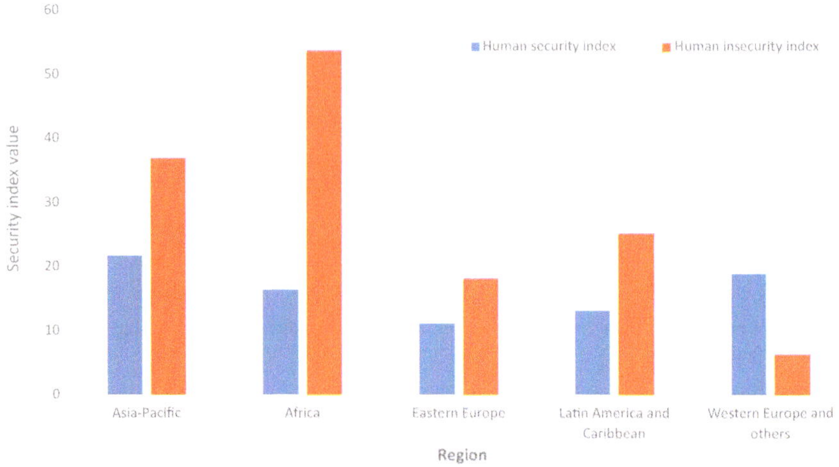

Fig. 4.3 Human security in Asia and the rest of the world (*Source* Data from Hastings [2011] and Werthes et al. [2011])

than capture human security as defined widely. Werthes et al. (2011) also argued that the human insecurity index represents human security that is compatible with the United Nations Trust Fund for Human Security (United Nations 2018), a general understanding that human security can be best measured by its absence rather than by its presence (UNDP 1994) and that presenting it as 'human security' would not provide a better picture of human insecurity that is more important for policy focus and as deterrence for countries being ranked as insecure.

To assess the level of human security in Asia and how it compares with those in other countries, the human security index as reported by Hastings (2011) and the human insecurity index as reported by Werthes et al. (2011) are presented Fig. 4.3. The grouping of countries in the figure follows the UN classification of regional groups (United Nations 2014b), which means that the 'Western Europe and Others' group includes the North American countries of the United States and Canada and 27 of the western European countries. The Asia-Pacific region ranks first in human security (see Fig. 4.3; higher values indicate better human security), showing marginally better performance than 'Western Europe and Others' and second in human insecurity (can be interpreted as 4th for human security on this classification) where 'Western Europe and Others' is ranked least in terms of human insecurity. The marginal better performance of Asia in terms of human security index could be attributed to higher human security values for west Asian countries such as Kuwait and Oman. As a result, six countries in Asia recorded human security index values of more than 0.4, while only one country in 'Western Europe and Others' had a value of more than 0.4 (0.455 for Luxembourg). If these six countries are excluded from Asia-Pacific, the region ranks second after 'Western Europe and Others' with an average value of 0.17. Countries such as Bangladesh, Cambodia, Mongolia, Syria, and Vietnam recorded the lowest human security index of 0.1 or less.

In terms of human insecurity, only three countries in the Asia-Pacific region recorded human insecurity index values of less than 10, while 80% of the countries in the Western Europe and Others group had values less than 10. In the Asia-Pacific region, 55% of the countries recorded a human insecurity index less than 40, and five countries had more than 60 (Afghanistan, Bangladesh, Myanmar, North Korea, and Timor Leste). The higher insecurity conditions in Asia are attributed to more countries facing higher economic insecurity, moderate level of food insecurity, moderate level of health insecurity, and moderate level of environmental insecurity. These countries specifically have recorded relatively lower levels of political insecurity, which was stated as an 'Asian phenomenon' by the authors as large parts of Asia suffer from political insecurity in the years of the assessment (Werthes et al. 2011). Japan is the only Asian country that recorded the lowest insecurity conditions, largely attributed to its early inroads into human security as its political and policy landscape was impacted by the human security approaches relatively earlier than many other countries. Interestingly, Japan received '0' for food insecurity and environmental insecurity, even though the country imports a significant proportion of its food from those countries that are negatively affected by climate change and other insecurities. The difference in the two index values and trends across regions and countries indicates that these two indices are not consistent with each other.

4.6 Limitations of Human Security as a Policy Imperative

Human security concepts and approaches have certainly made their dent into the international discourse and international policy processes and frameworks, including sustainable development goals and climate change. However, it is yet to infiltrate national and subnational levels where the actual impact is to be felt. Even though human security has pushed forward as an overarching policy framework for global development, the human security approaches are not far from criticism and limitations when it comes to making impacts at the policy and institutional levels. The limitations are on the side of human security concepts and on the side of the existing policy and institutional mechanisms. Limitations on the side of the human security principles, largely, have been attributed to human security approaches that have reinforced existing policy frameworks rather than challenged them (Chandler 2008). Part of the criticism of human security also lies in it being seen as a policy agenda in itself (Fukuda-Parr and Messineo 2012). This criticism emerges from seeing human security as an end in itself rather than as a means to achieve international security agenda. Human security has also been criticized as too narrow, too broad, and too vague by various scholars (Fukuda-Parr and Messineo 2012), which may be hindering its practical utility in formulating policies and in aligning institutional actions. While traditional security was seen as a narrow concept, human security was viewed as overly expansive and vague (Owen 2008). In addition, views that human security is contextual in nature (and hence generalization even at a national scale could vastly be inappropriate) can practically limit developing measurement tools that can work in wide circumstances within which human security needs can be assessed with certainty (Werthes et al. 2011).

There are also limitations with respect to what institutions and policies can engage in, the most important one being the lack of proper understanding of human security concepts and approaches. This prevents institutions from adopting human security principles in their businesses. Compared with other security principles, including food security and traditional security, human security is a relatively new advent. Often global in scope, it has largely failed to capture the imagination of those institutions that often work in specific domains and that may find sub-tenants of human security (i.e., food security, energy security, etc.) more relevant.

As an outcome of policies and actions, the human security concept faces a challenge as policies are often promulgated and implemented at the national level, are often broad in scope and reach, and tend to generalize and normalize the situations to be addressed, whereas human security could manifest in different forms under different circumstances. Hence, formulating human security-specific policies could be daunting. A way forward could be to integrate human security principles into various sectoral cross-cutting areas where policies, which often ignore the human security dimension, already exist.

4.7 Conclusion

The paper discussed how policies and institutions can help the cause of human security. It was argued that human security has two dimensions in the context of policies and institutions—i.e., it can be viewed both as a policy framework and as a policy outcome. Institutions, both formal and informal, play a vital role in both dimensions of human security. Human security provides a comprehensive framework for policies and institutions in the context of increasing disasters and climate change impacts as it emphasizes social cohesion and tries to achieve environmental and economic outcomes that current developmental approaches have failed to achieve. The evidence suggests that Asian countries are not doing well in terms of human security, as depicted by the high human insecurity, mostly due to political instability and related insecurity. The only high-performing country in the Asian region is Japan, whose good showing was mainly attributed to the early impacts of human security approach on the political agenda of the country and to better performance in terms of handling food and environmental insecurity.

On the positive side, human security principles have already made a significant dent at the international level and have started influencing the national developmental discourse. Several elements of human security have already been incorporated into programs and are increasingly gaining attention at the national and subnational levels. Approaches such as SSE and community-based participatory risk reduction measures are able to provide practical ways of tapping into some of these human security principles and can be treated as a way forward for mainstreaming human security into development and risk reduction.

However, there is still a long way to go—human security must find its place in the mainstream developmental discourse in general and risk reduction in particular. In order for human security to move from the stage of 'a principle and a concept' to a practical framework, there is a need for more empirical work to see how individual and community securities are affected by various drivers and pressures and how policies can address them comprehensively. The current measures of human security are incremental advancements of the human development index and related measures and hence are yet to fully capture the multidimensional nature of human security so they can be applied for practical purposes. As these measures have poor representation of policy and institutional indicators, no significant adoption occurred. There is no evidence that the measures presented in this chapter are employed for practical purposes. One possible limitation is the very limited progress in gaining conceptual clarity of what constitutes human security and of how these understandings can be systematically distilled into a measuring system. One possible way forward in this area could be to advance the work on measuring human security, which has already started but has not gained momentum, and pilot-test these measures in real-world scenarios in which policies and institutions work. There is a need to develop a robust 'test' that can identify if the proposed measures can reliably ascertain human security.

Acknowledgements The author is grateful that part of this research was supported by the Environment Research and Technology Development Fund (2-1801) of the Environmental Restoration and Conservation Agency of Japan and by the IGES Strategic Research Fund 2020.

References

ActionAid (2015) Inclusive community-based disaster risk reduction: guidance on good practice in the Ayeyarwaddy Delta, Myanmar. ActionAid Myanmar, Yangon, Myanmar

ADB (Asian Development Bank) (2017) Integrated participatory development and management of irrigation program. ADB, Manila, Philippines

Adelphi (2017) Factbook of environment, conflict and cooperation (ECC) platform [Online]. Available at https://factbook.ecc-platform.org/. Accessed 11 Dec 2017

Adger W, Pulhin JM, Barnett J, Debelko GD, Hovesrud GK, Levy M, Spring UO, Vogel CH (2014) Human security. In: C. Field et al. (eds) Climate change 2014: impacts, adaptation, and vulnerability. Part A: Global and sectoral aspects. Contribution of Working Group II to the Fifth Assessment Report of the Intergovernmental Panel on Climate Change. Cambridge, United Kingdom: Cambridge University Press, pp 755–791

Agrawal A, Kononen M, Perrin N (2009) The role of local institutions in adaptation to climate change. The World Bank, Washington, DC, USA

Akiyama N (2004) Human security at the crossroad: human security in the Japanese foreign policy context. Institute for Peace Science, Hiroshima University, Hiroshima, Japan

Alonso J, Garcimartín C (2009) The determinants of institutional quality: more on the debate. Nottingham, UK: Centre for Research in Economic Development and International Trade, University of Nottingham

Anderson JE (1984) Public policymaking: an introduction, 2nd edn. Houghton Mifflin, Boston, USA

Atwood B (2012) The benefits of transparency in development. OECD, Paris, France

Azulai M, Bandiera O, Blum F, Kleven H, Ferrara EL, Padró G, Tejada CM (2014) State effectiveness, growth and development. International Growth Centre, London, UK
Barker D (2013) Oligarchy or elite democracy? Aristotle and modern representative government. New Polit Sci 35(4):547–566
Baumgartner F, Jones B (2002) Theoretical beginnings. University of Chicago Press, Chicago, USA
Brauch HG (2011) Concepts of security threats, challenges, vulnerabilities and risks. In: Brauch H et al (eds) Coping with global environmental change, disasters and security. Springer, Berlin, Germany, pp 61–106
Caballero-Anthony M (2002) Overview of health and human security case studies. In: Noda PJ (ed) Health and human security: moving from concept to action. Fourth Intellectual Dialogue on Building Asia's Tomorrow. Tokyo, Japan: Japan Center for International Exchange, pp 21–44
Cairney P (2016) The politics of evidence-based policy making. Palgrave Macmillan, UK
Carstens A (2005) The role of transparency and accountability for economic development in resource-rich countries (pers. commun., International Monitory Fund, 27 Jan 2005)
Chandler D (2008) Human security: the dog that didn't bark. Secur Dialogue 39:4
Chestrad International (2013) Fact sheet on health and human security approach. Chestrad International, Ibadan, Nigeria
Considine M (2005) Making public policy. Polity Press, Cambridge, UK
Dimitrov RS (2020) Empty institutions in global environmental politics. Int Stud Rev 22(3): 626–650
Eldering M (2010) Measuring human (in)security. Human Secur Perspect 7(1):17–49
Elliott L (2015) Human security/environmental security. Contemp Polit 21(1):11–24
European Commission (2017) The EU approach to development effectiveness [Online]. Available at https://ec.europa.eu/europeaid/policies/eu-approach-aid-effectiveness_en. Accessed 8 Dec 2017
FAO (Food and Agriculture Organization) (2016) Human security and food security. FAO, New York, USA
Finel BI, Bartolf C (2009) Climate security index. American Security Project, Washington, DC, USA
Fontanel J, Corvaisier-Drouart B (2014) For a general concept of economic and human security. In Bellais R (ed.). The evolving boundaries of defence: an assessment of recent shifts in defence activities (Contributions to Conflict Management, Peace Economics and Development, vol. 23). S.l.: Emerald Group Publishing Limited, pp 75–96
Fukuda-Parr S, Messineo C (2012) Human security: a critical review of literature. Centre for Research on Peace and Development, Leuven, Belgium
Gandhi V, Herath G (2007) Determinants of institutional success for water in India: results from a study across three states. Australian Agricultural and Resource Economics Society, Queenstown, New Zealand
Gemenne F, Barnett J, Adger W, Dabelko G (2014) Climate and security: evidence, emerging risks, and a new agenda. Clim Change 123(1):1–9
Gentle P, Race RT, Alexander K (2013) Changing role of local institutions to enable individual and collective actions for adapting to climate change. The International Association for the Study of the Commons, Kitafuji, Japan
Gerston L (2010) Public policy making: process and principles. M.E. Sharpe, USA
Ghai D (1997) Economic globalization, institutional change and human security. United Nations Research Institute for Social Development, Geneva, Switzerland
GGGI (Global Green Growth Institute) (2011) Social dimensions of climate change: adaptation, water, energy and green employment. Seoul, Republic of Korean, GGGI
Gomez OA, Gasper D (2016) Human security: a thematic guidance note for regional and national human development report teams. United Nations Development Program, New York
Graham C (2002) Strengthening institutional capacity in poor countries: shoring up institutions, reducing global poverty. The Brookings Institution, Washington, DC, USA
Guadagno L (2016a) Human mobility in a socio-environmental context: Complex effects on environmental risk. In: Sudmeier-Rieux Manuela K, Penna Michel F, Gaillard J (eds) Identifying

emerging issues in disaster risk reduction, migration, climate change and sustainable development. Springer, Switzerland, pp 13–31
Guadagno L (2016b) Human mobility in the sendai framework for disaster risk reduction. Int J Disaster Risk Sci 7(1):30–40
Güiza F, Simmons P, Burgess J, McCall M (2016) Chronic institutional failure and enhanced vulnerability to flash-floods in the Cuenca Altadel Río Lerma. Mexico Disasters 40(1):112–133
Habib A, Shahidullah M, Ahmed D (2012) The Bangladesh cyclone preparedness program: a vital component of the nation's multi-hazard early warning system. In: Maryam G (ed) Institutional partnerships in multi-hazard early warning systems: a compilation of seven national good practices and guiding principles. Springer, Dhaka, Bangladesh, pp 29–62
Harvard Kennedy School (2016) The transparency for development initiative. Cambridge, USA: Harvard Kennedy School [Online]. Available at https://www.r4d.org/resources/transparency-development-initiative/. Accessed 15 Nov 2018
Hastings D (2010) Data & maps: HSI Version 2 & a "classic" HDI. Paper presented at the International Symposium on Geoinformatics for Spatial Infrastructure Development in Earth and Allied Sciences, Hanoi, Vietnam
Hastings DA (2011) The human security index: an update and a new release [Online]. Available at http://www.humansecurityindex.org/. Accessed 11 Dec 2017
Henry P (2016) Institutional bias. In: Nelson T (ed) Handbook of prejudice, stereotyping and discrimination. Sage, New York, USA, pp 426–440
Holzmann R, Sherburne-Benz L, Tesliuc E (2003) Social risk management: the World Bank's approach to social protection in a globalizing world. The World Bank, Washington, DC, USA
Hudson B (2015) Institutional preconditions for policy success. Tulane Law Rev 89:50
IISD (International Institute for Sustainable Development) (2011) Adaptive policy making [Online]. Available at http://www.iisd.org/climate/vulnerability/policy.asp. Accessed 8 Dec 2017
IISD (International Institute for Sustainable Development), TERI (The Energy and Resources Institute) (2006) Designing policies in a world of uncertainty, change, and surprise: adaptive policy-making for agriculture and water resources in the face of climate change. IISD and TERI, Manitoba, Canada
ILO (International Labor Organization) (2016) Social and solidarity economy [Online]. Available at http://www.ilo.org/global/topics/cooperatives/areas-of-work/WCMS_546299/lang–en/index.htm. Accessed 14 Dec 2017
Ingram H, Smith S (1993) Public policy for democracy. The Brookings Institution, Washington, DC, USA
Izumi T, Shaw R (2012) Role of NGOs in community-based disaster risk reduction. In: Shaw R (ed) community-based disaster risk reduction. Bingley, UK, Emerald Group, pp 35–54
JICA (Japan International Cooperation Agency) (2010) JICA's human security approach: features and case studies [Online]. Available at https://www.jica.go.jp/english/publications/jica_archive/brochures/pdf/human.pdf. Accessed 11 Dec 2012
Karlsson-Vinkhuyzen S, Jollands N (2015) Human security and energy security: a sustainable energy system as a public good. SSRN [Online], p 21. Available at https://papers.ssrn.com/sol3/papers.cfm?abstract_id=2562594. Accessed 10 Oct 2018
Kaunda J (2005) Effective states and engaged societies: capacity development for growth, service delivery, empowerment and security in Africa. The World Bank, Washington, DC, USA
Kay A (2006) The dynamics of public policy: theory and evidence. Edward Elgar, Northampton, MA, USA
Khan M (2007) Governance, economic growth and development since the 1960s. UN Department of Economic and Social Affairs, New York, USA
Kumbamu A (2017) Building sustainable social and solidarity economies: place-based and network-based strategies of alternative development organizations in India. Comm Dev 49(1):18–33
Lassa JA (2010) Institutional vulnerability and governance of disaster risk reduction: macro, meso and micro scale assessment. Rheinischen Friedrich-Wilhelms-Universität, Bonn, Germany

Lonergan S, Gustavson K, Carter B (1999) The index of human insecurity. Global Environmental Change and Human Security Program, s.l.
Manning M, Petit M, Easterling D, Murphy J, Patwardhan A, Rogner H, Swart R, Yohe G (2004) Proceedings of the IPCC workshop on describing scientific uncertainties in climate change to support analysis of risk and of options. Maynooth, Ireland: National University of Ireland
McDonald M (2013) Discourses of climate security. Polit Geogr 33:42–54
Mehta P (2006) From state sovereignty to human security (via institutions?). Nomos 47:259–285
MercyCorps (2009) Community-based disaster risk reduction: good practices. MercyCorps Nepal, Lalitpur, Nepal
Ministry of Foreign Affairs of Japan (2014) Human security [Online]. Available at http://www.mofa.go.jp/policy/human_secu/index.html. Accessed 8 Dec 2017
Muggah R, Krause K (2008) A true measure of success? The discourse and practice of human security in Haiti. Whitehead J Diplomacy Int Relat 57(2):153–181
Niskala M (2008) Human security: consolidation of peace and good governance [Online]. Available at http://www.ifrc.org/en/news-and-media/opinions-and-positions/speeches/2008/human-security-consolidation-of-peace-and-good-governance/. Accessed 11 Dec 2017
OECD (Organisation for Economic Cooperation and Development) (2011) Fourth high level forum on aid effectiveness [Online]. Available at http://www.oecd.org/dac/effectiveness/fourthhighlevelforumonaideffectiveness.htm. Accessed 8 Dec 2017
OECD (Organisation for Economic Cooperation and Development) (2014) OECD reviews of risk management policies: boosting resilience through innovative risk governance. OECD Publishing, Paris, France
Oels A (2013) Rendering climate change governable by risk: from probability to contingency. Geoforum 45:17–29
Ofreneo R, Hega M (2016) Women's solidarity economy initiatives to strengthen food security in response to disasters: insights from two Philippine case studies. Disaster Prev Manage 25(2):168–182
Owen T (2008) Measuring human security: methodological challenges and the importance of geographically referenced determinants. In: Liotta PH, Mouat DA, Kepner W, Lancaster J (eds) Environmental change and human security: recognizing and acting on hazard impacts. The Netherlands: Springer, p 478
Paksoy H (2006) Identities: how governed, who pays?. S.l.: Entelequia Revista Interdisciplinar, Texas, USA
Parkurst J (2017) The politics of evidence: from evidence-based policy to the good governance of evidence. Routledge, Oxon, UK
Peterson G, Leo GAD, Hellmann JJ, Janssen MA, Kinzig A, Malcolm JR, O'Brien KL, Pope SE, Rothman DS, Shevliakova E, Tinch RRT (1997) Uncertainty, climate change, and adaptive management. Conserv Ecol 1(2):4
Prabhakar S (2017) Approaches for financial inclusion and their efficacy for vulnerability reduction: evidences, bottlenecks and opportunities. In: Mohammed E, Uraguchi Z (eds) Financial inclusion for poverty alleviation: issues and case studies for sustainable development. Abingdon, UK: Routledge, p 260
Ragan R (2013) Institutional sources of policy bias: a computational investigation. J Theor Polit 25(4):467–491
Rasheed F (2015) Microfinance institutions and human security in Muslim majority countries: achievements and challenges. In: Hasan S (ed) Human security and philanthropy. Springer, New York, USA, pp 205–227
Raungratanaamporn I, Pakdeeburee P, Kamiko A, Denpaiboon C (2014) Government-communities collaboration in disaster management activity: investigation in the current flood disaster management policy in Thailand. Proc Environ Sci 20:658–667
Results for Development (2017) Transparency for development: Do transparency and accountability interventions improve health outcomes?. Results for Development, Washington, DC, USA

Rothstein B, Teorell J (2005) What is quality of governance? A theory of impartial political institutions. United Nations Division for Public Administration and Development Management, Göteborg, Sweden

Sachs J, McArthur JW, Schmidt-Traub G, Kruk M, Bahadur C, Faye M, McCord G (2004) Ending Africa's poverty trap. T. Brookings Papers on Economic Activity, vol 1, pp 117–240. Washington, DC, USA: The Brookings Institution

Siba E (2008) Determinants of institutional quality in sub-Saharan African countries. University of Gottenburg, Göteborg, Sweden

Smith A (1776) Wealth of nations, 1st ed. London, UK: Strahan & Cadell

SolidarityNYC (2016) Solidarity economy: the basics [Online]. Available at http://solidaritynyc.org/#/resources/. Accessed 12 Mar 2018

The Fund for Peace (2017) Fragile states index: methodology cast framework. The Fund for Peace, Washington, DC, USA

Torjman S (2005) What is policy?. Caledon Institute of Social Policy, Ottawa, Canada

UN (United Nations) (2014a) Human security approach [Online]. Available at http://www.un.org/humansecurity/human-security-unit/human-security-approach#a6. Accessed 08 Dec 2017

UN (United Nations) (2014b) United Nations regional groups of member states [Online]. Available at http://www.un.org/depts/DGACM/RegionalGroups.shtml. Accessed 15 Mar 2018

UN (United Nations) (2018) United Nations trust fund for human security [Online]. Available at https://www.un.org/humansecurity/. Accessed 15 Mar 2018

UN (United Nations) (2009) Human security in theory and practice: an overview of human security concept and the United Nations Trust Fund for Human Security. New York, USA: UN

UNDP (United Nations Development Program) (1994) Human development report 1994. UNDP, New York, USA

UNESCWA (The United Nations Economic and Social Commission for West Asia) (2014) Social and solidarity economy as a tool for social justice. Beirut, Lebanon: UNESCWA

UNRISD (United Nations Research Institute for Social Development) (2014) Social and solidarity economy and the challenge of sustainable development: executive summary of a TFSSE position paper. UNRISD, Geneva, Switzerland

UNRISD (United Nations Research Institute for Social Development) (2017a) Feminist analysis of social and solidarity economy practices: views from Latin America and India [Online]. Available at http://www.unrisd.org/feminist-sse. Accessed 14 Dec 2017

UNRISD (United Nations Research Institute for Social Development) (2017b) Social and solidarity economy for the SDGs: spotlight on the social economy in Seoul [Online]. Available at http://www.unrisd.org/unrisd/website/projects.nsf/(httpProjects)/C99640DFC9CD198FC12580EC005B3D00?OpenDocument. Accessed 14 Dec 2017

Wall K (2014) Human security and the citizen-state relationship. In: Wall K, Aulin J, Vogelaar G (eds) Empowerment and protection: Stories of human security. The Global Partnership for the Prevention of Armed Conflict, The Hague, the Netherlands, pp 106–123

Wallis J, Dollery B (1999) Market failure and government intervention. In: Wallis J, Dollery B (eds) Market failure, government failure, leadership and public policy. Palgrave Macmillan, London, UK, pp 9–31

Wang J, Wang Y, Li S, Qin D (2016) Climate adaptation, institutional change, and sustainable livelihoods of herder communities in northern Tibet. Eco Soc 21(1):5

Werthes S, Heaven C, Vollnhals S (2011) Assessing human insecurity worldwide: the way to a human (in)security index. Institute for Development and Peace, Duisburg, Germany

Weimer D, Vining A (1992) Policy analysis: concepts and practice. Prentice Hall, Upper Saddle River, NJ, USA

World Bank (2017) Worldwide governance indicators [Online]. Available at http://info.worldbank.org/governance/wgi/index.aspx#reports. Accessed 8 Dec 2017

Yamada K (2015) Added value of the human security approach to natural disasters and climate change. Japan International Cooperation Agency, Miyagi, Japan

Dr. S. V. R. K. Prabhakar works on mainstreaming climate change adaptation and disaster risk reduction into development, adaptation metrics, adaptive policies, risk insurance, regional adaptation planning for transboundary climate risks, loss and damage, vulnerability assessment, and training needs assessment in South and Southeast Asian countries. He has been Coordinating Lead Author for the UNEP Geo-6 Asia report, Lead Author for the HIMAP Report, and Contributing Author to the Fifth and Sixth Assessment Reports of the IPCC Working Group II. Prabhakar obtained Ph.D. in Agronomy from IARI, India. He possesses more than 17 years of experience in participatory research and development with international and national research and development organizations including IGES, Kyoto University, UNDP, NIDM, CIMMYT-RWC, IARI, and ICRISAT.

Chapter 5
Satisfying the Basic Needs: The Material Aspect of Human Security

Lorena L. Sabino and Juan M. Pulhin

Abstract Humans have basic needs to satisfy, which are widely expressed as adequate food, shelter, clothing, education, clean water and air, sanitation, and good health, among others. These universal physiological needs are at the core of the material aspect of human security. Meeting these needs is a requisite for survival and essential for human well-being. Human insecurity occurs when these basic needs are not met. Failure to satisfy them is unjust and a serious denial of basic human rights and security. Recent literature points out that unsatisfied basic needs may not solely be attributed to poverty and lack of access to basic services but also to pressing environmental factors such as climate change impacts and disaster risks. This chapter contends that the provision of basic needs, which forms the very core of human security, will be progressively threatened in many parts of Asia as climate changes and disaster risks escalate. It further argues that climate change adaptation (CCA) and disaster risk reduction (DRR) strategies have been employed by many Asian countries, although there is yet no comprehensive assessment of their effectiveness in satisfying the human basic needs and security of the poor and vulnerable sector. The chapter concludes that it is crucial to mainstream CCA and DRR in national and local development agenda particularly in the less economically developed and more vulnerable countries in Asia, with the end view of protecting the basic needs and security of the poor in the context of changing climate and increasing incidence of disasters.

Keywords Basic needs · Climate change impacts · Disaster risks · Human security · Poor · Vulnerable sectors

L. L. Sabino (✉) · J. M. Pulhin
Department of Social Forestry and Forest Governance, College of Forestry and Natural Resources (CFNR), University of the Philippines Los Baños, Laguna, Philippines
e-mail: llsabino@up.edu.ph

J. M. Pulhin et al. (eds.), *Climate Change, Disaster Risks, and Human Security*, Disaster Risk Reduction, https://doi.org/10.1007/978-981-15-8852-5_5

5.1 Introduction

The concept of human security was popularized at the global level with the publication of the 1994 Human Development Report (HDR) of the United Nations Development Programme. Traditionally, the notion of human security is regarded against threats of nuclear holocaust, which is more related to nation-states than to people (UNDP 1994). In the twentieth century, the meaning and content of security have been increasingly contested by scholars and practitioners (MacFarlane and Khong 2006). Human security concepts have been mainstreamed into development discourse, which shifted from a territorial-centered to a human-centered concept of security (Kumssa and Jones 2010). With the passing of the Cold War, new concepts of security emerged that addressed not only the military realities of the contemporary world but also the political, economic, and social realities (Synder 2011). Likewise, the concept of human security is now being applied to the threat of global poverty in the form of drugs, health, climate change, illegal migration, and terrorism (UNDP 1994).

Generally, security is the condition of being protected from or not exposed to danger (Barnett 2003). The UNDP characterized human security in two aspects: "freedom from fear and freedom from want" (UNDP 1994). Freedom from fear means security from chronic threats such as hunger, disease, repression, and even extensive environmental damage. On the other hand, freedom from want is akin to protection from sudden and hurtful disruptions in the patterns of daily life, whether at home, on the job, or in the community (UNDP 1994). Kumssa and Jones (2010) referred to freedom from fear as protection of individuals from violent conflicts and from denial of civil liberties, including freedom of expression and belief, whereas freedom from want was emphasized as satisfying the basic needs of individuals such as food, shelter, clothing, and gainful employment. The ultimate objective of human security is the "protection of the vital core of all human lives in ways that enhance human freedom and human fulfillment" (CHS 2003). It is an intrinsically people-centered rather than state-centered approach.

In the context of climate change and disaster risk, policies, and opportunities as means of exercising human freedom and human fulfillment should be given much importance in order to move away from insecurity. It means building peoples' capability and capacity to overcome foreseen and prepare for unforeseen threats to human survival. The UNDP (1994) emphasizes that people are secured if they are able to take care of themselves and have the opportunity to meet their own most essential needs and earn for their living. Human insecurity emerges from the interaction of multiple factors (Adger et al. 2014). Jonas and Burgess (2011) indicated that insecurity originated not only from civil conflict, failed states, poverty, disease, and small arms but also due from disasters brought about by climate change.

The climate change phenomenon is recognized as a major human security issue that poses a serious global threat (Dankelman et al. 2008). With the constant changes in climate, human security might progressively be threatened as livelihoods are undermined (Adger et al. 2014). The impacts of climate change include destruction of ecosystems, occurrence of extreme weather events, changes in hydrological cycles, and sea level rise. These impacts lead to loss of livelihood and worsen poverty. As a result, people are deprived of their basic needs, which lead to hunger, water scarcity, deteriorating health conditions, involuntary displacement and migration, and increased incidence of violence (Cameron 2011). Moreover, it will significantly threaten the security of individuals, the security of states and nations, as well as the global security of future centuries (Adger 2010). Finally, as livelihoods are undermined, vulnerability increases due to lower capacity to prepare, to cope, and to recover from shocks and stresses (DFID 2004).

This chapter examines how climate change and disaster risks in Asia threaten the satisfaction of human basic needs, which forms the core of human security as well as some of the key strategies adopted to address the problem. The chapter is divided into five sections. Following this brief introductory section, Sect. 5.2 discusses the concept of human basic needs as the foundation of human security. Section 5.3 elaborates how climate change impacts and disasters risks undermine the satisfaction of human basic needs. Section 5.4 presents illustrative examples of key strategies on climate change adaptation (CCA) and disaster risk reduction (DRR) in selected Asian countries relevant to satisfying basic needs and human security. The last section on conclusion highlights the need to mainstream CCA and DRR in development agenda to protect the basic human needs and security of the poor in an era of changing climate and increasing incidence of disasters.

5.2 The Concept of Human Basic Needs

This section introduces some of the general notion of human needs and as well as the basic needs of human being which form the core of the material aspect of human security.

5.2.1 Notion of Human Needs

Understanding the concept of human needs is an essential component of visualizing what the future would look like, particularly concerning the welfare of the people. The notion of need is usually viewed as the difference between what is (present/actual condition) and what ought to be (desired situation). The gap between the two is what is referred to as "needs" (Leagans 1964). It is a notion of necessity, although the concept of need is broad.

The need concept was first proposed by the psychologist Abraham Maslow (1943). In his paper "A theory of human motivation," he categorized human needs into five levels of hierarchy of prepotency (lower to upper), which are common to all people

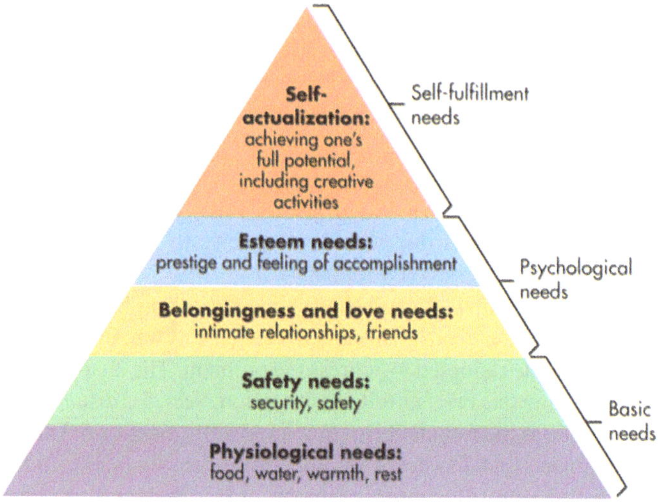

Fig. 5.1 Abraham Maslow's hierarchy of human needs (McLeod 2017)

(Fig. 5.1). Maslow's hierarchy of human needs includes physiological needs (food, water, shelter, etc.), the safety needs (protection, security), the love needs (family, affection), the esteem needs (achievement, status), and the need for self-actualization (personal growth and fulfillment). It is assumed that the lower level (physiological needs) should first be satisfied before the next level can be fulfilled.

For simplicity and practicality, Leagans (1964) came up with three categories of human needs: physical needs, which also refer to physiological needs (food, clothing, housing, etc.); social needs (status, affection, belonging, etc.); and integrative needs (relating oneself to something bigger). He also classified needs into two groups: felt needs (recognized) and unfelt needs (unrecognized). Felt needs serve as the motivating forces. Likewise, Alderfer (1969) simplified the five stages of Maslow's human needs theory into three: existence, relatedness, and growth. In contrast, it does not assume satisfaction at the lower level for the emergence of higher order needs.

As people grow, human needs may change through time, but the physiological aspect or basic needs will remain constant. Basic needs are universal and serve as keys for survival. Failure to satisfy the basic needs can cause human insecurity. It can also affect human personal development. For instance; chronic hunger due to insufficient food as a result of poverty or environmental stress can lead to health issues, negative conditions, and even unproductivity. Satisfaction of basic needs forms the very core of the material aspect of human security.

5.2.2 *Basic Needs as the Core of the Material Aspect of Human Security*

Basic needs are essentials required for survival and well-being and are, therefore, basic human rights. They include personal commodities such as adequate food, shelter, clothing, and sound sleep as well as public services such as provision of clean water, good sanitation, clear air, transportation, health care, employment, and education, among others (ODI 1978). Basic needs encompass the physical and non-physical elements required for human growth and development, including those things humans naturally strive to attain (Marker 2003). Such basic needs for living was iterated in the United Nations Universal Declaration of Human Rights of Article 25 (2015), which states that "Everyone has the right to a standard of living adequate for the health and well-being of himself and of his family, including food, clothing, housing and medical care and necessary social services, and the right to security in the event of unemployment, sickness, disability, widowhood, old age or other lack of livelihood in circumstances beyond his control."

Food is a fundamental need. Satter's hierarchy of food needs demonstrates that getting enough food is the most basic foundation of human needs (Satter 2007). The human body needs enough nutrients to keep it going, growing, and glowing. Fresh and healthy foods make life productive and happy.

Likewise, shelter is also a basic need because it is a place for rest, privacy, safety, and security. Everyone has the right to adequate housing. Shelter is a place of protection from cold, damp, heat, rain, wind, or other threats to health, structural hazards, and disease vectors (UN Office of the High Commissioner for Human Rights 2009). Shelter is required to provide security, personal safety, and protection from any disasters and to build resistance to ill health and disease. It is a critical element for human survival and vital for human dignity, to sustain family and community life, and to enable affected populations to recover from the impact of disasters (The Sphere Project 2011). It is a basic necessity during disasters. It can only be achieved if better preparedness is in place (Pothiawala 2015).

Similarly, clothes are also one of the necessities in life (Kaplan and Okur 2008). They provide protection and comfort to the body as well as a reflection of social status. Likewise, clean water is a fundamental need in sustaining life. Water is needed for hydration, hygiene, and sanitation. It is also needed in food preparation and is essential in agricultural food production. Provision of clean water makes people safe and healthy. It is essential for good health and well-being. According to the World Health Organization (2003), human beings cannot live without water. It plays essential roles for the body to function very well and strengthen the immune system and help eliminate waste matter. Human beings also need clean air for survival.

On the other hand, provision of non-physical human needs like education, employment, health care, and others enable people to have healthy, productive, and meaningful life. Education is very important because it plays a vital role in improving the value and excellence of one's life. Also, a person having a degree can get a satisfying job that pays well. Education and employment remove people from poverty and enable the person to afford the necessities in life (Campbell 2006).

Satisfying the basic needs, however, is very challenging. There are many factors hindering their satisfaction such as poverty and lack of access to basic services. To some extent, unsatisfied basic needs are due to pressing environmental factors like climate change and disaster impacts.

5.3 How Climate Change Impacts and Disaster Risks Undermine the Satisfaction of Human Basic Needs in Asia

The countries located in the Asia Pacific region are extremely vulnerable to the adverse impacts of climate change (ADB 2011, 2017). The 2018 global climate risk index showed that, for a 20-year period (1997–2017), 60% of the 10 countries most affected by climate risk are in Asia. They include Myanmar, the Philippines, Bangladesh, Pakistan, Vietnam, and Thailand (Eckstein et al. 2017). Majority of these countries are from Southeast Asia where vulnerability arises from the fact that a large proportion of their populations, settlements, as well as industrial and infrastructure sectors is concentrated along coastlines and riverine areas, which are generally high-risk locations (Gupta et al. 2014; Hijioka et al. 2014).

Table 5.1 shows the impact profiles of some countries in Asia due to climate-related extreme events. These climate change impacts include sea-level rise, droughts, floods, flashfloods, landslides, tropical cyclones, and wildfires (World Bank Group 2018). Also, increasing temperature extremes and significant variations in precipitation patterns are common issues in the region (ADB 2009; Hijioka et al. 2014).

Across Asia, the extreme climate events will have an increasing impact on human health, security, livelihood, and poverty (Hijioka et al. 2014). These impacts will also have significant effects on different sectors, for example, agriculture, water resources, coastal, forestry, and other natural resources (Amadore et al. 1996) and could significantly undo previous achievements of economic development and affect quality standard living and future growth (ADB 2017).

Climate change is just one of the components of disaster risk. Disaster risk is expressed as the potential loss of life, health status, livelihood, assets, and services from disasters in a given period of time (UNISDR 2009). As such, climate change impacts exacerbate disaster risk, affecting the vulnerable sectors and vulnerable populations. Within the context of human security, climate change impacts have serious implications on the satisfaction of basic human needs and development of the people (Marquina 2010). Food, water, shelter, livelihood sources, health, and even education are the basic human needs that will be undermined because of climate change impacts and disaster risk.

Table 5.1 Extreme climate impact profiles of some countries in Asia

Country	Impact							
	Sea level rise	Drought/desertification	Floods	Flash floods	Land-slides	Tropical cyclones/storms	Storm surges	Wild fires
Afghanistan		✓	✓	✓	✓	✓		
Armenia		✓	✓			✓		
Azerbaijan		✓	✓		✓	✓		
Bahrain		–	–	–	–	–	–	–
Bangladesh		✓	✓	✓		✓		
Bhutan			✓		✓			
Brunei								✓
Cambodia	✓	✓	✓			✓		
China		✓	✓		✓	✓		
East Timor	✓	✓	✓			✓		
Georgia		✓	✓			✓		
Hongkong			✓		✓	✓		
India	✓	✓	✓		✓	✓	✓	✓
Indonesia	✓	✓	✓		✓			✓
Iran		✓	✓					
Iraq		✓	✓					
Israel		✓	✓			✓		✓
Japan			✓		✓	✓		
Kazakhstan		✓	✓		✓			
Kuwait		✓	✓	✓		✓		

(continued)

Table 5.1 (continued)

Country	Impact							
	Sea level rise	Drought/desertification	Floods	Flash floods	Land-slides	Tropical cyclones/storms	Storm surges	Wild fires
Kyrgyzstan			✓		✓			
Laos		✓	✓			✓		
Lebanon		✓	✓			✓		
Malaysia		✓	✓		✓	✓		✓
Mongolia		✓				✓		
Myanmar			✓		✓	✓		
Nepal		✓	✓		✓			
North Korea			✓		✓	✓		✓
Oman			✓			✓		
Pakistan	✓	✓	✓		✓	✓		
Papua New Guinea	✓	✓	✓		✓	✓		
Philippines	✓	✓	✓	✓	✓	✓	✓	✓
Qatar		✓	✓					
Russia		✓	✓		✓	✓		✓
Saudi Arabia		✓	✓	✓				
Singapore	✓		✓			✓		
South Korea			✓		✓	✓		
Sri Lanka		✓	✓		✓	✓		
Syria		✓	✓		✓	✓		

(continued)

Table 5.1 (continued)

Country	Impact							
	Sea level rise	Drought/desertification	Floods	Flash floods	Land-slides	Tropical cyclones/storms	Storm surges	Wild fires
Taiwan			✓		✓	✓		
Tajikistan		✓	✓		✓			
Thailand		✓	✓		✓	✓		
Turkey			✓					
Turkmenistan		✓	✓	✓				
United Arab Emirates	–	–	–	–	–	–	–	–
Uzbekistan		✓	✓		✓			
Vietnam		✓	✓			✓		

Sources
ADB (2009)
Hijioka et al. (2014)
EM-DAT (CRED-Center for Research on the Epidemiology of Disasters). International Disaster Database. http://www.emdat.be
World Bank Group (2018)
https://www.preventionweb.net/english/countries.php
https://www.nccs.gov.sg/climate-change-and-singapore/national-circumstances/impact-of-climate-change-on-singapore

5.3.1 Impacts on Food Security and Livelihood

Food is among the most basic human needs (Millman and Kates 1990). Asia is responsible for two-thirds of the world's agricultural food production (Mendelsohn 2014). Agriculture is essential for most developing countries. It is a major source of livelihood, national income, and food. However, the agriculture sector is one of the most vulnerable sectors to climate change impact. Changes in climate patterns could lead to changes in the planting season (FAO 2008) and its associated impacts on water availability, occurrence of pests, and incidence of diseases. Extreme weather events are all likely to affect substantially the potential of agricultural production (Zhai and Zhuang 2009). Climate change acts as a hunger risk multiplier, which may aggravate food insecurity (Krishnamurthy et al. 2015).

Climate change impacts affect the four dimensions of food security: food availability, food accessibility, food utilization, and food system stability (FAO 2008). Krishnamurthy et al. (2015) indicated that food availability through variation in yield can be affected by changes in temperature and precipitation patterns. Changes in production will result in inadequate food supply and household's access to food. As such, dietary options are limited and could potentially lead to increased cases of malnutrition. Food utilization and food stability are also affected as climate change lessens the sufficiency of food production and people's livelihood.

In most countries in Asia, the observed impact is declining yield in corn and rice production as attributed to rising temperature, reduction of rainy days, and El Niño (IPCC 2007). With climate change, both producers and consumers will be affected. It affects agricultural crops, yields of production, livestock, and fisheries (Teng et al. 2015).

The drought episodes in the Philippines, for instance, had devastating impacts not only on agricultural production and food security but also on water resources, thereby placing the country's socioeconomic condition at risk. Central to the drought effects was the deficit in water supply for irrigation, which resulted in agricultural land becoming cracked and unproductive. Pest infestation also became a common problem among farmers. Consequently, decline in harvest severely affected the farmers' income, which led to hunger.

Historically, Hilario et al. (2009) reported an estimated damage of PHP 4.09 billion or 71% in agricultural production (rice, corn, vegetables, and fruit trees) in the Philippines during the 1991–1992 drought episode, of which Mindanao was severely affected. The 1997–1998 episode, on the other hand, had devastated about 74,000 ha of agricultural land located in 18 provinces in the country, resulting in a death toll of 74 and millions of people becoming hungry (Interagency Regional Analysts Network 2015). The most recent drought experienced in the Philippines was in 2015–2016 where hunger due to drought resulted in social unrest in Kidapawan City, Southern Philippines. Around 6,000 farmers rallied and 99 people were reported injured, while three farmers died in the commotion that ensued during the rally. It had severely affected 181,687 farmers and 224,834 ha of agricultural land planted with rice, corn, and high-value crops (International Federation of Red Cross and Red

Crescent Societies 2016). The agriculture sector in the Philippines could be severely affected by the changes in temperature and rainfall regimes as well as by extreme events such as floods or droughts. There could be shortfalls in agricultural crop production in the absence of effective and timely interventions (PAGASA 2011).

The World Bank Group climate change knowledge portal (2018) indicated that urban centers in Indonesia such as Jakarta, Medan, and Bandung are greatly exposed to floods, landslides, and sea-level rise as well as drought events. In 1997, drought resulted in water shortages, which adversely impacted 426,000 ha of rice, and reduced yields of coffee, cocoa, and rubber. It also caused forest fires. In Afghanistan, floods and droughts brought the greatest and most frequent climate change impacts that particularly affect the country's agriculture and water sector and, as such, pose a threat to income, livelihood, and food security among 85% of the Afghan population. Over 4 million Afghans were affected by drought between 2000 and 2003. On the other hand, changes in precipitation patterns and spring snowmelt will likely result in flooding and flash floods, which could affect the economy and poverty reduction efforts in Afghanistan. The United States Agency for International Development (2016) indicated that about 36% of the population in Afghanistan lives in poverty. By 2060, annual average temperature will increase up to 4.0 °C from 1.4 °C. The projected hotter and drier condition will affect agricultural production and water resources. Also, the frequency of drought events will affect livestock production. These phenomena will likely lead to a competition for resources, resulting in starvation.

The World Bank Group (2018) also indicated that Bangladesh and Cambodia are two of the most disaster-prone countries in Southeast Asia frequently affected by floods and flash floods as well as cyclones and droughts. In Bangladesh, nearly 70% gets flooded during heavy monsoon season, while floods along the Mekong River in Cambodia often result in disasters. Also, sea level rise could pose a significant threat to marine coastal areas in Cambodia due to storm surges, high tides, beach erosion, and sea water intrusion. These climate risks have devastating impact on food security since the economy of both Bangladesh and Cambodia is based predominantly on agriculture, forestry, and fishing. Meanwhile, floods, storms, and droughts are common climate impacts, affecting millions of people and causing economic losses in Lao PDR. In 1977, a severe drought affected more than 3 million people. Around 188,000 households in Lao PDR are at risk of food insecurity caused by drought, which affects soil moisture and changes the country's production pattern. The heavy flood in 1992 had caused economic damage of more than USD 21 million. Tropical cyclones have affected the country's economy and its people who have low adaptive capacity.

Sea level rise threatens to inundate coastal communities in India and can cause more severe storm surges. As such, agricultural production and biodiversity of mangrove and wetland ecosystems are vulnerable to rising seas. In 2002–2003, a drought event affected around 300 million people and more than 50% of the cropland, thereby affecting the country's food security. It is estimated that agricultural losses in India will be over USD 7 billion by 2030, but this could be reduced if climate adaptation policies are employed. Meanwhile, rainfall in Kuwait tends to be erratic. Drought occurrence is aggravated by low annual rainfall, while heavy storms cause floods and flash floods, which affect the agriculture sector. In Nepal, droughts are becoming more frequent especially during the winter months, posing a challenge for hydropower generation. Monsoon rainfall below normal condition has reduced rice cultivation by 21–30% (World Bank Group 2018).

The drought events contribute to agricultural challenges, food shortages, and food insecurity in Timor Leste (Fanzo et al. 2017). Agricultural pest infestation also poses a significant threat in agricultural productivity, particularly in the western part of the country. In 2007, about 4,500 ha of rice and corn had been affected. The locust outbreak and drought event caused a 30% decline in crop production, resulting in a food shortage (WFP 2007; World Bank Group 2018). About one-third of the population in Timor Leste experienced food shortages and many people were forced to eat wild root crops to survive. Almost half of the children below 5 years old were chronically malnourished (WFP 2006).

Generally, hunger can result from climate-induced disasters such as droughts, floods, landslides, and pest infestation that reduce food production as well as limit the distribution of food and wealth. Hunger leads to low levels of energy, decreased immunity, and increased susceptibility to poor health (WFP 2007).

Increased frequencies of floods and droughts due to climate change generally affect agricultural production, which often leads to higher food prices and cost of living, thereby exacerbating rural poverty in many parts of Asia (Hijioka et al. 2014).

5.3.2 *Impacts on Water Quality and Quantity*

Water plays a significant role in economic development because of its use in industrial and agricultural production (ADB 2009). Water scarcity is expected to be a major challenge for most Asian countries as a result of increased water demand by their rapidly growing population and lack of good management of their water resources (Hijioka et al. 2014).

Cases of flooding and avalanche due to the continued melting of glaciers in the Himalayan region are projected to increase in the next 2–3 decades. The coastal areas in Asia and its heavily populated delta regions will likely experience more flooding due to the rise in sea level and river flooding (Sharma and Sharma 2008).

As the economy of Southeast Asia is highly dependent on agriculture and water resources, it is expected that this region will be seriously affected by the impacts of climate change in 2050 (IPCC 2007). Likewise, about 40% of the global population, particularly in Asia, will be living in river basins that experience severe water stress due to the changing climate (UNISDR 2015).

The rates of evaporation as well as transpiration increase with increasing temperature, which, in turn, affects the quantity and quality of water available for agricultural production and human consumption. Irregular precipitation patterns affect the quantity of water used for storage, power generation, and irrigation. El Niño years cause reduced stream flows while La Niña years bring more frequent and intense rainfall, resulting in excessive runoff and water flows, which, in turn, cause severe erosion of river banks and sedimentation of transported soils in water reservoirs. Sedimentation reduces the volume of water stored by water reservoirs for future use. On the other hand, sea level rise can result in salt water intrusion into freshwater resources and aquifers, further aggravating water shortage in Southeast Asia (ADB 2009).

In Armenia, water resources are already limited. Since the early 1900s, glacial volume has declined by 50%. The rate of evaporation increases with rising temperatures, reducing winter snowpack and spring runoff. It is projected that, by 2030, aggregate river flow will decrease by 11.9% and by 37.8% by 2100 as a result of increasing temperatures and decreasing rainfall (USAID 2017).

Similarly, India's water resources already face challenges as the country is greatly affected by the changing climate, particularly in the states of Rajasthan, Punjab, and Haryana, due to unsustainable consumption of groundwater for irrigation as well as for other uses, thereby depleting this valuable resource. This has long-term implications on the amount and quality of water available (World Bank Group 2018).

Water is vital not only for economic use but also for human personal consumption and sanitation. The most immediate pathway through which climate change may affect health is water (ADB 2011). Climate change may, however, affect sanitation systems and lower the efficiency of local sewerage systems, leading to increased concentrations of pathogens in raw water supply. Human health depends on an adequate supply of potable water, but changes in rainfall patterns may reduce the water available for drinking and washing. Water shortage may force people to use poorer quality sources, which are often contaminated. All these factors could result in an increased incidence of diarrheal diseases (WHO 2000). Likewise, Hijioka et al. (2014) stated that more heavy rains and rising temperature will increase the risk not only of diarrheal diseases but also dengue fever and malaria, among others.

5.3.3 *Impacts on Human Health*

Changes in climate are expected to modify and even magnify the current burden of diseases significantly challenging the public health sector (ADB 2011). It is expected to exacerbate public health issues by increasing the incidence of certain water-, food-, and vector-borne diseases (World Bank Group 2018). More cases of diarrheal diseases, dengue fever, malaria, cholera, typhoid fever, and dysentery will be observed due to increased frequencies of intense rains and rise in temperature in Asia (Hijioka et al. 2014).

In Lao PDR, for instance, disease outbreaks of malaria, diarrhea, dysentery, dengue fever, and cholera have been registered. In 1994, a major cholera outbreak killed 500 people and affected 8,000. Similarly, in Afghanistan, the rise in cases of malaria, typhoid, cholera, and diarrhea was associated to flooding resulting from heavy rainfall and snowmelt (World Bank Group 2018; USAID 2016).

On the other hand, vector- and water-borne diseases associated with changing climate, particularly from drought and floods due to glacier melt, have been found

Table 5.2 Effects of climate change impacts and their outcomes related to human health

Effect of climate change impacts	Human health outcome
• Drought reduces food availability in populations that are highly dependent on agricultural productivity (crop yields and fish stocks) • Drought reduces water availability for hygiene • Drought increases the risk of forest fires	• Starvation, malnutrition, under nutrition, and diarrheal and respiratory diseases
• Flooding disrupts water supply and sanitation systems and may damage transport systems and health care infrastructure • Floods may provide breeding sites for mosquito vectors and lead to disease outbreaks • Floods may increase post-traumatic stress disorders • Floods, landslides, and wind storms cause death and injuries	• Water- and food-borne diseases (infectious diseases) • Mental disorders • Deaths and injuries
• Higher temperatures shorten the development time of pathogens in vectors and increase the potential transmission to humans • Each vector species has specific climate conditions (temperature and humidity) necessary to be sufficiently abundant to maintain transmission	Mosquito-, tick- and rodent-borne diseases (such as malaria, dengue, tick-borne encephalitis, and Lyme diseases)

Sources Kovats et al. (2003), ADB (2011)

to be increasing in Nepal. It is already evident that malaria, Kala-azar, Japanese encephalitis, and water-borne diseases such as typhoid and cholera are commonly experienced (Regmi et al. 2008). Table 5.2 present the effects of climate change impacts on human health.

5.3.4 *Impacts on Shelter*

Extreme climate events such as increased frequencies and intensities of precipitation will likely cause damage to houses due to flooding and landslides. Tropical Cyclone Haiyan (Typhoon Yolanda), which battered the Philippines in 2013, for instance, caused damage to a total of 1,140,332 houses (Table 5.3) (NDRRMC 2013).

In early February of 2008, two active phases of extreme monsoonal storm activity associated with La Niña produced localized winds, floods, and landslides, which destroyed 3,600 houses across all the districts of Timor Leste. Likewise, 48 settlements in southern Kazakhstan were inundated during the same period, forcing 13,000 people to leave their homes. Flooding causes widespread displacement (World Bank Group 2018).

Table 5.3 Number of damaged houses due to Tropical Cyclone Haiyan (Typhoon Yolanda) (NDRRMC 2013)

Region	Areas covered	Totally damaged	Partially damaged	Total
I V-A	Laguna, Batangas, and Quezon	34	806	840
I V-B	Oriental and Occidental Mindoro, Marinduque, Romblon, and northern Palawan	11,611	22,202	33,813
V	Albay, Catanduanes, Camarines Norte, Camarines Sur, Masbate, and Sorsogon	2,088	10,324	12,412
VI	Aklan, Antique, Iloilo, Capiz, Guimaras, Negros Occidental	229,326	253,023	482,349
VII	Northern Cebu, Bohol, Negros Oriental, Siquijor	62,840	48,479	111,319
VIII	Tacloban City, Eastern Samar Leyte, Western Samar, Biliran, Southern Leyte	244,550	248,306	492,856
X	Misamis Oriental, Misamis Occidental, Bukidnon, Camiguin and Lanao del Norte	2	18	20
XI	Compostela Valley and Davao Oriental	11	8	19
XIII	Dinagat Islands, Agusan del Norte, Surigao del Norte, Surigao del Sur	466	6,238	6,704
Total	Philippines	550,928	589,404	1,140,332

5.3.5 *Impacts on Infrastructure and Social Services*

Infrastructure is also affected by extreme weather conditions. For example, tropical Cyclone Haiyan (Typhoon Yolanda) brought considerable damage to roads, bridges, seaports, airports, water supply systems, drainage systems, agricultural and irrigation facilities, telecommunication facilities, as well as government and school buildings in the Philippines (NDRRMC 2013).

The World Bank Group (2018) indicated that, in Lao PDR, the floods in 2005 and 2006 caused damage to the irrigation system amounting to more than USD 5 million. In Bangladesh, the cyclone that struck the coastal areas in November 1970 caused more than 300,000 deaths and USD 2.5 billion worth of damage to property. Tropical Cyclone Vivienne, which occurred in February 2005, disrupted oil and gas production in Timor Leste. In 1999, Tropical Cyclone Orissa caused over 10,000 people to lose their lives and damaged buildings, lifeline infrastructure, and economic assets along many coastal districts. A 50 cm rise in sea level, combined with land subsidence in Jakarta Bay, Indonesia, could permanently inundate densely populated areas in Jakarta and Bekasi with more than 270,000 households. Both landslides and floods in Papua New Guinea directly affected human safety and access to health centers by blocking major communication and transportation arteries. Increasing frequencies of intense storms with accompanying landslides can destroy public infrastructure such as hospitals and roads, preventing delivery of health services to vulnerable populations in Papua New Guinea.

5.4 Selected CCA and DRR Strategies to Satisfy Basic Needs and Enhance Human Security

The converging goal of both CCA and DRR is to reduce the risk of climate change impacts (UNISDR 2010). DRR is done through systematic efforts of analyzing and managing the causal factors of disasters. It also involves reduction of exposure to hazards and vulnerability of people as well as improvement of preparedness to extreme events (UNISDR 2009). On the other hand, CCA means adjustment of both human and natural systems to actual or expected climate and its effects (IPCC 2014).

Different CCA and DRR strategies (Table 5.4) are developed among Asian countries in response to the challenge of mitigating climate change impacts that undermine human basic needs as material aspect of human security. For instance, in areas of food security, in particular the agriculture sector, implementation of an early warning system (EWS) as CCA and DRR strategies are the commonalities among some Asian countries, particularly in the Philippines, Afghanistan, and Indonesia. Accordingly, an EWS is a vital component to effectively reduce disasters, especially in countries that regularly face a wide range of climate-related hazards and frequently experience complex disasters (ESCAP 2015). EWS is essential in ensuring emergency preparedness and in mobilizing assistance (FAO 2001). It facilitates to act before disasters, mitigate, and prevent the impacts (FAO 2017).

According to the Climate Change Commission (2019), the institutional sources of risk information in the Philippines, are fragmented, absence of coordination and some duplication of functions accross government agencies. For example, the Philippine Atmospheric, Geophysical, and Astronomical Services Administration (PAGASA) is tasked to monitor hydro-meteorological hazards and risks, whereas the Philippine Institute of Volcanology and Seismology (PHIVOLCS) looks at seismic activity-related hazards and risks. The preparation of hazard maps is under the National Mapping and Resource Information Authority (NAMRIA), while the Mines and Geosciences Bureau (MGB) monitors landslide and flood hazards/risks. Accordingly, this existing setup for EWS is deemed inefficient and ineffective. As a result, the adaptive capacities of local communities in the Philippines are still weak in spite of the fact that the country has a comprehensive regulatory Philippines for DRM (Swiss NGO 2014).

Table 5.4 Illustrative examples of CCA and DRR strategies in selected Asian countries to satisfy basic needs and to enhance human security

Area	Strategies	Country
Food security	• Early warning system. Information on ENSO and typhoons provided by PAGASA to farmers • Introduction of drought-resistant varieties • Crop diversification • Changes in cropping calendar and crop rotation	• Philippines
	• Early warning system and hazard mapping. This is done by establishing critical risk information in key sectors to help citizens adapt and increase their resilience to climate-related impacts in the future	• Afghanistan
	• Cultivation of vegetable crops in floating gardens (using aquatic weeds as base). This cheap and sustainable technology is one way of improving the food resources of the people (Source: Schumacher Centre for Technology and Development, n.d.) • Research on and dissemination of drought-tolerant crop varieties • Storage mechanisms • Crop diversification • Agronomic manipulations such as shifting planting dates, using short-duration crop cultivars	• Bangladesh
	• Use of drought-resistant crops and early-maturing crop varieties can increase crop yield and productivity. • Crop diversification can lessen reliance on agriculture as the sole livelihood source	• Cambodia
	• Adopting crop varieties and plants that can reduce land degradation. This practice improves food security by planting different crop types/varieties and controlling the impact of droughts	• Iraq
	• Investment in both drought-tolerant and salt-tolerant crops • Crop diversification • Improved early El Niño warning systems	• Indonesia
	• Using indigenous and locally adapted plants combined with crop varieties adapted to drier variable conditions	• Kuwait

(continued)

Table 5.4 (continued)

Area	Strategies	Country
	• Change in cropping patterns/calendars • Improve information on climate variability and seasonal climate forecasting to reduce production risk	• Lao PDR
Livelihood security	• Livelihood protection in ecologically fragile areas and protection of vulnerable socioeconomic groups • Construction of multipurpose cyclone shelter centers for people and of raised platforms for livestock • Livelihood support and diversification systems given to coastal communities	• Bangladesh
	• Improving farmers' knowledge and access to weather information to enable them to carry out agricultural activities	• Cambodia
	• Facilitate knowledge exchange on climate science between institutions and small farmer groups to improve institutional management • Public awareness-raising campaigns on climate change impacts and adaptation • Livelihood diversification into off-farm activities	• Indonesia
	• Strengthening the capacity of farmers to implement improved water and agricultural techniques	• Kuwait
	• Improve farmers' knowledge about proper use of weather information	• Lao PDR
Water quality and quantity	• Implementing agro-forestry and adopting soil and water conservation measures. The practices help stabilize the slopes and reduce soil erosion and runoff (The World Bank Group 2018; Wangpakapattanawong et al. 2017)	• Afghanistan • Philippines • Indonesia • Vietnam • Lao PDR
	• Modernization of existing irrigation schemes and demand management aimed at optimizing physical and economic efficiency in use of water resources and recycled water in water-stressed areas • Rainwater harvesting, creation of water reservoirs, low cost filtration system in salinity-affected areas	• Bangladesh

(continued)

Table 5.4 (continued)

Area	Strategies	Country
	• Improved water use (water harvesting, small-scale irrigation, etc.) in drought-prone areas. • Establishing community-based, participatory watershed management and conservation programs to restore degraded watersheds	• Cambodia
	• Development of improved water collection and storage facilities	• Indonesia
	• Rehabilitation and reconstruction of water management and monitoring networks • Promotion of non-conventional water resources such as wastewater reuse and water harvesting to increase water supply • Improvement of water quality and upgrading of the water distribution network • Maintaining and improving the efficiency of existing pumping stations and sewerage networks, while meeting the required environmental standards for processed water prior to pumping to drainage points • Providing safe drinking water to acutely vulnerable populations • Reconstruction and rehabilitation of water treatment and distribution infrastructure	• Iraq
	• Promoting water- and energy-saving programs through public awareness campaigns • Implementing water-saving programs such as rationing to reduce current water wastage • Developing and implementing national policies aimed at discouraging water wastage and enforcing regulations and policies that promote water saving and conservation • Increasing the use of safe treated wastewater for agricultural irrigation • Improving agricultural and landscape drainage systems to curb waterlogging and salinization	• Kuwait
	• Investing in water monitoring and information systems as a basis for more efficient and equitable use of water resources in all sectors	• Lao PDR
Human health	• Better irrigation water management to reduce mosquito breeding sites • Improved disease/vector surveillance and monitoring • Education, training, and awareness campaigns on public health issues	• Bangladesh

(continued)

Table 5.4 (continued)

Area	Strategies	Country
	• Implementation of community-based health education programs to increase awareness and improve personal hygiene and enhance environmental health management • Introduction of a malaria surveillance system, coupled with improved methods of vector control for health workers and communities • Providing training programs to build manpower capacity to improve the provision of health extension services at local levels • Supporting health research and community health services	• Cambodia
	• Supporting vitamin, mineral, and micronutrient supplementation to vulnerable groups. Improving the nutritional status of the population by providing technical and logistical support to food safety laboratories • Increasing access to potable water in both urban and rural areas and improving sanitation services in rural areas • Improving government-led emergency response teams through capacity and disaster preparedness training • Improvement of sanitation services to vulnerable areas to avert water-borne diseases	• Iraq
	• Enhanced infectious disease control programs (vaccines, vector control, case detection, and treatment)	• Lao PDR
Settlement/shelter	• Monitoring of internal and external migration and providing support for rehabilitation • Improving the education and work skills of migrants and financing resettlement and rehabilitation	• Bangladesh
	• Climate-proof housing design	• Lao PDR
Infrastructure and social services	• Restoring and rehabilitating flood protection infrastructure, such as dikes in order to sustain communities during periods of drought and to supplement domestic and irrigation supply • Establishing grain banks • Construction of small check dams and rainwater harvesting schemes	• Cambodia

(continued)

Table 5.4 (continued)

Area	Strategies	Country
	• Development and deployment of more efficient irrigation techniques and infrastructure (such as troughs, sprinklers, and trickle irrigation) for the purpose of improving farming practices and supporting infrastructure	• Indonesia
	• Developing alternative sources of energy, such as solar and wind energy to supplement electricity production and for use in the water desalination process	• Kuwait
	• Improvement of flood control facilities, such as pumping stations, and protection of existing irrigation systems • Creation of multipurpose reservoirs, dams, and water-impounding systems • Development of integrated management plans for river basins and water catchment areas. • Rehabilitation of damaged irrigation and drainage facilities	• Lao PDR

Source World Bank Group (2018). http://sdwebx.worldbank.org/climateportal/countryprofile/home.cfm

Thus, the Climate Change Commission felt the need to strengthen and improve the coordination system and cooperation among agencies and stakeholders by establishing a national platform for integrating all risk- and hazard-related information and impact-based forecasting. Combining science and technology with on-the-ground local participation as well as with integrated regional and national strategies and policies make an effective EWS to climate change and disaster risk. This will serve as an avenue for building resilience among Asian countries (ESCAP 2015). In addition, collaboration is the basic element of an effective EWS (FAO 2001). FAO (2015) suggested that countries should continuously conduct monitoring and evaluation and find ways to improve the EWS to ensure effectiveness and efficiency.

Likewise, the use of drought-resistant varieties and crop diversification practices are common among countries such as the Philippines, Bangladesh, Cambodia, and Indonesia. Drought-resistant agricultural crops are able to endure heat stress; they grow, survive, and yield satisfactorily under extreme heat conditions (Zhang 1992). Moreover, practicing crop diversification suppresses pest occurrences and reduces the spread of pathogens. Such practices can improve resilience and increase economic benefits in the agriculture sector. Crop diversification is a cost-effective method to combat the negative impacts of climate change on agriculture (Lin 2011).

Changes in cropping calendar and use of indigenous, locally adapted, and early maturing crop varieties are also practiced in the Philippines, Bangladesh, Cambodia, Iraq, Kuwait, and Lao PDR. Cultivation of vegetable crops in floating gardens and raised beds is common in Bangladesh. Lao PDR, Cambodia, and Indonesia are currently improving their climate variability information and seasonal climate forecasting to reduce production risks.

Meanwhile, some of the CCA and DRR strategies for human health security include implementing better irrigation management systems as these reduce mosquito breeding sites. Likewise, more awareness campaigns on climate change are being done in Asian countries.

On the other hand, adoption of agroforestry technologies and soil and water conservation measures is common, particularly in the Philippines, Afghanistan, Philippines, Vietnam, and Lao PDR. These practices ensure good water management, particularly water quality and quantity. It stabilizes the slopes and reduces soil erosion and runoff from steep slopes. Also, rainwater harvesting, establishment/rehabilitation of water treatment facilities, and implementing community-based watershed management are common among Asian countries.

As to settlements/shelters in Lao PDR, investing in climate-proof house design has been observed particularly among those affected by climate change Also, some Asian countries are engaged in resettlement and rehabilitation projects.

Nonetheless, no comprehensive assessment is done to gauge the effectiveness and efficiency of specific CCA and DRR in the context of human security.

5.5 Conclusion

The Asian region is extremely vulnerable to adverse climate change impacts. As such, human basic needs are undermined, resulting in hunger, water scarcity, deteriorating health conditions, and displacement and migration increases. In the absence of effective and timely interventions, human security might progressively be threatened as global climate change and disaster risks escalate.

Some countries in the Asian region have been implementing significant CCA and DRR strategies. However, a more comprehensive assessment of the effectiveness of these strategies at the regional scale still needs to be done. Nevertheless, climate change should not be treated independently of other development concerns. It is crucial to mainstream CCA and DRR into national and local development agenda particularly in the less economically developed and more vulnerable countries in Asia, with the end in view of protecting basic human needs and security of the poor in the context of changing climate and increasing incidence of disasters.

References

Adger WN (2010) Climate change, human well-being and insecurity. New Polit Econ 15(2): 275–292. http://dx.doi.org/10.1080/13563460903290912. Accessed 4 Dec 2015

Adger WN, Pulhin JM, Barnett J, Dabelko GD, Hovelsrud GK, Levy M, Oswald Ú, Spring, Vogel CH (2014) Human security. In: Climate change 2014: Impacts, adaptation, and vulnerability. Part A: Global and sectoral aspects. Contribution of Working Group II to the Fifth Assessment Report of the Intergovernmental Panel on Climate Change [Field CB, VR Barros, DJ Dokken, KJ Mach, MD Mastrandrea, TE Bilir, M Chatterjee, KL Ebi, YO Estrada, RC Genova, B Girma, ES Kissel, AN Levy, S MacCracken, PR Mastrandrea, and LLWhite (eds)]. Cambridge University Press, Cambridge, United Kingdom, and New York, NY, USA, pp 755–791

Alderfer CP (1969) An empirical test of new theory of human need. Org Behav Hum Perform 4(2): 142–175. https://www.sciencedirect.com/science/article/pii/003050736990004X. Accessed 28 Mar 2018

Amadore L, Bolhofer WC, Cruz RV, Feir RB, Freysinger CA, Guill S, Jalal JK, Iglesias A, Jose A, Leatherman S, Lenhart S, Mukherjee S, Smith JB, Wisniewski J (1996) Climate change vulnerability and adaptation in Asia and the Pacific: workshop summary. Water Air Soil Pollut 92(1–2): 1–12. http://research3.fit.edu/sealevelriselibrary/documents/doc_mgr/403/Asia_Pacific_Vulnerability_&_Adaptation_Workshop_Summary_-_Amadore_et_al_1996.pdf. Accessed 28 Mar 2018

ADB (Asian Development Bank) (2009) The economics of climate change in Southeast Asia: a regional review. https://www.adb.org/sites/default/files/publication/29657/economics-climate-change-se-asia.pdf. Accessed 28 Mar 2018

ADB (Asian Development Bank) (2011) Accounting for health impacts of climate change. https://www.adb.org/sites/default/files/publication/28976/heath-impacts-climate-change.pdf. Accessed 28 Mar 2018

ADB (Asian Development Bank) (2017) A Region at risk: the human dimensions of climate change in Asia and the Pacific. https://www.adb.org/sites/default/files/publication/325251/region-risk-climate-change.pdf. Accessed 28 Mar 2018

Barnett J (2003) Security and climate change. Global Environ Change 13: 7–17. https://ac.els-cdn.com/S0959378002000808/1-s2.0-S0959378002000808-main.pdf?_tid=08be4610-1066-11e8-8a90-00000aab0f02&acdnat=1518489285_73219c851a8f847bba23511117497beb. Accessed 25 Mar 2018

Cameron E (2011) Development, climate change and human rights. From the margins to the mainstream? Social Development Working Paper of the World Bank Social Development, World Bank, Washington, DC. http://siteresources.worldbank.org/EXTSOCIALDEVELOPMENT/Resources/244362-1164107274725/3182370-1253125009973/Development-Climate-Change-Human-Rights.pdf. Accessed 27 Mar 2018

Campbell B (2006) Why education is important. http://nfl.packers.com/fan_zone/contests/badger_mutual_essay_contest/essay061119.pdf. Accessed 28 Mar 2018

CHS (Commission on Human Security) (2003) Human security now. https://reliefweb.int/sites/reliefweb.int/files/resources/91BAEEDBA50C6907C1256D19006A9353-chs-security-may03.pdf. Accessed 23 Mar 2018

Climate Change Commission (2019) Executive Brief: The Philippine National Climate Change Action Plan, Monitoring and Evaluation Report 2011–2016. Manila, Philippines. https://climate.gov.ph/files/The%20Philippine%20NCCAP%20M&E%20Executive%20Brief%20[v2].pdf.

Dankelman I, Alam K, Ahmed WB, Gueye YD, Fatema N, Kutin RM. (2008) Gender, climate change and human security: lessons from Bangladesh, Ghana and Senegal. http://www.gdnonline.org/resources/WEDO_Gender_CC_Human_Security.pdf. Accessed 27 Mar 2018

DFID (Department for International Development) (2004) The impact of climate change on the vulnerability of the poor. http://www.unpei.org/sites/default/files/PDF/resourceefficiency/KM-resource-DFID-impact-climatechange-vulnerability.pdf. Accessed 28 Mar 2016

Eckstein D, Künzel V, Schäfer L (2017) Global climate risk index 2018. Who suffers most from extreme weather events? Weather-related loss events in 2016 and 1997 to 2016. https://germanwatch.org/en/download/20432.pdf. Accessed 28 Mar 2018

ESCAP (Economic and Social Commission for Asia and the Pacific) (2015) Strengthening regional multi-hazard early warning systems. https://www.unescap.org/sites/default/files/pre-ods/E_CDR4_2.pdf. Accessed 29 Mar 2018

Fanzo J, Boavida J, Bonis-Profumo G, McLaren R, Davis C (2017) Timor-Leste strategic review: progress and success in achieving Sustainable Development Goal 2. https://docs.wfp.org/api/documents/WFP-0000015583/download/. Accessed 28 Mar 2018

FAO (Food and Agriculture Organization of the United Nations) (2001) Report of the FAO Asia-Pacific Conference on Early Warning, Prevention, Preparedness and Management of Disasters in Food and Agriculture. http://www.fao.org/docrep/005/AC120E/AC120e00.htm#TopOfPage. Accessed 23 Mar 2018

FAO (Food and Agriculture Organization of the United Nations) (2008) Climate change and food security: a framework document. http://www.fao.org/forestry/15538-079b31d45081fe9c3dbc6ff34de4807e4.pdf. Accessed 28 Mar 2018

FAO (Food and Agriculture Organization of the United Nations) (2015) Enhancing early warning capabilities and capacities for food safety. Training handbook, 1st ed. http://www.fao.org/3/a-i5168e.pdf. Accessed 23 Mar 2018

FAO (Food and Agriculture Organization of the United Nations) (2017) Global early warning—early action report on food security and agriculture. http://fscluster.org/sites/default/files/documents/a-i7149e.pdf. Accessed 23 Mar 2018

Gupta H, Kohli RK, Ahluwalia AS, Pal P (2014) Climate change and biodiversity: analysis of the policy framework. Environ Policy Law 44(6): 539–544. https://www.researchgate.net/profile/Ravinder_Kohli2/publication/277631896_Climate_change_and_biodiversity-Analysis_of_the_policy_framework/links/556ee27708aefcb861dbbca7.pdf. Accessed 28 Mar 2018

Hijioka Y, Lin E, Pereira JJ, Corlett RT, Cui X, Insarov GE, Lasco RD, Lindgren E, Surjan A (2014) Asia. In: Climate change 2014: impacts, adaptation, and vulnerability. Part B: Regional aspects.

Contribution of Working Group II to the Fifth Assessment Report of the Intergovernmental Panel on Climate Change [Barros VR, CB Field, DJ Dokken, MD Mastrandrea, KJ Mach, TE Bilir, M Chatterjee, KL Ebi, YO Estrada, RC Genova, B Girma, ES Kissel, AN Levy, S MacCracken, PR Mastrandrea, and LL White (eds)], Cambridge University Press, Cambridge, United Kingdom and New York, NY, USA, pp 1327–1370. Accessed 28 Mar 2018

Hilario F, de Guzman R, Ortega D, Hayman P, Alexander B (2009) El Niño southern oscillation in the Philippines: impacts, forecasts, and risk management. https://dirp3.pids.gov.ph/ris/pjd/pidspjd09-1enso.pdf. Accessed 28 Mar 2018

Inter-Agency Regional Analysts Network (2015) 2015/16 El Nino Event. October 2015. www.iris-france.org/wp-content/uploads/2017/…/IARAN-El-Nino-October-2015.pdf

IFRC-RCS (International Federation of Red Cross and Red Crescent Societies) (2016) Information Bulletin no. 1 GLIDE no. DR-2016-000017-PHL. http://www.ifrc.org/docs/Appeals/16/IBPHdr290416.pdf. Accessed 28 Mar 2018

IPCC (Intergovernmental Panel on Climate Change) (2007) Summary for policymakers. In: Parry ML, Canziani OF, Palutikof JP, van der Linden PJ, Hanson CE (eds) Climate change 2007: impacts, adaptation and vulnerability. Contribution of Working Group II to the Fourth Assessment Report of the Intergovernmental Panel on Climate Change, Cambridge University Press, Cambridge, UK, pp 7–22

IPCC (Intergovernmental Panel on Climate Change) (2014) Annex II: glossary. In: Mach KJ, Planton S, von Stechow C (eds) Climate change 2014: synthesis report. Contribution of Working Groups I, II, and III to the Fifth Assessment Report of the Intergovernmental Panel on Climate Change [Core writing team: RK Pachauri and LA Meyer (eds)], IPCC, Geneva, Switzerland, pp 117–130. https://www.ipcc.ch/pdf/assessment-report/ar5/syr/AR5_SYR_FINAL_Glossary.pdf. Accessed 30 Mar 2018

Jonas G, Burgess J (2011) Human security. In: Synder CA (ed) Contemporary security and strategy. Palgrave Macmillan, London, pp 89–105

Kaplan S, Okur A (2008) The meaning and importance of clothing comfort: a case study for Turkey. J Sens Stud 23: 688–706. https://doi.org/10.1111/j.1745-459X.2008.00180.x. Accessed 11 Mar 2018

Kovats RS, Bouma MJ, Hajat S, Worrall E, Haines A (2003) El Niño and health. The Lancet, 362(9394): 1481–1489. https://doi.org/10.1016/s0140-6736(03)14695-8

Krishnamurthy PK, Lewis K, Kent C, Aggarwal P (2015) Climate impacts on food security and livelihoods in Asia: a review of existing knowledge. http://documents.wfp.org/stellent/groups/public/documents/newsroom/wfp281745.pdf. Accessed 28 Mar 2018

Kumssa A, Jones JF (2010) Climate change and human security in Africa. Int J Sustain Dev World Ecol 17(6):453–461. https://doi.org/10.1080/13504509.2010.520453

Leagans JP (1964) A concept of need. J Coop Ext. https://www.joe.org/joe/1964summer/1964-2-a4.pdf. Accessed 28 Mar 2018

Lin BB (2011) Resilience in agriculture through crop diversification: adaptive management for environmental change. BioScience 61(3): 183–193. https://academic.oup.com/bioscience/article/61/3/183/238071. Accessed 23 Mar 2018

Marker S (2003) Unmet human needs: beyond intractability. In: Burgess G, Burgess H (eds) Conflict information consortium, University of Colorado, Boulder. https://www.beyondintractability.org/essay/human_needs. Accessed 28 Mar 2018

MacFarlane SN, Khong YF (2006) Human security and the UN: a critical history. United Nations Intellectual History Project. https://zodml.org/sites/default/files/%5BS._Neil_MacFarlane%5D_Human_Security_And_the_UN_A_.pdf. Accessed 25 Mar 2018

McLeod SA (2017) Maslow's hierarchy of needs. www.simplypsychology.org/maslow.html. Accessed 19 July 2019

Marquina A (2010) Global warming and climate change. Prospects and policies in Asia and Europe. Energy, Climate, and the Environment Series

Maslow AH (1943) A theory of human motivation. Psychol Rev 50(4): 370–396. http://citeseerx.ist.psu.edu/viewdoc/download?doi=10.1.1.318.2317&rep=rep1&type=pdf. Accessed 28 Mar 2018

Mendelsohn R (2014) The impact of climate change on agriculture in Asia. J Integr Agric 13(4): 660–665. https://ac.els-cdn.com/S2095311913607017/1-s2.0-S2095311913607017-main.pdf?_tid=937cb76e-0cce-11e8-87b9-00000aacb35f&acdnat=1518094381_19d246f3391d15384bbd4cbeadd1e8dc. Accessed 28 Mar 2018

Millman S, Kates RW (1990) Toward understanding hunger. In: L.F. Newman et al. (eds.). Hunger in history: Food shortage, poverty, and deprivation, Blackwell, Cambridge, MA. http://rwkates.org/pdfs/a1990.06.pdf. Accessed 28 Mar 2018

NDRRMC (National Disaster Risk Reduction and Management Council) (2013) Final report re effects of Typhoon Yolanda (Haiyan). NDRRMC update. http://ndrrmc.gov.ph/attachments/article/1329/FINAL_REPORT_re_Effects_of_Typhoon_YOLANDA_%28HAIYAN%29_06-09NOV2013.pdf. Accessed 28 Mar 2018

ODI (Overseas Development Institute) (1978) Basic needs. Briefing Paper. https://www.odi.org/sites/odi.org.uk/files/odi-assets/publications-opinion-files/6616.pdf. Accessed 28 Mar 2018

PAGASA (Philippine Atmospheric, Geophysical, and Astronomical Survey Administration) (2011) Climate change in the Philippines. https://web.pagasa.dost.gov.ph/index.php/climate-change-in-the-philippines. Accessed 28 Mar 2018

Pothiawala S (2015) Food and shelter standards in humanitarian action. Turkish J Emergency Med 15 (Supplement 1): 34–39. https://doi.org/10.5505/1304.7361.2015.98360. Accessed 11 Mar 2018

Schumacher Centre for Technology and Development (n.d.) Floating gardens in Bangladesh. http://www.fao.org/climatechange/17849-0e277b46b31f98942e6bc81bb22319243.pdf. Accessed 23 Mar 2018

Regmi BR, Pandit A, Pradhan B, Kovats S, Lama P (2008) Climate change and health in Nepal: capacity strengthening in the least developed countries for adaptation to climate change. http://pubs.iied.org/pdfs/G02664.pdf. Accessed 28 Mar 2018

Satter E (2007) Hierarchy of food needs. GEM No. 447. J Nutr Educ Behav 39: S187–S188. http://www.jneb.org/article/S1499-4046(07)00091-7/pdf. Accessed 11 Mar 2018

Sharma BR, Sharma D (2008) Impact of climate change on water resources and glacier melt and potential adaptations for Indian agriculture. https://www.researchgate.net/publication/228431195_Impact_of_climate_change_on_watr_resources_and_glacier_melt_and_potential_adaptations_for_Indian_agriculture. Accessed 18 July 2019

Sphere Project (2011) Humanitarian charter and minimum standards in disaster response. http://ocw.jhsph.edu/courses/RefugeeHealthCare/PDFs/SphereProjectHandbook.pdf. Accessed 11 Mar 2018

Swiss NGO (2014) Disaster risk reduction and management in the Philippines. http://www.drrplatform.org/images/DocPub/RiskAssessment.pdf. Accessed 28 Mar 2018

Synder CS (2011) Contemporary security and strategy, Palgrave Macmillan, London. https://www.macmillanihe.com/resources/sample-chapters/9780230241503_sample.pdf. Accessed 25 Mar 2018

Teng PS, Anthony MC, Tian G, Lassa JA (2015) Impacts of climate change on food production: options for importing countries. Policy brief. https://www.rsis.edu.sg/wp-content/uploads/2015/05/PB150529_Impact-of-Climate-Change-on-Food-Production.pdf. Accessed 28 Mar 2018

UNISDR (United Nations International Strategy for Disaster Reduction) (2009) UNISDR terminology on disaster risk reduction. https://www.unisdr.org/files/7817_UNISDRTerminologyEnglish.pdf. Accessed 28 Mar 2018

UNISDR (United Nations International Strategy for Disaster Risk Reduction) (2010) Climate change adaptation and disaster risk reduction: institutional and policy landscape in Asia and the Pacific. https://www.unisdr.org/files/17250_t0236000045500climatechangeadaptati.pdf. Accessed 28 Mar 2018

UNISDR (United Nations International Strategy for Disaster Reduction) (2015) Global assessment report on disaster risk reduction. https://www.preventionweb.net/english/hyogo/gar/2015/en/gar-pdf/GAR15_Pocket_EN.pdf. Accessed 28 Mar 2018

UNDP (United Nations Development Programme) (1994) Human development report 1994, Oxford University Press, New York. http://hdr.undp.org/sites/default/files/reports/255/hdr_1994_en_complete_nostats.pdf. Accessed 25 Mar 2018
UN (United Nations) (2015) Universal Declaration of Human Rights of Article 25, p 25 (UDHR). http://www.un.org/en/udhrbook/pdf/udhr_booklet_en_web.pdf. Accessed 11 Mar 2018
United Nations Office of the High Commissioner for Human Rights (2009) Fact Sheet no. 21. The human right to adequate housing. http://www.ohchr.org/Documents/Publications/FS21_rev_1_Housing_en.pdf. Accessed 11 Mar 2018
USAID (United States Agency for International Development) (2016) Climate change risk profile: Afghanistan. https://www.climatelinks.org/sites/default/files/asset/document/2016%20CRM%20Fact%20Sheet%20-%20Afghanistan_2.pdf. Accessed 28 Mar 2018
USAID (United States Agency for International Development) (2017) Climate risk profile in Armenia. https://www.climatelinks.org/sites/default/files/asset/document/2017_USAID_Climate%20Change%20Risk%20Profile_Armenia.pdf. Accessed 28 Mar 2018
Wangpakapattanawong P, Finlayson R, Öborn I, Roshetko JM, Sinclair F, Shono K, Borelli S, Hillbrand A, Conigliaro M (eds) (2017) Agroforestry in rice-production landscapes in Southeast Asia: a practical manual. Food and Agriculture Organization of the United Nations Regional Office for Asia and the Pacific World Agroforestry Centre Southeast Asia Regional Program. http://www.fao.org/3/a-i7137e.pdf. Accessed 24 Mar 2018
World Bank Group (2018) Climate risk and adaptation country profiles. Climate Change Knowledge Portal. http://sdwebx.worldbank.org/climateportal/countryprofile/home.cfm. Accessed 10 Feb 2018
WFP (World Food Programme) (2006) Annual report. https://www.wfp.org/sites/default/files/wfp_AR06_lowres.pdf. Accessed 28 Mar 2018
WFP (World Food Programme) (2007) World hunger series 2007. Hunger and health. https://www.wfp.org/sites/default/files/World_Hunger_Series_2007_Hunger_and_Health_EN.pdf. Accessed 28 Mar 2018
WHO (World Health Organization) (2000) Climate change and human health: impact and adaptation. http://apps.who.int/iris/bitstream/handle/10665/66776/WHO_SDE_OEH_00.4.pdf;jsessionid=E3D5380A38DB0E1E59E4529C576C664F?sequence=1. Accessed 28 Mar 2018
WHO (World Health Organization) (2003) The right to water. http://www.who.int/water_sanitation_health/en/righttowater.pdf. Accessed 11 Mar 2018
Zhai F, Zhuang J (2009) Agricultural impact of climate change: a general equilibrium analysis with special reference to Southeast Asia. ADBI Working Paper 131. Tokyo: Asian Development Bank Institute. http://www.adbi.org/workingpaper/2009/02/23/2887.agricultural.impact.climate.change. Accessed 28 Mar 2018
Zhang H (1992) Water use of two corn varieties, drought resistant and drought sensitive. Retrospective theses and dissertations. https://lib.dr.iastate.edu/cgi/viewcontent.cgi?article=11164&context=rtd Accessed 23 Mar 2018

Websites

Prevention Web. The Knowledge platform for disaster risk reduction. https://www.preventionweb.net/english/countries.php
Impact of Climate Change on Singapore. https://www.nccs.gov.sg/climate-change-and-singapore/national-circumstances/impact-of-climate-change-on-singapore

Dr. Lorena L. Sabino is Assistant Professor with the Department of Social Forestry and Forest Governance, College of Forestry and Natural Resources, University of the Philippines Los Baños (UPLB). She obtained her doctorate degree in Environmental Science with minors in Community Development and Social Forestry at UPLB. In her dissertation, she used a human security perspective in assessing how the vulnerability to climate variability and change is linked to household livelihood security. She is currently involved in climate change and disaster risk-related projects, sustainable mangrove rehabilitation, vulnerability, and human security research projects.

Dr. Juan M. Pulhin is full Professor and former Dean of the College of Forestry and Natural Resources, University of the Philippines Los Baños, Philippines. He was Lead Author of the 2007 Fourth Assessment Report of the Intergovernmental Panel on Climate Change (IPCC), Coordinating Lead Author of the Human Security Chapter of the 2014 IPCC Fifth Assessment Report and currently Lead Author of the Asia chapter of the IPCC Sixth Assessment Report. He has been a four-time a visiting professor at The University of Tokyo in 2007, 2010, 2016 and 2017 and a visiting scholar at Waseda University, Tokyo in 2018. He has written and edited 11 books and published more than 100 technical papers and book chapters on natural resources policy and governance, and on the human dimension of climate change and disaster risks. He is a fellow, invited expert, and council member of several international organizations and a member of the National Academy of Science and Technology (NAST), Philippines.

Chapter 6
Natural Resource Base as a Foundation of Human Security

Rose Jane J. Peras, Juan M. Pulhin, Dixon T. Gevaña, and Makoto Inoue

Abstract The natural resource base of Asia constitutes a salient foundation for human security as it provides households and communities with various ecosystems, goods, and services. These resources include but are not limited to forests, fishery, agricultural land, pasture, and fresh water. Natural disasters and climate change pose potent threats to each of these resources and the goods and services that humanity derives from them. In this chapter, we apply the sustainable livelihood framework to synthesize various literature and link the natural resource base to livelihood and human security. First, we review how "vulnerability context" affects "livelihood assets" or the extent to which various natural disasters and climate change is frustrating the role that the natural resource base plays for human security in Asia. Second, we examine how "transforming structure and process" could be improved to create a "livelihood strategy" for better "livelihood outcomes." Different innovative initiatives and approaches are being undertaken at various scales to prevent and cope with the threat of natural disasters and climate change. Third, we look at some promising practices in Asia to highlight the need to sustain the natural resource-base and livelihood in order to achieve human security. We conclude that a more integrated and participatory approach in natural resource governance is crucial to preserve the natural resources and ensure benefits for resource-dependent communities. These are prerequisites for achieving human security in the region in the context of challenging climate and increasing impacts of disasters.

Keywords Sustainable livelihood · Natural resource base · Human security

R. J. J. Peras (✉) · J. M. Pulhin · D. T. Gevaña
Department of Social Forestry and Forest Governance, College of Forestry and Natural Resources (CFNR), University of the Philippines Los Baños, Laguna, Philippines
e-mail: rjperas@up.edu.ph

M. Inoue
Faculty of Human Sciences, Waseda University, Saitama, Japan

J. M. Pulhin et al. (eds.), *Climate Change, Disaster Risks, and Human Security*, Disaster Risk Reduction, https://doi.org/10.1007/978-981-15-8852-5_6

6.1 Introduction

The condition of a country's natural resource base is an important indicator for attaining sustainable development. The concept of sustainable development, as popularized under the 1992 Rio de Janeiro United Nations Conference on Environment and Development (UNCED), underscores the indispensable role of natural resources to meet the needs of the present and future generations. Similarly, the UN recognizes the critical role of conservation, protection, rehabilitation, and sustainable use of various types of natural resources to achieve the 2030 sustainable development goals.

While natural resource abundance does not necessarily measure the country's economic development (Broad 1995), sustaining its natural capital can be a foundation for human security. In the context of climate change, human security has been defined as "a condition that exists when the vital core of human lives is protected, and when people have the freedom and capacity to live with dignity" (Adger et al. 2014). In Asia, where millions of people are dependent on natural resources for livelihood and environmental services, the nexus between natural resources and human security is not hard to imagine, considering the vulnerability of these resources to the changing climate and the increasing incidence of disasters.

For almost three decades, the popularity of the sustainable development framework has been well-embraced. Khagram et al. (2003) reiterated the human–environment nexus that emphasizes the role of natural resource conservation and human improvement in advancing the sustainability debate. They contend that efforts to protect nature will fail unless they simultaneously advance the cause of human betterment, while efforts to improve people's lives will fail if they fail to conserve or enhance the essential resources and life support systems.

In this chapter, we examine the link between natural resource base, sustainable livelihood, and human security in Asia vis-à-vis the challenges that are associated with climate change and disaster risks. We use the sustainable livelihood framework (SLF) as an analytical tool to describe this nexus. The chapter is divided into four major parts. First, we provide an overview of the key concepts and relationships among natural resources, livelihood, and human security, followed by a short description of the modified sustainable livelihood framework that describes this relationship. In the second part, we discuss the current status of Asia's natural resource base and the vulnerability context of its livelihood assets. We then delve into livelihood strategy transformation to achieve better livelihood outcomes in part three. In part four, we present some of the promising practices in livelihood strategy transformation in Asia. We conclude that a more integrated and participatory approach in natural resource governance is vital to sustain the natural resources and ensure benefits for resource-dependent communities. These are pre-requisites for achieving human security in the context of challenging climate and increasing adverse impacts of disasters.

6.2 Linking Natural Resource Base, Sustainable Livelihood, and Human Security

How are natural resource base, sustainable livelihood, and human security linked? This chapter argues that attaining human security will necessitate sustainable local livelihood that will not compromise nature's ability to provide its goods and services.

Natural resources largely include soil, water, air, and all living things that are vital for human survival. However, in many less developed and developing countries in Asia, these natural resources are usually poorly managed and continuously being degraded. Overexploitation of natural resources resulted in the decline of ecosystem productivity and resilience, making people and places vulnerable to floods, droughts, famines, epidemics, etc. (Chand and Gartia 2016). Future challenges confronting natural resource base is compounded by the threat of climate change, known as a global risk multiplier (Nautiyal et al. 2016).

The SLF was adopted in this chapter to establish the close link between livelihood, natural capital, and human security. Many studies on vulnerability have already used this framework (e.g., Mwakubo and Obare 2009; Hahn et al. 2009; Ekblom 2012; Shah et al. 2013; Amos et al. 2015; Panthi et al. 2015; Peras et al. 2017). SLF is primarily relevant in understanding the vulnerability context of communities in relation to the impacts of climate change, particularly in analyzing the key elements of livelihood systems and the factors that affect them. Some scholars have explored the effects of climate change and poverty nexus on livelihood (Reed et al. 2013; Pandey and Jha 2011). Similarly, the disruption of ecological and land use systems caused by climate change further compromises the availability of food and water, hence, ultimately affecting livelihood (Recanati et al. 2017; Reed et al. 2013; Maharjan and Issahaku 2014).

Further, the framework promotes better understanding of people's access to livelihood assets, which include human, natural, financial, physical, and social. Access to these assets is analyzed in relation to the context of livelihood, institutional and social processes, and livelihood strategies. However, this model is also criticized for its limited scope–i.e., the analysis of people's access to capital assets and the resulting flow of services, neglecting the overall assets' stocks and associated services. For the "strong sustainability" advocates, livelihood can only be made sustainable if the critical levels of natural capital are maintained (Reed et al. 2013).

The SLF covers a wide range of advantages. More importantly, it provides the basis for understanding the contribution of livelihood strategies to building the adaptive capacity of people in coping with climatic change by diversifying economic activities (livelihood sources) to enhance future climate resiliency.

The framework (Fig. 6.1) was originally proposed in 1999 by the United Kingdom's Department for International Development (DFID) to synthesize various relevant literature and to link the natural resource base to livelihoods and human security in the context of changing climate and increasing incidence of disasters in Asia. We highlight the role of natural capital in our discussion more than the other types of

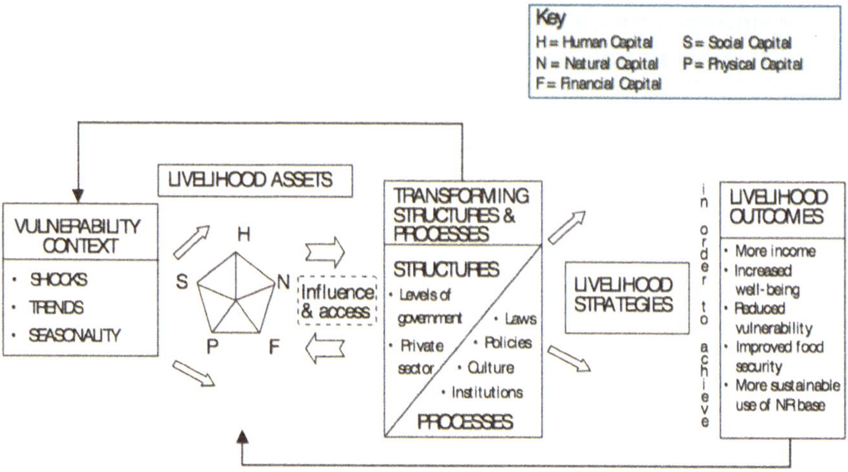

Fig. 6.1 Sustainable livelihood framework (*Source* Adapted from DFID 1999)

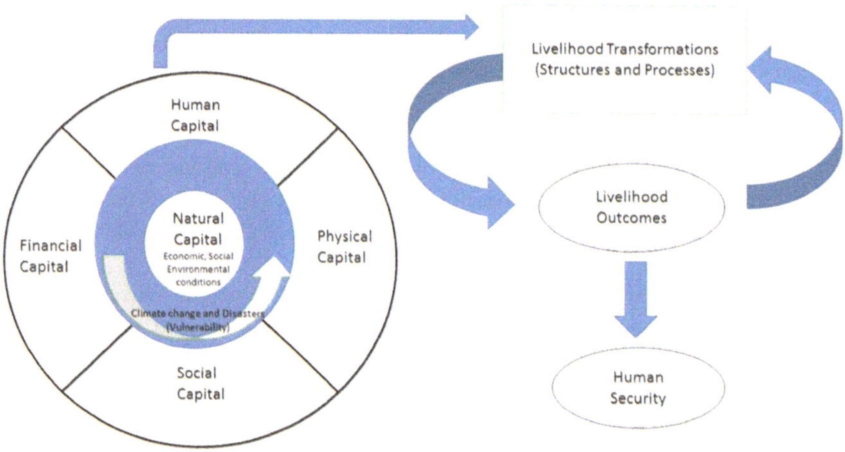

Fig. 6.2 Modified sustainable livelihood framework

capital assets to illustrate a point (Fig. 6.2). We argue that, in most developing countries of Asia where millions of populations are dependent on natural resources for livelihood and survival, sustainable livelihood strategies would have to be anchored on maintaining a "critical level of natural resource base" and thereafter enhance human security. Such critical level of resource base will vary from one country to another, depending on the country's specific economic, socio-cultural, political, and environmental conditions, including the future challenge brought about by climate change.

The asset pentagon in Fig. 6.1 is no longer prominent in the modified version of SLF. The emphasis is now on natural capital as a foundation for achieving human security. Natural capital is placed at the center of all the assets because a livelihood strategy that will be developed will need some critical or healthy level of natural resource base for it to be sustainable. Reed et al. (2013) posit that the overall stocks of the assets and associated services are oftentimes ignored due to the SLF's focus on people's access to capital assets and the resulting flow of services from which benefit can be derived. It also follows that the degree of people's access to the different capital assets will largely depend on the availability of the natural capital. Hence, limited natural resources could imply greater competition and conflict over people's access.

The Tonle Sapla Great Lake is a good example of the continuing pressure that human population has on a lake with already depleted natural resource base. Overfishing led to a decline in fish catch and smaller fish size (time constraint to regenerate). External market demand for turtles and snakes has likely depleted wildlife biodiversity. Further, the widespread agricultural development using chemical inputs, unmanaged household waste disposal, and planned infrastructure development around the lake have further degraded the natural ecosystem (Bonheur and Lane 2002). Pandey and Jha (2011) further reinforced the argument in their longitudinal analysis of rural indigenous mountain community in Srinagar, Pauri Garhwal, Uttarakhand, India. Their study underscored the importance of natural resources in enhancing social-ecological resilience—i.e., through diversifying income sources mixed with strong efforts to conserve forest resources. The multiple stressors governing the flow of different assets in and out of an ecosystem are essential for developing effective climate change initiatives (Reed et al. 2013). Hence, the evolution of the livelihood strategy is tailored to the different economic, social, and environmental conditions affecting the critical level of natural resources present.

The transformation of a livelihood strategy is also influenced by structures and processes. The different levels of governance system (local, regional/district, national, international), presence of civil society groups, private sector, policies, laws, and institutions have an equal share of influence in this transformation. These structures and processes create assets through the following: (1) infrastructure development, (2) technology generation, (3) strengthening local institutions, (4) determining access through ownership rights and regulating access to common resources, and (5) issuance of policies (DFID 1999). Overall, livelihood transformation can be viewed as a two-way process that undergoes a trial-and-error or an iterative process (IPCC 2014) to achieve a positive outcome and eventually gain a more secure future.

6.2.1 State of Asia's Natural Resource Base

The 2015 Global Forest Assessment reported that the global forest area is just under 4 billion ha. This area declined by 129 million ha (3.1%) during the period 1990–2015. At present, the world's forests house more than 75% of the world's terrestrial biodiversity (FAO 2015).

Asia's land area in 2010 comprised agriculture (50%), forest (20%), and other land (30%) use classes (FAO 2016). The Asia-Pacific region has a total forest area of 740.383 million ha or 18.36% of the global forest area (Table 6.1) (FAO 2011). For Asia, the total forest area is around 550.999[1] million ha. Although the forest area in the Asia Pacific region increased at 0.19% for 2000–2010, the forest area in Southeast Asia declined at a rate of −0.41% for the same period. This decline is compensated for by the increasing forest area coverage in South and East Asia in the same period, with a combined annual change of 1.03%.

The 2030 Agenda for Sustainable Development envisions the contribution of forests to food security and improved livelihood, which increases community resilience in finding integrated approaches to land use management and climate change adaptation and mitigation. The socio-economic development of the rural poor is largely dependent on the products and services provided by the forest (FAO 2015).

Forests have the potential to promote food security by contributing to rural livelihood and poverty alleviation goals from income derived from the production of forest goods and environmental services. The diversity of forest plants and animals is a source of food for millions of people. Likewise, wood fuel for cooking and water sterilization supports about 2.4 billion people (FAO 2015).

On a broader picture, the booming population adds to the problem of natural resource degradation. In Asia alone, population is currently set at 4.53 billion,

Table 6.1 Basic information in the Asian region

Land area (1,000 ha)	Population 2008				GDP 2008	
	Total (1,000)	Density (population/km^2)	Annual growth rate (%)	Rural (% of total)	Per capita (PPP)(US $)	Annual real growth (%)
3,093,763	4,075,307	132	1.1	59	6,070	3.0
Extent of forest 2010			Annual change rate (%)			
Forest area (1,000 ha)	% of land area	Area per 1,000 people (ha)	1990–2000		2000–2010	
			(1,000 ha)	(%)	(1,000 ha)	(%)
592,512	19	145	−595	−0.1	2,235	0.4

Source Adapted from FAO (2011)

[1] Asia forest cover = total Asia Pacific—Oceania.

accounting for about 60% of the global estimate. In a span of 10 years, the population in the Asian region increased to almost 500 million (Table 6.2). Population density also increased from 132/km^2 in 2008 to 146/km^2 (379 people/mi^2) in 2018. Roughly 50.4% are situated in the rural areas as compared with 59% in 2008. Urban areas are also densely populated, hence resulting in greater number of people exposed to climate-relatedimpacts and natural disasters.

The benefits derived from the coastal marine ecosystem are countless. Products such as wood, fish, and other marine resources generate income and feed a large coastal population of this region (Figs. 6.3 and 6.4). Asia has the world's largest and longest coastline, spanning 39,022 miles. In Asia Pacific alone, coastal population amasses to at least 1.4 billion, thus making the coastal area the most densely populated zone. Temperate to tropical coastal ecosystems also support major fisheries and industries. Further, tourism provides local communities with alternative sources of income. Ecological services such as flood control and providing habitat for rich and endangered coastal marine biodiversity are also directly and indirectly benefiting people on the coasts.

Biodiversity is essential for human well-being. It goes beyond provisioning for material welfare and livelihood; this includes achieving human resiliency and security. Many rural communities in Asia have benefited from the exploitation of these rich resources. However, their slow degradation has been observed as population became denser. Habitat loss beset poor ecosystem functioning, hence exacerbating the already impoverished communities who are heavily dependent on natural resources.

In Southeast Asia, biodiversity shares as much as 20% of the global terrestrial and marine flora and fauna. The region, however, is listed as a global hotspot due to high endemism and serious biodiversity loss. Deforestation rates were also observed to be highest in this region, reflective mainly of anthropogenic activities. The significant drivers of biodiversity decline include: (1) tree-plantations and deforestation as some countries have already lost over half their original forest cover; (2) hunting and trade for food, medicine, ornamentation, and as a status symbol; (3) mining of limestone and various minerals, which resulted in habitat fragmentation, leakage of heavy metals, and destruction of limestone karsts; (4) reservoir construction, wetland drainage, fires, pollution, invasive species, diseases; and (5) climate change. Further, it was observed that 30% of the species are already at increasing risk of extinction, whereas approximately 15–40% of the ecosystems are being affected by climate change. Climatic factors affect habitat and organisms through extreme changes in temperature and precipitation. Adaptation of these organisms (e.g., species migration) is further hampered as human has deprived them of space due to various land use changes.

Table 6.2 Forest area in Asia and the Pacific, 1990–2015

Subregion	Area (1,000 ha)				Annual change (1,000 ha)			Change rate (%)		
	1990	2000	2010	2015	1990–2000	2000–2010	2010–2015	1990–2000	2000–2010	2010–2015
East Asia	209,198	226,815	254,626	257,047	1,762	2,369	1,309	0.81	1.0	0.52%
South & Southeast Asia	319,615	298,645	295,958	292,804	−2,097	−269	−630	−0.68%	−0.14%	−0.21%
Western & Central Asia	39,309	40,452	42,944	43,511	114.29	249.19	113.43	0.29%	0.60%	0.26%
Total Asia	568,121	565,912	589,405	593,362	−221	2,349	791	−0.04%	0.41%	0.13%
Total Oceania	176,825	177,641	172,002	173,524	82	−564	304	0.05%	−0.32%	0.18%
Total Asia−pacific	744,946	743,553	761,407	766,886	−139	1,785	1,095	0.10%	0.09%	0.31%
World	4,128,269	4,055,602	4,015,673	3,999,134	−7,267	−3,993	−3,308	−0.18%	−0.10%	−0.08%

Source http://www.fao.org/forest-resources-assessment/past-assessments/fra-2015/en/
Adapted from FAO (2011)

Fig. 6.3 Mangrove-dependent rural community in Banacon Island, Bohol, Philippines

Fig. 6.4 Mangrove-dependent rural community in Iloilo, Philippines

6.2.2 The "Vulnerability Context" of Asia's "Livelihood Assets"

Natural resources shield local communities from different climate-related and natural disaster risks. Tree plantations located either in state-owned or private land serve as investment and insurance to reduce current vulnerability from food scarcity whenever their livelihood system is disrupted by climate-related stresses and disasters. In CARAGA region, Philippines, planted trees are well-known to serve as "bank deposits," not only addressing climate-related risks but also as "time deposit" for the college education of household members (Fig. 6.5). However, communities that are highly dependent on natural resources are likely to be severely affected as climate-related disasters ravage these investments. Therefore, maintaining some critical level of natural capital in view of the changing climate is central to the livelihood sustainability debate (Reed et al. 2013).

Turner et al. (2003) define vulnerability as the degree to which a human social and/or ecological system will be affected by some form of hazards. Hazards or perturbations, which include tropical cyclones, floods, landslides, volcanic eruptions, earthquakes, and tsunamis, are commonly referred to as a dangerous phenomenon, substance, human activity, or condition that may cause loss of life, injury or other health impacts; damage to property; loss of livelihood and services; social and economic disruption; or environmental damage. It usually involves climate-related physical events or trends or their physical impacts (IPCC 2014). Landslides and floods are best examples of hazards that pose great havoc to natural resources.

In 2006, a small village in Southern Leyte, Philippines, experienced a massive landslide that took numerous lives. In addition, the long dry spell (drought/El Niño) in 1997–1998 has destroyed marine and coastal ecosystems with the massive bleaching of the coral reefs. However, vulnerability can also be seen positively as Gallopín (2006) saw the local community's emergence from poverty. The community members took on vulnerability to their advantage by attaining some degree of livelihood transformation to get them out of poverty.

Fig. 6.5 Smallholder *Falcata* plantations in CARAGA, Philippines

Agriculture is one of the major economic sectors that are highly affected by climate change. In Asia, about 81% of the population still depend on agriculture for livelihood (Hijioka et al. 2014). In the Gaza Strip in Middle East Asia, water has significantly declined from 51 m^3/year per person (in 2005) to 33 m^3/year per person (2017). This is brought about by drought, driven by the deterioration of water resources, which led to a severe decline in agricultural production (Recanati et al. 2017). About 2 million people were reported to have been affected by this event.

The terrestrial systems are also vulnerable to climate change impacts. Some of these include shifts in phenologies, growth rates, and distribution of plant species and permafrost degradation. The projected climate change in the twenty-first century will further increase these observed impacts (IPCC 2014). Drought has been responsible for tree mortality, loss of forest biomass, diminishing size of water bodies, reduction of water quality, and loss of fish species. Alamgir et al. (2014) underscored the effects of climate change on the Thotne Khola community forest in Nepal. Records show that temperature has increased by 0.06 degree annually from 1977 to 1994. Shifts in precipitation patterns, glacier recession, less predictable timing of onset of monsoon seasons, and increasing incidence of severe storms, landslides, and droughts threaten the terrestrial ecosystems at alarming rates. Higher summer temperatures likewise led to high mortality of important trees such as coppiced *Schima wallichii* and *Castanopsis indica* and massive pest infestations. The fruiting *Myrica esculenta* was also affected by higher temperatures, thus losing revenues for farmers who are growing this for fruits.

In spite of the essential values that coastal ecosystems provide, Asia's coasts are also under threat. Coastal vulnerability worsens as the impacts and likelihood of flooding, storm surges, and sea level rise also increase through time. Sea level rise is expected to cause increased rates of coastal erosion. Areas of mangroves, salt marshes, and seagrass beds continue to decline, while the entry of saltwater, combined with rising sea level, will make coastal freshwater swamps and marshes more vulnerable. Coral reefs will experience increased widespread damage in the twenty-first century from warming and ocean acidification. There is high confidence that marine biodiversity will likely increase in temperate areas with the expansion of warm water species northward. However, marine biodiversity in the tropics may "decline if thermal tolerance limits are exceeded" (IPCC 2014).

In view of mangrove forests alone, deforestation remains a pressing problem. In the Southeast Asian region, this ecosystem covers about 4.5 million km^2 of land, which harbors 39% of the world's mangrove biodiversity (Giesen et al. 2006). Deforestation and degradation rate was observed to be significant, with about 2.1% loss over the past decade (Richards and Friess 2016; Gevaña et al. 2018). The major driver of mangrove loss was aquaculture development, which contributed to 30% loss in forest cover (Gevaña et al. 2019). This was followed by rice farming with a recorded loss of 21.7%. They also found out that mangrove conversion to oil palm plantation has emerged as a significant threat. Shedding-off of coastal forests was observed to have resulted in more severe impacts on human lives and natural resources, particularly during storm surges and tsunami events (Pulhin et al. 2017).

Extreme flood events uncovered the many lapses of livelihood systems, including access to alternative livelihoods, sustainability of existing livelihoods, and accumulated household debts, which led to the migration. The perpetual vulnerability of Tonle Sap Lake of Cambodia to flooding portrays a good example of this problem.

6.3 Livelihood Strategy Transformation for Better Livelihood Outcomes

Scholars on human dimension of climate change have underscored resiliency (or resilience thinking) as the unspoken yet ultimate aim of sustainable livelihood (SL). Sustainable livelihood focuses on how people's capabilities, assets and activities, and transformed structures and processes could result in better outcomes such as increased income, enhanced well-being, and improved food security. According to Adger (2000), livelihood stability is a subset of social resilience, which implies that a stable livelihood is essential for the social system to be resilient. However, social resilience is still a neglected concept from the practice theory standpoint (Speranza et al. 2014).

Livelihood strategies build people's resilience (Ekblom 2012). However, the challenges posed by climate change may dismiss this claim. IPCC (2014) calls for transformational pathways where a change in the fundamental attributes of natural and human systems is needed. It includes strengthening, alterations, or alignment of paradigms, goals, or values in the promotion of sustainable development goals. More specifically, transformations in political, economic, and socio-technical systems enhance mitigation and adaptation responses. In the context of climate change, significant transformations may lead to climate-resilient development pathways, which will require an understanding of risks, adaptive management, learning, innovation, and leadership. These transformational pathways include actions, strategies, and choices that reduce climate change impacts with the assurance that risk management and adaptation can be implemented and sustained.

Many of the initiatives at the global and national levels are focused on sustaining climate resiliency, but majority of the actions are localized. Evidence shows that failure to address the effects of emerging climate stressors has already eroded the basis for sustainable development, which includes, among others, household income, food security, health, education, housing, and livelihood assets. There is thus an urgent need at present for action toward climate-resilient pathways to materialize in the future (IPCC 2014).

Strategies and actions toward climate-resilient pathways equally intend to improve livelihood, social and economic well-being, and responsible environmental management. Transformation of livelihood strategies is necessary to address poverty reduction goals that are envisioned toward climate resiliency. The ability of the transformed livelihood system and the sustainability of the natural resource base to cope with future impacts of climate change are vital for enhancing human security. It can be done iteratively from monitoring, evaluation, learning, innovation to contingency planning (IPCC 2014).

The poor rural population are undergoing household-level adjustments in investments and productive activities in order to transform their livelihood systems. However, the outcomes of the poor's livelihood systems are generally influenced by structural and policy processes. In Srinagar, district of Pauri Garhwal, Uttarakhand, India, transformative institutional change has been the major concern for impoverished families to better cope with climate change impacts (Pandey and Jha 2011). The national program on forest restoration was noted critical for livelihood-related decisions of local communities. In Yan'an, China, the implementation of national soil erosion control policy has provided for the restoration of grasslands as well as the afforestation of degraded farmlands. Between 1999 and 2012, the forest area and forest cover of this place has increased by more than 610,000 ha. However, cultivated areas had declined by 50%, which then led to problems of food security and farm income. To counter such policy impact, the local residents pursued the construction of warping dams (see Sect. 6.4 for details). The local communities used the policy to their advantage by transforming land resources (i.e., eroded soil and sedimentations) to increase their area of cultivation (Zhang et al. 2018), thereby effectively addressing the issue.

Eder (2011), on the other hand, delved into environmental reconfigurations of the upland and coastal areas in the Philippines to highlight the various pressures and responses inflicted on the human–environment nexus. The influence of structures and processes on livelihood transformation is evident in the three different environments involving natural resources: resource-depleted environments, restricted-use environments, and off-limits environments. The strict implementation of the Northern Sierra Madre National Park (NSMNP) policies since 1997 has limited the Agta's (indigenous people of Sierra Madre) livelihood options, forcing them to seek other income sources to meet their market needs. Traditionally, the Agta people are forest gatherers, who collect and sell rattan, swiftlet nests (for a soup delicacy), and other commercially valuable forest products. Some of them are also upland farmers and small-scale loggers. However, the park's zoning system has placed their domain under a strict protection zone, allowing them to utilize resources in the area only for "traditional resource use." The Agta's traditional resource utilization was restricted to subsistence activities, ignoring the longtime importance of the need to earn more and become more resilient.

Shifting cultivation has been discouraged in the whole of Southeast Asia due to its contribution to deforestation and soil erosion. The abandonment of this practice as a national policy has either positive or negative implications. In Palawan, Philippines, a group of farmers benefitted in the process as they transitioned from upland tillers to lowland irrigated rice and agroforestry farmers. This has been the case for those who have sufficient capital assets and alternative livelihood options. However, the other group, composed of poorer farmers, suffered as they retreated to less accessible upland areas where shifting cultivation can be allowed (Eder 2011).

6.4 Promising Practices on Livelihood Strategy Transformation in Asia

Rural communities create livelihood strategies as a response to environmental challenges, including climate change and natural disasters. Livelihood strategies are transformed to positively adapt and produce better livelihood outcomes that will be sustainable in the long run. Livelihood is sustainable if it enables the natural resource base to cope with and recover from stresses and shocks. Generally, sustainable livelihood strategies promote human development, thereby ensuring human welfare and ecosystem health at present and in the future. A livelihood strategy interdependence on ecological, social, and economic systems implies the need for a stable ecosystem. This section provides selected examples of promising practices on livelihood transformations currently employed by rural communities, coupled by structural and political processes that either constrain or facilitate livelihood outcomes. Representative examples from each of the natural resources, like forest and biodiversity resources, land (soil) and water resources, and mangroves and coastal resources, are briefly discussed to illustrate efforts toward livelihood transformations in the region.

6.4.1 Forest and Biodiversity Resources

6.4.1.1 Reducing Emissions from Deforestation and Forest Degradation (REDD+)

Incorporating a broader range of forest values in decision-making processes is what comprises a resilient and durable economy. This entails a significant reduction in deforestation and forest degradation activities and a net increase in global forest area, hence, the birth of Reducing Emissions from Deforestation and Forest Degradation (REDD+). This is a recent initiative at the global level endorsed by individual governments, civil society, and the private sector, which moves for the sustainable management of forests, emphasizing the conservation and enhancement of forest carbon stocks. Considered as a "game changer," REDD+ is a financial mechanism that reduces deforestation through carbon emission trading. It also recognizes the non-market value of forests (FAO 2012).

REDD+ sites are generally characterized by higher rates of deforestation and poor governance. In Asia, countries undergoing REDD+ readiness activities include Cambodia, Indonesia, Laos, Malaysia, Nepal, Sri Lanka, and Vietnam (Table 6.3). Meanwhile, about 64 partner countries are being supported by the UN-REDD Program across Africa, Asia-Pacific, and Latin America and the Caribbean. There are 12 countries in Asia that benefitted from this facility with support intended for its national program and targeted for particular work area (outcome of the SNA or the Support to National REDD+ Action: Global Programme Framework 2011–2015).

Table 6.3 Asian REDD + countries with REDD + readiness activities

Country	REDD + projects	Other readiness initiatives	Forest cover	Deforestation rate	FCPF[a]	UN REDD
Cambodia	4	12	High	High	OK	OK
Indonesia	30	45	High	High	OK	OK
Laos	7	16	High	Low	OK	
Malaysia	0	5	High	Low		OK
Nepal	2	8	Low	Low	OK	OK
Sri Lanka	0	5	High	High		OK
Vietnam	4	28	High	Low	OK	OK

[a]Forest Carbon Partnership Facility (FCPF) Programme
Source https://theredddesk.org/countries

Bangladesh, India, Myanmar, Pakistan, and the Philippines are partner countries in addition to the seven countries already mentioned in Table 6.3.

Shrestha et al. (2014) show the case of Nepal as it implemented the Community Forest Management (CFM) REDD+ Pilot project where successful sustainable forest management has been promoted by local communities. This REDD+ project revealed enhanced forest vegetation growth rates through improved community forest management. This is evident even after project implementation with the conduct of regular and frequent meetings and with increased representation and participation of women and socially marginalized communities from community forest user groups (CFUGs). Furthermore, CFUG management improved with frequent meetings, open bank accounts, transparent financial records, and performance of targeted activities for marginalized groups. Co-benefits through improved livelihood and institutional and technical capacities within communities are reflected in the study site. The best practices promoted the advancement of REDD+ implementation by creating awareness, doing proper planning with baseline data, establishing institutional structures, regular monitoring and evaluation, and supporting communities' own ability to organize and manage their forests by addressing the livelihood concerns of the poor and socially marginalized. The successful implementation of the national level REDD+ program entails addressing livelihood concerns and problems of inequality and exclusion with the achievement of target emission reduction.

Similarly, the study of Traedal and Vedeld (2018) revealed the critical implications of their findings on REDD+ and REALU (Reducing Emission from Agriculture and Land Use) as they examine the multiple values of forest lands in the household economies in Bac Kan Province in northern Vietnam. The study identified a variety of forest land uses so households can generate livelihood income. This is consistent with findings from the study of Groom and Palmer (2012) in China where forest lands are an important livelihood asset for the poorer household economies.

6.4.1.2 Payments for Environmental Services (PES) Scheme

Payments for Environmental Services (PES) is a financial mechanism or scheme where the one who protects environmental/ecosystem service is paid. The relationship is established between the buyer (user) of the service and the seller (provider) of that service. Earlier implementers of the PES scheme primarily focused on enhancing natural resource management. Groom and Palmer (2012) examine the potential impacts of the Sloping Lands Conversion Programme (SLCP), China's national-level PES scheme, on livelihood, especially income and poverty alleviation. SLCP is mainly focused on reforestation to supply watershed services. Findings showed that households at the 25th percentile of income distribution had increased by RMB 331 per capita (50% increase). Likewise, the program was noted to have a positive impact on poverty alleviation, where only a few households with income below RMB 2,000 were accounted for. Poverty alleviation was measured by comparing economic conditions before and after program implementation. Since SLCP is a reforestation scheme, there is no assurance that households could be prevented from returning to agriculture in the long term. This implies that incentivizing forest conservation will require livelihood options that will complement the improvement of carbon sinks.

6.4.1.3 Community-Based Initiatives for Natural Resource Management

Community-based natural resource management (CBNRM) has various names: it is called community-based forest management (CBFM) in the Philippines and Indonesia, forest user groups in Nepal, joint forest management (JFM) in India, and community forestry in Myanmar, Bangladesh, and Vietnam. CBNRM is an institutional strategy that aims to attain the community's desired outcomes along with sustainable natural resource conservation. Apart from improving the natural resources, CBNRM also reduces their vulnerability to the impacts of climate-related risks. Apparently, the condition and extent of CBNRM will always have implications on local livelihood and human resilience (Ming'ate et al. 2014).

CBFM has been implemented in the Philippines over the past three decades. Illegal logging, timber poaching, and kaingin making as livelihood sources have been replaced by a more sustainable employment of the agroforestry system (Peras et al. 2015). The common interests and aspirations of forest community development espoused in CBFM distill lessons among community members to transform environmentally destructive livelihood sources (Fig. 6.6). CBFM also strengthens forest protection activities of people's organizations such as the Young Innovators for Sustainable Environmental Development Association (YISEDA) (Fig. 6.7) in central Philippines. Mohammed et al. (2015) cited the potential of CBFM as a transformation strategy in managing coupled socio-ecological systems, highlighting the cases of the Philippines and Bangladesh.

Fig. 6.6 The agroforestry system applied in the CBFM area of YISEDA, Southern Leyte

Fig. 6.7 The natural capital asset of YISEDA as a result of strengthened community-based forest management activities

6.4.2 Land and Water Resources

Land and water resources are particularly important for Asia. The quality of land and water resources is highly affected by unsustainable human activities. Poor land management has serious impacts: soil erosion and desertification, overgrazing resulting in grassland degradation, large-scale utilization of fertilizers and pesticides affecting soil quality, and land contamination brought about by industrial activities. Issues such as water crisis, water pollution, water conflicts, and water politics, among others, were observed to have worsened through time (AASA 2011). These problems are rooted in the increasing demand for food, water, and energy.

6.4.2.1 Warping Dams

China is confronted by the scarcity of arable land for food production. The vast track of marginal and degraded land from decades of unsustainable use, coupled with climate change impacts, is an opportunity unraveled. A successful case study of land transformation gave way to the development of an agricultural livelihood system that improved the food security of the local populace and promoted environmental conservation (Zhang et al. 2018).

Gully development control or gully stabilization in Yan'an City led to the construction of warping dams (Fig. 6.8). A warping dam is a local mitigating measure that addresses food insecurity caused by the limited area of cultivated land. This reduction in cultivated land is a national policy measure to control soil erosion by converting farmland into grassland and forest land. However, from the socioeconomic point of view, the policy was deemed unacceptable to the local population whose food production system was threatened by the drastic decline in cultivated area. Hence, the evolution of warping dams. The trapped sediments in the dams created artificial fields used for cultivating suitable crops. Advanced and efficient farming techniques were applied. This is one example of adapting to a transformed landscape to come up with a more sustainable livelihood strategy. However, the lack of maintenance in the warping dams endangers their sustainability, which may threaten food security and climate resiliency (Zhang et al. 2018).

Similarly, in the Philippines, the local communities of Pantabangan-Carranglan watershed saw an opportunity to adapt to drought in the watershed's reservoir by planting drought-resistant onions when the dam water recedes during dry season.

Fig. 6.8 A warping dam in Yan'an City (*Source* Adapted from Zhang et al. 2018)

This livelihood strategy is just one of the many adaptation strategies employed in the area (Peras et al. 2008).

6.4.3 *Mangroves, Coastal, and Marine Resources*

The marine ecosystems in Asia also play a vital role in highlighting the relationship between land and the sea. Asia's long coastlines harbor more than 50% of the human population directly dependent on marine resources, including mangroves and coastal reefs, for survival. Mangrove resources provide timber and non-timber forest products; serve as coast protector in the face of climate-related problems and disasters, which include, among others, sea level rise, tsunamis, etc.; and provide fish and shellfish. Like other natural resources, mangroves are threatened and severely damaged by industrial development and infrastructure. However, livelihood adaptation strategies are also identified to meet present and future challenges.

6.4.3.1 Community-Based Mangrove Restoration and Management

Few studies have investigated the factors that could lead to long-term community-based mangrove restoration and management. Damastuti and Groot (2017) demonstrated the effectiveness of various community-based mangrove management (CBMM) practices toward sustainable mangrove management in four coastal villages in central Java, Indonesia (Sriwulan, Bedono, Timbulsloko, and Surodadi). Among the factors that showed greater impact were type of community participation, level of organizational and economic assistance from external institutions, magnitude of rehabilitation project, and the time selected for rehabilitation and maintenance of strategies applied. The strategies employed in each village had resulted in different outcomes. The most successful was the one in Surodadi village where efficient resource utilization and much improved local livelihood were observed. Bedono is less successful in terms of livelihood support systems but showed effectiveness in extending and maintaining rehabilitated areas. Hence, to better promote sustainability, it was further suggested that best practices be combined with (1) additional external assistance both scientifically and technologically, (2) diversification of income, (3) institutional reinforcement, and (4) continuous monitoring of local institutions.

6.4.3.2 Construction of Raised Agricultural Fields

In developing new agricultural fields in Xinghua City, Jiangsu Province took advantage of the natural wetland that is present in the area. The province is popularly known for its traditional raised agricultural fields, declared as a heritage site in 2014 by

Fig. 6.9 Examples of raised agricultural fields in China (*Source* Adapted from Zhang et al. 2018)

UN-FAO (Fig. 6.9). This 600-year-old traditional practice evolved because of population pressure and severe and frequent flooding. Although flooding has decreased significantly due to the construction of water conservation facilities, these wetland raised fields have been retained to cater to the expansion of other cultivated land. The raised fields or small elevated platforms were formed from excavated (dirt/soil) drainage canals of the flooded area located at the edges of the lakes, coastal areas, and swamps. The raised field is 2–4 m above the water level and is 30 × 50 m wide with a spacing of 2–10 m between fields. Vegetables are the primary crops grown in these fields, particularly taro, chives, and oilseed rape. The establishment of vegetable processing enterprises in the 1960s led to the development of a specialized industry that produces dehydrated vegetables for export to Great Britain, Japan, Korea, and other countries. Apart from vegetable production, aquaculture (e.g., freshwater fish and shrimp production) along irrigation canals was observed to have grown. Furthermore, tourism has been an additional income-generating source for the local community as people continue to come to see these ancient landscapes.

6.5 Conclusion

Human security is threatened by the changing climate, coupled with the continuing decline of the natural resource base. Local communities create livelihood strategies in response to these environmental challenges. However, livelihood strategies remain vulnerable to climate-related risks. The employment of the modified sustainable livelihood framework highlights the nexus between natural resource base, livelihood, and human security. The natural capital asset is at the heart of this framework, providing the foundation for human security. Depleting the natural resource base could result in chaos as competition and conflict over people's access to the limited resource base could escalate. Hence, maintaining some healthy level of natural resource base is pursued to ensure livelihood sustainability.

Livelihood strategies are being improved and transformed for people to positively adapt and produce better outcomes, which make them resilient to the changing times. Therefore, for a livelihood strategy to become sustainable, it must be pressed toward the twin goal of improving human resilience and ensuring the stability of the natural resource base. The modified sustainable livelihood framework clearly portrays such nexus by underscoring the natural resources' role in sustaining livelihoods, the need for designing livelihood systems that will improve the availability and quality of these resources, and the importance of aligning livelihood strategies (including policies, institutions, and markets) to promote local development and resilience. Case studies from Asia suggest that there remain huge challenges in bridging these elements. Wrong policies and poor governance hamper the ability of poor rural communities to conserve their natural resources and effectively implement suitable and sustainable livelihood options. A number of promising practices were discussed, suggesting some potential pathways to overcome these challenges. Distillations from such practices reveal that a more integrated and participatory approach in natural resource governance is crucial to sustain the natural resources and to ensure benefits for resource-dependent communities, thereby advancing human security in the context of a challenging climate and increasing risks of disasters.

References

AASA (Association of Academies of Science in Asia) (2011) Towards a sustainable Asia: natural resources. South Korea: AASA. 65 p

Adger WN (2000) Social and ecological resilience: are they related? Progr Hum Geogr 3:347–364. https://doi.org/10.1191/030913200701540465

Adger WN, Pulhin JM, Barnett J, Dabelko GD, Hovelsrud GK, Levy M, Oswald Spring Ú, Vogel CH (2014) Human security. In: Field CB, Barros VR, Dokken DJ, Mach KJ, Mastrandrea MD, Bilir T, Chatterjee M, Ebi KL, Estrada YO, Genova RC, Girma B, Kissel ES, Levy AN, MacCracken S, Mastrandrea PR, White LL (eds) Climate change 2014: Impacts, adaptation, and vulnerability. Part A: Global and sectoral aspects. Contribution of Working Group II to the Fifth Assessment Report of the Intergovernmental Panel on Climate Change. Cambridge University Press, Cambridge, United Kingdom and New York, NY, USA, pp 755–791

Alamgir M, Pretzsch J, Turton SM (2014) Climate change effects on community forests: finding through user's lens and local knowledge. Small-scale For 13(4):445–460
Amos E, Akpan U, Ogunjobi K (2015) Households' perception and livelihood vulnerability to climate change in a coastal area of Akwa Ibom State. Nigeria. Environ Dev Sustain 17:887–908. https://doi.org/10.1007/s10668-014-9580-3
Bonheur N, Lane BD (2002) Natural resources management for human security in Cambodia's Tonle Sap Biosphere Reserve. Environ Sci Policy 5(1):33–41
Broad R (1995) The political economy of natural resources: case studies of the Indonesian and Philippine forest sectors. J Dev Areas 29(3):317–340
Chand D, Gartia R (2016) Climate change and natural capital: some veiled issues on sustainable livelihood in agriculture sector. In: Nautiyal S, Schaldach R, Raju KV, Kaechele H, Pritchard B, Sreenivasa Rao K (eds) Climate change challenge (3C) and social-economic-ecological interface-building: exploring potential adaptation strategies for bio-resource conservation and livelihood development. Springer International Publishing, Switzerland, pp 261–278
DFID (Department for International Development) (1999) Sustainable livelihoods guidance sheets, Numbers 1–8. DFID, London (also available on www.livelihoods.org)
Damastuti E, Groot RD (2017) Effectiveness of community-based mangrove management for sustainable resource use and livelihood support: a case study of four villages in Central Java. Indonesia J Environ Manage 203:510–521
Eder JF (2011) Environmental reconfigurations and livelihood transformations in rural Southeast Asia. Resource management. In: Rural livelihoods. Regions and cohesion 1(3): 145–164. https://doi.org/10.3167/reco.2011.010307
Ekblom A (2012) Livelihood security, vulnerability and resilience: a historical analysis of Chibuene. AMBIO, Southern Mozambique. https://doi.org/10.1007/s13280-012-0286-1
Giesen W, Wulffraat S, Zieren M, Scholten L (2006) Mangrove Guidebook for Southeast Asia. Dharmasarn Co., Ltd.: FAO and Wetlands International
FAO (Food and Agriculture Organization of the United Nations) (2011) State of the world's forests. FAO, Rome
FAO (Food and Agriculture Organization of the United Nations) (2012) State of the world's forests. FAO, Rome. 60 p
FAO (Food and Agriculture Organization of the United Nations) (2015) Global Forest Assessment 2015. FAO, Rome. 253 p
FAO (Food and Agriculture Organization of the United Nations) (2016) State of the world's forests. FAO, Rome
Gallopín GC (2006) Linkages between vulnerability, resilience, and adaptive capacity. Global Environ Change 16(3):293–303
Gevaña D, Camacho L, Pulhin J (2018) Conserving mangroves for their blue carbon: insights and prospects for community-based mangrove management in Southeast Asia. In: Makowski C, Finkl C (eds) Threats to mangrove forests. Coastal Research Library. vol 25. Springer Nature, pp 579–588
Gevaña D, Pulhin J, Tapia M (2019) Fostering climate change mitigation through a community-based approach: carbon stock potential of community-managed mangroves in the Philippines. In: Krishnamurthy RR, Jonathan MP, Srinivasalu S, Glaeser B (eds) Coastal management: global challenges and innovations. Elsevier, Oxford, UK, pp 271–282
Groom B, Palmer C (2012) REDD + and rural livelihoods. Biol Conserv 154:42–52
Hahn M, Riederer AM, Foster SO (2009) The livelihood vulnerability index: a pragmatic approach to assessing risks from climate variability and change—a case study in Mozambique. Global Environ Change 19:74–88
Hijioka, Y, Lin E, Pereira JJ, Corlett RT, Cui X, Insarov GE, Lasco RD, Lindgren E, Surjan A (2014) Asia. In: Climate Change 2014: impacts, adaptation, and vulnerability. Part B: Regional Aspects. Contribution of Working Group II to the Fifth Assessment Report of the Intergovernmental Panel on Climate Change [Barros VR, Field CB, Dokken DJ, Mastrandrea MD, Mach KJ, Bilir TE, Chatterjee M, Ebi KL, Estrada YO, Genova RC, Girma B, Kissel ES, Levy AN, MacCracken S, Mastrandrea PR, White LL (eds)]. Cambridge University Press, Cambridge, United Kingdom and New York, NY, USA, pp 1327–1370

IPCC (Intergovernmental Panel on Climate Change) (2014) Climate change. Working Group III: mitigation of climate change. Cambridge University Press, Cambridge
Khagram S, Clark WC, Raaad DF (2003) From the environment and human security to sustainable security and development. J Hum Dev 4(2):289–313
Maharjan KL, Issahaku ZA (2014) Chapter 1. Communities and livelihood strategies: an overview. In: Maharjan KL (ed) Communities and livelihood strategies in developing countries. Springer, Japan
Ming'ate FLM, Rennie H, Memon A (2014) Potential for co-management approaches to strengthen livelihoods of forest dependent communities: a Kenyan case. Land Use Policy 41:304–312
Mohammed AJ, Inoue M, Peras RJ, Nath TK, Jashimuddin M, Pulhin JM (2015) Transformation strategy for managing coupled socio ecological systems: case studies from Bangladesh and the Philippines. Small Scale For 15(2):213–227. https://doi.org/10.1007/s11842-015-9318-6
Mwakubo SM, Obare GA (2009) Vulnerability, livelihood assets and institutional dynamics in the management of wetlands in Lake Victoria watershed basin. Wetlands Ecol Manage 17:613–626. https://doi.org/10.1007/s11273-009-9138-6
Nautiyal S, Schaldach R, Raju KV, Kaechele H, Pritchard B, Rao KS, Sreenivasa Rao K (eds) (2016) Climate change challenge (3C) and social-economic-ecological interface-building: exploring potential adaptation strategies for bio-resource conservation and livelihood development. Springer International Publishing, Switzerland
Pandey R, Jha SK (2011) Climate vulnerability index–measure of climate change vulnerability to communities: a case of rural Lower Himalaya, India. Mitig Adapt Strat Glob Change 17:487–506. https://doi.org/10.1007/s11027-011-9338-2
Panthi J, Aryal S, Dahal P, Bhandari P, Krakauer NY, Pande VP (2015) Livelihood vulnerability approach to assessing climate change impacts on mixed agro-livestock smallholders around the Gandaki River Basin in Nepal. Reg Environ Change. https://doi.org/10.1007/s10113-015-0833-y
Peras RJ, Pulhin JM, Lasco RD, Cruz RVO, Pulhin FB (2008) Climate variability and Extremes in the Pantabangan-Carranglan watershed, Philippines: assessment of impacts and adaptation practices. J Environ Sci Manage 11(2):14–31
Peras RJ, Pulhin JM, Inoue M (2015) Local stakeholder's assessment of community-based forest management and the implications of REDD plus implementation in the Philippines. Asia Life Sci 24(1):349–381
Peras RJ, Pulhin JM, Inoue M (2017) Vulnerability of community-based forest management to climate variability and extremes: emerging insights on the contribution of REDD+. Small Scale For 16(2):249–274
Pulhin J, Gevana D, Pulhin F (2017) Community-based mangrove management in the Philippines: experience and challenges in the context of changing climate. In: Gupta R, Shaw R (eds) Participatory mangrove management in a changing climate. Tokyo, Japan, Springer Nature, pp 247–262
Recanati F, Castelletti A, Dotelli G, Melià P (2017) Trading off natural resources and rural livelihoods: a framework for sustainability assessment of small-scale food production in water-limited region. Adv Water Resour 110:484–493
Reed MS, Podesta G, Fazey I, Geeson N, Hessel R, Hubacek K, Letson D, Nainggolan D, Prell C, Rickenbach MG, Ritsema C, Schwilch G, Stringer LC, Thomas AD (2013) Surveys: combining analytical frameworks to assess livelihood vulnerability to climate change and analyse adaptation options. Ecol Econ 94:66–77
Richards D, Friess D (2016) Rates and drivers of mangrove deforestation in Southeast Asia, 2000–2012. Proc Natl Acad Sci 113(2):1–2
Shah KU, Dulal HB, Johnson C, Baptiste A (2013) Understanding livelihood vulnerability to climate change: applying the livelihood vulnerability index in Trinidad and Tobago. Geoforum 47:125–137
Shrestha S, Karky BS, Karki S (2014) Case study report: REDD+ pilot project in community forests in three watersheds of Nepal. Forests 5:2425–2439. https://doi.org/10.3390/f5102425
Speranza C, Wiesmann U, Rist S (2014) An indicator framework for assessing livelihood resilience in the context of social-ecological dynamics. Global Environ. Change 28:109–119

Traedal LT, Vedeld P (2018) Cultivating forests: the role of forest land in household livelihood adaptive strategies in the Bac Kan Province of northern Vietnam. Land Use Policy 73:249–258

Turner BL II, Matson PA, McCarthy JJ, Corell RW, Christensen L, Eckley N, Hovelsrud-Broda GK, Kasperson JX, Kasperson RE, Luers A, Martello ML, Mathiesen S, Naylor R, Polsky C, Pulsipher A, Schiller A, Selin H, Tyler N (2003) Illustrating the coupled human–environment system for vulnerability analysis: three case studies. Proc Natl Acad Sci USA 100(14):8080–8085

Zhang J, Chunxiong H, Li C, Shixiong C (2018) Improving food security in China by taking advantage of marginal and degraded lands. J Cleaner Prod 171:1020–1030

Dr. Rose Jane J. Peras is Professor and current Chair of the Department of Social Forestry and Forest Governance, College of Forestry and Natural Resources, University of the Philippines Los Baños. She has more than 20 years of experience in research and development and 15 years in forestry education and public service. Her research involvements in forestry, community-based forest management, integrated watershed management plan development, human dimension of climate change, disaster risk reduction/climate change adaptation, and forest governance led her to author/co-author 29 technical publications. She earned her Bachelor of Science in Forestry and Master of Science in Forestry from the University of the Philippines Los Baños and Ph.D. in Forest Governance from The University of Tokyo.

Dr. Juan M. Pulhin is full Professor and former Dean of the College of Forestry and Natural Resources, University of the Philippines Los Baños, Philippines. He was Lead Author of the 2007 Fourth Assessment Report of the Intergovernmental Panel on Climate Change (IPCC), Coordinating Lead Author of the Human Security Chapter of the 2014 IPCC Fifth Assessment Report and currently Lead Author of the Asia chapter of the IPCC Sixth Assessment Report. He has been a four-time a visiting professor at The University of Tokyo in 2007, 2010, 2016 and 2017 and a visiting scholar at Waseda University, Tokyo in 2018. He has written and edited 11 books and published more than 100 technical papers and book chapters on natural resources policy and governance, and on the human dimension of climate change and disaster risks. He is a fellow, invited expert, and council member of several international organizations and a member of the National Academy of Science and Technology (NAST), Philippines.

Dr. Dixon T. Gevaña is a faculty member of the Department of Social Forestry and Forest Governance, College of Forestry and Natural Resources, University of the Philippines Los Baños (UPLB). He teaches courses related to forest sustainability and conducts researches on coastal forest management and climate change mitigation. Most of his publications delved on the role of mangroves in addressing climate change. He obtained both his undergraduate (BS Forestry) and master's (MS Community Development) degrees from University of the Philippines, and doctorate degree in Forest Environmental Science from Seoul National University, South Korea.

Dr. Makoto Inoue is Professor at the Faculty of Human Sciences, Waseda University, Japan and former Professor at The University of Tokyo. He serves as Dean of Graduate School of Human Sciences, Waseda University since September 2020. His specialty is environmental sociology, forest governance, and Southeast Asian studies. He has written and edited/co-edited 34 books, and published more than 140 original articles in peer-reviewed international and national journals on the social aspects of forestry and the environment. He serves as President of The Japanese Association for Environmental Sociology (June 2019–present), Vice President of Japanese Society for Current Anthropology (January 2020–present), Vice President of Society for Environmental Economics and Policy Studies (April 2018–March 2020), Councilor of The International Association for the Study of the Commons (January 2019-present), a board member of various academic societies in Japan, and an editorial board member of several international journals.

Chapter 7
Agriculture and Food Security in Asia

Jose Ma. Luis Montesclaros and Paul S. Teng

Abstract Despite declining contributions to Asia's gross domestic product (GDP) and generation of jobs, agriculture continues to play an important role in preventing famine and reducing chronic undernourishment in the region. Its ability to ensure food security outcomes, however, is complex as hunger persists even if more food is available than is actually consumed, on a per-capita basis. Equal attention needs to be given to physical and economic access to food, as well as the way food availability and access are translated into nutrition outcomes. Climate change has played an important role in shaping agricultural performance, which links back to all of the crucial perspectives in food security. While climate change impacts on yields are expected to continue, and even worsen in significant parts of the region, these are compounded by competing demands for scarce water resources, rural-urban migration, food prices, and the growing demand for food in the long term. Regional approaches to food security will need to be transformed to keep up with the pace at which climate impacts are changing landscapes for food security in the region and globally.

Keywords Food security · Adaptation · Agriculture · Technology · Asia

7.1 Role of Agriculture to Ensure Food Security in the Asian Region

Agriculture plays an important role in Asia's food security. As income levels have improved, hunger has been reduced across the region. While the economic role of agriculture has fallen, in terms of share of GDP and contributions to job creation, other important roles of agriculture include the ability to provide greening in cities. It is also important in averting repeats of famines in the past resulting from combinations of natural hazards (droughts, floods) and state policies and in preventing food-related causes of conflict and political instability. Apart from this, agriculture

J. Ma. L. Montesclaros (✉) · P. S. Teng
S. Rajaratnam School of International Studies, Nanyang Technological University, Singapore, Singapore
e-mail: ismontesclaros@ntu.edu.sg

J. M. Pulhin et al. (eds.), *Climate Change, Disaster Risks, and Human Security*, Disaster Risk Reduction, https://doi.org/10.1007/978-981-15-8852-5_7

plays an important part in the region's cultural heritage, although the declining share of the labor force in agriculture raises questions on whether this can be sustained.

Despite its declining contribution to national GDP in many middle- and high-income Asian countries (ADB 2016a), agriculture remains an important economic activity in the lower income countries and generally continues to be a significant source of rural employment. Asia is home to approximately 60% of the world's population (UN FAO 2017a), but it only has 34% of its arable land, (UN FAO 2017b) and the Asia-Pacific as a whole has only 36% of its freshwater resources (United Nations (UN) 2017). This, together with the estimated 300 million smallholder farmers (International Fund for Agricultural Development (IFAD) 2011), means that food production, especially of staples such as rice, relies largely on small land holdings whose performance is heavily dependent on seasonal weather and on the varied capacities of these smallholder farmers to deal with unexpected disturbances to their production.

Today, Asia as a whole produces more agricultural commodities than it imports. Out of the total food supply in the region (total imports + total production), 91% of meat, 87% of cereals, 98% of vegetables, 99% of eggs, 82% of fish, and 86% of spices are produced domestically (UN FAO 2017c). At the country level though, there is still high dependence on imports for household commodities. India imports 46% of its total supply of vegetable oils (UN FAO 2017d), while Indonesia and Malaysia are among the world's largest exporters of vegetable (palm) oil, together making up 87% of global exports for palm oil in 2016 (in tonnage) (International Trade Centre (ITC) 2017). Most crucial among these are cereals, which make up the majority (57%) of Asian diets (UN FAO 2017e).[1] Figure 7.1 shows that Malaysia is still 73% import-dependent for cereals, while Brunei and the Maldives are entirely import-dependent (100%).

Given per capita income growth, hunger across all regions has fallen since 1990. Based on data from the Asian Development Bank (ADB), over 707 million people in the Asia-Pacific have been brought out of poverty since the start of the millennium (ADB 2016a). However, as Fig. 7.2 shows, some regions have experienced recent increases in the share of the hungry, especially Southeast Asia (e.g., Myanmar increased from 14.4 to 16.9%) and Western Asia (e.g., Yemen increased from 25.2 to 28.8%) (UN FAO 2017e).

The agri-food picture in Asia has also changed much in response to changes in its number of middle-class people and their rising economic power at the household level. Foremost of this is the increased consumption of animal protein, which has resulted in Asia being now the main importer of the world's supply of soybeans and corn. China alone imports 70–80 million tons of soybean each year to meet the needs of its animal industry, projected to increase to up to 112 million tons by 2023 (Gale et al. 2015).

An emerging trend in Asia is that of farming in and around cities (urban and peri-urban agriculture), contributing complementary food sources as well as greenery for urban environments. Agriculture also provides green spaces within the core of urban

[1]Note: Data based on 3-year average of 2009–2011, the latest available data for Asia.

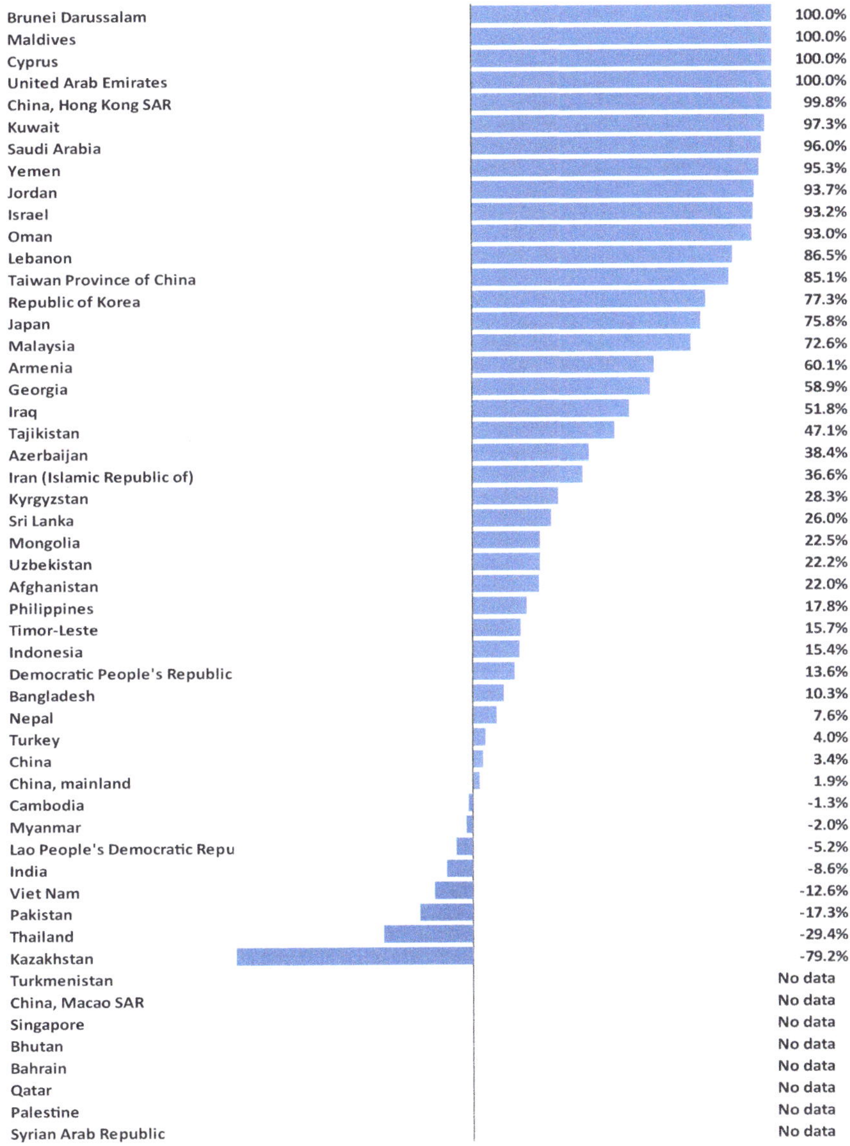

Fig. 7.1 Cereal import dependency ratio (%) (3-year average) in Asia. Countries with negative cereal dependency ratios are those that are net exporters of cereals. Data were not available for some countries in the UN FAO website. Latest data available for the countries above are for the period 2011–13 (UN FAO 2017e)

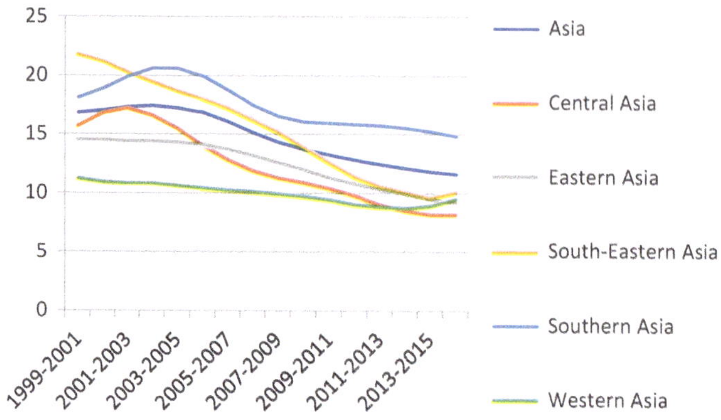

Fig. 7.2 Prevalence of undernourishment (%) (3-year average) (UN FAO 2017e)

centers, as a means toward sustainable landscape planning with not only environmental but also cultural and socioeconomic benefits (Carlet et al. 2017). In Seoul, such a movement started to take shape in the mid-2000s, with the cultivation of edible vegetables by urban dwellers in unused land, after a key policy was introduced in 2005 that offered tax benefits to farmers (Gee-hyun 2014). These offer cities some remedy from the heat-island effect, wherein densely populated areas are a few degrees warmer than their surroundings (Teng and Escaler 2014). Having agriculture enhances the ability of cities to cope with this, thereby also reducing energy costs for cooling (Hiemstra et al. 2017).

Historically, major parts of Asia have suffered from famines triggered by natural weather disruptions. The Bengal Famine in India in 1943 took at least 1.5 million lives and possibly even more than 3 million lives (data are scarce as the government at that time had an interest in making it appear that no famine was occurring) (Mukerjee 2014). Natural weather disruptions contributed by shrinking the *aman* rice harvests (during winter) to 83% of the average over the previous 5 years (aman harvests make up 72% of the yearly harvest [Sen 1981]). In the 1970s, 30 million lives (Kung and Lin 2003) were claimed in China, alongside combinations of natural disruptions, including flooding in the Yellow River in 1959 and droughts, typhoons, and disease infestations in 1960/61. In addition, the Cultural Revolution or the Great Leap Forward initiated by Mao Zedong led to reallocation of scarce food resources, with two-thirds of the exports of Chinese provinces diverted to urban centers like Beijing, Tianjian, and Shanghai and the industrial province, Liaoning (Kung and Lin 2003).

7.2 The Link Between Agricultural Performance and Food Security in Asia

Given the importance of agriculture and food security to the region, we now frame the challenge of food security.

Food security is defined by the Food and Agriculture Organization of the United Nations as "(a) situation that exists when all people, at all times, have physical, social and economic access to sufficient, safe, and nutritious food that meets their dietary needs and food preferences for an active and healthy life" (UN FAO; IFAD; World Food Programme (WFP) 2015).

The section goes deeper into the four key interconnected facets underpinning food security as identified in this definition, in particular, food availability, economic food access, physical food access, and food utilization. It provides the links between each of these facets and food security, supported by evidence from recent literature. Why, for instance, does a significant share of the world's population still remain hungry even as more calories of food are already produced than is required (on a per-capita basis)?

The lenses provided in this section are key to understanding the pathways by which climate change influences food security, which are discussed in succeeding sections.

7.2.1 *Agricultural Performance and Food Availability*

The first facet, food availability, is defined as having sufficient food available to meet growing demand. This is the most basic/fundamental level of analysis as hunger cannot be solved without a sufficient quantity of food produced to meet consumption needs.

Food availability is measured based on the amount of food produced, in comparison to total demand, as measured in calories of food supplied/needed per person. This is defined as the ratio of available food (in calories) to the average dietary energy requirement. Based on this indicator, Asia has an average dietary energy supply adequacy (ADESA) of 118% (3-year average from 2014 to 2016); in other words, it has excess food supply of close to one-fifth of the average dietary energy requirement. Subregions East Asia, Western Asia, and Central Asia, and Southeast Asia have ADESAs greater than the regional average (127, 121, and 120, and 119%, respectively), while South Asia has average ADESA below Asia, at 110% (UN FAO 2017f).

There are two sources of food, in particular, imports and domestic production. On one hand, there are a few economies that, even without significant agriculture, are still able to secure food through trade. Singapore, for instance, only produced 26% of the eggs it consumes, 8% of fish, and 8% of leafy vegetables in 2015 (Agri-Food and Veterinary Authority, Singapore 2015).

However, because all imports draw from the global pool of commodities for export, it is important that production as a whole keeps up with demand of a global population of more than 9.77 billion by 2050 (United Nations Department of Economic and Social Affairs 2017). For majority of countries in Asia, domestic production still makes up approximately 80% of total consumption, making the performance of the agricultural sector critical (UN FAO 2017c).

Next, production is driven by the space allocated for production and the yield or amount of harvest for every unit of space planted/allocated for production. The production base for agriculture has already been shrinking over the past decades. Extensive land-based agriculture has been traditionally done in rural areas. However, World Bank data show that the share of rural populations has declined from 66% in 1961 to 46% in 2016 (World Bank 2017). This can be attributed to the low value added per worker in agriculture. While this has increased from USD 650 to USD 1,200 per worker since 1980, it has never been anywhere close to value add per worker in industry, which ranged from USD 11,000 to USD 14,000, and in services, which ranged from 15,000 to 18,000 in services (World Bank 2017).

As a result, the challenge of increasing yields has been crucial to increasing production amid limited, and even shrinking, land areas for it. Research by the Global Rice Science Partnership (GRiSP) has shown that in rice alone, yields would need to grow by 1–2% annually to keep up with growing demand and amid unlikely expansion of areas allocated for rice production beyond 2020 (GRiSP 2013).

7.2.2 Agricultural Performance and Economic Food Access

Regardless of the amount of food available through either imports or production, people may still not be able to consume the required amount of food if they cannot afford it. Economic access is a key factor, wherein wages and the support given by governments need to provide the people with sufficient purchasing power to obtain needed quantities of food.

Nobel Laureate Amartya Sen has shown how misalignment of prices, incentive policies, and government and private sector behavior may prevent economic access to food (Sen 1976). He compared India's domestic price of rice, a basic staple in that country, with prices of other commodities which he used as a basis for wages of workers in other sectors (i.e., cloth, umbrellas, fish, milk, haircuts). If total revenues from the other sectors are not growing at the same pace as food prices, then wage rates in those sectors will shrink relative to food prices. Explaining the Great Bengal Famine, Sen's data show the price of rice increasing by more than 500% from December 1941 to June 1942, whereas the prices of commodities in other aforementioned sectors were declining. In such a case, it was natural that people could be less able to afford rice (Sen 1981).

On the other hand, food prices may increase for several reasons. First is if there is a shortage in the commodity, such as during disruptions in production (e.g., droughts, floods). Food prices may also increase if the cost of producing or of acquiring food through imports increases or if the price of goods from international trade becomes more expensive, such as during shortages.

In countries where there are monopolies over inputs to production, such as fertilizers or seeds, there can also be deliberate or insufficient capacity to increase the amount of inputs available. This brings up the price of inputs to the benefit of input producers, thereby increasing input costs. The other way the prices of production inputs increase is when factors such as climate change reduce rice yields, requiring farmers to adapt by making use of more fertilizers or better technology.

In both cases, producers would then increase the price of food sold to remain profitable, to the detriment of consumers, among which the poorest may not be able to purchase them.

Last, when international prices of traded food increase, a "pass through" effect occurs such that domestic prices increase (Durevall and van der Weide 2014). This is when local producers decide to increase the amount of a commodity they sell in international markets to make larger profits at those higher prices and there develops a shortage in domestically available rice.

7.2.3 Agricultural Performance and Physical Food Access

Next, even if food is available and affordable globally, the countries that need it may not always have physical access to it, especially if it needs to be transported from foreign sources. During the 2007–2008 crisis, countries like Thailand and India banned the export of rice in order to meet domestic demand (Timmer 2009). Trade routes may also be impeded, such as in the Middle East, where Qatar has suffered from transport route inaccessibility from neighboring countries (Atkinson 2017).

There can also be delays in transporting food, such as during the port strikes in Brazil in May 2017 wherein potentially 2,000 containers of cargo, which include perishables like food, were held up at ports. A 2-day port strike would require additional 2–3 days just for clearing cargo at the ports (Ward 2017). These delays could lead to spoilage of vegetables, eggs, and even chilled products, unless appropriate storage mechanisms are used. A study by the European Food Safety Authority (EFSA) Panel on Biological Hazards (BIOHAZ) requested by the European Commission, for instance, shows an increase in food-spoiling bacteria if meat is stored at 7 degrees for 11 days (BIOHAZ 2016).

Next, even within countries, physical access to food may be limited in the event that suppliers practice hoarding behavior. The rationale behind this is that, during times of food price inflation, suppliers anticipate higher prices of food in the future. This leads them to stock up on the rice, waiting for prices to increase, before finally selling their stocks at higher prices. The impact of this is that, until prices are high enough, suppliers of food may not release as much rice as is needed by the market (Timmer 2009).

7.2.4 Agricultural Performance and Food Utilization

Last is food utilization, which involves the way by which food addresses hunger and brings about desirable nutritional and health outcomes among the populace.

On one hand, the lack of available food drags down performance in food utilization. Table 7.1 shows that the highest level of undernourishment can be seen in the Democratic People's Republic of Korea (40.8%), Tajikstan (30%), and Afghanistan (99%), where the ADESAs are all below 100%. Undernourishment is measured by the share of the population with daily per capita consumption (in calories) below the dietary energy requirement (UN FAO 2017f). On the other hand, having sufficient food available is no guarantee that undernourishment will be alleviated. Specifically, the rest of Asia has ADESAs greater than 100%, but undernourishment is still above 0% in those countries. This is especially so in the case of Yemen, Iraq, Timor-Leste, and Sri Lanka, where more than a fifth of the national population is undernourished.

Available food may not contain the needed nutrients that allow for productive and healthy living, given concerns of food safety, nutrition, and unhealthy consumption habits.

Food safety was touched on earlier. Food-borne diseases that emerge from improper feeding practices or excessive use of plant chemicals can worsen malnutrition, and thus impact on food utilization, especially among the poor (Trench et al. 2011). These sprout in different parts of the supply chain, from improper food-growing practices, to spoilage during transport and storage before reaching consumers, to food stocking and storage behavior in households, and finally to consumption habits of households.

Data from the World Health Organization (WHO) show the extent of food-borne diseases in the region. Table 7.2 shows the number of disability-adjusted life years per 100,000 population across the major regions of the WHO, which include Asian countries, for food-borne diseases.[2] It shows that Southeast Asia suffers the most from invasive infectious disease agents, which include hepatitis A virus, *Listeria monocytogenes*, *Mycobacterium bovis*, *Salmonella paratyphi A*, and *Salmonella typhi*, among others. It also shows that after Africa (AFR) and the Eastern Mediterranean region (EMR), Southeast Asia (SEAR) has the largest vulnerability to diarrheal disease agents, which include *E. coli* (enteropathogenic, enterotoxigenic, and Shiga toxin-producing) and non-typhoidal *Salmonella enterica*, among others (UN WHO

[2]Regions that include Asian countries are (1) Western Pacific Region (WPR) A: Australia, Brunei Darussalam, Japan, New Zealand, Singapore; (2) WPR B: Cambodia, China, Cook Islands, Fiji, Kiribati, Lao People's Democratic Republic, Malaysia, Marshall Islands, Micronesia, Federated States of, Mongolia, Nauru, Niue, Palau, Papua New Guinea,Philippines, Republic of Korea, Samoa, Solomon Islands, Tonga, Tuvalu, Vanuatu, Viet Nam; (3) Southeast Asian Region (SEAR) B: Indonesia, Sri Lanka, Thailand, Timor-Leste; (4) SEAR D: Bangladesh, Bhutan, Democratic People's Republic of Korea, India, Maldives, Myanmar, Nepal; (5) Eastern Mediterranean Region (EMR) B: Bahrain, Cyprus, Iran, Islamic Republic of, Jordan, Kuwait, Lebanon, Libyan Arab Jamahiriya, Oman, Qatar, Saudi Arabia, Syrian Arab Republic, Tunisia, United Arab Emirates; (6) EMR D: Afghanistan, Djibouti, Egypt, Iraq, Morocco, Pakistan, Somalia, Sudan, and Yemen (UN WHO 2017c).

Table 7.1 Average dietary energy supply adequacy and prevalence of undernourishment in Asia (UN FAO 2017f)

	Average dietary energy supply adequacy or ADESA (%) (3-year average)	Prevalence of undernourishment (%) (3-year average)		Average dietary energy supply adequacy or ADESA (%) (3-year average)	Prevalence of undernourishment (%) (3-year average)
Asia	**118**	**12**	**Southeastern Asia**	**119**	**10**
Central Asia	**120**	**8**	Brunei Darussalam	125	No data
Kazakhstan	137	No data	Cambodia	112	15
Kyrgyzstan	120	6	Indonesia	123	8
Tajikistan	97	30	Lao People's Democratic Republic	106	17
Turkmenistan	120	6	Malaysia	127	No data
Uzbekistan	117	6	Myanmar	112	17
Eastern Asia	**127**	**9**	Philippines	117	14
China	129	10	Singapore	No data	No data
China, Hong Kong SAR	132	No data	Thailand	113	10
China, Macao SAR	132	6	Timor-Leste	102	27
China, mainland	129	10	Viet Nam	122	11
China, Taiwan Province of	123	5	**Western Asia**	**121**	**10**
Democratic People's Republic of Korea	88	41	Armenia	121	4
Japan	112	No data	Azerbaijan	131	No data
Mongolia	106	20	Bahrain	No data	No data
Republic of Korea	135	No data	Cyprus	107	5
Southern Asia	**110**	**15**	Georgia	118	7
Afghanistan	99	23	Iraq	110	28

(continued)

Table 7.1 (continued)

	Average dietary energy supply adequacy or ADESA (%) (3-year average)	Prevalence of undernourishment (%) (3-year average)		Average dietary energy supply adequacy or ADESA (%) (3-year average)	Prevalence of undernourishment (%) (3-year average)
Bangladesh	109	15	Israel	160	No data
Bhutan	No data	No data	Jordan	132	4
India	109	15	Kuwait	140	No data
Iran (Islamic Republic of)	129	6	Lebanon	125	5
Maldives	120	9	Occupied Palestinian Territory	No data	No data
Nepal	121	8	Oman	123	6
Pakistan	110	20	Qatar	No data	No data
Sri Lanka	113	22	Saudi Arabia	139	4
			Syrian Arab Republic	No data	No data
			Turkey	158	No data
			United Arab Emirates	124	4
			Yemen	101	29

2017c), unlike the Western Pacific Regions (WPR) and the European Region (EUR) (the suffixes after the region represent different levels of disease burden, with "A" indicating very low child and adult mortality, and "D" indicating high child and adult mortality).

In this regard, agricultural performance in terms of safety in operations could make a big impact in addressing food-borne diseases. One among the emerging challenges in this regard is that of antimicrobial resistance, where diseases develop resistance to the very antibiotics which were meant to treat them. This results when antibiotics are overused, for the purpose of promoting faster growth of livestock; when antibiotics are applied wholesale to entire flocks; or when antibiotics are used to treat infections that recur (Wall et al. 2016).

Next, nutrition security requires that sufficient amounts of needed nutrients are consumed by the populace. Linking to the earlier facets, the implications of lack of access to food are strongly felt in children. Twenty percent of children under five in 19 developing countries in the region suffer from stunting (Asian Development

Table 7.2 Food-borne disability-adjusted life years per 100,000 (UN WHO 2017a)

	SEAR B	SEAR D	WPR A	WPR B	EMR B	EMR D
Diarrheal disease agents	1,649	1,676	119	322	2,180	2,243
Invasive infectious disease agents	724	630	51	252	168	282
Helminths	121	176	27	574	29	115
Chemicals and toxins	0	110	0	0	0	15
All hazards	2,592	2,666	202	1,169	2,404	2,710

	AFR D	AFR E	AMR A	AMR B	AMR D	EUR A	EUR B	EUR C
Diarrheal disease agents	2,633	2,502	90	298	304	133	78	96
Invasive infectious disease agents	226	222	52	153	179	45	68	73
Helminths	368	344	10	129	637	2	15	20
Chemicals and toxins	0	3	1	0	0	18	1	17
All hazards	3,266	3,107	156	597	1,146	205	168	210

Bank (ADB) 2016a). Anemia among children under five is highest in Pakistan (61%), Yemen (59.3%), and India (59%), as of 2011 (UN FAO 2017f). On average, there is also a food deficit of 90 kcal/person per day in Asia, and India shows a 25% prevalence of iodine deficiencies, which is half of those in Kyrgyzstan (50.5%) and Mongolia (52.8%), as of 2006 (latest available data) (UN FAO 2017f).

Last, given the rapid pace of urbanization occurring globally, with more than two-thirds of the world expected to live in cities by 2050, it is important to note that diets and food consumption behavior of the populace are changing rapidly. The majority of urban dwellers do not have the capacity to grow their own food, and lifestyles fuel a trend toward consuming processed, "fast" food. This leads to greater obesity, with evidence that, in 2010–2014, obesity increased more than 20% inIndonesia, Malaysia, the Philippines, Singapore, Thailand, and Vietnam (Food Industry Asia 2016). Obesity brings with it health risks, which include coronary heart diseases, ischemic stroke, diabetes, as well as risks of cancer (including breast, colon, and prostate cancer) (UN WHO 2017b). Increasingly, international groups such as the UN FAO and the International Food Policy Research Institute (IFPRI) are advocating that proper nutrition in this "food utilization" part of food security is important to ensure healthy societies by reducing the incidence of non-communicable diseases, notably exemplified by diabetes. Agriculture as a means to produce food with improved nutritive value is also the main thrust of successful programs such as "Harvest Plus," which has produced new sweet potato varieties with higher Vitamin A (Jones and de Brauw 2015).

7.3 Agricultural Performance, Climate Change, and Natural Disasters

Agricultural performance has been shown to impact on different facets of food security. This section explores the linkages between climate change impacts and agricultural performance, based on historical evidence as well as scholarly interpretations of such.

For a continent that holds approximately 60% of the world's population (UN FAO 2017g), Asia only has 34% of the world's arable land (UN FAO 2017a), and 36% of its freshwater resources (UN 2017), both essential for agriculture. Eighty-seven percent of the world's smallholder farmers also reside in the Asia-Pacific, making it a region with one of the lowest per farmer agricultural land (Thapa 2009). All this underlies the importance of stability in the agricultural sector if food and nutrition security is to be maintained. It is in this context that the impacts of climate change, both in the long term (yields declining) and short term (reduced production due to droughts/floods) need to be considered.

7.3.1 Changes in Crop Yield Growth

While yields continue to grow in Asia, it can be seen that they have not grown as fast as they did in previous periods. In particular, when we compared the growth rate in yields in 1990–2010 with the previous 20-year period, 1970–1990, we found that yield growth in cereals was slower in the 1990–2010 period. The same applies to coarse grain, fiber crops, jute, oilcakes, oils, and pulses (Fig. 7.3).

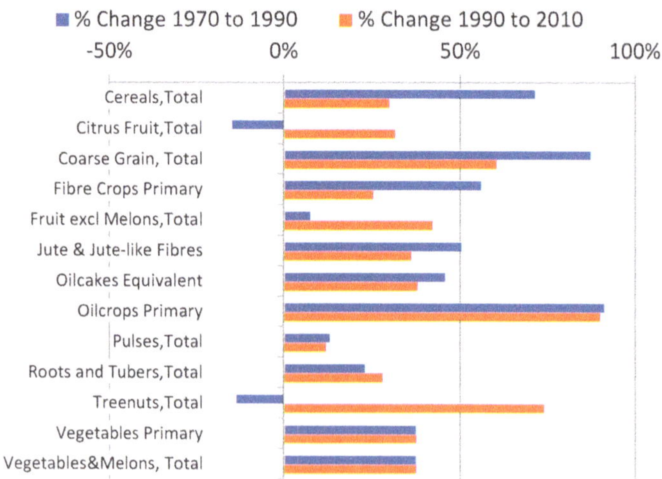

Fig. 7.3 Changes in crop yield growth in Asia (UN FAO 2017h)

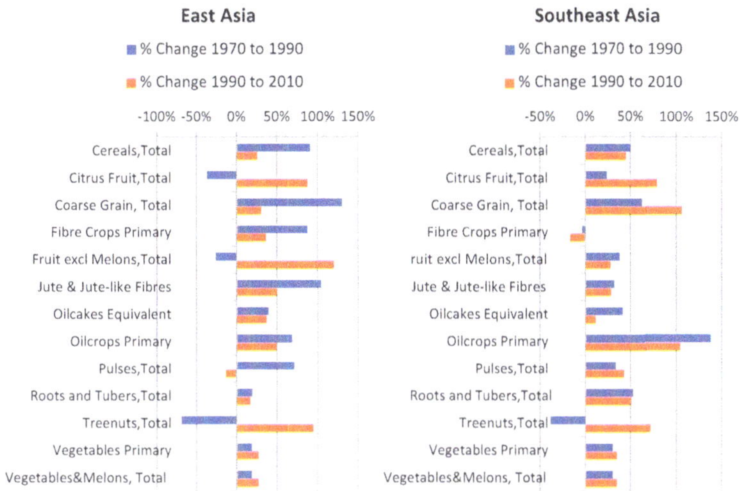

Fig. 7.4 Changes in crop yield growth in East and Southeast Asia (UN FAO 2017h)

Disaggregating yield growth trends by region, the figures show the changes in yields across East, Southeast, South, and Western Asia. In East Asia, there are consistent trends of declines in the growth rate in yields in cereals, coarse grains, and roots, among others. In contrast, though, perennial plants, such as fruits, seem to show trends of increasing yields. In addition, vegetable production yields grew faster in this subregion.

In Southeast Asia, coarse grains grew faster in the 1990–2010 period than in the 1970–1990 period, indicating, to some extent, technology transfer from East Asia. Moreover, yield growth for vegetables as well as perennial plants accelerated in 1990–2010 (Fig. 7.4).

Next, unlike in East and Southeast Asia, vegetable production yield growth decelerated in 1990–2010 in South and Western Asia. Moreover, citrus fruits in South Asia have been growing more slowly, while yields in pulse production have been growing faster in Western Asia (Fig. 7.5).

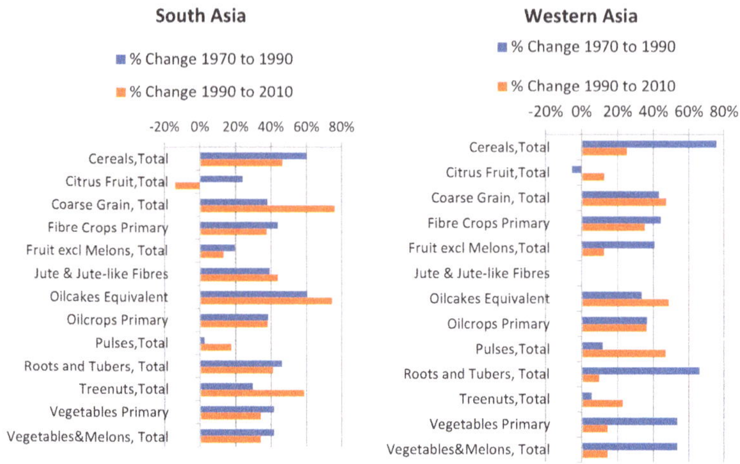

Fig. 7.5 Changes in crop yield growth in South and Western Asia (UN FAO 2017h). Data are not available for jute and jute-like fibers in Western Asia

7.3.2 Potential Contributors to Decline in Yield Growth: Observed Changes in Temperature, Precipitation, Droughts, and Floods

Declines in yield growth across Asia's subregions can be interpreted as indications of different approaches to improving agriculture, impacts of climate change in the region, and measures to adapt to climate change. While there is a myriad of possible combinations of practices, climatic conditions, and adaptation measures, scientists have nonetheless tried to uncover links between changing climatic conditions and agricultural performance.

Historically, there have been observed increases in warm days (and decreases in cold days) across the region. Spatially varying trends in heavy precipitation have also been observed in Central and East Asia, while heavy precipitation has been seen to increase in some regions of North Asia, and to decrease in West Asia. Last, spatially varying trends in dryness and droughts have been observed in North, Central, and Southeast Asia, with a tendency to increase in East Asia (Field et al. 2014).

We now go through each of these factors, beginning with temperature. Since 1910, the warmest land temperature anomalies in Asia were seen in 2007 (3.44 °C); in fact, 7 out of the top 10 warm land temperature anomalies since 1910 have happened in the past 30 years. Moreover, since 1990, there were 3.5 times more warm anomalies than there were cold anomalies (NOAA National Centers for Environmental Information (NCEI) 2017a).

Crop production is highly sensitive to temperature, with extreme cold temperatures at night time and extreme hot temperatures in day time being damaging to plant growth. For instance, a controlled experiment at the International Rice Research

Institute (IRRI) farm in Los Baños, Philippines, showed a 1% increase in night time temperature during the dry season, causing a 10% fall in rice yield (Peng et al. 2004). A study in India with 292 different rice cultivars likewise showed that, with every additional growing degree hour (GDH) in temperatures above 33 °C, spikelet sterility increased by 0.26 percentage point (Bheemanahalli et al. 2016). The impacts are not all negative, as the latest assessment by the Intergovernmental Panel on Climate Change (IPCC) shows, but negativeimpacts were cited to be more common than positive impacts (Field et al. 2014).

The next potential contributors to declining yields are disruptions from droughts and floods, as agriculture is dependent on a stable supply of water. At the same time, submergence caused by floods can impede the growth of plants.

There have been spatially varying trends in dryness and drought in North, Central, and Southeast regions of Asia, based on the IPCC's regions (Field et al. 2014). Impacts of floods and droughts have been devastating in the region over the past decade, given erratic rainfall patterns (Hijioka et al. 2014). As a snapshot, substantive parts of the Association of Southeast Asian Nations (ASEAN) in 2015 were affected by both droughts (492,000 ha, mostly in Thailand and Indonesia) and floods (81,000 ha in Brunei, Cambodia, Indonesia, Laos, Malaysia, Myanmar, and the Philippines), to the detriment of rice paddy farming activity (ASEAN Plus Three Food Security Information System 2016). Changing patterns of precipitation, especially in the case of rainfed crops, in both cases of under- and oversupply of water (droughts and floods, accordingly), are exemplified in El Niño and La Niña events.

7.3.3 Potential Contributors to Decline in Yield Growth: Pests and Diseases

Different climatic variables may allow for greater likelihood of pests and diseases in agriculture (UN FAO 2008).

One example is rice leaf blast, which can kill plants at both early stages (seedling, tillering) as well as reduce grain yield by reducing the number of grains due to neck blast (IRRI 2017). Temperature changes have been found to have a significant impact on disease occurrence in warmer areas in cool subtropical zones and in cooler areas in humid tropics/warm humid subtropics, based on data in China, Japan, Korea, the Philippines, and Thailand (Luo et al. 1995).

Diseases likewise may develop resistance to changes in temperature or the plants themselves that are affected by diseases may be more susceptible to infection when plants are slow to develop their own defense mechanisms (with less moisture, and especially during droughts) (Elad and Pertot 2014). To provide a snapshot of ASEAN in 2015, Thailand experienced as much as 74,000 ha of land damaged by pests and 37,000 ha damaged by diseases, while Malaysia experienced damage of 37,000 ha and close to 20,000 ha, from both causes, respectively (ASEAN Plus Three Food Security Information System 2016).

7.3.4 Potential Contributors to Decline in Yield Growth: Limits to Productivity and Yields Posed by Soil Nutrients

Apart from direct impacts of temperature and precipitation changes on plant growth, there are also indirect impacts when combined with the impacts of climate change on nutrients in the soil.

On one hand, increasing carbon dioxide (CO_2) implies the potential for more CO_2 fertilization, wherein plants absorb the carbon in the soil and, as a result, grow faster. However, flooding-caused soil erosion over time may cause either a loss in soil nutrients or an oversupply of them, both of which are undesirable. For instance, a 2-year experiment in Kyoto shows that while elevated CO_2 concentrations have the potential to enhance yields, this impact can be limited by temperature conditions as well as the amount of nitrogen (N) in the soil (Nakagawa et al. 1993).

N and phosphorus (P) are crucial for the process of photosynthesis (converting sunlight to produce sugars from water and CO_2), the lack of which leads to slower plant growth and yield losses (Brevik 2013). But too much N deposited in soils, beyond the capacity at which it can be absorbed by the plants, allows for the growth of invasive grass, which absorbs the nutrients and blocks sunlight, at the expense of the slower growing plants (Settele et al. 2014).

7.3.5 Societal Impacts

Climate impacts on agriculture have been both long term (temperature changes) and sudden (droughts, pests, diseases) in nature. These have had pronounced impacts on the global community, taking into account the interrelationships between food availability and physical and economic access.

In 2007–2008, for instance, when droughts disrupted India's rice farming activity, the country government had to ban the export of rice to meet domestic consumption needs. Because rice is thinly traded, with a small amount sold on international trade relative to the total amount of production, and with roughly 80% of exports concentrated in just five countries, rice prices increased significantly during this period (Timmer 2009). As rice prices increased, Thailand and Vietnam likewise discouraged the export of rice, with the intent of waiting for prices to rise further before selling their grains, thereby contributing to an upward spiral of price increases (Timmer 2009).

Worst hit by the combined production and trade disruptions were the poor, who could barely make ends meet even under normal prices. Price increases caused 100–200 million people to fall below the poverty line, and 63 million more people to become undernourished (World Bank Group 2013). In 2010–2011, more modest food price inflation occurred in maize (73%), sugar (76%), and soybean oil and palm oil (54%), from June to December 2010, but these nonetheless are estimated to have led 44 million more people to live on less than $1.25 a day (Ivanic et al. 2011). These in turn led to food riots in over 60 countries, globally (*The Economist* 2009).

7.4 Evaluation of Projected Climate Change Impacts on Agriculture Going Forward

Looking forward, climate change paints a pessimistic picture of suitable temperatures and precipitation, which are needed for agricultural production. Water demand is likewise expected to increase from other sectors, while the trend of rural to urban migration is not helpful. There is also a chance that agricultural research, which has the potential to sustain yields amid these changes, may not receive sufficient support in the future, given market barriers that prevent the entry of much needed technologies. The failure to implement technologies that can arrest climate impacts can cause declining yields in key staples, such as rice, wheat, and maize, and potential decline in per capita calorie consumption. These impacts may be worsened by continued rural-to-urban migration, which has historically had a poverty angle to it, with urban poverty increasing amid urban migration.

7.4.1 Projections on Climate Change Moving Forward

A 34% increase in global population is expected by 2050 and, combined with income growth, this leads to an estimated 70% growth in global demand (UN FAO 2015). Projections on future production environments for meeting this demand are not very encouraging, however. Based on IPCC's subregions, heavy precipitation is likely in North Asia, East Asia, and Southeast Asia, while both frequency and intensity of heavy precipitation are projected to increase in South Asia. More hot days are projected across all subregions as well. Last, there are still inconsistent signals of change that can be generalized to project future trends in dryness and drought at the level of the subregions (Field et al. 2014).

7.4.2 Analysis of IPCC Projections on Climate Change: Impacts on Agriculture and Prices

Changes in climate will have particular impacts on yields in agricultural production in the region. Estimates from IFPRI show 8–14% declines in rice yields in South Asia (SA), and East Asia and the Pacific (EAP), although yields in Europe and Central Asia (ECA) decline by less than 1%. In the case of wheat production, EAP is expected to increase in yields, by close to 2%, but massive declines of 43–50% are expected in SA and ECA. Last, maize production yields are expected to decline by 9–19% in SA and by 19–38% in ECA, but there is a range of uncertainty on whether impacts will be positive or negative in EAP, ranging from an increase of 9% to a decline of 13% (Nelson et al. 2009).

The implications of these, combined with global demand growth and changes in dietary patterns, are that prices of agricultural commodities could increase further by 2050, in rice (by 113–121%), wheat (by 171–194%), and maize (by 148–153%) (Nelson et al. 2009).

7.4.3 Water: Trade-Offs Between Agricultural and Non-agricultural Uses of Water

Most of agricultural water use for crops is from rainfed agriculture. However, with forecast instability in precipitation, irrigation will be in greater demand; the share of agricultural areas equipped with irrigation increased from 10.6% in 1960 to 14.2% in 2014, and the increase is more pronounced in South Asia, from 13.8% in 1961 to 35.1% in 2014, close to a threefold increase (UN FAO 2017a).

This will not be without its challenges, as growing population and income will also lead to competing uses of water, across agricultural products and between agricultural and non-agricultural uses, i.e., energy/industry purposes. Today, 80% of the region's water is utilized for agricultural purposes, with the remaining 20% being used for non-agricultural purposes, including industry and energy production (ADB 2016b). Agricultural demand for water is expected to increase by 55%, from 3,500 km^3 in 2000 to 5,500 km^3 in 2050, to be accompanied by a 400% increase in water demand for manufacturing purposes, a 140% increase for electricity generation purposes, and a 130% increase for domestic water use (OECD 2012).

7.4.4 Impacts on Agricultural Businesses Along the Supply Chain

According to the director general of IFPRI and the former president of the International Fund for Agriculture and Food (IFAD), "A commitment to treat smallholder farms as viable businesses is key to unlocking the sector's potential to contribute to the broader development agenda" (Nwanze and Fan 2016: 14). This statement is powerful, as smallholders occupy 53% of agricultural land globally (Graeub et al. 2016).

If trends of climate change translate to increased costs of production and lower productivity to meet higher costs, and with more occurrences of destabilizing extreme weather events, it is possible that more farmers, in particular smallholders, may exit the agriculture scene in favor of other sectors.

A downward spiral is also possible, if farmers resort to over extraction of water/improper use of agrochemicals to make up for yield declines from climate change. Doing so would not only be economically suboptimal, as there is a limit to the amount of agrochemicals that can be efficiently applied in the soil; it could also lead to further degradation of soil quality and ecosystem services, as well as water pollution resulting from soil erosion (Nwanze and Fan 2016).

7.4.5 *Impacts on Rural-Urban Migration and the Future of Cities*

The impacts of climate change are worsened when coupled with rapid migration from rural to urban areas. Forty-eight percent of Asians live in cities, but this is projected to increase to 64% by 2050 (United Nations Department of Economic and Social Affairs, Population Division 2015). The problem is that, alongside the increase in urbanization from 38% in 1990 to 43% in 2008, urban poverty also increased from 15.7 to 21.9% in the same period (ADB 2014).

If trends continue, we might see greater urban poverty, which worsens economic access to food, especially since the urban poor do not have the same access to land for farming, and thus have few alternative food sources during rapid food price inflation, such as during the 2007–2008 crisis when international food prices doubled, with cereal prices more than doubling (World Bank Group 2013). Parallel to this is the challenge of physical access to food as a more urbanized Asia will mean that more food is transported from rural to urban areas.

7.4.6 *Changing Locations of Food Production*

The changing dynamics of temperatures and precipitation, among others, have the potential to shape the landscape of areas suitable for crop production. In China, spatial trends have been found in areas suitable for planting sugarcane, which provides up to 90% of sugar sources, under different representative concentration pathways used for testing climate change scenarios. Mean temperature in the coldest quarter, the range of annual temperatures, and minimum temperature in the coldest month were found to be the top three significant factors shaping future areas suitable for crop production. Predictions, moving forward, are for areas in Guangxi, Yunnan, and Guangdong to have more suitable planting areas in the future (Mi et al. 2017).

Apart from the impacts on locations for crop production, climate change also has led to mass migrations of diverse species of wildlife. Essentially, temperature changes lead to changes in availability of nutrients, plant availability, and ultimately, changes in abundance of fish (Roessig et al. 2004). Temperate regions, globally, are predicted to see northward shifts in Pacific salmon, with decreasing populations and changes in size; for instance, sockeye salmon is expected to shift toward the sub-Arctic; in tropical regions, specific species of sea bass and snappers are also predicted to shift to new areas (Roessig et al. 2004; Ding et al. 2017).

7.4.7 *Impacts on Global Food Supply and Demand*

Negative impacts of climate change can impede the potential of the agricultural sector to catch up with the other sectors in terms of profitability and contribute to further migration of workers away from rural agriculture.

Looking at both growing populations and supply changes in the future, a key indicator is the amount of daily per capita calories available. Simulations by IFPRI (Nelson et al. 2009: 11) show that, in the absence of carbon fertilization effects, daily per capita calorie availability could improve from 2,424 kcal/day in 2000 to 2,660 kcal/day by 2050 in the case of South Asia, but that climate change could reduce this to 2,226 kcal/day by 2050, undoing the benefits of development over the period. Similar impacts are seen in the other regions, which include Asian countries (East Asia and the Pacific, Middle East and North Africa, and Europe and Central Asia).

7.5 Policy Areas to Consider in Light of Postulated Future Impacts of Climate Change on Agriculture

Global and regional approaches to food security will need to be transformed at a pace that is congruent with the manner that the landscape for food security is evolving, given climate change and related factors. A few issues and solutions are suggested here to address the gap in penetration of new technologies in areas where they can make an impact; to mitigate risks of these technologies; to encourage sustainable water consumption behavior; and to draw participation of more actors in the supply chain as well as the youth in cities.

7.5.1 *Develop Coherent Climate Change Adaptation Strategies in Collaboration with the Private Sector*

A number of technologies and approaches for addressing climate change have already been identified. A survey published by the Boston Consulting Group and AgFunder, an online investment platform for agriculture, shows that big data are the top priority among executives (Walker et al. 2016). The FAO's Adapt program has also prescribed examples such as sustainable and climate-smart production practices for crops, livestock, forestry, and fisheries; addressing maladaptive practices such as the promotion of monoculture over biodiversity; multidisciplinary research on climate impacts across sectors/activities; and developing climate monitoring and impact assessment capacities at the national/subnational level (UN FAO 2011). The onus is now on governments to help mainstream these practices and to draw collaboration among the sectors that impact on/are affected by climate change. The policy areas that follow give further details on how adaptation can be implemented.

7.5.2 Encourage Site-Specific Experimentation for New Cultivars of Crops to Support Climate-Smart Agriculture

Climate-smart crop varieties or crop varieties that have increased resilience to droughts, floods, and temperature changes constitute one of the best tools for smallholder farmers to combat climate change effects. Conventional breeding has allowed for new varieties of crops that maintain crop yields amid stressors such as prolonged droughts, submergence (by allowing stomatal closing during times of submergence), and temperature fluctuations, all of which are predicted to increase in frequency.

The IRRI, for instance, has worked to develop rice varieties that can withstand prolonged submergence and drought and has assessed rice varieties and rice innovations that are crucial in meeting future rice consumption needs of a growing population (GRiSP 2013). An outstanding example of a climate-smart crop variety is the submergence-tolerant rice variety Swarna-Sub1, now planted extensively in eastern India, which contains the *sub1* gene that allows rice plants to tolerate submergence of up to 15 days (Singh et al. 2013). Different topographies (slopes, depressions, summits) under different farming systems and management practices influence crop biomass (Ladonia et al. 2016).

To maximize the use and effectiveness of new varieties, more experimentation will need to be permitted and conducted on conventionally bred crops, thereby bridging information gaps on the suitability of crops in different local farming systems in target countries. Findings on yield benefits to crops and impacts on environments and ecosystems in particular areas can then serve as bases for regulating the distribution and deployment of these crops and for providing recommended management practices that maximize yield improvement in various growing environments (Montesclaros et al. 2019).

7.5.3 Develop Science-Based Policies and Regulatory Systems for Novel Technologies (e.g., Biotechnology, Genetically Modified Organisms, and Gene-Edited Varieties)

Beyond conventional breeding, novel technologies such as genetic modification and gene editing have been leveraged to develop new traits for agronomic and consumer benefits. Biotechnologies such as genetic engineering offer powerful tools to create new plant types that conform to the requirements of climate-smart agriculture.

The challenge lies in whether research and development (R&D) on these technologies can be scaled. This is in part because of dogmatic, rather than science-based, interpretations of biosafety provisions surrounding agricultural technologies such as biotechnology (Kent 2004). There is presently a lack of comprehensive public understanding of the nature of genetic modification, which has led to negative perceptions such as those on health risks (toxicity, allergenicity, genetic hazards); the potential for insects and weeds to become impervious to the pest- and herbicide-resistant properties of crops; uncertain ecosystem impacts; and the potential for antibiotic traits to transfer from targeted organisms to pathogenic bacteria, making it resistant to antibiotics (Zhang et al. 2016). It should be noted that there is currently no credible scientific evidence to confirm these perceptions, as all the aforementioned concerns are addressed by governments before any biotechnology-crop is approved for release, and the history of over 20 years of use by approximately 18 million farmers worldwide has shown a strong record of safety (ISAAA 2016).

Uncertainties and the lack of public understanding have led to regulatory delays in approving the entry of genetically modified (GM) rice varieties in rice-producing countries and have further raised the uncertainty on whether investment returns can be made from these R&D efforts. There is also lack of coordinated regulatory capacity and failure to reach resolutions in addressing regulatory issues (Smyth 2017). These cause delays averaging 3.7 years for novel products introduced before 2002, and the time period has increased to 5.6 years for products in 2011 (McDougall 2011).

This raises questions on the extent to which the private sector will continue to invest in research in developing technologies that can boost production yield in the long term. Given the growing demand in the region, assessment of novel technologies will need to be revised in two ways (Teng 2008). First, they will need to be more science-based, relying on data. This can be helped by developing biosafety frameworks for assessment of risks, similar to those developed by the South African Committee for Genetic Experimentation (Van der Walt 2013). Alongside this, the net benefits of GM crops will also need to be surfaced much more through research and given due attention in assessing whether risks indeed do outweigh benefits to society.

7.5.4 Implement Targeted Water Pricing Policies and Encourage Research into Enabling Technologies

Today, there are calls for improving water pricing policies to boost conscientious and efficient use of water.

In response to overconsumption and wasteful use of water, some have recommended increasing water prices based on scarcity to curb behavior (OECD 2012). On one hand, higher water prices lead to higher production cost. Given that demand for agricultural products is less sensitive to price changes compared with that for other commodities, producers may pass on the higher cost to consumers through higher agricultural prices, leading to less economic access to food.

On the other hand, if water prices are not increased, then inefficient use of water in energy, food, or industrial production may lead to continued overuse. Today, there is still a wide gap in water use efficiency. For every cubic kilometer of water consumed, Marshall Islands and Tuvalu produce only USD 2,100 billion worth of industrial output and Cook Islands, Nauru, Palau, and Tajikstan produce only USD 5,500 billion, while countries such as Japan and Singapore produce USD 50 billion or more (ADB 2016b, c).

A delicate balance thus needs to be maintained if water sufficiency and food security are to be both achieved. Key to this will be identifying which wasteful behaviors need to be curbed by which actors and targeting corrective price measures at those. Research in addressing challenges in measuring water use, identifying appropriate prices, and comparing rates across countries (Garrido and Calatrava 2013) will be critical to implementing targeted corrective price policies.

7.5.5 Identify and Promote Inclusive Models for Viable Distribution of New Crop Varieties and Technologies

Because of the complex nature of climate change, solutions to it necessitate the use of layers of technologies that complement one another. For instance, weeds pose a significant challenge to crop growth and may significantly reduce yield. As it takes hundreds of farmer-hours to manually remove weeds, or alternatively, large amounts of energy to mechanically remove them, one solution is to apply herbicides to kill the weeds. However, herbicides may also kill the crops they are meant to protect, and this leads to the need for seeds that are immune to herbicide effects. At the same time, as certain challenges become more pronounced, such as fluctuations in temperature and in precipitation, more adaptive traits will need to be integrated into the seeds.

The cost of using layers of technologies, however, may be prohibitively high, barring smallholder farmers with limited financial means from accessing them. This prevents society from increasing production to meet growing demand, and it prevents farmers as well from being more profitable, given the rapid pace of environmental change. In Asia, which is estimated to have about 87% of the world's smallholders, this is an important consideration as these smallholders are both food producers and consumers.

More exploration is needed on new types of business models that can make agriculture more inclusive, as well as on enablers for these models. One of these is through farmer-owned businesses, which are formalized/legally incorporated farming cooperatives (Teng and Oliveros 2016). This provides farmers with greater leverage for obtaining credit and drawing investments, which can be used to fund new technologies (Montesclaros et al. 2019). There is also growth in financial technologies and platforms, such as AgFunder, that link investors with new agribusinesses or with agribusinesses wanting to venture into new technologies (Walker et al. 2016).

A step forward is to explore appropriate methods of financing procurement of inputs, alongside investment models that allow smallholders to have shares in the company. This is because uncertainties in farming incomes (e.g., from droughts/floods, price changes, pests, and diseases) affect their reliability as borrowers. As new models of crop insurance are developing in the private sector, farmers may be educated on the merits of these initiatives to provide protection to farmers as they begin to be integrated into the private/financial sector (Kalra 2013). For instance, vertical integration, where different functions (sourcing for seeds and inputs/growers, cultivation of crops/growing of livestock, transport, storage, marketing/distribution) are integrated into one company, has reduced transaction costs (Kissoly et al. 2016: Murdoch Commission 2015).

7.5.6 Assist Farmers in Adapting to Supply Chain Uncertainties/Changes

Supply chain risks are not uncommon. There are uncertainties in demand for products and future prices for those products, as well as in the supply and prices of production inputs. Supply chains are also being transformed, with booming supermarkets and less reliance on informal means of selling produce (e.g., wet markets). These imply greater challenges to cash flows, especially when supermarkets delay their payments to producers.

Farmers will need to be orientated on how to navigate this changing landscape. Education initiatives on managing inventories and cash flows may be implemented by governments, international organizations, and farmer cooperatives, not only in disseminating information, but also in paying licenses for using related computer applications (Teng et al. 2019). Alternatively, production planning may be offered as a service to farmers by the private sector.

7.5.7 Promote Urban Agriculture and Related Technologies and Professions

While it would not be wise to conceive of cities producing all their own food, agriculture in urban settings can nonetheless do more in addressing food security concerns and helping urban dwellers mitigate some of the effects of climate change on food supply and prices (Teng 2012). One way to do this is through the practice of urban and peri urban agriculture. With more people living in cities, governments may provide more room for allowing households to discover its merits and for entrepreneurs to expand their businesses in the agriculture industry.

There are already technologies that can boost yields, helping make urban production a viable business option moving forward. One innovation can be found in indoor agriculture, in which growing environments (light, temperature, water, nutrients) are manipulated in order to boost yields (Kozai et al. 2015). This requires the use of sensors to monitor the environments and variable rate input devices, which allow for controlling the amount of fertilizers and other inputs used. There are also crop analytic softwares that use data from the sensors to identify the optimal amount of inputs to use. Beyond these, there is potential for more sharing of information across farmers to provide recommendations that encompass a wider range of growing environments and experiments (Teng 2017). Complementing these are developments in vertical farming that boost productivity as many times as there are layers of crops.

However, the newness and capital-intensive nature of these technologies increase the level of uncertainty of whether it would indeed be viable to invest in them. Model cases and best practices in business planning may be disseminated, as was done by the city of San Francisco, wherein the government developed a financial model for urban agriculture that could be customized by farmers and "farmtrepreneurs" for corporate planning purposes, in order to address uncertainties in the financial viability of urban farming (Ganguly et al. 2011). This was subsequently adapted to an Asian country's context in assessing the viable vegetable production quantity for indoor farms in government-tendered land in Singapore (Montesclaros et al. 2018). Zoning policies, for industrial space and building use, may also be adjusted to allow for prudent application of indoor agriculture initiatives.

Smallholders may not be expected to own entire large-scale indoor farms, also known as "plant factories," which employ all these technologies (Kozai et al. 2015). They may nonetheless benefit from some component technologies, such as environment sensors (especially for plants grown on rooftops) and drones that can deliver foliar fertilizers or pesticides more effectively.

For this to happen, though, these technologies will need to be affordable and accessible. This can be realized if markets for supply/provision of these technologies are competitive, with barriers to entry eradicated, so that these technologies can be made available at competitive market prices to smallholders. Alternatively, smallholders may offer their services as input providers (e.g., converting their farms to produce biofertilizers or biopesticides) or produce a vegetable mix that complements what is being grown in indoor farms. Examples include growing plants like climbing beans, which require more space, such as the angled bean, *Psophocarpus tetragonolobus* (Lepcha et al. 2017); they may also produce flowers that can be consumed as vegetables, such as the climber Yellow Cowslip creeper, *Telosma cordata* (National Parks Board Singapore 2013). By leveraging on these, the produce from modern plant factories can be augmented for improved nutrition vegetable packages sold to consumers. A sector or cluster approach can be used for defined urban areas so that a mixture of high-tech controlled-environment plant factories and open-environment farms can coexist to collectively form a sectoral supply chain with multiple players (Montesclaros and Teng 2018).

Governments and/or the private sector may also aid smallholders by providing them with friendly loan terms/interest rates. Educational grants may be provided to encourage more youth to see their future in agriculture, given growing demand globally. Support may also be given to training initiatives that allow related industries to customize their range of services provided in order to support agriculture, including supply chain management, product prototyping, IT adaptation (e.g., internet of things or IOT in agriculture) (Montesclaros et al. 2019), quality assessment, and other professions/industries whose relevance to agriculture is yet to be discovered.

References

ADB (Asian Development Bank) (2014) Urban poverty in Asia. ADB, Manila (Philippines). http://www.adb.org/publications/urban-poverty-asia. Accessed 29 August 2017

ADB (Asian Development Bank) (2016a) Key indicators for Asia and the Pacific 2016. ADB, Manila (Philippines). https://www.adb.org/publications/key-indicators-asia-and-pacific-2016. Accessed 29 August 2017

ADB (Asian Development Bank) (2016b) Asian water development outlook 2016. ADB, Manila (Philippines). https://www.adb.org/sites/default/files/publication/189411/awdo-2016.pdf. Accessed 28 August 2017

ADB (Asian Development Bank) (2016c). Asian water development outlook 2016: description of methodology and data. ADB, Manila (Philippines). https://www.adb.org/sites/default/files/publication/222676/awdo-2016-methodology-data.pdf. Accessed 28 August 2017

Agri-Food and Veterinary Authority, Singapore (2015) The food we eat. http://www.ava.gov.sg/explore-by-sections/food/singapore-food-supply/the-food-we-eat. Accessed 29 August 2017

ASEAN Plus Three Food Security Information System (2016) ASEAN early warning information no. 16. ASEAN Plus Three Food Security Information System. http://www.aptfsis.org/index.php/publications-m/ewi-report-sub-a. Accessed 15 August 2017

Atkinson S (2017) Qatar row: economic impact threatens food, flights and football. BBC News, 5 June. http://www.bbc.com/news/business-40156029. Accessed 29 August 2017

Bheemanahalli R, Sathishraj R, Tack T, Nalley LL, Muthurajan R, Jagadish KSK (2016) Temperature thresholds for spikelet sterility and associated warming impacts for sub-tropical rice. Agric For Meteorol 221:122–130. https://doi.org/10.1016/j.agrformet.2016.02.003

BIOHAZ (2016) Growth of spoilage bacteria during storage and transport of meat. EFSA J 14(6):4523. https://www.efsa.europa.eu/en/efsajournal/pub/4523. Accessed 28 August 2017

Brevik EC (2013) The potential impact of climate change on soil properties and processes and corresponding influence on food security. Agriculture 3:398–417. https://doi.org/10.3390/agriculture3030398

Carlet F, Schilling J, Heckert M (2017) Greening U.S. legacy cities: urban agriculture as a strategy for reclaiming vacant land. Agroecol Sustain Food Syst. https://doi.org/10.1080/21683565.2017.1311288

Ding Q, Chen X, Yu W, Chen Y (2017) An assessment of fishing down marine food webs in coastal states during 1950–2010. Acta Oceanol Sin 36(2):43–50. https://doi.org/10.1007/s13131-017-1003-5

Durevall D, van der Weide R (2014) Importing high food prices by exporting: rice prices in Lao PDR. In: Policy research working paper (poverty and inequality team, development research group, world bank group) (7119). http://documents.worldbank.org/curated/en/894041468278696193/pdf/WPS7119.pdf. Accessed 29 August 2017

Elad Y, Pertot I (2014) Climate change impacts on plant pathogens and plant diseases. J Crop Improv 28(1):99–139. https://doi.org/10.1080/15427528.2014.865412

Field CB, Barros VR, Mach KJ, Mastrandrea MD, van Aalst M, Adger WN, Arent DJ, Barnett J, Betts R, Bilir TE, Birkmann J, Carmin J, Chadee DD, Challinor AJ, Chatterjee M, Cramer W, Davidson DJ, Estrada YO, Gattuso J-P, Hijioka Y, Hoegh-Guldberg O, Huang HQ, Insarov GE, Jones RN, Kovats RS, Romero-Lankao P, Larsen JN, Losada IJ, Marengo JA, McLean RF, Mearns LO, Mechler R, Morton JF, Niang I, Oki T, Olwoch JM, Opondo M, Poloczanska ES, Pörtner H-O, Redsteer MH, Reisinger A, Revi A, Schmidt DN, Shaw MR, Solecki W, Stone DA, Stone JMR, Strzepek KM, Suarez AG, Tschakert P, Valentini R, Vicuña S, Villamizar A, Vincent KE, Warren R, White LL, Wilbanks TJ, Wong PP, Yohe GW (2014) Technical summary. In: Field CB, Barros VR, Dokken DJ, Mach KJ, Mastrandrea MD, Bilir TE, Chatterjee M, Ebi KL, Estrada YO, Genova RC, Girma B, Kissel ES, Levy AN, MacCracken S, Mastrandrea PR, White LL (eds) Climate change 2014: Impacts, adaptation, and vulnerability. Part A: Global and sectoral aspects. Contribution of working group II to the fifth assessment report of the intergovernmental panel on climate change. Cambridge University Press, Cambridge, United Kingdom and New York, NY, USA, pp 35–94. http://www.ipcc.ch/report/ar5/wg2/. Accessed 29 August 2017

Food Industry Asia (2016) South-East Asia is getting fatter, faster. https://foodindustry.asia/south-east-asia-is-getting-fatter-faster. Accessed 29 August 2017

Gale F, Hansen J, Jewison M (2015) China's growing demand for agricultural imports. Econ Inf Bull 136. Economic Research Service, U.S. Department of Agriculture. https://www.ers.usda.gov/webdocs/publications/43939/eib-136.pdf?v=42058. Accessed 29 August 2017

Ganguly S, Kujac P, Leonard M, Wagner J, Worthington Z (2011) Lively'Hood farm financial analysis: a feasibility study of commercial urban agriculture in San Francisco. San Francisco Department of the Environment. https://sfenvironment.org/download/financial-modeling-for-livelyhood-farm-a-hypothetical-urban-farm. Accessed 29 August 2017

Garrido A, Calatrava J (2013) Agricultural water pricing: EU and Mexico. OECD Publishing, Paris. https://www.oecd.org/france/45015101.pdf. Accessed 29 August 2017

Gee-hyun S (2014) City agriculture sprouts in Korea. The Korea Herald, 28 March. http://www.koreaherald.com/view.php?ud=20140328001435. Accessed 28 August 2017

Graeub B, Chappell J, Wittman H, Ledermann S, Kerr R, Gemmill-Herren B (2016) The state of family farms in the world. World Dev 87:1–15. https://doi.org/10.1016/j.worlddev.2015.05.012. Accessed 29 August 2017

GRiSP (Global Rice Science Partnership) (2013) Rice almanac, 4th edn. International Rice Research Institute, Los Baños (Philippines). http://irri.org/resources/publications/books/rice-almanac-4th-edition. Accessed 22 February 2018

Hiemstra JA, Saaroni H, Amorim JA (2017) The urban heat island: thermal comfort and the role of urban greening. In: Perlmutter D et al. (eds) The urban forest. Future City 7. https://link.springer.com/chapter/10.1007/978-3-319-50280-9_2. Accessed 29 August 2017

Hijioka Y, Lin E, Pereira JJ, Corlett RT, Cui X, Insarov GE, Lasco RD, Lindgren E, Surjan A (2014) Asia. In: Barros VR, Field CB, Dokken DJ, Mastrandrea MD, Mach KJ, Bilir TE, Chatterjee M et al. (eds) Climate change 2014: impacts, adaptation, and vulnerability. Part B: regional aspects. Cambridge University Press, Cambridge, UK and New York, NY, USA, pp 1327–1370. http://www.ipcc.ch/report/ar5/wg2/. Accessed 22 February 2018

IFAD (International Fund for Agricultural Development) (2011) Viewpoint: smallholders can feed the world. IFAD, Rome. https://www.ifad.org/documents/10180/ca86ab2d-74f0-42a5-b4b6-5e476d321619. Accessed 29 August 2017

IRRI (International Rice Research Institute) (2017) Blast (leaf and collar)—rice knowledge bank. http://www.knowledgebank.irri.org/index.php?option=com_zoo&task=item&item_id=808&Itemid=739. Accessed 16 August 2017

ISAAA (International Service for the Acquisition of Agri-Biotech Applications) (2016) Global status of commercialized biotech/GM crops: 2016. In: ISAAA brief no. 52 – 2016. ISAAA, Ithaca, NY. https://www.isaaa.org/resources/publications/briefs/52/download/isaaa-brief-52-2016.pdf. Accessed 4 January 2018

ITC (International Trade Centre) (2017) List of exporters for the selected product_product: 1511 palm oil and its fractions, whether or not refined (excluding chemically modified). http://www.trademap.org/(S(u5ribc55210oue55dz31wrrn))/Country_SelProduct_TS.aspx?nvpm=1|||||1511|||4|1|1|1|2|2|1|1|2|2|1|1. Accessed 28 August 2017

Ivanic M, Martin W, Zaman Z (2011) Estimating the short-run poverty impacts of the 2010–11 surge in food prices. In: Policy research working paper, agriculture and rural development team, development research group. The World Bank. http://documents.worldbank.org/curated/en/560951468330321207/pdf/WPS5633.pdf. Accessed 28 August 2017

Jones KM, de Brauw A (2015) Using agriculture to improve child health: promoting orange sweet potatoes reduces diarrhea. World Dev 74:15–24. https://doi.org/10.1016/j.worlddev.2015.04.007

Kalra A (2013) Partnering for food security in emerging markets. Swiss Re Sigma 1/2013. http://media.swissre.com/documents/sigma1_2013_en.pdf. Accessed 6 January 2018

Kent L (2004) What's the holdup? addressing constraints to the use of plant biotechnology in developing countries. AgBioForum 7(1&2):63–69. http://www.agbioforum.org/v7n12/v7n12a12-kent.pdf. Accessed 6 January 2018

Kissoly L, Faße A, Grote U (2016) The integration of smallholders in agricultural value chain activities and food security: evidence from rural Tanzania. Food Secur. https://doi.org/10.1007/s12571-016-0642-2

Kozai T, Niu G, Takagaki M (2015) Plant factory: an indoor vertical farming system for efficient quality food production, 1st edn. Academic Press, Cambridge, MA, USA. https://www.elsevier.com/books/plant-factory/kozai/978-0-12-801775-3. Accessed 22 February 2018

Kung JK, Lin JY (2003) The causes of China's Great Leap Famine, 1959–1961. Econ Dev Cult Change 52(1):51–73. https://doi.org/10.1086/380584. Accessed 29 August 2017

Ladonia M, Basir A, Robertson P, Kravchenko A (2016) Scaling-up: cover crops differentially influence soil carbon in agricultural fields with diverse topography. Agric Ecosyst Environ 225(1):93–103

Lepcha P, Egan AN, Doyle JJ, Sathyanarayana N (2017) A review on current status and future prospects of winged bean (*Psophocarpus tetragonolobus*) in tropical agriculture. Plant Foods Hum Nutr 72(3):225–235. https://doi.org/10.1007/s11130-017-0627-0

Luo Y, TeBeest DO, Teng PS, Fabellar NG (1995) Simulation studies on risk analysis of rice leaf blast epidemics associated with global climate change in several Asian countries. J Biogeogr 22:633–678. https://www.jstor.org/stable/2845969. Accessed 28 August 2017

McDougall P (2011) The cost and time involved in the discovery, development and authorisation of a new plant biotechnology derived trait: a consultancy study for Crop Life International. Consultancy Study for Crop Life International by P McDougall, Midlothian. https://croplife.org/wp-content/uploads/pdf_files/Getting-a-Biotech-Crop-to-Market-Phillips-McDougall-Study.pdf. Accessed 22 February 2018

Mi CR, Zu Q, He L, Huettmann F, Jin N, Li J (2017) Climate change would enlarge suitable planting areas of sugarcanes in China. Int J Plant Prod 11(1). http://ijpp.gau.ac.ir/article_3315_6aa492f9a465a3d20b3f52db393f8446.pdf. Accessed 29 August 2017

Montesclaros JML, Liu S, Teng PS (2018) Scaling commercial urban agriculture to meet food demand in Singapore: an assessment of the viability of leafy vegetable production using plant factories with artificial lighting in a 2017 land tender (First tranche). In: NTS report no. 7.

Singapore, RSIS Centre for NTS Studies. https://www.rsis.edu.sg/wp-content/uploads/2018/02/NTS-Report-No.-7-%E2%80%93-Scaling-Up-Commercial-Urban-Agriculture-to-Meet-Food-Demand-in-Singapore_-February-2018.pdf. Accessed 1 March 2018

Montesclaros JML, Teng PS (2018) Ensuring a successful Singapore urban food cluster. NTS Insight, No. IN18-02. S. Rajaratnam School of International Studies (RSIS) Centre for Non-Traditional Security (NTS) Studies, Singapore. http://www.rsis.edu.sg/wp-content/uploads/2018/02/NTS-Insight-_-Ensuring-A-Successful-Singapore-Urban-Food-Cluster-_-February-2018-__.pdf. Accessed 1 March 2018

Montesclaros JML, Babu SC, Teng PS (2019) IoT-enabled farms and climate-adaptive agriculture technologies: Investment lessons from Singapore. IFPRI Discussion Paper 1805. Washington, DC: International Food Policy Research Institute (IFPRI). https://doi.org/10.2499/p15738coll2.133079.

Mukerjee M (2014) Bengal Famine of 1943: an appraisal of the Famine Inquiry Commission. Econ Polit Weekly 49(11):71–75. https://www.academia.edu/6351378/The_Bengal_Famine_of_1943. Accessed 28 August 2017

Murdoch Commission (2015) Operation flood: India's dairy revolution through cooperatives. In: Food security, trade and partnerships: towards resilient regional food systems in Asia. Second Murdoch commission final report. Murdoch Commission. http://www.murdoch.edu.au/Murdoch-Commission/Commissions/Second-Murdoch-Commission/_document/Reports/Murdoch_Commission_Report_2015_English_Fullreport.pdf. Accessed 6 January 2018

Nakagawa H, Hone T, Nakano J, Kim HV, Wada K, Kobayashi M (1993) Effects of elevated CO_2 concentration and high temperature on the growth and development of rice. J Agric Meteorol 48(5):799–802. https://www.jstage.jst.go.jp/article/agrmet1943/48/5/48_5_799/_pdf. Accessed 4 September 2017

National Parks Board Singapore (2013) *Telosma cordata* (Burm. f.) Merr.—NParks Flora and Fauna Web. https://florafaunaweb.nparks.gov.sg/special-pages/plant-detail.aspx?id=1515. Accessed 5 January 2018

Nelson GC, Rosegrant MW, Koo J, Robertson RD, Sulser T, Zhu T, Ringler C et al. (2009) Climate change: impact on agriculture and costs of adaptation. In: IFPRI food policy report. IFPRI, Washington, DC. http://www.ifpri.org/publication/climate-change-impact-agriculture-and-costs-adaptation. Accessed 28 August 2017

NOAA (National Oceanic and Atmospheric Administration) National Centers for Environmental Information (NCEI) (2017a) Climate at a glance: global time series, July 2017. https://www.ncdc.noaa.gov/cag/time-series/global/asia/land/1/1/1910-2017. Accessed 15 August 2017

Nwanze K, Fan S (2016) Climate change and agriculture: Strengthening the role of smallholders. In: Global food policy report. IFPRI, Washington, DC, pp 13–22. http://www.ifpri.org/publication/climate-change-and-agriculture-strengthening-role-smallholders. Accessed 22 February 2018

OECD (Organization for Economic Cooperation and Development) (2012) OECD environmental outlook to 2050: the consequences of inaction. OECD Publishing. http://dx.doi.org/10.1787/9789264122246-en. Accessed 28 August 2017

Peng S, Huang J, Sheehy JE, Laza RE, Visperas RM, Zhong X, Centeno GS, Khush GS, Cassman KG (2004) Rice yields decline with higher night temperature from global warming. Proc Natl Acad Sci USA 101:9971–9975. http://www.pnas.org/content/101/27/9971.full. Accessed 28 August 2017

Roessig JM, Woodley CM, Cech JJ, Hansen LJ (2004) Effects of global climate change on marine and estuarine fishes and fisheries. Rev Fish Biol Fisheries 14:251–275. https://link.springer.com/article/10.1007/s11160-004-6749-0. Accessed 28 August 2017

Sen A (1976) Famines as failures of exchange entitlements. Econ Polit Weekly 11:(31/33). Special number: population and poverty. https://www.jstor.org/stable/i398911. Accessed 28 August 2017

Sen A (1981) The great Bengal Famine. In: Poverty and famines: an essay on entitlement and deprivation. Clarendon Press, Oxford, pp 52–85. http://staging.ilo.org/public/libdoc/ilo/1981/81B09_608_engl.pdf. Accessed 28 August 2017

Settele J, Scholes R, Betts R, Bunn S, Leadley P, Nepstad D, Overpeck JT, Taboada MA (2014) Terrestrial and inland water systems. In: Field CB, Barros VR, Dokken DJ, Mach KJ, Mastrandrea MD, Bilir TE, Chatterjee M, Ebi KL, Estrada YO, Genova RC, Girma B, Kissel ES, Levy AN, MacCracken S, Mastrandrea PR, White LL (eds) Climate change 2014: impacts, adaptation, and vulnerability, Part A: global and sectoral aspects. In: Contribution of working group II to the fifth assessment report of the intergovernmental panel on climate change. Cambridge University Press, Cambridge, UK and New York, NY, USA, pp 271–359 http://www.ipcc.ch/pdf/assessment-report/ar5/wg2/WGIIAR5-Chap4_FINAL.pdf. Accessed 28 August 2017

Singh US, Dar MH, Singh S, Zaidi NW, Bari MA, Mackill DJ, Collard BCY, Singh VN, Singh JP, Reddy JN, Singh RK, Ismail AM (2013) Field performance, dissemination, impact and tracking of submergence-tolerant (Sub1) rice varieties in South Asia. SABRAO J Breed Genet 45:112–131. http://eds.b.ebscohost.com/eds/pdfviewer/pdfviewer?vid=0&sid=1527aeae-2738-4854-87ff-85bb5f1fb186%40pdc-v-sessmgr01. Accessed 4 January 2018

Smyth S (2017) Genetically modified crops, regulatory delays, and international trade. Food Energy Secur 6(2):78–86. http://onlinelibrary.wiley.com/doi/10.1002/fes3.100/full. Accessed 22 February 2018

Teng PS (2008) An Asian perspective on GMO and biotechnology issues. Asia Pac J Clin Nutr 17(1):237–240. https://www.ncbi.nlm.nih.gov/pubmed/18296345. Accessed 31 August 2017

Teng PS (2012) Food security: cities as part of the solution and not the problem. RSIS Commentary. https://www.rsis.edu.sg/rsis-publication/nts/1804-food-security-cities-as-part/#.WaPF5z6g-M8. Accessed 28 August 2017

Teng PS (2017) Knowledge-intensive agriculture: the new disruptor in world food? RSIS Commentary No. 124, 23 June 2017. https://www.rsis.edu.sg/wp-content/uploads/2017/06/CO17124.pdf. Accessed 6 January 2018

Teng PS, Escaler M (2014) Urban food security and urban agriculture in Asia: cities as part of the solution. RSIS Commentary. https://www.rsis.edu.sg/rsis-publication/nts/1804-food-security-cities-as-part/#.WaO8tT6g-M8. Accessed 28 August 2017

Teng PS, Oliveros J (2016) The enabling environment for inclusive agribusiness in Southeast Asia. Asian J Agric Dev 13(2). http://searca.org/ajad/read-articles/13-view-article?aid=688. Accessed 22 February 2018

Teng PS, Montesclaros JML, Hulme R, Powell A (2019). The Evolving Singapore Agrifood Ecosystem. NTS Insight, no. IN19-03, August 2019. Singapore: Centre for Non-Traditional Security Studies (NTS Centre), S. Rajaratnam School of International Studies (RSIS), Nanyang Technological University, Singapore. http://hdl.handle.net/11540/11207.

Thapa G (2009) Smallholder farming in transforming economies of Asia and the Pacific: challenges and opportunities. In: Discussion paper prepared for the side event organized during the 33rd Session of the IFAD's Governing Council, 18 February 2009. International Fund for Agriculture and Development. https://www.ifad.org/documents/10180/a194177c-54b7-43d0-a82e-9bad64b76e82. Accessed 28 August 2017

The Economist (2009) Whatever happened to the food crisis? 2 July. http://www.economist.com/node/13944900. Accessed 28 August 2017

Timmer P (2009) Rice price formation in the short run and the long run: the role of market structure in explaining volatility. In: Working papers (Center for Global Development) (172). https://www.cgdev.org/sites/default/files/1422136_file_Timmer_Rice_Price_FINAL.pdf. Accessed 29 August 2017

Trench PC, Narrod C, Roy D and Tiongco M (2011) Responding to health risks along the value chain. In: 2020 conference paper 5. 2020 conference: leveraging agriculture for improving nutrition and health. International Food Policy Research Institute, Washington, DC. https://microlinks.org/sites/microlinks/files/resource/files/IFPRI%20-%20Responding%20to%20Health%20Risks%20Along%20the%20Value%20Chain.pdf. Accessed 3 January 2018

UN (United Nations) (2017) Water for Life decade—Asia and the Pacific. http://www.un.org/waterforlifedecade/asia.shtml. Accessed 29 August 2017

UN FAO (United Nations Food and Agriculture Organization) (2008) Climate-related transboundary pests and diseases: technical background document from the expert consultation held on 25 to 27 February 2008. UN FAO, Rome. http://www.fao.org/3/a-ai785e.pdf. Accessed 28 August 2017
UN FAO (United Nations Food and Agriculture Organization) (2011) FAO-adapt—FAO's framework programme on climate change adaptation. UN FAO, Rome. www.fao.org/docrep/014/i2316e/i2316e00.pdf. Accessed 22 February 2018
UN FAO (United Nations Food and Agriculture Organization) (2015) How to feed the world in 2050. http://www.fao.org/fileadmin/templates/wsfs/docs/expert_paper/How_to_Feed_the_World_in_2050.pdf. Accessed 29 August 2017
UN FAO (United Nations Food and Agriculture Organization) (2017a) Annual population. http://www.fao.org/faostat/en/#data/OA. Accessed 29 August 2017
UN FAO (United Nations Food and Agriculture Organization) (2017b) Land use. http://www.fao.org/faostat/en/#data/RL. Accessed 29 August 2017
UN FAO (United Nations Food and Agriculture Organization) (2017c) Food balance sheets. http://www.fao.org/faostat/en/#data/FBS. Accessed 2 August 2017
UN FAO (United Nations Food and Agriculture Organization) (2017d) The future of food and agriculture–trends and challenges. Rome. http://www.fao.org/3/a-i6583e.pdf. Accessed 28 August 2017
UN FAO (United Nations Food and Agriculture Organization) (2017e) 'Food security indicators' FAOStat database. http://www.fao.org/economic/ess/ess-fs/ess-fadata/en/#.WkxHJlWWaM8. Accessed 3 January 2018
UN FAO (United Nations Food and Agriculture Organization) (2017f) Suite of food security indicators—FAOStat database. http://www.fao.org/faostat/en/#data/FS. Accessed 29 August 2017
UN FAO (United Nations Food and Agriculture Organization) (2017g) Annual population—FAOStat database. http://www.fao.org/faostat/en/#data/RL. Accessed 28 August 2017
UN FAO (United Nations Food and Agriculture Organization) (2017h) Crops—FAOStat database. http://www.fao.org/faostat/en/#data/QC. Accessed 29 August 2017
UN FAO (United Nations Food and Agriculture Organization); IFAD (International Fund for Agricultural Development); WFP (World Food Programme) (2015) The state of food insecurity in the world 2015—glossary of selected terms. UN FAO. http://www.fao.org/hunger/glossary/en/. Accessed 2 September 2017
UN WHO (United Nations World Health Organization) (2017a) Food borne disease burden. https://extranet.who.int/sree/Reports?op=vs&path=/WHO_HQ_Reports/G36/PROD/EXT/FoodborneDiseaseBurden. Accessed 29 August 2017
UN WHO (United Nations World Health Organization) (2017b) Obesity—situation and trends. http://www.who.int/gho/ncd/risk_factors/obesity_text/en/. Accessed 29 August 2017
UN WHO (United Nations World Health Organization) (2017c) WHO estimates of the global burden of food borne diseases. http://apps.who.int/iris/bitstream/10665/199350/1/9789241565165_eng.pdf. Accessed 28 August 2017
United Nations Department of Economic and Social Affairs (2017) World population prospects: the 2017 revision, key findings and advance tables. Population Division. United Nations Department of Economic and Social Affairs, New York. https://esa.un.org/unpd/wpp/Publications/Files/WPP2017_KeyFindings.pdf. Accessed 2 September 2017
United Nations, Department of Economic and Social Affairs, Population Division (2015) World urbanization prospects: the 2014 revision (ST/ESA/SER.A/366). https://esa.un.org/unpd/wup/publications/files/wup2014-report.pdf. Accessed 6 January 2018
Van der Walt WJ (2013) After 33 years of biotech communication and 21 years of GM crop testing and adoption, where to now for South Africa? In: ISHS acta horticulturae 974: II genetically modified organisms in horticulture symposium. https://doi.org/10.17660/actahortic.2013.974.3
Walker D, Kurth T, Van Wyck J, Tilney M (2016) Lessons from the frontlines of the agtech revolution. AgFunder and Boston Consulting Group. https://www.bcg.com/publications/2016/pro

cess-industries-building-materials-strategy-lessons-frontlines-agtech-revolution.aspx. Accessed 6 January 2018

Wall BA, Mateus A, Marshall L, Pfeiffer DU (2016) Drivers, dynamics and epidemiology of antimicrobial resistance in animal production. United Nations Food and Agriculture Organization, Rome. http://www.fao.org/3/a-i6209e.pdf. Accessed 28 August 2017

Ward R (2017) Two-day strike disrupts Brazilian ports. JOC, 16 May. http://www.joc.com/port-news/south-american-ports/port-santos/two-day-strike-causes-disruptions-brazil-shippers_20170516.html. Accessed 28 August 2017

World Bank (2017) World development indicators. http://data.worldbank.org. Accessed 28 August 2017

World Bank Group (2013) The world bank group and the global food crisis: an evaluation of the world bank group response. http://ieg.worldbankgroup.org/Data/reports/food_crisis_eval_1.pdf. Accessed 28 August 2017

Zhang C, Wohlhueter R, Zhang H (2016) Genetically modified foods: a critical review of their promise and problems. Food Sci Hum Wellness 5:116–123. https://doi.org/10.1016/j.fshw.2016.04.002

Jose Ma. Luis Montesclaros is Associate Research Fellow with the Centre for Non-Traditional Security Studies (NTS Centre), S. Rajaratnam School of International Studies (RSIS), Nanyang Technological University, Singapore (NTU). He conducts policy analysis with dynamic models of food security, economics, agriculture, and climate change in ASEAN, and has provided inputs to Singapore's Inter-Ministry Committee on Food Security (IMCFS). As consultant, he co-authored the World Bank's "2014 ASEAN Equitable Development Monitor Report". Holding a master's degree in Public Policy from the Lee Kuan Yew School of Public Policy (LKYSPP), National University of Singapore (ASEAN Scholar), and a BS Economics degree from the University of the Philippines Diliman, he was among LKYSPP's "Leaders of Tomorrow", 2014 St. Gallen Leadership Awards (Switzerland).

Paul S. Teng is Adjunct Senior Fellow (Food Security) at the NTS Centre, RSIS, NTU, and concurrently Managing Director of the National Institute of Education (NIE) International Pte Ltd. He has over thirty years of experience in developing countries from positions at the World Fish Centre, the International Rice Research Institute, and Monsanto Company. Paul has conducted research on new agri-technologies, science-based entrepreneurship, food security and sustainable development, with recent focus on urban food security and agtech. Professor Teng has been recognized with an Honorary Doctor of Science by Murdoch University, Australia; the Eriksson Prize in Plant Pathology and as a Fellow of the American Phytopathological Society; and The World Academy of Sciences. He has published over 250 technical papers.

Chapter 8
Cultural Dimensions of Human Security

Clarissa D. Ruzol and Juan M. Pulhin

Abstract This chapter explores the cultural dimensions of human security by problematizing the nature of knowledge on climate, climate change, and human security, including its inherent relation with power and varying perceptions across cultures in Asia. The analysis is based on an extensive review of studies that attempt to capture the underlying cultural factors of vulnerability and adaptation. Grounded on an anthropological perspective, the chapter covers the profoundness of culture and its importance in climate change and disaster research from its most popular element of indigenous knowledge and practices to material structures and technology across time and space, different ways of thinking, and ethical considerations. Anthropological approaches that academics and practitioners could use to address human security are also discussed. These simply refer to the set of viewpoints that can be used to frame the analysis of cultural dimensions of human security as informed by theories and practice in anthropology. We argue that adding the concept of human security to climate change and disaster research brings the discourse into a realm foregrounding the human experience to motivate a more powerful political action.

Keywords Culture · Human security · Knowledge · Climate change · Risk perception

8.1 The Importance of Culture and the Convenience of Ignoring It

Various institutions, predominantly from the academe and international organizations, have been raising efforts to emphasize the central role of culture in facing

---

C. D. Ruzol (✉)
Department of Anthropology, London School of Economics and Political Science, London, United Kingdom
e-mail: clarissa.ruzol@gmail.com

J. M. Pulhin
Department of Social Forestry and Forest Governance, College of Forestry and Natural Resources (CFNR), University of the Philippines Los Baños, Laguna, Philippines

J. M. Pulhin et al. (eds.), *Climate Change, Disaster Risks, and Human Security*, Disaster Risk Reduction, https://doi.org/10.1007/978-981-15-8852-5_8

many of the world's pressing environmental problems today. Yet, in international platforms on climate change, disaster risks, and human security, the significance of culture remains in the margins of the dominant discourse and politics. If any, the role of culture is being conveniently and superficially encompassed by its elements such as "indigenous knowledge and practices" in climate change adaptation and disaster risk reduction or as experiential evidence for climate variability and its impacts as validated by or to inform scientists. These are indeed very important but delimiting the study of culture into traditional/indigenous knowledge alone undermines the profoundness of what and how culture affects people of different contexts. Another popular dialogue is that lived experiences under the new normal presents uncertainties beyond the scope of local, traditional, and indigenous knowledge (Puri 2015; Adger et al. 2014). Such dichotomy between the indigenous and the scientific knowledge reinforces the notion that indigenous knowledge, as part of culture, is static when in actuality, it is being constructed by dynamic, emerging particular human–environment interactions. In hindsight, Merton (1973) argues that the superiority of scientific knowledge is being favored precisely because it liberates the truths of nature from its distinctive socio-cultural setting.

The role of culture needs to be given more devotion especially that policies and institutional actions that are unequivocally based on the scientific understanding of climate and disasters still remain ineffective in strengthening human security against impacts of climate change (Barnes et al. 2013). And the incorporation of indigenous knowledge and practices in climate change adaptation and disaster risk reduction is limited by institutional support, such as funding in promotion and implementation (Balay-As et al. 2018). Culture significantly matters in all aspects of human affairs but to ignore it seems to be more convenient than to actually understand what it is.

Culture is a central concept in anthropology. It has been the anthropologist's core explanatory instrument to recognize universal patterns of what makes us human and our particularities in how we perceive the world and manage our own societies. In a culturally diverse region such as Asia, assuming that members of disparate societies and cultures think and perceive the same way about their environment, the climate, disaster risks, and even notions of "security" is fallacious. While modern science may have attempted to globalize and standardize not just technology but knowledge itself, it is culture that determines how people would respond and communicate the information. Even within nation states, irrespective of level of economic development or political status, there are variations in cultures and peoples' worldviews. National policies aim to homogenize many nation states in pursuit of an identity distinguishable from or to emulate others, which tend to ignore cultural diversity.

By looking into culture, one can have a deep and holistic understanding of how people are vulnerable to environmental pressures and the effective ways on how to adapt to them. Hoffman (2015b) explains that, in the context of hazards, risks, and disaster recovery, the perception of reality and the environment is cross-cultural, including "how a particular people calculate peril, experience catastrophes, and recover from them, or do not recover, or do not protect themselves." Culture is the complex whole that signifies meanings in kinship and social networks, gender roles and identity, management of time, use of language, perception of color, emotions, arrangement in space, subsistence, division of labor, prestige and status, power relations, ways of knowing, land use, place attachments, child raising, laws, rules, belief in higher spiritual beings, religion, the unseen world, death, afterlife, and many more. These elements of culture dictate peoples' receptions and responses to ideas and practice in human security, climate change adaptation, and disaster risk management.

There have been efforts to elaborate the role of culture in climate change and disaster research especially by disciplines that adapted the concept or as Hoffman (2015b) termed it "co-opted" it, such as by journalism, geography, environmental science, sustainability science, and development studies. Using a cultural lens to climate change and disaster research challenges the prevailing conventional approach to science and policy. In particular, there is difficulty in integrating comprehensive cultural analysis of contexts of the human condition across spatial, political, and temporal scales when climate and disasters are predominantly viewed as an apolitical, ahistorical, predictable external force outside and separate from the human world, usually at the regional and global levels. A form of co-opting the concept of culture mainly manifests in the growing popularity of research on indigenous knowledge and practices for climate change adaptation and disaster risk reduction in a rather facile way. As Hoffman (2015b) argues, co-optation has "defanged" culture as a holistic explanatory tool. It delimits the dynamic and comprehensive nature of culture into aspects that could fit in the popular scientific discourse.

If climate change and disaster risks are not simply problems of the physical but a social, cultural, and perceptual one, cultural analysis should encompass science itself and the nature of knowledge that reproduces it. Many scholars have already written about how climate science continues to be independent of the culturally laden human experience in its production and reception (Demeritt 2001; Jasanoff 2010; Lahsen 2010, 2015). In the Intergovernmental Panel for Climate Change assessment reports that established climate change as a global phenomenon, for example, the methodological process detached knowledge from the meaning of the subjective, embedded experience of human actors engaged with nature (Jasanoff 2010), while values were hardly accounted for (Duraiappah 2010).

Incorporating the concept of human security into climate change and disaster research brings the discourse into a realm closer to the human experience. This can propel onto the mainstream the studies on the cultural dimension of climate change and disasters by social sciences and humanities as well as motivate more powerful political actions. The following sections will give detail to what it means and what it takes to analyze the cultural dimensions of human security in Asia through an anthropological perspective by citing cases and examples from literature.

8.2 Knowledge and Views on Climate Change and Human Security

8.2.1 Climate and Climate Change

In spite of the breadth and depth of scientific evidence for anthropogenic climate change, debate remains polarized with many still unconvinced of its reality, while a smaller minority believe it is a deliberate conspiracy (Barnes et al. 2013; Hoffman 2015a). The perplexing phenomenon of polarized views, one claiming unequivocal climate change and the other disproving it as a hoax, can be attributed to the communication and reception of science that are mainly being acted upon by culture and institutions. Institutional narratives about climate change and disaster risks limit our understanding of local and regional impacts that may impede long-term adaptation and undermine social vulnerabilities (Sherpa 2014). National policies and actors are immanently designed to generalize and perpetuate the authoritative discourse from the dominant language of science that emanated from the West.

In Pharak, Nepal, there is a significant difference between the institutionalized narrative of climate change (*jalvayu parivartan* in Nepali language) described in terms of global warming, melting of glaciers, and glacial lake outburst flood from the Sherpa's experienced impacts, namely, changes in snowfall and rainfall patterns and increased temperatures that they compare with past experiences and direct effects on their crops (Sherpa 2014). Consequently, recurring warnings about the melting glaciers and the flood were causing anxiety and panic among the Sherpa even when nothing has happened:

> Every time we begin to forget about the threats from glacial lake outburst, then comes news of yet another study through the radio and television, and this has been happening over and over again for more than 15 years now… Instead of having to fear death like that again and again, we would rather die once if the lake really bursts out one day. (Navin Singh Khadka 2012 as cited in Sherpa 2014)

The example from Pharak, Nepal, illustrates the disconnect between scientific knowledge and lived experiences of the Sherpa. We can differentiate the nature of these two knowledge systems: scientific knowledge is based on calculated predictions of select observations, while local knowledge is constructed from life-long experiences, which anticipates future scenarios. When scientific forecasts attempt to produce a localized knowledge for the awareness of the villagers, localizing from the perspective of climate science means downscaling the model by simply calculating it. Krauss (2009) argues that it is not as "simple" as calculating but instead localizing is a world of complex uncertainties, ramifications, and unexpected "social" context—a challenge for both scientists and ethnographers. In Pharak, Nepal, climate change perception among the Sherpa is being constructed through different knowledge systems based on their socioeconomic backgrounds, such as gender, age group, occupation, and residence in on-route or off-route villages (Sherpa 2014). What the

Sherpa of Nepal perceive is not separate from their view of nature and the environment, in particular the meanings, attached to sacred places such as the Sagarmatha (Mount Everest) and the distinction between natives, migrants, and tourists, unlike that of the scientist's whose epistemological premise relies on the objective detachment of self from its subject of study.

In Maldives, there is skepticism in the news of sea level rise and islanders tend to focus more on events they personally experience:

> If you ask local people about fuel or food subsidy, they would be interested to talk about it, as it affects their daily life, but climate change-induced sea level rise is not a concern as it is occurring so far ahead. They do not know and don't understand what TV or newspapers have to say about this. (as cited in Arnall and Kothari 2015)

People's temporality, or how people sense and experience time, also determines their responses and reception of the science of climate change and disasters. In contrast with the 50- and 100-year models of climate change impacts by scientists, the local people of Maldives, particularly the non-elites according to Arnall and Kothari (2015), are concerned about what the impacts of climate change might be on their children but often did not articulate the issue in a more distant future. As a result, the people are not worried of sea level rise and are confident of their adaptation measures, which is otherwise being publicized by the temporally extensive view of experts as an "urgent" matter. The same holds true for many farmers in Asia and Africa who have rather short-term horizon and appreciation of climate change impact studies. For instance, farmers would care less about changes in precipitation and increase in temperature in the distant future but care more about present weather and changes in seasonal rainfall. This is an example of how culture interprets science affecting peoples' attitude in adaptation. In this case, there should be a dialogue between groups of people to integrate the different perspectives.

Climate change discourse is not just about the communication or localization of science, but of the science itself (Hoffman 2015a), more specifically the nature of knowledge that is intertwined with ethics and values. Western science rationalizes the physical world to be fundamentally orderly and predictable. While science rationalizes the environment by coming up with universal truths and replicable technology to control it, many cultures do not necessarily follow the divide between human and nature. In the Chinese and Japanese relational epistemology, the truth can be found within yourself by becoming a part of the universe, which is natural. In essence, the *natural* does not signify "nature" as it is commonly understood in modern times, i.e., outside of, but "from itself" (Tucker 2003). For example, consistent with their belief that the north-south orientation is the best and most fortunate orientation, Chinese traditional courtyard houses follow a north-south orientation "to increase solar gain and ambient air temperature in the winter, and decreasing it in the summer" (Soflaei et al. 2017). Scholars of Japanese history and nation building in early twentieth century such as Julia Thomas also describe Japan's modern nationhood as conscious self-creation reinscribing the nation as natural, i.e., "naturalizing the nation" and subsequently "nationalizing nature" (Thomas 1998).

Another important matter to consider is that the production of knowledge on climate differs through time. We argue that the twentieth-century framing of climate that affects the science being communicated about warming is insufficient in addressing climate change and, hence, its cultural elements should be given more emphasis. Cultural constructs about the climate can also be learned from narratives in the colonial past. For instance, the understanding and experience of climate by the European colonizers were also used to justify their dominion over "degenerate" populations exposed to tropical climates such as the case of colonial Bombay, India (Adamson 2012). This disregard of the insiders' point of view is comparable to how the climate change narrative is being communicated by institutions today. The North and South, terms ubiquitous in the language of international climate change policies, are disconcertingly comparable to the narrative in the colonial parts of Asia, including India, the Philippines, and Indonesia. One way for European colonial powers to make sense of the world was by division into two zones based on the interaction of climate, health, and race[1]: "temperate" and "torrid" (Kenny 1995). The temperate zone located in the North (particularly Northern European) was salubrious to health, producing races that are "strong," "industrious," and "intelligent" with "high morals," while the climate in the South was "lethargic," "effeminate," and "indolent" (Livingstone 1991; Harrison 2010). The science and knowledge about climate, in this case, has been used as one of the justifications for European imperialism in Asia. Similar to the North-South divide during the colonial past in Asia, the North-South[2] divide in climate change has significant implications on how science is framed, produced, applied, communicated, and interpreted (Blicharska et al. 2017). The North-South knowledge divide deprives the scientific community of considerable intellectual wealth, influences research priorities and commitment, and narrowly frames approaches and paradigms that are devoid of certain cultural settings and perspectives (Karlsson et al. 2007). The North-South divide in climate change becomes economic and political with the North being associated with the developed countries and the South, the developing ones. Climate change particularly puts pressure on the rising economies in Asia (to emulate the North) that continue to struggle in managing its environmental and socioeconomic problems, such as from pressures of urbanization, population growth, and globalization.

[1]Race is used as it is used in the original citation to portray the colonial context.

[2]The 'North' refers to countries that are members of the OECD (Organisation for Economic Co-operation and Development) or are classified as high-income economies by the World Bank. These are largely, but not exclusively, countries in Europe, North America, East Asia, and Australasia. The term 'South' is used to refer to countries classified as upper-middle income, lower-middle income, or low-income economies, which are mostly, located in the rest of Asia, Africa, and Latin America (*World Bank Country and Lending Groups*, Accessed 28 May 2016. URL: http://go.nature.com/2gpbHhv).

8.2.2 Human Security

King and Murray (2001) claim that human security replaces the traditional ideas of state (military and economic) security and human development toward a people-centered paradigm. Although there is no coherent definition of what human security is, it has considerably articulated the concepts of social justice, dignity, and human rights in the 1990s, which are notably closer to the human experience than the prevailing dialogue on climate change at the time. In one of the frameworks of human security, Nef (1999) proposed the five dimensions of human security based on his analysis of the elements of the world system: ecology/environment, economy, society, polity, and culture. The inclusion of culture in a framework that ought to be applied universally at least considered cultural diversity for policy, particularly foreign policy.

Human security is based on the cultural notions of "safety," "danger," and "risk," particularly in the context of climate change and disasters. What ties these constructs together is how humans perceive nature. What is generally considered as "dangerous" or "risky" is out there, outside of the human body, while being "safe" denotes a condition of being protected from or not exposed to danger. Understanding how different cultures in Asia perceive "nature" is one way to understand how they cope with new dangers and risks. For example, in Thailand, the word used for "nature" today in Thai language is *thammachaat*, a combination word borrowed from Sanskrit with *thamma*, or *dhamma*, meaning "truth" (particularly the teachings of the Buddha) and *chaat*, meaning rebirth from the concept of reincarnation (Darlington 2003). Applying this adopted concept of "nature," for example, to the forest, sanitizes the danger and risk, termed by Stott (1991) as "barbaric" qualities of the wild. Another example is among the Orang Asli of Malaysia. According to the Orang Asli, the forest is perceived as the parent providing care for them, its children, hence, "killing" the forest is tantamount to killing one's parent (Gomes 2012). In this sense, the forest is not a "barbaric wild" that puts the person at risk but a sanctuary that signifies parental care.

Indigenous and traditional knowledge on coping and adaptation is formed through experiences from the past, which have endured dangers and risks. Distinctively, many indigenous peoples have a historical consciousness that is spatialized rather than temporalized, such as among the Ilongot of the Philippines (Rosaldo 1980) and the Orang Asli of Malaysia (Gomes 2012). This means that history is being conceptualized not through events in time, but where it "took place," that is, it is inscribed in the landscape, including events of disasters and places of hazards. This sense of place influences the social construction of risk and its connections to survival and security (Bankoff et al. 2015). While risk perception varies based on the culture's notion of time and space, anthropology has one of many ways to analyze risk perception, that is, using cultural theory.

The cultural theory of risk has been one of the most significant contributions of anthropology to contemporary issues today. Conceptualized by anthropologist Mary Douglas and later on with political scientist Aaron Wildavsky (see Douglas 1970,

1982), the risk framework categorizes social organizations according to the interplay of internal structure and social cohesion, namely, as individualist, hierarchist, fatalist, and egalitarian. The hierarchists, for example, respect authority and initiate solutions through policy, while individualists prioritize self-interest. The egalitarians participate through their strong views on social cohesion and volunteerism, while the fatalists are not likely to participate in social movements but still abide by the rule of law. The cultural theory of risk has been valuable in understanding perceptions and insights for natural resource management, environmental change, and management of conflicting views (Hoogstra-Klein et al. 2012; Sharp et al. 2015; McEvoy et al. 2017; Blais-McPherson and Rudiak-Gould 2017; Ruzol et al. 2017). According to Kahan (2010), cultural cognition that influences these group values better explain differences in climate risk perceptions than individual attributes such as gender, ethnicity, income, educational level, political ideology, or personality type. On the other hand, much like any other elements of culture, this cultural grouping is fluid and individuals from one category can move to another by force or circumstance (Douglas 1970). The application of cultural theory to assessing human security in the time of climate change and disasters can provide valuable insights that climate scientists do not usually consider in their approaches.

8.3 Some Underlying Cultural Factors in the Context of Climate Change and Disasters

Human security as a social construct is determined by the mix of idiosyncratic and shared cultural factors. Cultural "factors" are used in this chapter to describe culture as a "tool" to help the readers grasp its different components and use the approach to form their own contextual knowledge about human security. It is important to note that these cultural factors comprise a holistic system of symbols that signify meaning to a particular group of people, and hence are interrelated to one another. The discussion in this section reflects this interrelatedness between cultural factors, in addition to the nature of knowledge and perception (discussed above). The examples cited here constitute the most recent and emerging discussion about the cultural dimension of human security in the context of climate change and disasters in some countries in Asia.[3]

The first cultural factor to be discussed here is the most popular in climate change and disaster research, that is, indigenous knowledge and practices. Indigenous knowledge refers to the ways of thinking of indigenous peoples that reflect their interactions with the environment. In coastal and island communities, heavy rainfall and strong winds can be predicted by observing changes in the clouds—changes in texture (thin

[3]The gender dimension of human security in the context of climate change is also one of the more popular topics, but this chapter will not cite examples of it. The development of gender theories has its own history that gave it its own prominence in international policy and research. It also created its own critiques among practitioners.

or thick), color (white, dark, yellow. or red), location (over mountains or the sea), and movement (to/from the coast), including speed (fast) and direction (vertical or horizontal); waves—changes in color (white), direction, and height (high); the wind, sun, and stars—the direction (usually east or west) and temperature (cold or warm) of winds, the position (high or low) and size (large or small) of the sun, and visibility (many or absent) and constellations of stars (Hiwasaki et al. 2014). It is a way to navigate the self across the different features of nature. Oral tradition as part of the knowledge system, including myths, legends, and poems, may also embed warnings and narratives about past disaster events that have occurred in the cultural landscape, for example, of the 2003 Bam earthquakes in Iran (Parsizadeh et al. 2015) and the 2004 Indian Ocean tsunami, prompting the Onge tribe to retreat to the high grounds (Kakoty 2018), as well as the stories of the *Smong* (tsunami in Devayan language) in Aceh, Sumatra. The *Smong* oral literature provided an early warning system to the Simeulueans of Aceh, alarming people to run away to a higher place after a major earthquake. The *Smong* was being communicated through a traditional lullaby called *Buai-buai*, the poem of *Nandong*, and reiterative narration among family and community members (Syafwina 2014).

Indigenous knowledge may also refer to traditional knowledge, recognizing that traditions continually change and evolve over time as cultural groups borrow, innovate, and adapt to the changing environment (Dudgeon and Berkes 2003). Indigenous and traditional knowledge has been used to devise structural designs, spatial planning, seasonal calendars, etc., which are adaptable to the climate and resilient to hazards. In the literatures we reviewed for this chapter, the adoption of indigenous knowledge for climate change adaptation and disaster management in India is well-researched. An example of this is the indigenous knowledge of *bandalling* for navigation enhancement in low-water periods, such as in the Brahmaputra River and the Ganges River. The *bandal* structures, a vertical bamboo screen mounted on a bamboo frame, also help create new agricultural land as sediment deposits along river banks while reducing risk from bank erosion and flooding (Zhang and Nakagawa 2018). *Bandal* structures do not always perform as desired and to ensure its effectiveness means to learn about flow structure, sediment transport properties, and morphology (Rahman et al. 2003; Zhang et al. 2010). Vernacular built environments in India, including houses and settlements, are considered resilient, having evolved over centuries of experimentation by the locals (Choudhary 2016).

On the other hand, Indian culture also generally practices inclusivism to integrate new knowledge into traditional ones in contrast to the Western culture, particularly the United States, where innovation and replacement of old concepts is the driving force for resilience (Schuler 2014). In the state of Assam in the Brahmaputra valley in northeast India, the incorporation of amphibious technology to traditional *Assam*-type house architecture can make communities resilient to flooding (Das and Mukhopadhyay 2018). The *Assam* houses are made of lightweight bamboo suitable for the hybrid amphibious design with a buoyant base that can float on floodwater.

Traditional ecological knowledge and practices sustain the subsistence livelihood of local communities. While indigenous peoples have been resilient and well-adapted to their environment, some are greatly being challenged by impacts of climate change.

In the permafrost regions of Siberia, the Sakha people have been successful pastoralists for the past 600–800 years (Takakura 2016). They have effectively controlled the landscape by draining lakes and waterlogged areas to maximize hay production or by holding water in times of drought (Crate et al. 2017). However, increased precipitation that results in the thawing of the permafrost and causing the death of inland forests threatens their traditional subsistence livelihood and cultural landscape (Takakura 2016; Crate et al. 2017). In his article, Takakura (2016) details that spring waterfloods (*saasky uu*) caused by ice jams are necessary for the growth of the pasture land, but the increase in the frequency and scale of black water floods (*khara uu*) causing damage to private property and livestock may exemplify the limits of the indigenous cultural adaptation of the Sakha people. Similarly, the people of Bangladesh are at risk of more intense rain-induced flooding but saw the potential advantage of the indigenous floating agriculture to support farming communities in waterlogged situations, which can also control invasive aquatic weeds (Chowdhury and Moore 2017).

The resilience of agricultural landscapes also depends on the concept of space arrangement in one's culture. In Bali, Indonesia, the concept of *Tri Hita Karana* (Fig. 8.1) that has been learned over the past thousand years prescribed the people's spatial distribution from micro to macro scales vertically and horizontally (Asmiwyati et al. 2015). In this concept, as Asmiwyati and colleagues explain (2015), the Balinese landscape reflects the harmony between *Parahyangan* (God) signified in sacred places such as the mountain, the forest, the temple, and sources of water like springs and lakes; the *pawongan* (people) settled in villages where the *Subak* (local irrigation system) can be found below the forest; and the *palemahan* (nature) that is being respected through sustainable cultivation in paddy fields and mixed gardens. The translocation of water necessary for the resilience of the terraced agricultural landscape is highly dependent on the communal *Subak* water temple that will course the water to the private terraced paddy fields. The *Subak* system has been a successful example of resilience in agriculture amidst the changing climate because it also unites farmers through obligatory communal commitments, communal labor assignments, and flexible labor management of on-farm and off-farm work (Lorenzen and Lorenzen 2010).

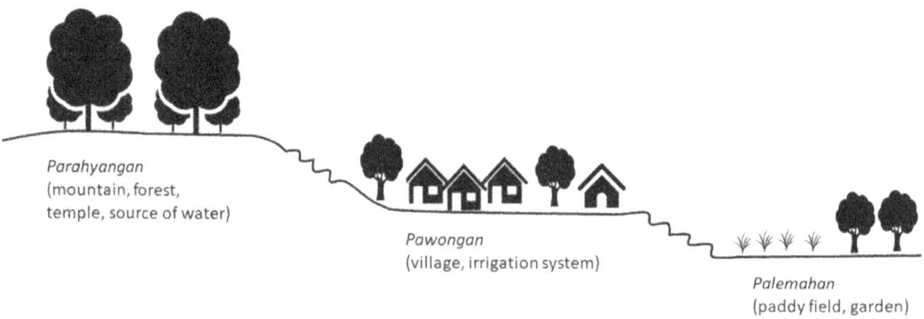

Fig. 8.1 Tri Hita Karana concept in Bali, Indonesia

Religion is another way for people to make sense of unexplainable phenomena and to cope with adverse life events, such as disasters. From a religious perspective, disasters can be understood as the people's relationship with God, a form of divine intervention, or punishment. For example, increased occurrence of extreme events is perceived by the Sherpa in Nepal as the consequence of diminishing religious faith and behaviors (Sherpa 2014). In the context of human security, climate change and occurrence of disasters are related to the Sherpa's cosmology. They also described that non-Sherpa migrants pollute sacred sites and increase religious pollution (Sherpa 2014). This angered local deities that could have brought heavy rains and flooding.

Many cultures in Asia are heavily influenced by Buddhism, Daoism, Confucianism, and Shintoism. These philosophies originated in East Asia, particularly China and Japan. There are about 507,237,000 Buddhists in Asia, the highest population in the world in 2015 (Johnson and Grim 2018). Buddhism is mutually related with trees, forests, and groves that play significant roles in the life of the Buddha, including his understanding, knowledge, and wisdom that Buddhists follow as the Noble Path (Sponsel and Natadecha-Sponsel 2003). Buddhism defines nature as the sum total of reality, all beings, and all things. The non-human elements of nature have the same moral standing as the humans that interact with it. Hence, the Buddhist perspective on taking action against climate change can be derived from the intrinsic value they put in all living and non-living things in nature and the interconnected causality and consequences of human actions.

In Thailand, the third country with the highest Buddhist population, Buddhism intersects with Thai spirit beliefs on the forest and the way they value it throughout the Thai cultural region (northern Thailand, Laos, southern Yunnan Province, China, parts of Myanmar, and northern Vietnam) (Pei 1985; Darlington 2003). This interplay between Buddhism and spirit beliefs provides the framework for the incorporation of cultural and religious concepts in (political) developmental and environmental efforts in Thailand (Darlington 2003), which is relevant to climate change adaptation and disaster risk management. Another example of a religious perspective on the changing climate is the Daoist climate ethics. There are two responsibilities that humans should do, according to Daoism as cited in Xia and Schönfeld (2011): (1) "to learn to become stewards of the biosphere, to shepherd life, and protect complexity; and (2) to refrain from further disruptions of the flow, to learn to become mitigators of climate change, to soften the impact, and to calm down the waves." In stark contrast with the science of the West that follows the anthropocentric view of nature, Buddhist and Daoist perceptions on climate change are influenced by their preferred way of knowing, that is, by being a part of nature itself.

In post-disaster events, resilience in Asian cultures is in many ways connected to the sense of dignity, a core concept in human security. Dignity is seen as a universal human right, according to the United Nations Universal Declaration of Human Rights, separating us from mere biological existence (Fiori et al. 2013). But Field (2017) argues that dignity is a value and idea different across cultures. After typhoon Haiyan (locally Yolanda) hit the Philippines, one of the strongest typhoon to make landfall in history, Filipino teachers viewed that wearing cosmetics one month after the typhoon reflect their dignity (*pagkatao* in Tagalog) and self-worth, boosting a community sense of recovery (Field 2017). In one of the interviews by the Department of Social Welfare and Development of the Philippines in 2013 (as cited in Field 2017), a teacher responded that

> *Gusto ko* Brother, *kapag nakita ako ng mga students ko sa pagsimula ng klase, maayos ang itsura ko. Gusto kong isipin nila na nakabangon na ako para sila rin, ma-inspire na bumangon* (Brother, I want to look good when my students will see me. I want them to think that I have recovered so that they will be inspired to also move on)

The Filipino community's sense of recovery after a disaster is also embedded in the *pakikipagkapwa* (treating other people as equal in moral standing) at *pagkakaisa* (solidarity). Especially in the Philippines, these values are reflective of the significant role of kinship and social networks, including the *barangay* (village), during recovery from typhoons. A strong bond of kinship and social network during events of disasters are valuable in accessing and mobilizing resources, establishing communication, mutual engagement, and maintaining social order when people have a sense of shared expectations and responsibilities from one another.

Strong community bonds help a community become resilient after a disaster event. It is one of, if not the most, reliable support a person can have, especially in remote villages, such as in the Harsil Valley in India. The built environment in Harsil Valley is designed to reinforce interdependence among members of the community. For example, open spaces between houses are multifunctional, fostering community bonding from women drying grains and apple pieces while chatting with one another to becoming vibrant living spaces for community worship and procession of gods and goddesses (Choudhary 2016).

Community bonds are crucial in severe situations when people are experiencing adverse life events. Weakening of community bonds has been cited as a reason for farmers affected with severe drought to take extreme steps such as committing suicide (Hall and Scheltens 2005; Hogan et al. 2012; Vins et al. 2015; Choudhary 2016). Support from family and social network establishes not just relationships between individuals but with society as a whole, a sense of belongingness and a purpose of existence. Similarly, people's experiences of depression, anxiety as well as of hope caused by losses from impacts of climate change varies across cultures. Termed as "ecological grief" in the recently published article by Cunsolo and Ellis (2018), it refers to climate change-driven mental and emotional responses, including mourning and the feeling of pain from losses attached to land and identity.

Indeed, community resilience is essential in climate change adaptation (Adger et al. 2012; Folke et al. 2010). Rituals and ceremonies play an important role in achieving community resilience in Asia. In Indonesia, Casson (2016) notes the importance of an emic (insider/member of the culture) understanding of climate change adaptation strategies and resilience, such as by analyzing the meaning behind the rat ceremony performed by village elders in the island of Flores. While the national government encourages the use of pesticide to get rid of rats in farms, the local agricultural department supports and prefers the performance of the rat ceremony. As quoted from an agricultural department official by Casson (2016: 24) during her fieldwork in Flores:

> This special ceremony sends the rats back to where they belong–the sea–and is the most successful approach to clearing rats from agricultural fields. In my experience, the ceremony guarantees that rats will not return to a field for at least five years. If pesticides are used [instead of the ceremony], the rats will return next year with anger.

Casson (2016: 24) further explains that most people in East Flores believe that "today's rats are descendants of ancient rats that aided the farmers' ancestors in a time of crisis." The farmers must ask the rat nicely to return to their home in the sea; they believe that using pesticide is not polite to the rats. The oral story about rats–that they were the navigators that directed the early people to their homeland in Flores–has been part of the reiterative storytelling from generation to generation. Hence, rats are not considered as random pests attacking the fields but misguided old friends (Casson 2016). In this case, solidarity does not only involve living harmoniously among members of the community but with nature itself. The ritual also reflects traditional ecological knowledge such as the association of rat behavior with the monsoon season and patterns of rainfall.

So far, the recent studies on the cultural dimensions of human security mentioned above and in the previous section have cited examples for the incorporation of scientific knowledge into indigenous and traditional knowledge for adaptation, challenges to traditional ecological knowledge and subsistence livelihood, space arrangement, temporality, cosmology and religion, sense of dignity, social bonds including kinship and community networks, rituals and ceremonies, implications to cross-cultural experiences of mental health, as well as some maladaptive options, i.e., to choose not to respond or to protect themselves from disasters. There is another set of studies that provides insights from early cultures in the past. In the Lake Daihai basin in Mongolia, the relatively mild and humid climate enabled the growth of the Laohushan culture (4500–4100 cal BP) favoring millet planting as a major food source (Xu et al. 2017). However, based on archaeological and climatic evidence from Xu and colleagues (2017), the Laohushan culture collapsed in 4100 cal BP due to the drop in temperature and changes in humidity, while other cultures south of the basin simultaneously flourished. Climate, particularly the changing precipitation in northwest China, affected agricultural production and hunting resources of the Majiayao culture (5300–4000 cal BP) according to evidence of variations in site distribution, site density, and subsistence strategies (Dong et al. 2013). There is also evidence that rapid change in climate (8300–8050 cal BP) in northeastern Hokkaido, Japan, caused the sudden appearance of sophisticated blade technology from the previous simple flake technology (Morisaki et al. 2018). There is valuable knowledge to be gleaned from the studies of the deep past, such as the influence of climate on the rise and collapse of cultures, settlement systems and human movements, material culture, subsistence, demographic change, and cultural change.

8.4 Anthropological Approaches to Human Security

Throughout this chapter, we have been discussing human security "in the context" of climate change and disasters. This is to situate the cultural dimensions of human security amidst the known impacts and the uncertainty of the changing climate. In anthropology, *context* is at the core of research. It is through context that culture (of one's own and of others) is being understood and interpreted. Context refers to the experiential world of the people belonging to the same culture, including that of the world of the researcher. For the sake of its relevance to human security, context can be framed as something that is actively evolving through time (diachronic) or dynamically interrelated with different elements of the culture at a specific moment in time (synchronic) and can be a product of a system of inequalities where the disenfranchised and disempowered challenge the hierarchical power structure (critical and feminist). The three approaches included here (diachronic, synchronic, critical and feminist) are not mutually exclusive of one another and we do not claim that they are exclusively anthropological.

8.4.1 Diachronic Approach

Transformative pathways of change and responses in adaptation have been gaining prominence in recent literature (see for instance Denton et al. 2014; O'Brien et al. 2014; Few et al. 2017). However, the temporal dimension of human-environment studies (including human-climate and human-disaster) is still poorly integrated in adaptation (Adamson et al. 2018). To understand human–climate and human–disaster interactions, it is not sufficient to focus on the present alone because the cultural dimensions of human security are embedded within its context and intricately long history. The diachronic approach to human security focuses on the evolutionary process of adaptation (and maladaptation) and knowledge production and the environmental history of the place. Context can be captured by understanding the circumstances of the past and its continuity into the future, that is, contextualizing through baselining.

Baselining can be referred to as the scientist's practice of gathering data at points in time in a specific place and comparing them across time. This also describes the way scientists localize knowledge on climate change. Localizing in this sense is also a way of standardizing the features and attributes of a particular group of people and place. Of course, there will always be data to collect from the set of assumptions selected by the scientist but the question remains, "What is the context?" As mentioned above, localizing is not as simple as calculating. Instead, using a diachronic approach captures the intricate social and cultural context of the place. Anthropologists understand context from the indigenous way of "localizing" and understanding from the emic point of view. For instance, the historical consciousness of many indigenous peoples (also mentioned above) is a way to capture context through time. Instead of localizing, it is a way of *particularizing*. Their experience and interactions with the environment as a way of adaptation happened in a particular place.

Historical particularism and cultural ecology, for example, can be used to frame the evolution of adaptation options of people through their interaction with the environment. The discipline of anthropology, together with archaeology, history, geography, and paleontology that have been guided by a diachronic approach in their studies on human–environment relationships, can significantly contribute to reconstructing the cultural dimensions of the past, present, and future of human security.

8.4.2 Synchronic Approach

Anthropologists traditionally use the synchronic approach in making sense of other peoples' mode of thought since they usually study cultures other than their own. Ethnography or the practice of writing about peoples exemplifies this intellectual tradition. A synchronic approach entails two important methodologies: (1) understanding how things are related by observing the society (village or ethnic group) as a whole and (2) examining the case study in relation to others. In contrast to the diachronic approach, synchronic analysis is interested in a particular moment in time, usually contemporary phenomena, such as human security in the time of climate change and disasters.

Conventionally, ethnography uses Clifford Geertz' "thick description" to understand meanings that people place in language, actions, material culture, institutions, etc., providing the cultural context, for example, to understand how politics fits together with gender structures, religion, or economics, or to compare the various adaptation options of nomadic groups across Asia amidst climate change, globalization, Westernization, and assimilation to the mainstream market economy. The anthropologist would be interested in a holistic perspective trying to capture into narratives, discourses, and various media the complexity of the system that is otherwise silenced or ignored by reductionist sciences. While surveys and interviews can be helpful, the anthropologists' preferred method of data collection and analysis is participant-observation. Participant-observation entails researchers to socialize and enculturate themselves in the context of the culture embedded in interactions, behaviors, non-verbal language, unspoken rules, and norms. Contrary to the scientific method that requires the scientist to observe a phenomenon in a controlled setting expecting to generate objective results, in participant-observation, knowledge is a construct created as a result of the researcher's interaction with the people in the study. The synchronic approach can effectively capture the cultural context of human security based on the people's experienced reality.

8.4.3 Critical and Feminist Approaches

The critical and feminist approaches can be diachronic, synchronic, or a combination of both but with emphasis on the role of power in a cultural context. Critical and feminist theories have undeniable commonalities that lean toward radicalism. Both critique the scientist's positivistic science, including that of the social scientists, that distance the researcher from the people in the study, not taking accountability of the results and not informing or even misinforming them of the aims of the research. Feminism, in particular, is a collective movement addressing inequality, including marginalization of social classes, disenfranchisement of indigenous peoples, misrepresentation, and pursuit of neoliberal agenda that reinforce the highly skewed power dynamics. According to critical and feminist theories, dominant power and knowledge structures direct the attention away from the root causes of vulnerability and constrain the link of adaptation to the decolonizing process. For instance, in vulnerability assessments, indices can measure the magnitude of hazards and socioeconomic capacities to adapt, but they alone do not provide an explanation as to *why* and *how* these present characteristics manifested in this peculiar way. This is because vulnerability assessments and adaptation policies were *framed*, as argued by Colette (2016), by scientists and decision makers themselves. This has consequently reduced the diversity of local contexts and concepts of vulnerability as well as associated terms like risk, susceptibility, and resilience, to a set of quantifiable variables.

In the rapidly growing field of climate change and disaster research that is unequivocally defined by its uncertainty and complexity, a critical view of the practice of science can help one see how hegemonic structures of power and knowledge cause adverse impacts, at times unintended, on the marginalized. In a quote from Balay-As et al. (2018: 22), a Kankanaey elder in the Philippines expressed both optimism and anxiety about scientific knowledge:

> All of us must be engaged in a continuing search for knowledge that will strengthen our warning systems for typhoons. This benefits no other than the people whose lives and livelihoods are threatened by the impacts of typhoons. However, we must ensure that this search for knowledge does not become oppressive to anyone. (Elder 2)

Through the critical and feminist approaches, human security becomes a topic on ethics and social justice, reminding us to consider the inequalities already in place. These approaches recognize social, economic, and political inequalities as the root causes of vulnerability to climate change and disasters and understand the meaning of the set parameters and assumptions for the people themselves. Using this approach, the goal is to dismantle the gap between the empowered and the disempowered and realize people's agential power to positively act for their own and the future generation's welfare. In Box 8.1, a critical approach is used to briefly analyze the root causes of vulnerability of the Calawit Tagbanua in Palawan, Philippines, by looking into the historical narrative of marginalization and entitlements.

Box 8.1: Case Study on the Human Security of the Calawit Tagbanua in Palawan, Philippines

Typhoon Haiyan/Yolanda

Typhoon Haiyan/Yolanda in November 2013 was the strongest that the Calawit Tagbanua experienced. It was most damaging to property and agricultural crops. The Calawit Tagbanua resorted to their own efforts to improve the structural integrity of their houses, particularly the roof, or to evacuation to a more stable shelter such as the elementary school. Farming households in the Calawit Tagbanua community also explored new opportunities to compensate for their low income through marine harvesting. Subsistence fishing was one, as well as fishing for high-value commodity such as the grouper, which is becoming a trend among other coastal Tagbanua communities in Palawan in response to declining yields, increasing debt, and the potential of a new export market (Dressler and Fabinyi 2011).

The story of displacement

Calauit Game Reserve and Wildlife Sanctuary
Photo source: Clarissa Ruzol (July 2014)

Today, many of the life ways of the indigenous Calawit Tagbanua have changed and the indigenous communities are in the continuous process of reasserting their culture and identity as the people of the *teeb ang surublien* (ancestral land and seas). In the recent past, the Calawit Tagbanua were driven out of Calauit Island to give way to the Calauit Game Preserve and Wildlife Sanctuary (Presidential Proclamation No. 1578) in 1976. There are mixed opinions on why the then Marcos dictatorial government funded the transport and creation of a sanctuary of exotic animals, including giraffes and zebras, in Calauit Island. But it was said to result in being the "first successful wildlife translocation experiment in Asia" (Palawan Council for Sustainable Development 2018).

A few years preceding the proclamation of the Calauit Island as a protected area in 1976, the then Bureau of Lands started to survey the island, announcing that the purpose was for landholding titling. Individual resettlement agreements were orderly followed and signed by the Calawit Tagbanua in 1973. In the beginning of the resettlement of the Calawit Tagbanua, villages Halsey and Burabod in the Leper Colony Reservation in Culion Island were identified as resettlement areas through Presidential Proclamation No. 1626 of 1977. Every Calawit Tagbanua family in the island was relocated to Culion Island with accounts of forceful displacement.

The case of the Calawit Tagbanua's displacement remained dormant during the first decade of their settlement in Culion Island. From 1986 to 1987, the Calawit Tagbanua, now organized as the Balik Calawit Movement (Return to Calawit Movement), began to resettle in Calauit Island and filed a complaint against the Conservation and Resource Management

Foundation, Inc.—the organization that handled the management of Calauit Island—and the state government to the Philippine Commission on Human Rights (PCHR). Petitions filed in court by the Calawit Tagbanua detailed that crops could not be grown and essential public utility services were absent in the resettlement areas in the neighboring island of Culion. Aside from the difficulty of sustaining livelihood in the resettlement areas, the displacement has physically detached the Calawit Tagbanua from places of cultural significance, ceasing the traditional ways of knowledge transfer and skill acquisition to the new generation. After a fact-finding mission, the PCHR recommended the (1) repeal of the proclamation of the protected area for being violative of the Calawit Tagbanua's Bill of Rights, as settlers of the island, and (2) the immediate return of the settlers to Calauit Island. Return to Calauit Island was documented to be successful in June 1987 until a Special Order was issued by the Secretary of the Department of Environment and Natural Resources (DENR) commanding the Calawit Tagbanua to vacate the protected area and return to the resettlement areas in Culion Island in July of the same year.

From 1988 to 1994, the Calawit Tagbanua persistently petitioned their case to permanently settle in Calauit Island and rescind the proclamation of the island as a protected area. Assessing the events within almost two decades since the Tagbanua's displacement, the Regional Trial Court has (1) identified the defendants who could stay at the island and those who needed to vacate and live in the resettlement areas; (2) ordered the DENR to procure another suitable relocation site; (3) demanded the DENR to deliver the expected quality of service in the resettlement areas to prevent petitioners from going back to Calauit Island and contribute to the disturbance and destruction of the Calauit Safari. In 2002, the Court of Appeals affirmed the assailed ruling and further disputed the Calawit Tagbanua's claim of ownership to lands of Calauit Island because of absence of proof presuming that the island, as a public domain, belongs to the state. Applications for consideration, including to the Office of the President, by the Calawit Tagbanua have ultimately been denied even with the support of *pro bono* counsels and a letter from the Bishop of the Apostolic Vicariate of Taytay, Palawan, to the Chief Justice requesting for reconsideration.

Not until 2008 did the Calawit Tagbanua gained legitimacy over the island when their Certificate of Ancestral Domain Title (CADT) was issued under Republic Act 8371 or the Indigenous Peoples Rights Act of 1991. The CADT covers 3, 683.2334 hectares of land and water, including barangays Calauit and Quezon of the municipality of Busuanga. The holder of the CADT are the Calawit Tagbanua as represented by the *Balik Calawit Movement*. The application for CADT as well as the formulation of the Ancestral Domains Sustainable Development and Protection Plan (ADSDPP) of the Calawit Tagbanua was supported by other Tagbanua communities in Calamianes, particularly by the organization *Saragpunta Tagbanua Calamian Inc.*, that have also been awarded of CADT prior. Through the initial efforts of the *Saragpunta* federation, the Calawit Tagbanua and the Balik Calawit Movement have been able to outsource technical and financial support from partners such as the province of Antwerp in Belgium and the University of the Philippines.

Negotiating power

The case of the Calawit Tagbanua shows evidence of how dispossession and eviction from the island threaten the human security of the indigenous peoples, while land entitlements that directly addressed some of the issues of marginalization reduced vulnerability. Based on the case of the Calawit Tagbanua, the reorientation and redistribution of power can be achieved through (1) an enabling national policy, that is, the Indigenous Peoples Rights Act, and (2) a combination of resistance to marginalization and conformity to the politics of the state. The transformative period in the history of the Calawit Tagbanua struggle, that is, when they have organized themselves as the Balik Calawit Movement, illustrated these two points. Presenting their indigeneity in terms perceptible by the state (Eder 2013) gained them legitimacy and a level position as a negotiating party to the issue. In particular, organizing into

an identifiable collective of people with definite and clear goals conforming to the necessities of issue-oriented institutions had made them recognizable not only to the state but to other institutional actors as well (Hirtz 2003). The Balik Calawit Movement has outsourced and mobilized resources and partnerships that helped them build a support network and represent the indigenous community in policymaking. For instance, the Calawit Tagbanua is now being represented by the Balik Calawit Movement in consultations for the formulation of the Local Climate Change Action Plan and Disaster Risk Reduction Management Plan of the municipality of Busuanga, among others. Accepting the reality that the Calauit Safari shall continue its existence on the ancestral land, the Calawit Tagbanua hence then coped by adjusting to the political dynamics and gaining "entitlements" that come with the CADT.
Data source: Supreme Court of the Philippines document G.R. No. 156022 dated 6 July 2015.

8.5 Conclusion

Anthropology has long been familiar with the intellectual interests that recently emerged from the concept of the Anthropocene (Orr et al. 2015), including human security, climate change, and disaster management. In this chapter, we argue that placing the human condition and experience at the core of the discourse is what human security could offer to improve the prevailing science of climate change and disasters. The incorporation of the concept of human security brings closer to people of different cultures the science of climate change and disasters as well as encourages the rethinking of how science is framed.

The cultural dimension of climate change and disaster risk reduction has been abstracted for the convenience of policy, governance, and management. In response, we describe the cultural dimensions of human security as a complex variety of ways of knowing—each with its own right—and its manifestations in the different tangible and intangible elements of culture. It has reiterated narratives about human–environment interactions alternative to the Western anthropocentric worldview, where people and nature are not seen as separate from one another or when people are not viewed as the superior species. It emphasized how people shape and are being shaped by their particular environment and the power structures that constrain them. Taking these particularities into consideration and learning from the insiders' point of view is what it means to understand the cultural dimensions of human security.

Finally, the notion of culture as place-specific is continuously being challenged by global drivers of change not just by climate change but by globalization, population movements, and the dominance of science and technology. We must take caution in our scientific practice of delimiting the dynamic and complex nature of culture and understand that, for many of the peoples in Asia, switching from one frame of knowing to another is commonplace and people have the agency to negotiate and assess how alternative options can be assimilated into their culture. Perhaps, this is how human security should be addressed, in a transformative cultural context toward a path for future possibilities.

References

Adamson G (2012) 'The languor of the hot weather': Everyday perspectives on weather and climate in colonial Bombay, 1819–1828. J Hist Geogr 38(2):143–154. https://doi.org/10.1016/j.jhg.2011.11.018

Adamson G, Hannaford M, Rohland E (2018) Re-thinking the present: the role of a historical focus in climate change adaptation research. Global Environ Change 48:195–205. https://doi.org/10.1016/j.gloenvcha.2017.12.003

Adger W, Barnett J, Brown K, Marshall N, O'Brien K (2012) Cultural dimensions of climate change Impacts and adaptation. Nat Clim Change 3:112–117. https://doi.org/10.1038/nclimate1666

Adger W, Pulhin J, Barnett J, Dabelko G, Hovelsrud G, Levy M, Vogel C (2014) Human security. In: Field CV (ed) Climate change 2014: impacts, adaptation, and vulnerability. Part A: global and sectoral aspects. Contribution of working group II to the fifth assessment report of the inter-governmental panel on climate change. Cambridge University Press, Cambridge and New York, pp 755–791

Arnall A, Kothari U (2015) Challenging climate change and migration discourse: different under-standings of timescale and temporality in the Maldives. Global Environ Change 31:199–206. https://doi.org/10.1016/j.gloenvcha.2015.01.011

Asmiwyati I, Mahendra M, Arifin N, Ichinose T (2015) Recognizing indigenous knowledge on agricultural landscape in Bali for micro climate and environment control. Procedia Environ Sci 28:623–629. https://doi.org/10.1016/j.proenv.2015.07.073

Balay-As M, Marlowe J, Gaillard J (2018) Deconstructing the binary between the indigenous and scientific knowledge in disaster risk reduction: approaches to high impact weather hazards. Int J Disaster Risk Reduct 30:18–24. https://doi.org/10.1016/j.ijdrr.2018.03.013

Bankoff G, Cannon T, Kruger F, Schipper L (2015) Exploring the links between cultures and disasters. In: Kruger F, Bankoff G, Cannon T, Orlowski F, Schipper L (eds) Cultures and disasters: understanding cultural framings in disaster risk reduction. Routledge, New York, USA, pp 1–16

Barnes J, Dove M, Lahsen M, Mathews A, McElwee P, McIntosh R, Yager K (2013) Contribution of anthropology to the study of climate change. Nat Clim Change 3:541–544. https://doi.org/10.1038/NCLIMATE1775

Blais-McPherson M, Rudiak-Gould P (2017) Strengthening inter-disciplinary and inter-ideological collaboration on REDD: a cultural theory approach. Global Environ Change 42:13–23. https://doi.org/10.1016/j.gloenvcha.2016.10.008

Blicharska M, Smithers R, Kuchler M, Agrawal G, Gutierrez J, Hassanali A, Mikusinski G (2017) Steps to overcome the North-South divide in research relevant to climate change policy and practice. Nat Clim Change 7:21–27. https://doi.org/10.1038/NCLIMATE3163

Casson S (2016) Ensuring community and agricultural resiliency to climate change: ceremonial practices as emic adaptive strategies. Future Food: J Food Agri Soc. http://www.futureoffoodjournal.org/index.php/journal/article/view/230/pdf

Choudhary P (2016) Vernacular built environments in India: an indigenous approach for resilience. In: Urban disasters and resilience in Asia. Elsevier, pp 269–286. https://doi.org/10.1016/b978-0-12-802169-9.00017-3

Chowdhury R, Moore G (2017) Floating agriculture: a potential cleaner production technique for climate change adaptation and sustainable community development in Bangladesh. J Cleaner Prod 150:371–389. https://doi.org/10.1016/j.jclepro.2015.10.060

Colette A (2016) The politics of framing risk: minding the vulnerability gap in climate change research. World Dev Perspect 1:43–48. https://doi.org/10.1016/j.wdp.2016.06.003

Crate S, Ulrich M, Habeck J, Desyatkin A, Desyatkin R, Fedorov A, Takakura H (2017) Permafrost livelihoods: a transdisciplinary review and analysis of thermokarst-based systems of indigenous land use. Anthropocene 18:89–104. https://doi.org/10.1016/j.ancene.2017.06.001
Cunsolo A, Ellis N (2018) Ecological grief as a mental health response to climate change-related loss. Nat Clim Change 8:275–281. https://doi.org/10.1038/s41558-018-0092-2
Darlington S (2003) The spirit(s) of conservation in Buddhist Thailand. In: Selin H (ed) Nature across cultures: views of nature and the environment in non-Western cultures. Springer Science + Business Media, Dordrecht, pp 129–145
Das S, Mukhopadhyay P (2018) Multi-hazard disaster resilient housing with bamboo-based system. Procedia Eng 212:937–945. https://doi.org/10.1016/j.proeng.2018.01.121
Demeritt D (2001) The construction of global warming and the politics of science. Ann Assoc Am Geogr 91(2):307–337. http://www.jstor.org/stable/3651262
Denton F, Wilbanks TJ, Abeysinghe AC, Burton I, Gao Q, Lemos MC, Masui T, O'Brien KL, Warner K (2014) Climate-resilient pathways: Adaptation, mitigation, and sustainable development. In: Field et al (eds) Climate change 2014: impacts, adaptation, and vulnerability. Part A: global and sectoral aspects. Contribution of working group II to the fifth assessment report of the intergovernmental panel on climate change. Cambridge University Press, Cambridge, UK, pp 1101–1131
Dong G, Wang L, Cui Y, Elston R, Chen F (2013) The spatiotemporal pattern of the Majiayao cultural evolution and its relation to climate change and variety of subsistence strategy during late Neolithic period in Gansu and Qinghai provinces, northwest China. Quat Int 316:155–161. https://doi.org/10.1016/j.quaint.2013.07.038
Douglas M (1970) Natural symbols: explorations in cosmology. Cresset, London
Douglas M (1982) Risk and culture: an essay on the selection of technical and environmental dangers. University of California Press, Berkeley
Dressler W, Fabinyi M (2011) Farmer gone fish'n? Swidden decline and the rise of grouper fishing on Palawan Island, the Philippines. J Agrarian Change 11(4):536–555. https://doi.org/10.1111/j.1471-0366.2011.00309.x
Dudgeon R, Berkes F (2003) Local understandings of the land: traditional ecological knowledge and indigenous knowledge. In: Selin H (ed) Nature across cultures: views of nature and the environment in non-Western cultures. Springer Science + Business Media, Dordrecht, pp 75–96
Duraiappah A (2010) Social sciences in intergovernmental processes. Paper presented at the changing nature, changing science symposium, Nagoya, Japan, 13 December 2010
Eder J (2013) The future of indigenous peoples in the Philippines: sources of cohesion, forms of difference. Philipp Q Culture Soc 41(3/4):273–294. https://www.jstor.org/stable/43854732
Few R, Morchain D, Spear D, Mensah A, Bendapudi R (2017) Transformation, adaptation and development: relating concepts to practice. Palgrave Commun 3:17092. https://doi.org/10.1057/palcomms.2017.92
Field J (2017) What is appropriate and relevant assistance after a disaster? Accounting for culture(s) in the response to Typhoon Yolanda. Int J Disaster Risk Reduct 22:335–344. https://doi.org/10.1016/j.ijdrr.2017.02.010
Fiori J, Dempsey B, Owen G (2013) Humanitarianism recast. In: Internal paper. Save the Children, London
Folke C, Carpenter S, Walker B, Scheffer M, Chapin T, Rockstrom J (2010) Resilience thinking: integrating resilience, adaptability and transformability. Ecol Soc 15(4):20. http://www.ecologyandsociety.org/vol15/iss4/art20/
Gomes A (2012) Alternative 'development': indigenous forms of social ecology. Third World Q 33:1059–1073. https://doi.org/10.1080/01436597.2012.681491
Hall G, Scheltens M (2005) Beyond the drought: towards a broader understanding of rural disadvantage. Rural Soc 15:348–358. https://doi.org/10.5172/rsj.351.15.3.348
Harrison M (2010) Medicine in an age of commerce and empire: Britain and its tropical colonies, 1600–1830. Oxford University Press, Oxford and New York

Hirtz F (2003) It takes modern means to be traditional: on recognizing indigenous cultural communities in the Philippines. Dev Change 34:887–914. https://doi.org/10.1111/j.1467-7660.2003.00333.x
Hiwasaki L, Luna E, Syamsidik Shaw R (2014) Process for integrating local and indigenous knowledge with science for hydro-meteorological disaster risk reduction and climate change adaptation in coastal and small island communities. Int J Disaster Risk Reduct 10:15–27. https://doi.org/10.1016/j.ijdrr.2014.07.007
Hoffman A (2015a) How culture shapes the climate change debate. Stanford University Press, Stanford
Hoffman S (2015b) Culture: the crucial factor in hazard, risk, and disaster recovery: the anthropological perspective. In: Collins A, Samantha J, Manyena B, Walsh S, Shroder J (eds) Hazards, risks, and disasters in society. Elsevier, Amsterdam, pp 289–305. https://doi.org/10.1016/b978-0-12-396451-9.00017-2
Hogan A, Scarr E, Lockie S, Chant B, Sylvia A (2012) Ruptured identity of male farmers: subjective crisis and the risk of suicide. J Rural Social Sci 27:118–140. https://search.proquest.com/docview/1429487606?pq-origsite=gscholar
Hoogstra-Klein M, Permadi D, Yasmi Y (2012) The value of cultural theory for participatory processes in natural resource management. Forest Policy Econ 20:99–106. https://doi.org/10.1016/j.forpol.2011.12.001
Jasanoff S (2010) A new climate for society. Theor Cult Soc 27(2–3):233–253. https://doi.org/10.1177/0263276409361497
Johnson T, Grim B (ed) (2018) World religion database. Brill, Leiden and Boston. https://www.worldreligiondatabase.org. Accessed 5 Apr 2018
Kahan D (2010) Fixing the communications failure. Nature 463:296–297. https://doi.org/10.1038/463296a
Kakoty S (2018) Ecology, sustainability and traditional wisdom. J Cleaner Prod 172:3215–3224. https://doi.org/10.1016/j.jclepro.2017.11.036
Karlsson S, Srebotnjak T, Gonzales P (2007) Understanding the North-South knowledge divide and its implications for policy: a quantitative analysis of the generation of scientific knowledge in the environmental sciences. Environ Sci Policy 10:668–684
Kenny J (1995) Climate, race and imperial authority: the symbolic landscape of the British hill station in India. Ann Assoc Am Geogr 85(4):694–714
King G, Murray C (2001) Rethinking human security. Polit Sci Q 116. https://doi.org/10.2307/798222
Krauss R (2009) Localizing climate change: a multi-sited approach. In: Falzon M (ed) Multi-sited ethnography. Ashgate, Abingdon, pp 149–164
Lahsen M (2010) The social status of climate change knowledge. WIREs climate change 1. https://doi.org/10.1002/wcc.27
Lahsen M (2015) Digging deeper into the why: cultural dimensions of climate change skepticism among scientists. In: Barnes J, Dove M (eds) Chapter 9. Climate cultures: anthropological perspectives on climate change. Yale University Press, New Haven and London, pp 221–248
Livingstone D (1991) The moral discourse of climate: historical considerations on race, place and virtue. J Hist Geogr 17(4):413–434
Lorenzen R, Lorenzen S (2010) Changing realities–perspectives on Balinese rice cultivation. Hum Ecol 39(1):29–42. https://doi.org/10.1007/s10745-010-9345-z
McEvoy J, Gilbertz S, Anderson M, Ormerod K, Bergmann N (2017) Cultural theory of risks as a heuristic for understanding perceptions of oil and gas development in Eastern Montana, USA. Extractive Ind Soc 4:852–859. https://doi.org/10.1016/j.exis.2017.10.004
Merton R (1973) The normative structure of science. In: Merton R, Storer N (eds) The sociology of science: theoretical and empirical investigations. University of Chicago Press, Chicago
Morisaki K, Kunikita D, Sato H (2018) Holocene climatic fluctuation and lithic technological change in northeastern Hokkaido (Japan). J Archaeol Sci 17:1018–1024. https://doi.org/10.1016/j.jasrep.2016.04.011

Nef J (1999) Human security and mutual vulnerability: the global political economy of development and underdevelopment. International Development Centre, Ottawa
O'Brien K, Eriksen S, Inderberg TH, Sygna L (2014) Climate change and development: adaptation through transformation. In: Håkon Inderberg T, Eriksen S, O'Brien K, Sygna L (eds) Climate change adaptation and development: transforming paradigms and practices. Routledge, Abingdon, UK, pp 273–289
Orr Y, Lansing J, Dove M (2015) Environmental anthropology: Systemic perspectives. Annu Rev Anthropol 44:153–168. https://doi.org/10.1146/annurev-anthro-102214-014159
Palawan Council for Sustainable Development (2018) The Calauit game preserve and wildlife sanctuary. In: Protected areas. http://pcsd.gov.ph/protected_areas/calauit.htm. Accessed 2 April 2018
Parsizadeh F, Ibrion M, Mokhtari M, Lein H, Nadim F (2015) Bam 2003 earthquake disaster: on the earthquake risk perception, resilience and earthquake culture–cultural beliefs and cultural landscape of Qanats, gardens of Khorma trees and Argh-e Bam. Int J Disaster Risk Reduct 14(P4):457–469. https://doi.org/10.1016/j.ijdrr.2015.09.011
Pei S (1985) Some effects of the Dai people's cultural beliefs and practices on the plant environment. In: Hutterer K, Rambo A, Lovelace G (eds) Cultural values and human ecology. University of Michigan, Michigan, pp 321–339
Puri R (2015) The uniqueness of the everyday: herders and invasive species in India. In: Barnes J, Dove M (eds) Chapter 10. Climate cultures: anthropological perspectives on climate change. Yale University Press, New Haven and London, pp 249–272
Rahman M, Nakagawa H, Khaleduzzaman A, Ishigaki T (2003) Flow and scour-deposition around bandals. In: Proceedings of the 5th international summer symposium, JSCE, Tokyo, pp 177–180
Rosaldo R (1980) Ilongot headhunting, 1883–1974: a study in society and history. Stanford University Press, Stanford
Ruzol C, Banzon-Cabanilla D, Ancog R, Peralta E (2017) Understanding water pollution management: evidence and insights from incorporating cultural theory in social network analysis. Global Environ Change 45:183–193. https://doi.org/10.1016/j.gloenvcha.2017.06.009
Schuler B (2014) Introduction: the dynamics of climate and the dynamics of culture. In: Environmental and climate change in South and Southeast Asia: how are local cultures coping? Brill, Leiden, pp 1–24
Sharp L, Macrorie R, Turner A (2015) Resource efficiency and the imagined public: insights from cultural theory. Global Environ Change 34:196–206. https://doi.org/10.1016/j.gloenvcha.2015.07.001
Sherpa P (2014) Climate change, perceptions, and social heterogeneity in Pharak, Mount Everest Region of Nepal. Hum Org 73:153–161. https://doi.org/10.17730/humo.73.2.94q43152111733t6
Soflaei F, Shokouhian M, Zhu W (2017) Socio-environmental sustainability in traditional courtyard housed of Iran and China. Renew Sustain Energy Rev 69:1147–1169. https://doi.org/10.1016/j.rser.2016.09.130
Sponsel L, Natadecha-Sponsel P (2003) Buddhist views of nature and the environment. In: Selin H (ed) Nature across cultures: views of nature and the environment in non-Western cultures. Springer Science + Business Media, Dordrecht, pp 351–371
Stott P (1991) Mu'ang and pa: elite views of nature in a changing Thailand. In: Chitakasem M, Turton A (eds), Thai constructions of knowledge. School of Oriental and African Studies. University of London, London, pp 142–154
Syafwina (2014) Recognizing indigenous knowledge for disaster management: *Smong*, early warning system from Simeulue Island, Aceh. Procedia Environ Sci 20:573–582. http://doi.org/10.1016/j.proenv.2014.03.070
Takakura H (2016) Limits of pastoral adaptation to permafrost regions caused by climate change among the Sakha people in the middle basin of Lena River. Polar Sci 10(3):395–403. https://doi.org/10.1016/j.polar.2016.04.003

Thomas J (1998) Naturalizing nationhood: Ideology and practice in early twentieth-century Japan. In: Minichiello A (ed) Japan's competing modernities: Issues in culture and democracy, 1900–1930. Honolulu: University of Hawai'i, p 114–132

Tucker J (2003) Japanese views of nature and the environment. In: Selin H (ed) Nature across cultures: views of nature and the environment in non-Western cultures. Springer Science + Business Media, Dordrecht, pp 161–183

Vins H, Bell J, Saha S, Hess J (2015) The mental health outcomes of drought: a systematic review and causal process diagram. Int J Environ Res Public Health 12(10):13251–13275. https://doi.org/10.3390/ijerph121013251

Xia C, Schönfeld M (2011) A Daoist response to climate change. J Global Ethics 7:195–203. https://doi.org/10.1080/17449626.2011.590279

Xu L, Liu Y, Sun Q, Chen J, Cheng P, Chen Z (2017) Climate change and human occupations in the Lake Daihai basin, north-central China over the last 4500 years: A geo-archaeological perspective. J Asian Earth Sci 138:367–377. https://doi.org/10.1016/j.jseaes.2017.02.019

Zhang H, Nakagawa H (2018) Validation of indigenous knowledge for disaster resilience against river flooding and bank erosion. In: Shaw R, Shiwaku K, Izumi T (eds) Science and technology in disaster risk reduction in Asia. Elsevier, London, pp 57–76

Zhang H, Nakagawa H, Baba Y, Kawaike K, Teraguchi H (2010) Three-dimensional flow around Bandal-like structures. Annu J Hydraulic Eng 54:175–180

Clarissa D. Ruzol is a student of MRes/PhD Anthropology at the London School of Economics and Political Science. They worked as Assistant Professor of Anthropology at the Department of Social Forestry and Forest Governance, University of the Philippines Los Baños from 2017 to 2020. They were a graduate fellow of the Asia Research Institute in the National University of Singapore in 2015 and had received grants from the University of Montana, the Southeast Asian Regional Center for Graduate Study and Research in Agriculture, the Swedish Institute, and the UNICEF in the Kyrgyz Republic to present their works in conferences. They have done research on topics such as cultural risk theory, social network analysis, local knowledge, environmental policy, and social impact assessment. They also write about structural violence, power, scientific knowledge, and infrastructure development in the Philippines.

Dr. Juan M. Pulhin is full Professor and former Dean of the College of Forestry and Natural Resources, University of the Philippines Los Baños, Philippines. He was Lead Author of the 2007 Fourth Assessment Report of the Intergovernmental Panel on Climate Change (IPCC), Coordinating Lead Author of the Human Security Chapter of the 2014 IPCC Fifth Assessment Report and currently Lead Author of the Asia chapter of the IPCC Sixth Assessment Report. He has been a four-time a visiting professor at The University of Tokyo in 2007, 2010, 2016 and 2017 and a visiting scholar at Waseda University, Tokyo in 2018. He has written and edited 11 books and published more than 100 technical papers and book chapters on natural resources policy and governance, and on the human dimension of climate change and disaster risks. He is a fellow, invited expert, and council member of several international organizations and a member of the National Academy of Science and Technology (NAST), Philippines.

Chapter 9
Population Movements and Human Security

Oscar A. Gómez

Abstract At the outset of global efforts to understand and confront climate change, migration was presented as a cautionary tale: a security threat compromising the stability of affected societies and surrounding countries, which justified mitigation action. Thirty years later, the negative view of migration persists, but it is much more nuanced and contested by a plethora of research that suggests a complex picture of intersections between climate change, disaster risk, and population movements. The chapter provides an overview of these intersections, particularly in the Asian context. The presentation is divided around three main themes. The first deals with the initial framing of climate change and disasters as a cause and migration as a threat. It describes the causal mechanisms proposed and how regional evidence does or does not support migration fears. The second theme introduces current efforts to reframe migration as a means to adapt to the effects of climate change. Evidence from existing experience using migration to confront environmental stress, individually by people themselves or through resettlement programs, is reviewed. The third theme refers to emerging issues derived from confronting climate change and disasters on a background of moving populations. Two issues receive specific attention: addressing the needs of migrant populations affected by disaster crises and the role of remittances in responding and recovering from them. The final section synthesizes the three themes, identifying the prospects of reducing human insecurity and providing security through migration, or despite it.

Keywords Environmental migration · Adaptation · Resettlement · Vulnerable populations · Security referent

---

O. A. Gómez (✉)
College of Asia Pacific Studies, Ritsumeikan Asia Pacific University, Beppu, Japan
e-mail: oagomez@apu.ac.jp

J. M. Pulhin et al. (eds.), *Climate Change, Disaster Risks, and Human Security*, Disaster Risk Reduction, https://doi.org/10.1007/978-981-15-8852-5_9

9.1 Introduction

At the outset of global efforts to understand and confront the threat of climate change, migration was presented as a cautionary tale. If international commitment to reducing emissions of greenhouse gases was not achieved, hordes of environmental refugees escaping from destruction triggered by weather disruptions would become a security threat, compromising the stability of their societies and surrounding countries, with global repercussions. Migration was thus mostly framed as a negative consequence of climate change, foreseen as the product of mainly fast-onset disasters. The refugee card was a strategy to convince major stakeholders to recognize the problem and act on its mitigation before it was too late; this, despite the fact that there was no evidence then about climate events actually triggering such effects. Thirty years later, the negative view of migration persists, but it is much more nuanced and contested by a plethora of research that suggests a complex picture of intersections between climate change, disaster risk, and population movements. In the present chapter, I provide an overview of these intersections, particularly in the Asian context.

Figure 9.1 shows the interactions addressed by this review. Area A is our main interest but limiting the review to only this would leave too much outside. On the one hand, area C, made of research on disasters not related to climate change and their overlaps with migration, is an important source of insights about the nature of the relation between the three topics. For example, it has been observed that the effects of tsunamis and storm surges can be very similar, as in the Philippines after the 2013 Typhoon Yolanda, so non-climatic disasters also deserve consideration. Experiences in evacuation and relocation related to disaster risk reduction (DRR) would also be of great interest to understand what is at hand when we talk about population movements. On the other hand, not all displacement related to climate

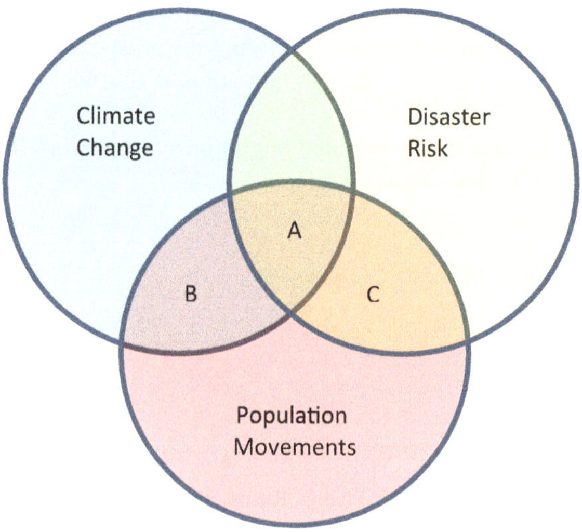

Fig. 9.1 Framework for analyzing the interactions between climate change, disaster risk, and population movements

change is connected to disasters, at least not to fast-onset ones. Much attention has been given to ecosystem deterioration resulting in migration in order to make up for lost livelihoods; as the review will show, this is indeed the mechanism of displacement that most certainly is going to be intensified by climate change and, thus, it must be part of the assessment. Displacement, expected following sea level rise, is also a recurrent topic in global literature that is not connected to disaster risk. In sum, area B also needs to be considered.

It is important to note that this is not an exhaustive revision and that sources included were filtered by the use of keywords such as "migration" and "displacement" in titles and abstracts. Thus, research that mentions migration but in which the topic is not central was not included. So, for example, there are only four papers on Google Scholar that cover the topic of migration and displacement after Typhoon Yolanda, and there is none about the 2008 Cyclone Nargis in Myanmar. Such dearth in the literature about even major events is recognized not only in Asia but globally. Indeed, it has been observed how the body of literature on climate change and migration is growing, but the topic is still understudied, fragmented, and in need of robust evidence (Raleigh et al. 2008; Bardsley and Hugo 2010; Piguet et al. 2011; United Nations Task Team on Social Dimensions of Climate Change 2011; Gray and Bilsborrow 2013). The chapter on human security of the 5th IPCC Assessment Report (Adger et al. 2014), aware of this limitation, points out the existence of a very significant new body of research in the 5 years preceding its publication, on which it based its review. Still then, the chapter only quotes about nine papers specific to migration in Asia. Fortunately, the research body has kept growing, with more regional work available as reports of the Asian Development Bank (ADB) (2012, 2017) hint at. Including the literature related to area C of Fig. 9.1 in the review complements the literature and adds other perspectives to the discussion. The IPCC (2012) special report on extreme events even uses examples of dam constructions to discuss about the role of relocations, which I briefly mention. Still, it is important to keep in mind that the evidence is limited and much more research is necessary to have robust conclusions.

Besides, as the ADB (2012: 54) suggests, "too often, research on climate-induced migration in Asia and the Pacific is conducted from industrialized countries, sometimes without field studies." In order to address this, in this review I give prominence to existing empirical research on each of the intersections covered. General presentations are only briefly introduced through the following sections—i.e., the human security chapter of the 5th IPCC report, as well as the ADB work on migration (2012) and its 2017 update—and should be directly read by those looking for a global perspective. Instead, the contribution of this regional review is to provide evidence supporting global narratives when possible and to contest them when necessary. I used general reports as baselines and expanded the literature review beyond what they already covered. Still, there are two final observations: first, the review is limited to works published in English, which may not reflect all the existing evidence being produced locally and, second, given the paucity of academic publications, think tank reports, and other gray literature from local sources are also included. Sanwal et al. (2017) suggest such a broader inclusion of information would be part of the new role of the IPCC supporting national implementation; still, conflicts of interest in non-peer-reviewed reports remain to be solved. Each of the fields intersecting in Fig. 9.1 is associated with different epistemic communities trying to maintain some domain over the international agenda through different framings of population movements and advocacy. This is easier to grasp by distinguishing existing narratives on migration and displacement as addressing different human security questions (Gomez et al. 2016). In the next section, I distinguish them and explain the structure of the review.[1]

9.2 The Structure of the Review: Population Movements and Human Security Questions

As the most populous region in the world, Asia is expected to be particularly affected by climate change effects, just as it is already vulnerable to disasters. In relation to population movements, "more than two-thirds of all new displacement associated with disasters in 2016 took place in East Asia and the Pacific, where 16.4 million incidents accounted for 68 percent of the global total" (IDMC 2017). South Asia accounts for an additional 3.6 million, increasing the share to 82.4% of all displacements. China, Philippines, India, and Indonesia are the most affected, while the top 10 also includes Japan, Bangladesh, Myanmar, and Sri Lanka; by size of population affected, Philippines, Sri Lanka, and Myanmar are also among the 10 most affected countries—see Fig. 9.2.

Therefore, from the outset of concern on climate change, the region received particular attention. The widely cited work of Homer-Dixon (1991) on conflict discussed the case of displacement of populations between Bangladesh and India as an example of social effects of climate change that may in the future result in confrontation. He also mentions Filipino fishermen as a tentative population due to become "environmental refugees." Elliot (2012) shows that such traditional connection between climate change, human security, and migration in terms of social instability and conflict is still relevant in Southeast Asia.

This initial narrative presented displacement as an unprecedented threat and, so, it required research to show that concern was well-founded—although, since it was only a precautionary tale, the plausibility of the story was the most important, what explains the dearth of research. For this, mechanisms linking climate change effects to displacement have to be singled out, as well as hot spots where such mechanisms would have the more significant effects. Particularly important is to determine whether the displacement is domestic or international, as they result in different types of social instability. Because these are effects without precedents, a study of previous or ongoing displacement is of limited support for forecasting, although good enough for affirming the feasibility of the threat. In fact, understanding the existing experience and the actual complexity underlying the intersections between climate change,

[1] This review draws substantially from my previous general review on climate change and migration (Gomez 2013a), expanding, deepening, as well as reevaluating observations made in that occasion.

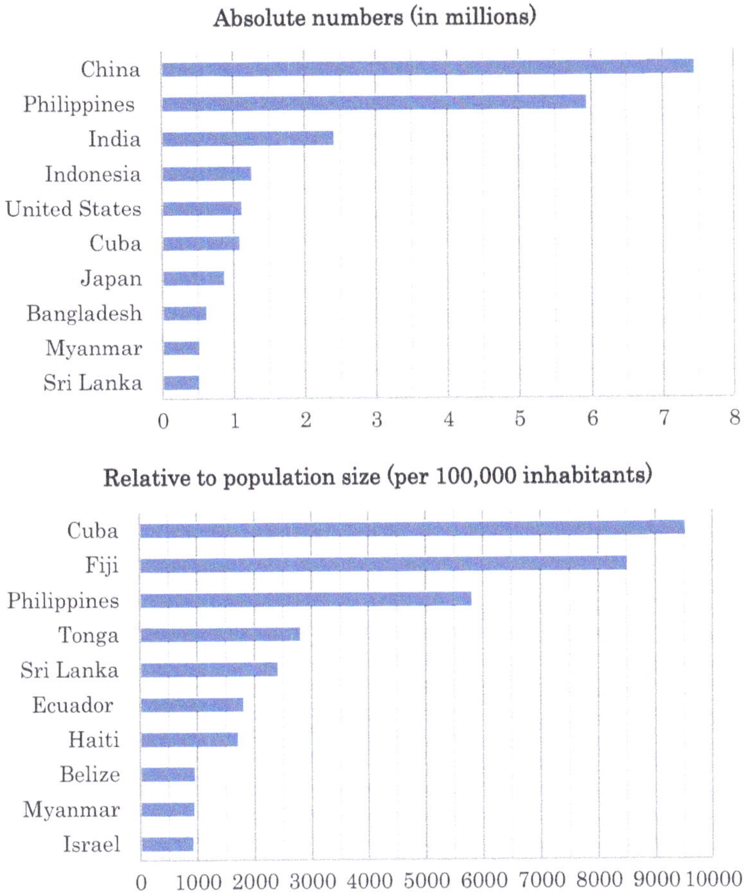

Fig. 9.2 Countries with the most new displacements by disasters in 2016 (*Source* IDMC [2017: 38])

disaster risk, and population movement will open the door for alternative narratives. I identify at least other two, which, in combination with the threat approach, are used to organize the review.

A second framing—i.e., population movements as a means to deal with environmental change—has slowly emerged through climate change discussions. There are multiple arguments defending migration as a means against climate change. The most basic one is the prehistoric evidence of humanity moving around the world pressured by environmental threats. Findley (1994) very early suggested that the alternative to act against droughts was finding ways to encourage migration. In that sense, migration as a means has become particularly important for two reasons: because of the change in emphasis from mostly mitigation of climate change toward more consideration to adaptation and because the evidence suggests that most of the migration will happen indirectly through effects on livelihoods, which is the phenomenon that Findley describes. As Black et al. (2011) suggest, "Migration may be the most effective way to allow people to diversify income and build resilience where environmental change threatens livelihoods."

Migration as a means to human security implies a different approach to its understanding. Causality becomes less important and the attention concentrates on understanding what kind of adaptation it is: a last resort or a useful alternative? Voluntary or involuntary? Much attention is given to actual patterns of displacement existing now, particularly in relation to rural-urban migration. Section four describing this framing also includes literature from disaster studies in relation to relocations, which is also seen as an example of population movement that could come in handy for dealing with future adaptation needs. Rural-urban migration and the disaster literature also coincide in the topic of maladaptation: the fact that population movements can also result in an increase of their vulnerability rather than the expected security. This is an important contribution since the attention to migration as adaptation that the IPCC recognizes (Adger et al. 2014) has already shown that in practice this may be a very limited, complicated option, as the Asian experience also suggests.

Maladaptation to climate change and disaster risk highlights the importance of not only considering migration as a threat or adaptation means but also looking at migrants themselves, both as menaced by threats—i.e., in security-speak, as referents of human security—and as providers of their own protection. In other words, there is a third possible framing of population movements that puts aside any causality questions and looks instead at how climate change and disaster risk affect migrants. This is a rather emerging narrative, so the evidence is even more limited than in the previous two framings, still it is possible to distinguish two topics that have received some attention: remittances as a source of support after disasters and dealing with migrant populations when disasters strike. As with the case of migration as a means, talking in terms of migrants opens the discussion to new alternatives but includes its own limitations, particularly when "migrant" is treated as a uniform identity. In the final section, I discuss the strengths and limitations of each of the framings in tandem, weighing general policy recommendations against the reviewed evidence from the region.

Before moving on to the first framing, it is worth noting how a good share of the literature reviewed is specific to climate change and migration research. Indeed, disaster studies have made limited emphasis on the topic. Handbooks such as Rodriguez et al. (2007) and Wisner et al. (2012) do not have dedicated chapters on migration, although traditional themes such as evacuation, sheltering, housing, and livelihoods are heavily related to the displacement that disasters generate. Wisner et al. (2012) include a section of five chapters on "Vulnerabilities and capacities" that does not include migrants, although they recognize that illegal immigrants and refugees are vulnerable groups requiring attention. Indeed, it is the chapter on climate change in the volume that presents a more significant comment on migration. This suggests that the topic is entering disaster studies through climate change just recently, at the level of a hypothesis which relevance needs to be confirmed and, thus, be taken with a grain of salt.

9.3 Disasters and Climate Change as Causes of Migration

The possible disruptions resulting from climate change were summarized by the UK government-sponsored Foresight project (2011) on migration and environmental change in the following six areas:

- sea level rise
- changes in tropical storm and cyclone frequency or intensity
- changes in rainfall regimes
- increases in temperature
- changes in atmospheric chemistry
- melting of mountain glaciers

Not all of these disruptions are relevant as direct causes of migration—e.g., the atmospheric chemistry—while several interact, generating different patterns of expected migration. At least three of such mechanisms can be distinguished (Martin 2009: 356; Gomez 2013a):

(1) intensification of natural disasters, such as hurricanes and cyclones that destroy housing and livelihoods and require people to relocate for shorter or longer periods;
(2) increased warming and drought that affect agricultural production, diminishing people's livelihoods, and access to clean water;
(3) rising sea levels that render coastal areas uninhabitable.

Martin mentions a fourth, hypothetical mechanism in which climate change-induced conflict results in displacement. I leave this aside because conflict is covered on a different chapter (please see Chapter 10)—and also because the composed causality makes it even harder to qualify.

Evidence from each of these mechanisms can be analyzed separately, although they are usually used in tandem in order to identify hot spots where displacement effects are due to be more visible. From Asia, the places considered usually comprise Bangladesh, the Philippines, China, the Indus Delta, and the Mekong Delta in Vietnam (ADB 2017). Surprisingly, there is little literature about climate-induced migration in the Philippines, despite disasters displacing up to 7 million people annually, as in 2013 when Typhoon Yolanda took place (IDMC 2015). Literature on the Indus Delta is similarly scarce.

There is also the "coastal approach" to visualize climate change vulnerability: as climate change effects due to sea level rise and tropical storms are due to hit populations living near the sea, those populations can be framed as becoming more vulnerable. This is supported by the fact that people, particularly in China and India, "have tended to move out of marginal dry and mountainous regions and to move broadly into coastal areas" (ADB 2017: 83). For instance, Seto (2011) identifies cities in nine mega-deltas in Asia that receive large numbers of migrants and that would be affected by climate change effects: the Yangtze River Delta, the Pearl River Delta, the Red River Delta, the Mekong River Delta, the Chao Phraya Delta, the Irrawaddy Delta, the Ganges-Brahmaputra Delta, the Indus River Delta, and the Tigris-Euphrates Delta. Cities on these deltas account for more than 350 million people, an undetermined share of whom would be exposed to floods, storms, and sea level rise.

Overall, as the ADB (2017: 82) points out, "existing projections of future climate change impacts on migration in Asia are limited in number, regional scope, and predictive power." Besides, as far as international migration caused by climate change and disasters goes, Obokata et al. (2014) find basically no research providing empirical evidence. From 31 papers they identified, only five looked at specific countries, while the rest made general approaches. Thus, they suggest that "international migration for obvious environmental reasons is not occurring in vast numbers." Indirect impacts are of influence, but there are no bases to estimate how important the phenomenon is. They also observe that more attention goes to migration within and from rural populations, while there is less about urban challenges. This is equally relevant for domestic displacement research (and most probably relates to the difficulty of doing such research).

9.3.1 Displacement Following Fast-Onset Disasters

Evidence from the region suggests that fast-onset disasters are not going to result in massive, destabilizing migration. Following Naik et al. (2007), Ewing (2012: 24) points out how the Indian Ocean Tsunami, "the largest single displacement event in modern regional history," displaced "nearly half a million people without leading to large-scale relocation beyond nearby areas." Naik et al. (2007: 58) suggest that migration resulting from the disaster may be a middle-term phenomenon, once families realize the impacts on livelihoods and the pace of economic recovery becomes evident. They also suggest that massive exodus may have been prevented by the degree of humanitarian assistance available. Certainly, the Indian Ocean Tsunami has been one of the better funded disaster management operations in history, but several other examples in the region also fail to show any massive displacement.

Apparently, the 2013 Typhoon Yolanda broke the Indian Ocean Tsunami's record of displacement with 4 million cases reported out of 14–16 million people affected (Yonetani and Yuen 2014). The numbers, nonetheless, have to be treated carefully, as Yonetani and Yuen show, because of lack of clarity on methods and definitions used and the fluidity of the numbers through the emergency. Despite these large numbers, Yonetani and Yuen (2013) and Sherwood et al. (2015) report that the vast majority of the displaced populations went back to their homes, some within hours. Major challenges following the emergency were those common to disaster research: recovering the lifeline of basic services, the standard of living, lost livelihoods, and addressing housing needs in a way that does not replicate past vulnerabilities—that is, building back better, relocating populations when necessary, and enforcing "No Dwelling Zones."

Paul (2005) presents the case of a tornado in north-central Bangladesh, close to the frontier with India, where fieldwork showed no major permanent displacement effects; only occasional migration linked to reconstruction efforts in other affected areas was reported. He maintains, as Naik et al. (2007) do, that aid from the government and non-government organizations plays a role in discouraging out-migration. Instead, the recovery process is seen as a livelihood opportunity.

In relation to floods, research examples from the Mekong Delta (Dun 2011) and Jakarta in Indonesia (Nurlambang 2012: 80) report no immediate effects. Thomas (2016) reports 1.7 million people displaced by floods in Myanmar in 2015, but the main permanent effects described are related to relocations and the effects on already internally displaced populations—perhaps because of lack of means for monitoring displacement. Instead, Dun (2011) identifies seasonal or permanent migration after repeated disasters as an adaptation behavior, a pattern that may be relevant to other periodic disasters. Islam and Hasan (2016) report not immediate but longer-term effects of displacement and continued vulnerability following Cyclone Aila in Bangladesh in 2009. Such pattern is closer to the slow-onset disaster mechanism, as described in the next subsection.

Finally, an outlier among the cases in the region is the experience after the 2011 Great East Japan Earthquake and Tsunami, particularly in relation to the nuclear reactors' meltdown in Fukushima prefecture. Around 110,000 persons were forced to evacuate because of the radiological emergency, plus an estimated 40,000 persons in areas not under mandatory evacuation who, nevertheless, decided to move, such as mothers with their children leaving husbands behind (Mosneaga 2015). Five years after the event, 70,000 persons are still to return, many of whom have restarted their lives elsewhere and may never go back. In other tsunami-affected areas, population decrease has also been noted, despite the government's ongoing efforts to revive communities. In this case, a common observation has been that the tsunami just accelerated ongoing trends of depopulation and gentrification of rural areas, common elsewhere in the country. All in all, while the fast-onset disasters do not appear to cause long term, destabilizing displacement, the case of the nuclear power plants suggests that other factors can result in exceptions to this trend.

9.3.2 Slow-Onset Effects and Livelihoods

While immediate displacement effects of fast-onset disasters are easier to discern, the causality from long-term effects of distress requires a different approach. In this case, there is no single cause, but several factors such as the economy and the household's social capital also become relevant in explaining displacement; therefore, methodologies to discern the relevance of environmental factors need to be more sophisticated, capturing a wider range of data and extracting their significance.

There have been several studies about the region showing the connection between environmental distress and displacement, as well as the limited effect of fast-onset disasters. Using a longitudinal survey conducted in rural Pakistan, Muller et al. (2014)

identify weather-related migration and suggest that it is heat stress, not flooding, what actually increases long-term migration. The effects are particularly relevant to men. Extreme heat stress is associated to migration of both land owners and the land- and asset-poor; still the magnitude and statistical significance for the latter is most pronounced. The authors suggest that owning land and assets may become a liability when considering migration as an adaptation strategy (briefly discussed in the next section).

A similar result is found by Bohra-Mishraa et al. (2014) using data from Indonesia. Following province-to-province movement of more than 7,000 households over a decade and a half, they also find that an increase in temperature is more significant than variations in rainfall and disasters in causing permanent out-migration. They point out that 25 degrees Celsius average temperature is the turning point for the effect in migration, while rainfalls over 2.2 m annually also are correlated with migration, although to a lesser extent. The mechanism of this migration is economical, affecting the value of household assets. The authors find only one natural disaster having consistent effects, although marginally significant, on migration: landslides. In a similar research about the Philippines, Bohra-Mishraa et al. (2017) identify an analogous pattern of migration, although in this case, typhoon activity also appears correlated to migration, through the effects on rice yields. Rainfall has no effect, which authors suggest may be related to the spread of irrigation.

Using a unique database following populations for 36 months on Chitwan Valley in Nepal, Massey et al. (2010) identify causal relations between environmental change and migration. They focus on declining land cover, rising times required to gather organic inputs, increasing population density, and perceived declines in agricultural productivity as environmental factors. They find a strong linkage between migration and some environmental factors but these are limited to local mobility, not long-distance one.[2] They dismiss fears of environmental refugees, pointing out the legacy of colonialism in such classification. On the other hand, they point out different patterns of movement, depending on caste and gender, which are of relevance from a human security perspective, as I suggest in Sect. 5.

Other research in the region includes Hutton and Haque (2004) and Rabbi et al. (2013), who identify displacement by riverbank erosion in Bangladesh, although this also tends to be short distance, reproducing vulnerabilities as exposure to environmental stress is maintained. They observe how "While the voluntary migrants from rural areas or small urban centers tend to choose to move to the capital city of Dhaka or to larger urban centers, most natural hazard-induced displacees attempt to stay within the vicinity of their rural place of origin." Elsewhere, Kumar and Viswanathan (2012) also find a linkage via agriculture between weather and migration in India.

Studying the case of Cambodian migration to Thailand in 2009 following flood, drought, and poor rainfall the previous year, Bylander (2016) provides evidence of international migration also taking place because of environmental factors, although still the numbers found were small. Factors associated with displacement were

[2]Using apparently the same data, Shrestha and Bhandari (2007) had found correlation on all types of migration, suggesting the difficulties of this kind of research.

reported drought in the previous year, reported poor rainfall in the previous rainy season, and household crop loss. The study presents multiple limitations related to the quantity and quality of the data, but it suggests possible exceptions to the observation of limited international migration so far observed.

Based on this evidence, it seems that the most relevant mechanism for migration caused by climate change and disasters is the indirect, middle- and long-term effects of environmental stresses, mostly inside countries. Since this is not a new mechanism creating hordes of refugees, but a process that has been going on for ages, as households and societies adapt to changing conditions, it should be analyzed in the context of present patterns of population movements (Hugo 2012).

9.3.3 Sea Level Rise

Sea level rise is one mechanism about which there is a lot of speculation but not much evidence. Its effects on displacement are divided into two major types: coastal effects and the case of small islands. About the former, besides the suggestions of Seto (2011) quoted above, Bangladesh often is mentioned because of the large extension of its low-lying coastal areas. Dun (2011) also mentions salt intrusion in the Mekong Delta areas in Vietnam as affecting livelihoods, but the causes are not environmental but present commercial uses. Evidence from these two countries suggests that effects are similar to those of slow-onset disasters; therefore, the adaptation approach may be more useful to understand the dynamics of population movements resulting from sea level rise in bigger countries.

Besides, there is only one major case of a small island country discussed in the regional literature, the Maldives. Martin (2009) describes existing considerations made by the government as part of the national adaptation program of action, which recognizes that relocation from vulnerable islands to less vulnerable ones may be necessary but takes no specific steps to prepare plans for action. Surveying local experts and 347 households, Stojanov et al. (2017) find that, while internal and external migration appears among the islanders' strategies to improve their welfare, this is rarely done because of environmental reasons. Climate change is perceived, yes, but other problems are seen as more pressing. The future need of migration is acknowledged, although a quarter of respondents still reject such idea and experts see it only as the last option.

Overall, Adger et al. (2014) find that protection measures against this climate change effect, such as dikes, are very likely to be implemented because of the high cost of not doing it, especially in major urban centers. The unsolved questions are when and how this is going to happen.

9.4 Population Movement as Means for Adaptation and Resilience

Migration as adaptation to climate change and disasters suggests in principle a positive experience. As a counter narrative against framing migrations as a threat, it suggests that population movements are a viable solution, as it was recognized through its inclusion in the 2010 Cancun Accord of the UN Framework Convention on Climate Change. This recognition and the increasing importance of adaptation have resulted in more academic attention to migration as a means. However, research ensuing from the adaptation approach shows that such migration is also very complex in practice and of limited success.

There are at least two major subtopics inside this narrative: one focusing on existing patterns of adaptation and another looking at the specific experience of planned relocations as adaptation. Studies on both subtopics have in common the question about whether the outcome of such migration does indeed result in less vulnerability to future effects of climate change and disasters. Still, I will present them separately.

9.4.1 The Role of Migration in Managing Risks

Changing the framing of migration from a threat to a means to deal with climate change and disaster risk brings along new questions. The issue of causation becomes less relevant, while the role of migration in managing risks associated with changing environmental conditions becomes more important (Warner and Afifi 2014). Particularly, Warner and Afifi (2014) stress that there is a need to make clear whether migration as adaptation is a positive measure of the ways in which populations deal with climatic stresses or if it is rather "a sign that in situ adaptation is decreasingly feasible."

Following a large-scale study reaching 2,000 respondents and 1,300 households in rural districts of eight countries around the world, Warner and Afifi (2014) suggest four different types of migration as adaptation, namely:

- Migration that improves household resilience
- Migration used to survive but not flourish
- Migration as an erosive coping strategy
- Migration not an option: trapped populations.

Their research focused on the risk of rainfall variability and food security, and only four of the countries surveyed were from Asia: Bangladesh (Etzold et al. 2014), India (Murali and Afifi 2014), Thailand (Sakdapolrak et al. 2014), and Vietnam (van der Geest et al. 2014). Still, the classification is general enough to be of use in understanding the intersection of migration, climate change, and disaster risk.

Migration contributing to household resilience was particularly found in Thailand (Sakdapolrak et al. 2014). The households surveyed did not show a direct link between rainfall and migration; still migration was one of several options to deal with stressors, making them better prepared for future problems. Indeed, it was the only place in Warner and Afifi's (2014) project where permanent migration as a positive strategy was detected. This could be because positive accounts of migration as adaptation are not easy to grasp through research looking into climate and disaster risks. In fact, migration is a socially accepted behavior, actively pursued in Bangladesh (Martin et al. 2014a; Ayeb-Karlsson et al. 2016) and Vietnam (Etzinger and Scholten 2016), not only in the face of deteriorating conditions but also during normal times. Households that flourish in part, thanks to migration, explain the decision to migrate on wishes for a better life, as in the Maldives above, not pressured by risks. Benefits in harsh time are rather a positive externality, the reason I introduce the specific case of remittances not in this section but in the following.

The second category, migration for survival but not flourish, was the most common pattern in Bangladesh (Etzold et al. 2014), India (Murali and Afifi 2014), and Vietnam (van der Geest et al. 2014). In these cases, migration is a major strategy to deal with climate variability, particularly on poor households, but it does not result in households improving their condition; indeed, it sometimes involves sacrificing schooling, for instance, that may perpetuate this pattern of adaptation. It can also be the case that, in search of economic opportunities, migrants remain exposed to environmental risks, as Nurlambang (2012) reports about vulnerable residents in Jakarta. It should be noted that evidence reviewed in Sect. 9.3.2 tends to fall into this category, so I do not discuss that literature again here to avoid repetition.

Evidence about the third and fourth categories of Warner and Afifi suggests there is a fine line between them that deserves further exploration. To start with, the third category, last-resort migration, was not identified in their Asian case studies, although the general observation of migrants moving into highly vulnerable places is usually mentioned (ADB 2012: 15; Foresight 2011). On the other hand, trapped populations, the fourth category, are identified through the existing research, but with a twist. While in principle they are pictured as the most vulnerable of the four categories, the practice seems to be less linear. This is because surveys show that the populations more reluctant to migrate are those who own land or assets that they are not willing to leave behind. For them, migration is the last resort, as in the case of the Maldives earlier described. This is also true in Pakistan (Muller et al. 2014), where migration may not be considered because of some asset ownership. In Vietnam, Melde et al. (2017) report about how those who responded that they had to stay and could not migrate belonged to the most affluent households; "it was actually the poorest who moved in the context of environmental degradation and hazards." Following personal livelihood histories in Bangladesh, Ayeb-Karlsson et al. (2016) find the same situation: migration included in the long run as an adaptation strategy against environmental impacts, but livelihoods and properties making populations vulnerable to fast-onset disasters. Family divisions after the Fukushima disaster may follow the same logic of husbands staying behind because of their properties and

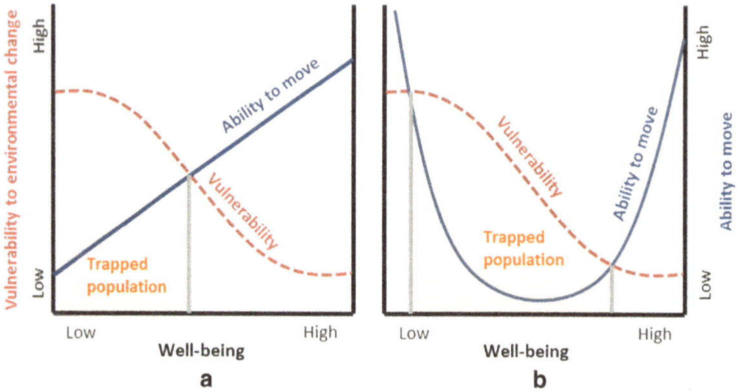

Fig. 9.3 Possible relationships between vulnerability to environmental change and mobility (*Source* Model A from Adger et al. [2014: 768]; model B elaborated by the author)

livelihoods. If this is the case, then the established relationship between vulnerability to environmental change and mobility may require revision, as presented in Fig. 9.3.

It should be kept in mind that there is also research that identifies trapped populations among the most economically vulnerable. In Bangladesh, Etzold et al. (2014) found that extremely poor families are not sensitive to rainfall variability because they do not depend on agriculture. These are usually households without male adults who could work as labor migrants and without access to migration networks. Such research supports the intuitive view elsewhere about non-migrants being actually the most vulnerable (e.g., Hugo 2012; Obokata et al. 2014), which is also embraced by the IPCC—see model A in Fig. 9.3. More research is needed to clarify how much well-being and ability to move are indeed proportionally correlated.

9.4.2 Resettlement Experiences

Resettlement experiences in Asia connected to disasters and climate change is limited (Hugo 2012); however, the ADB (2012: 36) affirms that there is considerable literature related to mega projects in the Asia-Pacific, which has been actively used to suggest principles for future climate-related resettlement (e.g., Tadgell et al. 2017a). In this sense, it is important to point out that resettlement caused by development is a sensitive issue connected to allegations of human rights abuses, making it a complicated topic to document and research. On the other hand, the economic rise of China has included multiple experiences on relocation (Rogers and Xue 2015), which can be of use informing future resettlement needs.[3] Moreover, the Chinese

[3]The transmigration scheme of Indonesia should also not be ignored (van Der Wijst 1985). Thanks to Professor Pulhin for pointing this out.

government has promised to move 1.4 million more people away from geological threats by 2020, adding to the 1.2 million economic migrants already resettled (Chiu 2017). While the applicability of this experience to other contexts in the region and elsewhere is not clear, it would be certainly worth careful observation.

Academic research is scarce, while existing examples highlight the many challenges of relocation. Rogers and Xue (2015) follow the case of a poverty alleviation-related resettlement in China's Shanxi Province. Their qualitative survey, covering 53 households and 18 local offices, suggests that without proper planning, resettlement can replicate vulnerability and result in maladaptation. Tan (2017) undertook a survey in Hongsibu, central Ningxia, which is the largest single environmental resettlement area in China, receiving up to 170,000 people between 2001 and 2010. The author finds that climate impacts are repeated in the relocation area, inducing households to pursue a new relocation; besides, households with low socioeconomic status and those with more local kinship ties were less likely to plan for relocation. Guo and Kapucu (2017), reviewing an experience in the Qinling Mountains, find as well that livelihoods are of great importance for the success of relocation, particularly because of dissimilar expectations between government officials and resettled households. The elderly and women are more vulnerable because of their disadvantage in terms of diversifying livelihoods; lack of education also plays a role. Besides, it has also been noted that relocations have evolved from rural-to-rural resettlement to rural-to-urban resettlement, making emphasis on small industry creation and employment (Tan 2008; de Sherbinin et al. 2011).

Another government that has undertaken specific relocation efforts is Vietnam. It is famous for a long-standing relocation program in the Mekong Delta in order to deal with floods, as well as for other purposes such as poverty reduction (Chun 2015; UNHCR et al. 2017). Its approach is centered on low-interest loans to which vulnerable populations have access in order to pay for the cost of relocation. Several authors who have studied the experience agree on the positive impact of this approach in reducing vulnerability to environmental effects, but they point out multiple complications (Melde et al. 2017). One problem is the decrease in income following relocation, which could compromise the ability to repay the loan (Chun 2015). Depending on the type of hazard that is being addressed, relocation cannot be undertaken close to the original site, thus affecting established livelihoods. van der Geest et al. (2014) points at restrictions related to raising animals in new locations as an example of negative impact. However, there seems to be a trade-off between the distance of the relocation site and the diversification of livelihood, since relocations near previous livelihood areas may not change the state of vulnerability against climate change and disaster effects. A key for success, as in the Chinese case, is a broader program addressing infrastructure and living conditions (Etzinger and Scholten 2016). It is also worth noting how in "the Mekong River Delta in Viet Nam, the relocation of communities led to subsequent migration patterns to urban areas" (Melde et al. 2017), effects that echo the evolution of Chinese efforts.

Elsewhere, UNHCR et al. (2017) point out the experience of the Myanmar government, including safeguards for relocated communities through its National Framework for Recovery after the 2015 floods and landslides. Such safeguards include consultation and participation through the whole process of deciding to relocate, planning and executing the relocation, as well as setting up mechanisms for monitoring and accountability. Unfortunately, Thomas (2016) affirms that these safeguards have not been put into practice. In the Philippines, Tadgell et al. (2017b) try to examine prospectively the resettlement of informal dwellings in Metro Manila, finding that the option is seen as a last resort, and much more attention is given to poverty reduction as a way to influence future risk avoidance behavior.

9.5 Migration in a World of Disasters and Changing Climate

Seeing migrants, not migration, as referents and possible providers of human security further broadens the spectrum of issues that appear in the intersection with climate change and disaster risk. As the most populated region of the world, any disaster or climate change effect will be felt both by Asian migrants elsewhere in the world and by migrants in Asia. ADB (2012: 2) quotes figures of 80 million Asian international migrants, 40–50 million Chinese diaspora, 20 million Indians living outside India, and 15 million Filipino foreign workers. Internal migration numbers are difficult to come by, but, for instance, ADB (2012) quotes figures of 210 million rural-urban migrants in China. The effects of disasters on tourism have been also a driving force behind a focus on migrants: about 309 million tourists arrived in Asia-Pacific in 2016, China being the fourth largest tourist destination in 2015 with 56.9 million international arrivals in 2015 (WTO 2017). These numbers suggest that migrant needs and behavior can add both to the problem or to the solution. Two issues receive specific attention: addressing the needs of migrant populations affected by disaster crises, be they caused or not by climatic variation, and the role of remittances in responding and recovering from them.

9.5.1 Migrants as Vulnerable Populations

One fundamental issue about approaching migrants as vulnerable populations is whether they are seen as a group or if they are disaggregated into specific identities. The latter is more common, particularly about the specific vulnerabilities related to gendered effects of disaster and displacement, as I already mentioned in studies on Nepal, Vietnam, China, and Japan. Additionally, Nellemann et al. (2011) suggest from observations in Nepal that disasters and environmental distress can result in gender-based violence and organized trafficking. Dun (2011) reports anecdotal evidence of human trafficking following floods in Vietnam, as well as cases of children being taken to the city in order to receive additional support. Etzinger and Scholten (2016) further add that relocation efforts in Vietnam result in men being able to continue their economic activities, while women were left without livelihoods. "The move did not lead to new income opportunities for women, making them potentially more vulnerable to future environmental shocks" (Melde et al. 2017: 92).

Research specific on other populations traditionally considered vulnerable, such as children, the elderly, ethnic minorities, and people with disabilities was not found. However, it should be observed that through the evolution of DRR, there has been growing attention to these populations, not only as subjects of research but as active participants in major events where regional and global agreement for action is produced, such as the 2015 Third UN World Conference on Disaster Risk Reduction in Sendai, Japan. Some of these have to do with displacement effects, particularly of evacuations. After the Great East Japan Earthquake, it was repeatedly noted how old residents would be distinctly affected by evacuations, changing the public health profile of disasters to increased needs on chronic diseases, as well as imposing new challenges for reconstruction. All of these populations are covered in respective chapters in the Wisner et al. (2012) handbook, but reviewing this literature is out of the scope of the present review.

There has been an effort to study "migrants" as a population group of concern, pioneered by Martin et al. (2014b) and followed by the Migrants in Countries in Crisis Initiative (MICIC). The MICIC is "a government-led effort co-chaired by the United States and the Philippines, [that] aims to improve the protection of migrants when the countries in which they live, work, study, transit, or travel experience a conflict or natural disaster."[4] Besides producing some guidelines, the initiative has included efforts to collect evidence from around the world that offers some insights about the experience in Asia. The report edited by Guadagno et al. (2017) includes 20 case studies of migrant inclusion practices into disaster risk management activities, some of which are from Asia, namely Bangladesh, Japan, Thailand, and Turkey. Testimonies from practitioners in Japan (Khovanova-Rubicondo and Kikuchi 2017; Parzniewski and Phillimore 2017) particularly point at using normal time activities for migrants' inclusion as an option to provide knowledge and engage them in disaster prevention. This includes training multilingual disaster support volunteers to help during emergencies. The case of the Filipino community of housewives in Kesennuma, a city heavily affected by the tsunami, shows the possibility of empowerment after the disaster. Elsewhere, Bravi and Schaur (2017) point at the vulnerability of illegal migrants accessing support after the Thailand floods in 2011 and, in general, the possible language barriers to access information. In relation to homeless migrants in Bangladesh, Clark-Ginsberg and Hunt (2017) introduce NGO activities tailored to the special needs of these populations. Finally, Wilson and Paradise (2017) estimate fatalities among refugees in the Turkish border with Syria in case of an earthquake, suggesting the need to include them in future planning against disasters.

These pieces remain at the level of reports in need of more formal research. Guadagno et al. (2017) recognize the need for evidence and recommend including the voices of the migrants more proactively. Adger et al. (2014: 768) report several studies showing how migrants more often live in more hazardous locations than

[4] From the MICIC homepage https://micicinitiative.iom.int/about-micic (accessed January 9, 2021).

long-term residents. For instance, a survey of migrant population in Shanghai by Wang et al. (2012) found less awareness about typhoons among migrants because of their lack of disaster experience and general ignorance about typhoons. The case of tourists' vulnerability (Steckley and Doberstein 2011) and how they can overwhelm local systems (Deebaj et al. 2011) was also discussed after the Indian Ocean Tsunami. That time, some 20,000 Swedes, 10,000 Britons, and 4,000 Finns were in the disaster area, resulting in multiple logistic complications. Therefore, there are reasons for further investigation about migrants as a vulnerable population.

However, it should be observed that different groups considered migrants are so dissimilar that treating them as a homogeneous group can be of no use. In Japan, the category international migrant includes tourists, spouses, international students, temporary workers, and permanent residents. Researchers who have tried to study these as a single population in the context of disaster management attest to their heterogeneity (Henry et al. 2012; Omura 2012; Yonekura 2012). I have done research about foreign students after the Great East Japan Earthquake (Gomez 2013b) through which it has been evident that, even for this subgroup, vulnerability is not uniform. Research about tourists also shows that their risk perceptions vary widely (Seabra et al. 2013) and that cultural differences among tourists from different countries are not as significant as differences among tourists from the same country (Reisinger and Crotts 2010). Their decisions may even change, depending on whether tourists are starting or ending their trips and whether they are staying at a hotel or with relatives (Drabek 1999).

Moreover, treating migrants as a group can also result in harm. Giving privileged support to migrants when they are not particularly vulnerable or the most affected population can reinforce social divisions existing in peaceful times—leaving aside it goes against humanitarian principles for relief provision. I also contend that governments with good intentions trying to protect populations abroad during disasters actually forcefully displace them while trying to appease domestic audiences, even at the cost of worsening the local situation (Gomez 2014). Nonetheless, international evacuations are becoming common place, as in Japan, Nepal after the earthquake in 2015, as well as conflicts in Libya and Yemen, so this topic also deserves more attention.

9.5.2 Migrants as Providers of Security: Remittances

On the bright side, research on migration usually points out some of the benefits of supporting networks once disasters strike, particularly in relation to remittances. While migrant workers suffer multiple difficulties in the target countries, they can become a source of strength, as hypothesized in Sect. 9.4.1. This is another topic about which there is much attention outside of the disaster studies epistemic community, which are just recently being incorporated.

The volume edited by Milan et al. (2016) includes several chapters on remittances in Pakistan and Nepal, suggesting small but positive effects that reduce vulnerability.

The United Nations' and the World Bank's (2010: 154–158) review of the literature provides a mixed picture: positive in as much as remittances are indeed resources communities can use for relief and reconstruction, but these do not necessarily reach all the population, neither are they a guarantee that results would reduce vulnerability. In India, Banerjee et al. (2017) identify positive effects that could be potentiated if further financial relevant literacy and skills training were provided to recipient households.

However, there are other limitations. Etzinger and Scholten (2016) note that support from remittances is not an immediate effect of migration, but that it depends on the migrants' success, so sending households may, for some time, instead send money to support migrants looking for opportunities away from the household. Besides, as the ADB (2012: 51) rightly points out, "remittances are private family income and should not be considered as a substitute for state investments, development aid, or international obligations to address climate change." In other words, remittances can be of great help but they are not a policy tool.

9.6 Conclusion: Between Evidence-Based Policies and Getting Ready for the Unthinkable

This review of intersections between climate change, disaster risk, and population movements in Asia showed increasing, although uneven, understanding of all its implications. In spite of the advance, the need for more research is clear and well-recognized, so any conclusions should be taken with caution. Findings are summarized in Table 9.1.

Evidence on different mechanisms resulting in displacement suggests that movement after slow-onset disasters is set to be the most important of the envisioned mechanisms, particularly heat stress, with mainly domestic effects. Not even the most catastrophic events have so far resulted in long-term displacement, although the undermining effects of repeated disasters on people's livelihoods do result in such a displacement. Sea level rise-related displacement may follow a similar pattern, while it is foreseen that, given its costs, mitigation infrastructure would be built before large impacts take place. The disaster in Fukushima prefecture, Japan, warns against possible exceptions to this trend.

Displacement after slow-onset disasters is not a new phenomenon but a practice as old as humanity. It is, therefore, better framed as an adaptation. Most of it takes place inside country borders, suggesting that dealing with this will not require radically different strategies from those already existing to deal with domestic population movements. In that sense, it is worth noting that most of the knowledge we have so far in this respect is about rural-to-urban migration, so there is a need for more research about the urban situation, for which new methodologies may be necessary (Tan et al. 2015). ADB (2017: 94) stresses the importance of cities, pointing out how

Table 9.1 Main findings of the review

	As a security…	Main findings
Population movements	Threat	• Emphasis on causality and dimension of the threat • Still a common framing • Fast-onset disasters not relevant causes • Slow-onset disasters/effects are important but mostly generate local displacement
	Means	• Emphasis on the effectiveness of migration • Observed in response to slow-onset effects • The existing understanding of people's "ability to move" may require revision • Need for more research on urban migration, as well as on resettlement
	Referent	• Emphasis on vulnerability, protection, and empowerment • Scarce but promising evidence on different groups of migrants as a vulnerable population (not "migrants" as a whole), as well as a source of protection—through remittances • Important approach to counter framing migration as a threat • Still, the migration approach has had no major impact in disaster studies

new vulnerabilities arise in relation to "air pollution, malfunctioning water supply, and worsening housing conditions," adding layers of complexity to the problem.

In this sense, research on the region highlights one particular complexity behind mounting vulnerability from climate change, disaster risk, and displacement: it is not necessarily the case that the poorest populations will be the most vulnerable to the specific effects of climate change and disasters. Evidence points out that it is those who possess land or assets who would be more reluctant to move out of harms' way until the last minute, if at all, while those without anything to lose will move more easily—although this movement would most probably just reproduce their vulnerability. This situation suggests that a two-pronged approach to protect people from disasters and climate change effects would be required: developing insurance instruments to protect those with land and assets, so more lives are saved from extreme disruptions, while making sure that poverty reduction strategies are informed by climate change adaptation and disaster reduction plans, so the population exposed to these risks stops growing. It is in this sense that attention to the urban environment becomes crucial, as moving to cities is one leading poverty reduction strategy.

With regard to relocation, the region has a lot of experience linked to development and many ongoing initiatives from which to draw lessons, as manifest in the UNHCR et al. (2017) work. Further research and lesson-sharing opportunities would be vital to ensure horizontal learning, keeping in mind that the sensitivity of the topic requires opening space for multiple actors, such as academia and civil society, to make sure that all voices are heard and that the best possible solutions are reached.

While the evidence suggests that the international level would not be particularly important, the uncertainty of future effects deserves some contingency preparation. Several global and regional processes discussing and promoting migration such as the Colombo process and the 2018 global compact on migration indicate that future global arrangements would incorporate concerns about climate change and disaster risk. Nonetheless, framing migration as a threat and the menace of backtracking on policies more open to population movements are not going to disappear any time soon, so continued efforts to keep migration-friendly arrangements are necessary. That said, it should be kept in mind that, so far, viewing disasters from the perspective of migration studies is still very new, and much more research is necessary to show that it can make a meaningful contribution to policy and practice. I highlight the fact that focusing on migrants as a vulnerable population may not be adequate because of how overstretched the category "migrant" will result. On the other hand, the agenda of remittances offers very positive prospects in the reframing of migration and promoting a positive view of migrants, but its usefulness as a policy tool for disaster risk reduction seems very limited. Therefore, while initiatives like the MICIC may help maintain a friendly environment, governance structures that make sure population movements' insights are properly included in policy are better organized around national authorities in charge of development and disasters. In the meantime, the parallel global agenda on migration should make sure that the positive externalities in dealing with disasters keep materializing in the years to come.

References

Adger WN, Pulhin JM, Barnett J et al. (2014) Human security. In: Field CB, Barros VR, Dokken DJ et al. (eds.) Climate Change 2014: Impacts, adaptation, and vulnerability. Part A: Global and sectoral aspects. Contribution of Working Group II to the Fifth Assessment Report of the Intergovernmental Panel on Climate Change. Cambridge University Press, Cambridge, pp 755–791

Asian Development Bank (2012) Addressing climate change and migration in Asia and the Pacific. ADB, Manila

Asian Development Bank (2017) A region at risk: the human dimensions of climate change and the Pacific. ADB, Manila

Ayeb-Karlsson S, van der Geest K, Ahmed I, Huq S, Warner K (2016) A people-centred perspective on climate change, environmental stress, and livelihood resilience in Bangladesh. Sustain Sci 11(4):679–694. https://doi.org/10.1007/s11625-016-0379-z

Banerjee S, Kniveton D, Black R, Bisht S, Das PJ, Mahapatra B, Tuladhar S (2017) Do financial remittances build household-level adaptive capacity? A case study of flood-affected households in India. Knomad Working Paper 18. The World Bank, Washington, DC

Bardsley DK, Hugo GJ (2010) Migration and climate change: Examining thresholds of change to guide effective adaptation decision-making. Pop Environ 32(2–3):238–262

Black R, Bennett SR, Thomas SM, Beddington JR (2011) Climate change: migration as adaptation. Nature 478(7370):447–449

Bohra-Mishraa P, Oppenheimera M, Hsiang SM (2014) Nonlinear permanent migration response to climatic variations but minimal response to disasters. Proc Natl Acad Sci USA 111(27):9780–9785. https://doi.org/10.1073/pnas.1317166111

Bohra-Mishraa P, Oppenheimera M, Cai R, Feng S, Licker R (2017) Climate variability and migration in the Philippines. Pop. Environ. 38(3):286–308

Bravi A, Schaur K (2017) Migrants in the 2011 floods in Thailand: Improving migrants' access to emergency and rescue services. In: Guadagno L, Fuhrer M, Twigg J (eds) Migrants in disaster risk reduction: practices for inclusion. Geneva and Strasbourg, IOM and Council of Europe, pp 71–74

Bylander M (2016) Cambodian migration to Thailand: The role of environmental shocks and stress. Knomad Working Paper 7. Washington, DC, The World Bank

Chiu J (2017) China to relocate millions of people away from disaster zones. IRIN News. Available at https://www.irinnews.org/feature/2017/04/11/china-relocate-millions-people-away-disaster-zones

Chun J (2015) Planned relocations in the Mekong Delta: A successful model for climate change adaptation, a aautionary tale, or both?. Brookings-LSE Project on Internal Displacement, Brookings Institution, Washington, DC

Clark-Ginsberg A, Hunt D (2017) We are human too! Concern worldwide's efforts to reduce risks for the homeless migrants of Dhaka, Bangladesh. In: Guadagno L, Fuhrer M, Twigg J (eds) Migrants in disaster risk reduction: practices for inclusion. Geneva and Strasbourg, IOM and Council of Europe, pp 27–31

Deebaj R, Castrén M, Öhlén G (2011) Asia tsunami disaster 2004: experience at three international airports. Prehospital Disaster Med 26(1):71–78

de Sherbinin A, Castro M, Gemenne F et al (2011) Preparing for resettlement associated with climate change. Sci 334:456–457

Drabek TE (1999) Understanding disaster warning responses. Soc Sci J 36(3): 515–523

Dun O (2011) Migration and displacement triggered by floods in the Mekong Delta. Int Migr 49(S1):e200–223. https://doi.org/10.1111/j.1468-2435.2010.00646.x

Elliot L (ed) (2012) Human security, climate change and migration in Southeast Asia. S. Rajaratnam School of International Studies, Singapore

Etzinger H, Scholten P (2016) Adapting to climate change through migration: A case study of the Vietnamese Mekong River Delta. IOM (International Organization for Migration), Geneva

Etzold B, Ahmed AU, Hassan SR, Neelormi S (2014) Clouds gather in the sky, but no rain falls. Vulnerability to rainfall variability and food insecurity in Northern Bangladesh and its effects on migration. Clim Dev 6(1):18–27. https://doi.org/10.1080/17565529.2013.833078

Ewing J (2012) Contextualising climate as a cause of migration in Southeast Asia. In: Elliot L (ed) Human security, climate change and migration in Southeast Asia. S. Rajaratnam School of International Studies, Singapore, pp 13–27

Findley SE (1994) Does drought increase migration? A study of migration from rural Mali during the 1983–1985 drought. Int Migr Rev 28(3):539–553

Foresight Project, Government Office for Science (2011) Migration and global environmental change: future challenges and opportunities. Government Office for Science, London

Gomez OA (2013a) Climate change and migration: A review of the literature. ISS Working Paper Series / General Series No. 572. International Institute of Social Sciences of Erasmus University (ISS), The Hague. Retrieved from http://hdl.handle.net/1765/50161

Gomez OA (2013b) Lessons from international students' reaction to the 2011 Great East Japan Earthquake: the case of the School of Engineering at Tohoku University. Int J Disaster Risk Sci 4(3):137–149

Gomez OA (2014) Forcing migration of globalised citizens. Forced Migr Rev 45:48

Gomez OA, Gasper D, Mine Y (2016) Moving development and security narratives a step further: human security in the human development reports. J Dev Stud 52(1):113–129. https://doi.org/10.1080/00220388.2015.1081176

Gray C, Bilsborrow R (2013) Environmental influences on human migration in rural Ecuador. Demography 50(4):1217–1241. https://doi.org/10.1007/s13524-012-0192-y
Guadagno L, Fuhrer M, Twigg J (eds) (2017) Migrants in disaster risk reduction: practices for inclusion. International Organization for Migration and Council of Europe, Geneva and Strasbourg
Guo X, Kapucu N (2017) Examining the impacts of disaster resettlement from a livelihood perspective: a case study of Qinling Mountains. China Disasters 42(2):251–274. https://doi.org/10.1111/disa.12242
Henry M, Kawasaki A, Meguro K (2012) Foreigners' disaster information gathering behavior after the 2011 Tohoku Earthquake (Part 3). Analysis of foreign students considering their post-disaster action. Seisan Kenkyu 64(4):497–503 (in Japanese)
Homer-Dixon TF (1991) On the threshold: environmental changes as causes of acute conflict. Int Sec 16(2):76–116
Hugo G (2012) Climate change and migration: some lessons from existing knowledge of migration in Southeast Asia. In: Elliot L (ed) Human security, climate change, and migration in Southeast Asia. Singapore, S. Rajaratnam School of International Studies, pp 45–59
Hutton D, Haque CE (2004) Human vulnerability, dislocation and resettlement: Adaptation processes of river-bank erosion-induced displacees in Bangladesh. Disasters 28(1):41–62
IDMC (Internal Displacement Monitoring Centre) (2015) Philippines: long-term recovery challenges remain in the wake of massive displacement. IDMC, Geneva
IDMC (Internal Displacement Monitoring Centre) (2017) Global report on internal displacement. IDMC, Geneva
IPCC (Intergovernmental Panel on Climate Change) (2012) Managing the risks of extreme events and disasters to advance climate change adaptation. A special report of Working Groups I and II of the IPCC [Field CB, Barros V, Stocker TF, Qin D, Dokken DJ, Ebi KL, Mastrandrea MD, Mach KJ, Plattner GK, Allen SK, Tignor M, Midgley PM (eds.)]. Cambridge University Press, Cambridge
Islam MR, Hasan M (2016) Climate-induced human displacement: a case study of Cyclone Aila in the south-west coastal region of Bangladesh. Nat Hazards 81(2):1051–1071. https://doi.org/10.1007/s11069-015-2119-6
Khovanova-Rubicondo K, Kikuchi A (2017) The diversity advantage is the answer: how local partnerships and engagement contribute to disaster risk reduction. In: Guadagno L, Fuhrer M, Twigg J (eds) Migrants in disaster risk reduction: practices for inclusion. Geneva and Strasbourg, IOM and Council of Europe, pp 23–26
Kumar KSK, Viswanathan B (2012) Weather variability and agriculture: Implications for long- and short-term migration in India. Working Paper No. 220. Centre for Development Economics, New Delhi
Massey DS, Axinn WG, Ghimire DJ (2010) Environmental change and out-migration: evidence from Nepal. Pop Environ 32(2):109–136
Martin SF (2009) Managing environmentally induced migration. In: Laczko F, Aghazarm C (eds) Migration, environment and climate change: Assessing the evidence. Geneva, International Organization for Migration, pp 353–384
Martin M, Billah M, Siddqui T, Abrar C, Black R, Kniveton D (2014a) Climate-related migration in rural Bangladesh: a behavioural model. Pop Environ 36(1):85–110
Martin SF, Weeransinghe S, Taylor A (2014b) Humanitarian crises and migration: causes, consequences and responses. Routledge, Abingdon
Melde S, Laczko F, Gemenne F (eds) (2017) Making mobility work for adaptation to environmental changes: results from the MECLEP global research. International Organization for Migration, Geneva
Milan A, Schraven B, Warner K, Cascone N (eds) (2016) Migration, risk management and climate change: evidence and policy responses. Springer, Heidelberg
Mosneaga A (2015) Restoring livelihoods after disasters: the case of Fukushima's nuclear evacuees. Policy Brief No. 2. United Nations University, Tokyo

Muller V, Gray C, Kosec K (2014) Heat stress increases long-term human migration in rural Pakistan. Nat Clim Change 4(3):182–185. https://doi.org/10.1038/NCLIMATE2103

Murali J, Afifi T (2014) Rainfall variability, food security and human mobility in the Janjgir-Champa district of Chhattisgarh state. India Clim Dev 6(1):28–37. https://doi.org/10.1080/17565529.2013.867248

Naik A, Stigter E, Laczcko F (2007) Migration, development and natural disasters: insights from the Indian Ocean Tsunami. International Organization for Migration, Geneva

Nellemann C, Verma R, Hislop L (eds) (2011) Women at the frontline of climate change: Gender risks and hopes. A rapid response assessment. United Nations Environment Programme, GRID-Arendal, Arendal

Nurlambang T (2012) Public policy matters on climate change and migration in Indonesia: the case of Jakarta City. In: Elliot L (ed) Human security, climate change and migration in Southeast Asia. S. Rajaratnam School of International Studies, Singapore, pp 74–83

Obokata R, Vernois L, McLeman R (2014) Empirical research on international environmental migration: a systematic review. Pop Environ 36(1):111–135

Omura M (2012) After an unprecedented large disaster, what happened with the scattered foreigners? In: Suzuki E (ed) East Japan Big Earthquake and foreign immigrants. Akashi Books, Tokyo, pp 34–55 (in Japanese)

Parzniewski S, Phillimore J (2017) Multilingual disaster support volunteer training in Japan: Participant observation of methods and practice in Toyama Prefecture. In Guadagno L, Fuhrer M, Twigg J. Migrants in disaster risk reduction: practices for inclusion. International Organization for Migration and Council of Europe, Geneva and Strasbourg, pp 36–40

Paul BK (2005) Evidence against disaster-induced migration: the 2004 tornado in north-central Bangladesh. Disasters 29(4):370–385

Piguet E, Pécoud A, de Guchteneire P (2011) Migration and climate change: an overview. Refugee Surv Q. 30(3):1–23

Rabbi H, Saifullah ASM, Sheikh MS, Sarker MMH, Bhowmick AC (2013) Recent study on river bank erosion and its impacts on land displaced people in Sirajgonj riverine area of Bangladesh. World J Appl Environ Chem 2(2):36–43

Raleigh C, Jordan L, Salehyan I (2008) Assessing the impact of climate change on migration and conflict. World Bank seminar on exploring the social dimensions of climate change. The World Bank, Washington, DC

Reisinger Y, Crotts JC (2010) Applying Hofstede's national culture measures in tourism research: illuminating issues of divergence and convergence. J Travel Res 49(2):153–164

Rodriguez H, Quarantelli EL, Dynes RR (2007) Handbook of disaster research. Springer, New York

Rogers S, Xue T (2015) Resettlement and climate change vulnerability: evidence from rural China. Glob Environ Change 35:62–69. https://doi.org/10.1016/j.gloenvcha.2015.08.005

Sakdapolrak P, Promburom P, Reif A (2014) Why successful in situ adaptation with environmental stress does not prevent people from migrating: empirical evidence from northern Thailand. Clim Dev 6(1):38–45. https://doi.org/10.1080/17565529.2013.826129

Sanwal M, Wang C, Wang B, Yang Y (2017) A new role for IPCC: balancing science and society. Global Policy 8(4):569–573. https://doi.org/10.1111/1758-5899.12470

Seabra C, Dolnicar S, Abrantes JL, Kastenholz E (2013) Heterogeneity in risk and safety perceptions of international tourists. Tourism Manage 36:502–510

Seto KC (2011) Exploring the dynamics of migration to mega-delta cities in Asia and Africa: contemporary drivers and future scenarios. Glob Environ Change 21(s1):s94–s107. https://doi.org/10.1016/j.gloenvcha.2011.08.005

Sherwood A, Bradley M, Rossi L, Guiam R, Mellicker B (2015) Resolving post-disaster displacement: Insights from the Philippines after Typhoon Haiyan (Yolanda). The Brookings Institution, Washington, DC and International Organization for Migration, Geneva

Shrestha SS, Bhandari P (2007) Environmental security and labor migration in Nepal. Pop Environ 29(1):25–38

Steckley M, Doberstein B (2011) Tsunami survivors' perspectives on vulnerability and vulnerability reduction: Evidence from Koh Phi Phi Don and Khao Lak. Thailand Disasters 35(3):465–487
Stojanov R, Duzi B, Nemec D, Prochazka D (2017) Slow-onset climate change impacts in Maldives and population movement from islanders' perspective brain. Knomad Working Paper 20. The World Bank, Washington, DC
Tadgell A, Doberstein B, Mortsch L (2017a) Principles for climate-related resettlement of informal settlements in less developed nations: a review of resettlement literature and institutional guidelines. Clim Dev 10(2):102–115. https://doi.org/10.1080/17565529.2017.1291401
Tadgell A, Mortsch L, Doberstein B (2017b) Assessing the feasibility of resettlement as a climate change adaptation strategy for informal settlements in Metro Manila, Philippines. Int J Disaster Risk Reduction 22: 447–457
Tan Y (2008) Resettlement in the three Gorges project. Hong Kong University Press, Hong Kong
Tan Y (2017) Resettlement and climate impact: addressing migration intention of resettled people in west China. Austr Geogr 48(1):97–119
Tan Y, Liu X, Hugo G (2015) Exploring relationship between social inequality and adaptations to climate change: evidence from urban household surveys in the Yangtze River Delta. China Pop Environ 36(4):400–428
Thomas A (2016) Accelerating threats from climate change: disasters and displacement in Myanmar. Refugees International, Washington, DC
United Nations and the World Bank (2010) Natural hazards, unnatural disasters: the economics of effective prevention. The World Bank, Washington, DC
United Nations Task Team on Social Dimensions of Climate Change (2011) The social dimensions of climate change. International Labour Organization, Geneva
United Nations High Commissioner for Refugees (UNHCR), Institute for the Study of International Migration (ISIM), International Organization for Migration (IOM) (2017) A toolbox: Planning relocations to protect people from disasters and environmental change. UNHCR, ISIM, and IOM, Geneva and Washington, DC
van der Geest K, Khoa NV, Thao NC (2014) Internal migration in the Upper Mekong Delta, Viet Nam: what is the role of climate-related stressors? Asia-Pacific Pop J 29(2):25–41
van Der Wijst T (1985) Transmigration in Indonesia: an evaluation of a population redistribution policy. Pop Res Policy Rev 4:1–30
Warner K, Afifi T (2014) Where the rain falls: evidence from eight countries on how vulnerable households use migration to manage the risk of rainfall variability and food security. Clim Dev 6(1):1–17
Wilson B, Paradise TR (2017) Refugee inclusion in earthquake casualty estimation: A case study in south-east Turkey. In: Guadagno L, Fuhrer M, Twigg J (eds) Migrants in disaster risk reduction: Practices for inclusion. International Organization for Migration and Council of Europe, Geneva and Strasbourg, pp 13–17
Wisner B, Gaillard JC, Ilan Kelman (2012) The Routledge handbook of hazards and disaster risk reduction. Routledge, Abingdon
WTO (World Tourism Organization) (2017) 2016 Annual report. WTO, Madrid
Wang MZ, Amati M, Thomalla M (2012) Understanding the vulnerability of migrants in Shanghai to typhoons. Nat Haz 60:1189–1210. https://doi.org/10.1007/s11069-011-9902-9
Yonekura R (2012) Media use and information behavior of foreigners in Japan during disaster. NHK Monthly Rep Broadcast Res 62(8):62–75 (in Japanese)
Yonetani M, Yuen L (2014) The evolving picture of displacement in the wake of Typhoon Haiyan: An evidence-based overview. Government of the Philippines, Department of Social Welfare and Development, IOM, IDMC, SAS, Manila

Dr. Oscar A. Gómez (Ph.D., Tohoku University) is Assistant Professor at the College of Asia Pacific Studies, Ritsumeikan Asia Pacific University. His main interest is in global governance and human security practice, with special emphasis on humanitarian crises. He was Research Fellow at

Japan International Cooperation Agency (JICA) Research Institute for almost 5 years and, before that, Japan Society for the Promotion of Science Postdoctoral Fellow at the Graduate School of Global Studies, Doshisha University. He co-wrote background papers for UNDP's Global Human Development Reports 2014, 2016, and 2020. He is a member of the Scientific Committee of the Humanitarian Encyclopedia project. Recent publications include co-edited volumes on the humanitarian-development nexus (Routledge) and human security norms in East Asia (Palgrave).

Chapter 10
Conflict, Development, and the Environment in Asia

Brendan Howe

Abstract In spite of dramatic progress in economic development and governance in Asia, major challenges to human security endure. In some cases, these have been exacerbated by national security and development policymaking. Among the most serious challenges are conflict, poverty, environmental degradation, and "natural" or "nature-induced" disasters. Each of these threatens human security. Through interaction with each other, however, they can also serve as insecurity multipliers. This chapter will focus on the intersections of these variables, using case material from several Asian countries. The first part of the chapter will consider conflict as a direct threat to human security, as a poverty multiplier, as a contaminator of the environment, and as a stimulus for state security policy prioritization that directly impacts upon the human security of the most vulnerable, as well as indirectly through resource allocation. Case material for this section will be drawn from the Democratic People's Republic of Korea (DPRK), Myanmar, and the Lao People's Democratic Republic (Lao PDR). The second part of the chapter will examine how insecurity at a state level can lead to a focus on national security and development models, which not only leads to environmental degradation but can also increase disaster vulnerability. It will further evaluate how mega-development practices can pose a direct threat to the most vulnerable, as well as indirectly through impact on the environment. The major case studies here will be Myanmar leading up to the devastating impact of Cyclone Nargis in 2008 and hydroelectric dams in Lao PDR and Malaysia.

Keywords Conflict · Development · Environment · Security · Vulnerability

For more detailed analysis of some of the issues and case studies addressed in this chapter, see Brendan Howe, *The Protection and Promotion of Human Security in East Asia* (Basingstoke: Palgrave Macmillan 2013). The author gratefully acknowledges research support provided by Heeseo Lee of the Graduate School of International Studies, Ewha Womans University.

B. Howe (✉)
Graduate School of International Studies, Ewha Womans University, Seoul, South Korea
e-mail: bmg.howe@gmail.com

J. M. Pulhin et al. (eds.), *Climate Change, Disaster Risks, and Human Security*, Disaster Risk Reduction, https://doi.org/10.1007/978-981-15-8852-5_10

10.1 Introduction

East Asia (including the sub regions of Northeast Asia and Southeast Asia) has sometimes been viewed as ranking among the most dangerous or conflictual regions on the planet, enduring colonial and Cold War legacies and a number of potential flashpoints (Calder and Ye 2010). International governance initiatives have focused on nonintervention, and order rather than justice; while domestic governance has emphasized national interest and strength in terms of security and economic growth. Thus, Asian states remain among the most ardent champions of Westphalian sovereignty (Acharya 2003). Not only are states considered the main referent object of security, but also security threats have been generally identified from the perspective of the state (Nishikawa 2009). Regional security arrangements reflect these perspectives, with Southeast Asia's "ASEAN Way" and Northeast Asia's "Shanghai Spirit" both emphasizing nonintervention, and Asian actors across the political spectrum resisting the Western interventionary interpretation of the Responsibility to Protect (R2P) and de-coupling it from the human security discourse (Howe 2013a).

In many ways, this regional security regime would seem to be successful. At the state-centric level of security reference, remarkable progress has been made in East Asia. In Southeast Asia, the Association of Southeast Asian Nations (ASEAN) regime has, despite historical, geopolitical, and societal animosities, prevented the outbreak of serious hostilities since the 1978 Vietnamese invasion of Cambodia. In Northeast Asia, since the armistice suspending the Korean War in 1953, there has been little physical manifestation of national rivalries beyond saber rattling and the occasional minor skirmish. With the possible exception of the People's Republic of China (PRC) and the Republic of China (ROC—Taiwan), as well as and the Democratic People's Republic of Korea (DPRK) and the Republic of Korea (ROK), states in the region no longer pose existential threats to each other. All seem to recognize a rational imperative to come to some sort of accommodation with one another in order best to secure collective goals and satisfy collective interests.

Throughout East Asia a premium is placed on economic development, with rapid success in this field combined with high levels of industrialization, urbanization, and modernization across the board. At the turn of the millennium, in a report on the increasingly networked global economy, "Emerging Asia" was highlighted by the US Central Intelligence Agency (CIA) as the fastest growing region, led by breakout candidates China and India, whose economies (at that time) already comprised roughly one-sixth or 16.67% of global GDP (CIA 2000). By 2017, China alone accounted for 14.84% (with India on 2.83%), and other regional economies, Japan (5.91%), South Korea (1.86%), Indonesia (1.16%), Thailand (0.53%), Hong Kong (0.42%), and Singapore (0.39%) (Gramer 2017). Yet, major challenges to human security endure in the region, some of which have even been exacerbated by the supposed successful developmental policies focusing on industrialization, urbanization, and modernization. Among the most serious challenges are conflict, poverty, environmental degradation, and "natural" disasters, which due to contributions of human agency discussed later, should perhaps be termed "natural-induced disasters." Each, individually, poses an increasing threat to the lives and well-being of the most vulnerable. Additionally, however, through interaction with each other, they can also serve as insecurity multipliers.

First and foremost, despite the decline of interstate wars, militarized international disputes continue, and intrastate conflicts and their legacies threaten directly the human security of the most vulnerable in numerous cases. This threat extends beyond the official cessation of hostilities through conflictual legacies such as contamination by the explosive remnants of war (ERW), socioeconomic disruption, governance failure, and environmental degradation. Conflictual operating environments also pose indirect threats to the human security of the most vulnerable. These include the securitization of governance, the normalization of state violence, contamination of the environment, and the diversion of resources to national defense projects. This avenue will be pursued in the first analytical section that follows. Case material for this section will be drawn from the Lao People's Democratic Republic (Lao PDR or Laos), the DPRK, and Myanmar.

Second, state insecurity and underdevelopment can stimulate unsustainable development practices that not only lead to environmental degradation, but can also increase disaster vulnerability, as well as lead to intrastate resource conflicts, all of which threaten the human security of the most vulnerable. Furthermore, national government policymaking, in response to the perceived double threat of poverty and insecurity, has impacted on the human security of vulnerable individuals and communities and upon the environment. Regional developmentalism and economic governance have been labeled "econophoria" (Buzan and Segal 1998). The related policy prioritization has contributed to remarkable patterns of economic growth, but it has also fed into ongoing and new human security challenges (Buzan and Segal 1998). For instance, the relocation of often already vulnerable and marginalized local populations to make way for developmental mega projects such as hydroelectric dams can further undermine human security (as described later). Infrastructure development can increase vulnerability to traffic accidents, to human trafficking, and to the degradation of the local environment in terms of drinkable water and breathable air. Hence, the prioritization of macroeconomic development in East Asia, has been described as assuming "cult-like status" (Christie and Roy 2001). This chapter will, therefore, further evaluate how state-centric, mega-development practices can pose a direct threat to the most vulnerable, as well as indirectly through impact on the environment. The major case studies in the second part will be Myanmar leading up to the devastating impact of Cyclone Nargis in 2008 and hydroelectric dams in the Lao PDR and Malaysia (Table 10.1).

All of these levels of security and insecurity are intricately linked in a nonhierarchical causality, with the potential to spill over across realms in any direction. The tendency of regional actors in East Asia to impose a security hierarchy on policymaking running from the top of this table to the bottom constitutes one of the central problems developed in the chapter. The following section expands on definitional challenges and the interaction between levels of security in theory and practice.

Table 10.1 Levels of security/insecurity and existential threats

Type of security	Main actors	Existential threats from	Referent objects	Issues
Traditional	States	States	States	Defense, deterrence, balance of power
Comprehensive security	Intergovernmental organizations (IGOs), states	Non-state actors, environment	States and communities	Water, food, environmental hazards, "natural" disasters, energy, terrorism, international crime
Environmental security	States, IGOs	States, multinational corporations (MNCs), communities, development	Ecospheres, biosphere, localized ecosystems	Climate change, global warming, sustainability, the Anthropocene, biodiversity, the global commons, pollution, consumption
Human security	IGOs, states, Non-governmental organizations (NGOs) international community	Environment, states and non-state actors	Individuals and vulnerable communities	Explosive remnants of war (ERW), responsibility to protect (R2P), peacekeeping operations (PKOs), humanitarian intervention shelter, food, water, stability, sustainability, "nature-induced" disasters, conflict transformation

10.2 Conflict, National Security, and Human Insecurity

Nontraditional security agendas have increasingly come to the fore in both academic and policy discourse and are often termed "new security challenges." The characteristics of such challenges include some, or all, of the following: a focus on non-military rather than military threats, transnational rather than national threats, and multilateral or collective rather than self-help security solutions (Acharya 2002; Waever 1995).

Human security is an emerging multidisciplinary paradigm for understanding global vulnerabilities at the level of individual human beings, incorporating methodologies and analyses from several research fields, including strategic and security studies, development studies, human rights, international relations, and the study of international organizations, existing at the point where these disciplines converge on the concept of protection (Howe 2012). The complexity of threats in people's daily lives now range from poverty, unemployment, drugs, terrorism to environmental degradation and social disintegration, and the concomitant obligations upon those who govern are summed up as a commitment to provide freedom from fear and freedom from want (UNDP 1994). Despite remaining distinct in terms of focus and referent objects, there is a close relationship between traditional and nontraditional security (NTS) approaches, and considerable spillover between them (Howe 2013b). All forms of security imply the existence of a referent object free from threats to its continued existence. Likewise, insecurity means that the referent object is not able to enjoy such freedom from threat. Vulnerabilities relate to the likelihood that the referent object(s) will be exposed to existential threats.

National insecurity (wherein those acting in its name perceive there to be existential threats to the state) may lead to human insecurity (existential threats to the lives of individuals) along various paths. It can divert resources from human development, which in turn focuses on "human flourishing" or improving the lives of people rather than the richness of the national economy (UNDP 2015). It can drain energy (Suh 2013). It can create a permissive political circumstance where national security is privileged at all costs (Unger 2012). Furthermore, it is likely to produce and perpetuate an operating environment within which the exceptional use of internal as well as external violence by the state becomes a permanent feature of the state (Suh 2013). The human costs of modern conflicts, whether interstate or intrastate, are borne, primarily, by the most vulnerable sections of society (Tirman 2015). Civilians can be directly targeted, used as human shields, or become the victims of "collateral damage" during conflicts.

Asia is certainly no stranger to the categories of international crimes listed under the R2P: genocide, war crimes, ethnic cleansing, and crimes against humanity. Indeed, the crimes against humanity, against their own people, perpetrated by the Khmer Rouge regime in Cambodia were of such magnitude as to have been termed "autogenocide" (Vittal 2001). Genocide of the more traditional or "purest" form (Shaw 2000), focusing on the elimination of ethnic and religious minorities perceived as posing a threat to the regime, was also perpetrated during this time. North Korea's crimes against its own people (detailed later) are of such a magnitude as to qualify as crimes against humanity, although perhaps not reaching the level of autogenocide (OHCHR 2014a). During the most recent interstate wars in the region, the Korean War in Northeast Asia, and the collection of related Southeast Asian conflicts involving the United States, China, Vietnam, Cambodia, and Laos, all sides have been accused of actions amounting to war crimes under contemporary definitions, as well as lasting environmental degradation (World Peace Foundation 2015a, b; Leaning 1993; Austin and Bruch 2000). Most recently, Myanmar has been accused of perpetrating a paradigmatic case of ethnic cleansing against the Rohingya Muslim minority (OHCHR 2017; also detailed later).

The legacies of conflicts can impact on the human security of the most vulnerable for years, decades, or even generations to come (Watson 2004). Postbellum threats to both life and well-being include the breakdown of law and order, the spread of disease due to refugee camp overcrowding, poor nutrition, infrastructure collapse, scarcity of medical supplies (although ironically often a proliferation of illicit drugs), and continued criminal attacks on civilian populations, unemployment, displacement, homelessness, disrupted economic activity, stagflation, and perhaps, most directly, ERW contamination (GICHD 2007; UNDP 2016). ERW include unexploded ordnance (UXO), landmines, and abandoned explosive ordnance (AXO). Negative effects include physical harm, amputation and death, psychological trauma, food insecurity, infrastructure limitations, and increased rebuilding costs.

10.2.1 Conflictual Legacies in Laos

During the Second Indochina War (1955–1975), four million large bombs, defoliants, herbicides, and more than 270 million submunitions were dropped over Laos (Landmine Monitor 2008). As a result, Laos is the most bombed country in history (Kingshill 1991; Cave et al. 2006). Civilian casualty levels were appallingly high (Cave et al. 2006). Many other vulnerable individuals became refugees, with 600,000 from a total population of just three million, fleeing their country between 1962 and 1971 (Paul 1971). For those who remained, living conditions were often deplorable (Zasloff 1970). Air-delivered submunitions are the most prolific form of ERW in Lao PDR (Handicap International 2004). Indeed, these have left Laos the world's most ERW-contaminated country (NRA 2008), with 15 of Laos' 17 provinces contaminated, and a quarter of all villages severely contaminated (UNDP 2008; NCCR 2008). UXO restricts access to usable land; increases the cost and time of development initiatives; inhibits access to shops, schools, and medical facilities; disrupts potential earnings from tourism, mining, and hydroelectric projects; and causes significant human casualties. Now, more than four decades since the cessation of the overt interstate conflict, there remains a close correlation between Laos' poorest districts and those of which were most bombed (Howe 2013a).

The costs of medical treatment and funerals for ERW casualties can push poor families deeper into poverty and destitution (Cave et al. 2006). As Tim Horner, former senior technical advisor to the National Regulatory Authority for the UXO Sector in Laos has explained, "The people hit hardest by UXO are often the poorest. Having to pay for treatment basically takes people from just being able to cope to being poverty stricken" (Howe and Sims 2011). Rural villagers commonly sell livestock to pay for medical treatment, yet in doing so they sacrifice vital family assets needed to stave off future financial hardship. Furthermore, the average cost of treatment for ERW injuries can be half the annual income of a rural family, and can thereby serve as a "poverty multiplier" (Handicap International 2004). On a broader scale, treatment and care of UXO survivors can be a burden for entire communities and a significant additional expenditure for Laos' overtaxed medical system (UNDP 2008).

ERW is perhaps the primary factor limiting social and economic development in Laos. When mapped against each other, there is a clear visual correlation between the prevalence of poverty and ERW contamination (Phoenix 2001; NCCR 2008). UXO contamination is considered an ongoing development concern in all UNDP development assistance programs in the Lao PDR. Likewise, the government's National

Socio-Economic Development Plan for 2006–2010 not only recognized UXO as a significant inhibitor to development, but also that it affects, in particular, already poor and vulnerable groups (Cave et al. 2006). As the population of Lao PDR gradually becomes more urbanized, in turn leading to population growth, there is an ongoing and growing need for more safe land. Hence, given the high levels of subsistence farming in the country, the inhibitory effect of ERW on land access represents a challenge to both the human security and the human development of significant sections of the populace.

10.2.2 National Insecurity, Underdevelopment, and Human Insecurity in North Korea

National insecurity can also contribute to an operating environment ill-suited to the promotion of safe havens, free from fear, want, and indignity for the most vulnerable sections of society. Successive administrations in Pyongyang have used the supposed hostile international operating environment within which the DPRK finds itself to foster a belief in the need for, or justification of, policies that further undermine human security among the most vulnerable of their citizenry. Foremost among these policies are *juche* (self-reliance) and *songun* (military first), as well as the successor policy under Kim Jong-un of *byungjin* (parallel development) of the economy and nuclear weapons. These misguided attempts at self-sufficiency, independence, and national strength-building have led to shortages, underdevelopment and more human insecurity. Attempts to promote energy sovereignty through privileging domestic sources and minimizing its reliance on trade contribute ultimately to the underproduction and inefficient use of energy.

Attempts to promote sovereignty in food policies, have had equally or even more disastrous results. In the 1980s, the growth rate of rice production began to slow, and when North Korea was hit in the 1990s by the combination of natural disasters and the collapse of the Soviet bloc, widespread hunger exploded into massive starvation and famine (Haggard and Noland 2005). In 2019 and 2020, bad weather, poor harvests, and the impact of the COVID-19 pandemic as well as government countermeasures, have exacerbated food insecurity in the DPRK. In May 2019, food rations, upon which about 70% of the population depend, were cut from 550 g to just 300 g per person following poor results in the year's early harvest, while drought, typhoons, and flooding impacted the late harvest that year (BBC 2019). 2020 has already been a record year for typhoons impacting the Korean Peninsula, following on from the historically long and heavy rainy season. Furthermore, the United Nations Food and Agriculture Organization (FAO) has warned that North Korea could face an even more serious food shortage due to the coronavirus pandemic, with over ten million North Koreans, some 40% of the population, being food insecure (KBS World 2020).

State security has deprived the people of resources and rights, which in turn have undermined trust in the government to alleviate the challenges faced by people in their everyday lives. This has led to greater regime insecurity, and therefore a greater emphasis on state security, as well as the deliberate destabilization of the international operating environment in order to create conditions, which justify national security policy prioritization. A vicious circle. Economic mismanagement, governance failure, negligence, and oppressive behavior by the state has had grave consequences for ordinary citizens, and the country remains deep in distress despite various flawed attempts at reform and the intermittent inflow of international aid (Howe 2013a). Hence, the "hostile" international environment is used to justify domestic security and governance policy. Indeed, so pervasive has this approach become that North Korea has even deliberately destabilized the international operating environment through brinksmanship. As a result, "the 'military first policy' becomes a useful tool both domestically and internationally" (Koga 2009).

Despite internal and external crises, severe economic stagnation, and political pressures from the international community, the regime has survived. At least, in part, this regime's survival has been purchased at the expense of human security and human rights of the citizens. The concept of "our own socialism," combined with a cult of personality, has contributed to a situation in which the directives of Kim Jong-eun, (and Kim Jong-il and Kim Il-sung before him), trump all laws and directly control and discipline the North Korean people, as well as their access to information (Lankov 2015). There is no room for rebellion even if a great number of people suffer at the hands of an evil authoritarian regime. And suffer they have. Grave human rights abuses in North Korea have increasingly attracted the attention of the international community, culminating in the February 2014 report of the UN special commission documenting "unspeakable atrocities" committed in the country. In late November 2014, the United Nations Third Committee—the Social, Humanitarian and Cultural Committee—took the initiative in demanding stronger actions toward DPRK by requesting that the United Nations Security Council refer the DPRK and the regime of Kim Jong-eun to the International Criminal Court (ICC) to face charge of crimes against humanity (OHCHR 2014b). Although the initiative did not make it to the ICC, the action by the HRC was at least a clear indicator of the seriousness of the issue.

Other policies deemed necessary for national security or regime survival, such as the continuation, proliferation, or expansion of its notorious prison camps, have further exacerbated the situation. In 2011, Amnesty International noted a significant expansion in the camps from the footprint of 2001, visible via satellite imagery (Amnesty International 2011). A 2014 white paper for the Korea Institute for National Unification (KINU) noted a lower but still horrific set of figures estimating the prison camps as holding approximately 80,000–120,000 people (Han et al. 2014). Meanwhile, spending on the military has diverted funds essential for the protection and promotion of human security in North Korea. This has also been the experience of the citizens of Myanmar.

10.2.3 National Security Prioritization in Myanmar

Martin Smith has described Myanmar as "a pre-eminent example of a post-colonial state subsumed in what development analysis describes as a 'conflict trap'" (Smith 2007). From the coup in 1962, until a partial retreat from the forefront of governance in 2011, a series of military juntas ruled the country, more often in conflict with internal opposition groups and ethnic insurgencies, than with external threats. Even after the replacement of military with civilian control, given national security challenges both real and perceived, the government has continued to prioritize policy platforms emphasizing national sovereignty, territorial integrity, and the national unity of diverse ethnic nationalities. This has continued under the leadership of Nobel Peace Prize laureate Aung San Suu Kyi and her National League for Democracy (NLD) administration, with the persecution of one of the most vulnerable groups of people in the country, the Muslim Rohingya (OHCHR 2017).

Since 2008, Myanmar's third Constitution has guaranteed the military 25% of parliamentary seats and the control of significant ministries and positions in the executive. Despite successive waves of people empowerment, first, in the immediate aftermath of the devastating cyclone Nargis (addressed in more detail later); second, with constitutional and governmental changes in 2010; third, with the breakthrough in negotiations between the government, the military, and the NLD in 2011; and most recently, with stunning election victory of the NLD in 2015, the governance structure remains best described as a mixed system of civilian and military government. Indeed, for Alistair Cook "the formal role of the military remains intact" (Cook 2017). There are even suspicions that there may be a military "shadow cabinet" waiting in the wings to resume control if their interests are too overtly challenged (Pe AZ, Panna Institute, 2017, pers. commun.). The national security focus and military dominance have distorted budgetary allocations, leaving little for human-centered development and resilience building. Direct military spending consistently totals around a quarter of the national budget; higher than that on health and education combined. For the 2015–2016 fiscal year, overall, 10.66% of the national budget was allocated for the four government ministries in the social sector, most responsible for human-centered policies: The Ministry of Health (3.65%), the Ministry of Education (6.79%), the Ministry of Labor, Employment and Social Security (1.15%), and the Ministry of Social Welfare, Relief and Resettlement (0.07%) (Win K, NLD Central Working Committee, 2016, pers. commun.).

In fact, the government has little control over the military. This in part explains the brutal treatment meted out since October 2016 by Myanmar's security forces to the mostly stateless Muslim Rohingyas in the country's Rakhine state. The OHCHR (2017) has reported "devastating cruelty" against the Rohingyas and "the very likely commission of crimes against humanity." While this may explain the NLD's inability to act to rein in the military, it does not absolve the leadership of their duty to speak out. Aung San Suu Kyi has responded to criticism that: "[W]e must work ourselves for our country's responsibilities, because we are the ones who best understand what our country needs" and that a UN inquiry was "not suitable for the situation of our country" (Selth 2017). Aung San Suu Kyi's office has accused international aid workers of helping "terrorists," a claim that has prompted fears for their safety and been condemned as dangerously irresponsible (Holmes 2017). Others fear that the NLD is disempowering NGOs and maintaining the national security/unity focus of previous administrations (Pe AZ, Panna Institute, 2017, pers. commun.). The threat posed by such a focus, to both human security and the environment, was graphically demonstrated by the impact and response to cyclone Nargis in 2008. This also segues into the second area of analysis.

10.3 Environmental Degradation, Unsustainable Development, and Human Insecurity

Within governance literature, human security, development, and poverty are readily understood as interrelated and connected in a complex causality (Howe 2013a). These linkages are even more apparent when it comes to consideration of environmental degradation and natural disasters. Natural disasters lead to human and economic losses with the potential to have a long-term impact on national economies, in turn leading to a new generation of vulnerable extreme poor. The extreme poor need resources to survive, and often resort to short-term desperate and unsustainable measures degrading the natural environment. This degraded environment in turn increases vulnerability to natural disasters (UNDP 2004; Ahrens and Rudolph 2006).

In this circular linkage, the poor are the most vulnerable when natural disasters occur, and human security issues are the most pronounced in areas of heaviest dependence on natural resources. Thus, by creating disaster-resilient communities, it is possible to enhance human security (Shaw 2006). It follows, therefore, that in order to break this vicious cycle, resilient communities, must be built from the bottom up rather than the top down, imposed through national security and development policy platforms. Furthermore, as families, neighbors, and local authorities are likely to be the first responders assisting those affected by natural disasters, local communities must not only be educated and trained but also empowered. Unfortunately, in the following case studies, at times due to a national unity and security focus legacy of conflict, regional governments have come up short in fulfilling these obligations.

10.3.1 Vulnerabilities in Myanmar

Myanmar is highly vulnerable to natural hazards. A total of 27 "natural" disasters, or as indicated earlier, "nature-induced" disasters, were recorded between 1980 and 2010, causing the death of approximately 140,000 people and affecting the lives and livelihoods of 3.9 million people, an average of 125,000 people a year (Disaster Risk

Reduction Working Group 2013). Cyclone Nargis, the most powerful cyclone to have struck Myanmar and that devastated parts of the country in May 2008, affected more than 50 towns, left more than 140,000 people dead, and 2.4 million people made more vulnerable by loss of property and livelihoods (UNEP 2009). Underlying governance failures and long-term trends leading up to Nargis had, however, left the country and its citizens even more vulnerable to the devastating impact of the storm than might otherwise have been the case.

As detailed above, due to military expenditure prioritization, Myanmar was particularly deficient in basic services which would contribute to disaster risk reduction (DRR). Resources earmarked for infrastructure improvement were, and continue to be diverted into macro-development projects, such as the construction of the new, isolated (and thus more secure) capital of Naypyidaw, gas pipelines like the Yadana Project, which takes natural gas from the Andaman Sea through a pipeline in southern Myanmar to Thailand and is the single largest source of revenue for the military (Global Security 2016), and hydroelectric dams (the impact of which is addressed below). By contrast, shortly before the cyclone hit, the national health system was ranked 190th out of 191 globally, and the government spent just 0.2% of GDP on health care (Mullany et al. 2008). The actual burden of responding to Nargis lay with the Ministry of Social Welfare, Relief and Resettlement (MSWRR) which received 0.003% of the national budget (Aung PT, ASEAN Committee on Women, 2016, pers. commun.). Local and national authorities had also refrained from building storm shelters and instituting other measures to increase resilience (Howe and Bang 2017). A lack of preparedness at the national government level was compounded by insufficient resilience education at the provincial level (Myint HT, NLD Central Working Committee, 2016, pers. commun.). As a result, necessary evacuations did not take place (Lwin and Maung 2011). Indeed, post-disaster analysis identified the main cause of the tragedy as being a lack of awareness of the danger, how to prepare, and when and where to evacuate (Shikada et al. 2012).

The United Nations Environment Programme (UNEP) noted in its case study on Learning from Cyclone Nargis, that because previously the government focused on development priorities rather than on natural resource sustainability, "[a]ttention was directed towards increasing production by harnessing more land area and building infrastructure such as embankments, without fully considering environmental impacts," thereby reaffirming the linkages between environment deterioration, disaster vulnerability, and human insecurity. The pre-disaster struggle for survival among the most vulnerable communities had further resulted in overexploitation of the environment and unsustainable natural resource management. As a result, these vulnerable communities lived in an insecure environment susceptible to catastrophic impacts from natural disasters. Senator U Kyi Win, Irrawaddy representative of the NLD Central Working Committee, places the blame for the magnitude of the disaster squarely on human impact on the environment, in particular the deforestation of mangroves which would previously have provided a natural barrier to storm surge (2016). The mangrove destruction had been caused by the expansion of fisheries and paddy fields (including a government-sponsored initiative for double cropping) at least in part for export; industrial salt production; the production of coke (for which the government granted an excessive number of licenses); and the direct use of mangrove wood. Poor forestry management was exacerbated by poor enforcement of what regulations were in place due to corruption, including bribery to escape punishment for infringements (Win K, NLD Central Working Committee, 2016, pers. commun.).

Furthermore, Cyclone Nargis's intensity and unusual direction were believed to be caused by global climate change (Lwin and Maung 2011). Figure 10.1 demonstrates how, due to global climate change, the path of Nargis deviated from that normally taken by cyclonic tropical storms in the region, impacting on a much more vulnerable part of the country, thereby causing much greater devastation and suffering.

Given that extreme weather events are occurring with increased frequency around the world, with its long-exposed coastline, Myanmar remains particularly vulnerable. Resilience remains weak in every region, even Naypyidaw (Khaing AT, Ministry of Livestock, Fisheries, and Rural Development, 2016, pers. commun.). According to the 2017 University of Notre Dame Global Adaptation Index (ND-GAIN 2017), Myanmar ranks 144 out of 181 countries in terms of vulnerability and 172 out of 191 for readiness. This is an improvement of the 2014 rankings, but it still shows serious vulnerability and lack of readiness. Along with Honduras and Haiti, Myanmar has been ranked as one of the three countries most affected by climate risk for the 20-year period from 1996 to 2015 (Kreft et al. 2017). Deforestation contributed to the severity of the 2015 floods and landslides, but it can also contribute to drought, the natural

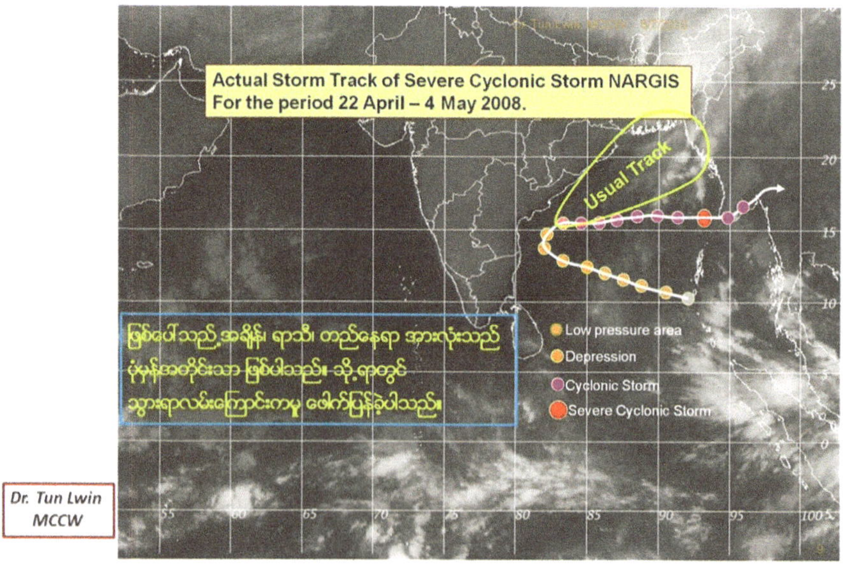

Fig. 10.1 Path of Cyclone Nargis (*Source* Tun Lwin, former director general of the Meteorology and Hydrology Department of Myanmar, now chairman and CEO of Myanmar Climate Change Watch. Interviewed by author in Yangon, January 22, 2016. Reproduced here by kind permission)

hazard seen by some as ultimately an even greater threat to Myanmar's resilience (Myint HT, NLD Central Executive Committee, 2016, pers. commun.).

ASEAN's Regional Integrated Multi-Hazard Early-warning System (RIMES), did notify Myanmar's Department of Meteorology and Hydrology about the impending storm on April 26 and continued to give updates until it made landfall 5 days later (Lwin T, Myanmar Climate Change Watch, 2016, pers. commun.). And during this period, the Director General of the Department of Meteorology and Hydrology contacted senior local and national government officials, and military commanders, with regular warnings and updates, as well as briefing official news outlets, but it was difficult and even dangerous to convince those in power that a terrible disaster, exacerbated by poor governance, was about to happen (Aye NY, Ministry of Social Welfare, Relief and Resettlement 2016, pers. commun.). An interview with the Director General was eventually published on May 2 but hidden inside the paper rather than on the front page and with a very toned-down message (Lwin T Myanmar Climate Change Watch, 2016, pers. commun.).

Due to security-related paranoia concerning domestic non-governmental actors, authorities clamped down on the activities of civil society first responders, ultimately arresting and imprisoning some of those who had done the most to respond to Nargis (Win K, NLD Central Working Committee, 2016, pers. commun.). Likewise, despite, within a week of landfall, twenty-four countries pledging USD 30 million of financial support, the regime was reluctant to accept foreign assistance, and ultimately only did so on the basis that it could control aid distribution, after which it was "slow to issue visas to foreign specialists and to allow aid into Myanmar" (Selth 2008). Indeed, the government did not allow international experts and cargo ships with aid goods into its territory for several weeks (Kapucu 2011). Even then, only "friendly" countries were allowed into Myanmar, while naval vessels loaded with aid supplies from the United States, Britain, and France were denied permission to dock or deliver supplies by helicopter (Selth 2008).

10.3.2 Vulnerabilities in Laos

The government of Lao PDR has likewise been concerned primarily with national security and development; or, according to Deputy Prime Minister of Foreign Affairs Thongloun Sisoulith, the need to safeguard political stability for national development (Sisoulith 2011). Being the only country in Southeast Asia faced with Collier's "poverty trap" of lacking direct access to the sea (Collier 2008), has presented barriers to Lao's participation in international trade since the sixteenth century (Pholsena and Banomyong 2006). Hence the government has focused, almost to the point of obsession, on transitioning from being "landlocked" to becoming "land-linked" or a crossroad for trade (Rigg 1998). In its attempts to achieve this geoeconomics objective, the Lao PDR has forged partnerships with neighboring countries. These include two major railway projects linking the country with Thailand, Vietnam, and China.

Giant Consolidated, a Malaysian company in charge of constructing the 220-kilometer railway linking Lao's western border with Thailand to Vietnam, finally began construction operations after minor derailments, which had postponed the groundbreaking ceremony. Laos has also been in negotiations to borrow 7.2 billion USD from China to fund the 430-kilometer railway stretching from the capital Vientiane to Vietnam and ultimately to southwestern China. China initially pulled out of the partnership after assessing that the project would be unprofitable, but, in October 2014, agreed once more to finance the project. Guan Huabing, China's ambassador to Laos, has noted that the project may boost bilateral ties between the two countries, characterizing it as "strategic" and "historic" (Gerin 2014). Furthermore, implementation of large-scale public infrastructure projects at the macro level, often in partnership with neighboring countries, has been the focus of the government's National Development Plan. Projects have typically been in the form of either (1) hydropower projects to generate electric power and revenue or (2) commercial exploitation of land to boost growth.

Energy-hungry Thailand has invested heavily in the Lao hydroelectric industry and also forms the major market for the export of surplus energy produced as a result of dam construction. As first out of 11 dams proposed to be built on the Lower Mekong, the Xayaburi Dam is situated between Laos and Thailand, in Xayaboury Province of northern Laos. Standing 810 meters tall, it is expected to generate 1,260 megawatts of electricity, most of which will be exported to Thailand. The total cost of building the dam is estimated to be 3.5 billion USD (International Rivers 2011). The second and most recent dam proposed to be built on the Mekong River Basin, the Don Sahong Dam, is to be situated on the Siphandone area of southern Laos, approximately 2 km north of the Laos–Cambodia border. The dam would stand between 30 and 32 meters high and generate 260 megawatts of energy for export to Thailand and Cambodia (ibid). The Government of Laos has also encouraged the development of rubber plantations in partnership with Vietnamese and Chinese interests. Further Chinese investment has come through the development of casinos and tourist infrastructure, and Vietnamese companies have partnered with the Government of Laos in logging ventures. Stuart-Fox (2009) has explored the importance of these international partnerships for Laos, terming them the "Vietnamese Connection" and the "Chinese Connection." Official sources report that roughly 1.1 million hectares or 5 percent of the country's arable land has been subjected to 2,600 land deals since 2010 (Heinimann and Messerli 2013). These land leases take the form of foreign investment projects brokered between the government and private enterprises.

Taken together with the econophoria of national developmentalism, however, these macroeconomic projects pursued with international partners, have the potential negatively to impact both human security and environmental considerations. Either human needs are neglected because of the prioritization of macroeconomic policies or they are sacrificed for the "collective good" (Howe and Park 2015). The Xayaburi Dam has been criticized for being a threat to aquatic biodiversity and fishery productivity as well as people's livelihood (International Rivers 2011). The dam will block the migration paths for 23–100 species of fish. By destroying the river's ecosystem, the dam will put at least 41 species of fish at the risk of extinction. More than 2,100 people will have to be resettled because of floods created by the dam and 202,000 people who meet their basic needs through fishing or cultivating rice on the riverbank's farmlands would face challenges to their livelihoods. The government has promised financial assistance to those who have had to resettle since 2013, with aid packages totaling 5 million kip (670 USD) per family per year. Families were granted up to 20 kilograms of rice per adult and up to 15 kilograms per child. The assistance and resettlement process was poorly managed, however, and could not replace access to river's resources, which provided a natural source of sustainable income and food (Vandenbrink 2013). The combination, therefore, of low-quality resettlement programs, environmental degradation, and a lack of viable livelihood options has directly contributed to further impoverishment of the most vulnerable.

Blockage caused by the Don Sahong Dam threatens the migration, feeding, and breeding patterns of the 201 species of fish living between Laos and the neighboring countries. For people living in the lower Mekong Basin, their livelihoods would be greatly affected because fish is a major source of food as well as revenue (International Rivers 2008). As for the Irrawaddy dolphins who inhabit the Laos–Cambodia border, the noise from constant bombing for rock excavation would damage the dolphins' sensitive hearing and their environment will be greatly altered because of increased boat traffic, changes in water quality, and habitat destruction (World Wildlife 2014; Ryan 2014).

The problems associated with land leases or "land grabs" are twofold: for ethnic minorities or indigenous people, their displacement puts them at risk of increased poverty and higher mortality; for land deals themselves, lack of transparency in the process makes accountability and enforcing land regulations increasingly difficult (MacLean 2014). For ethnic minorities and indigenous people living in rural areas, the land grabs continue to drive them off arable farmland without adequate consultation or compensation. As a result of which, mortality rates among the rural poor who have been forced to abandon their traditional livelihoods and move to urban areas, have been found to have risen 30% (UNDP 2001). Furthermore, faced with a lack of opportunity to conduct traditional farming practices, a lack of work skills, and inadequate health and education facilities, the resettled communities are driven into deeper poverty. Those who are forced to relocate from the Mekong River Basin as a result of climate change face similar challenges. Chapman and Van (2018) have noted that the 18 million inhabitants of the low-lying Mekong Delta in Vietnam are also some of the world's most vulnerable to climate change and that, over the preceding 10 years, around 1.7 million people migrated out of its vast expanse of fields, rivers, and canals while only 700,000 arrived. Likewise, according to Nguyen Huu Thien (2017), the Mekong Delta faces the three big challenges of climate change, unsustainable development, and hydropower plants.

These problems and vulnerabilities are exacerbated by the loose enforcement of existing land regulations to compensate displaced populations. As a communist country, Laos retains the right to forcibly seize or redistribute land without prior consent. Although a 2005 decree requires investors to compensate resettled villagers affected in full or in part at replacement cost, implementation has often been piecemeal or non-existent, with little negotiation taking place between the villagers and the government. This unlawful deprivation of land shows how the country's natural resources have been "captured by an elite growing spectacularly rich while one-third of the population lives on less than $0.61 a day" (Redd 2013).

While Laos, Myanmar, and North Korea are all conflict-affected states lying toward the bottom end of most development indices, it would be wrong to assume that only in such cases of challenged governance do we find these sorts of policy patterns. Similar threats to the environment and the human security of the most vulnerable, generated by regime insecurity and econophoric macroeconomic developmentalism, can be found throughout the region in more developed countries, from China, to Vietnam, to the final case study addressed in this chapter, Malaysia. All these countries share a history of insecurity and internal and external conflict, leading to policy prioritization of national political and economic strength and stability. Often, the environment and the human security of the most vulnerable sections of society are seen as the necessary and inevitable casualties of achieving these national goals.

10.3.3 Vulnerabilities in Malaysia

Malaysia achieved independence in 1957. Since that time the country has experienced rapid industrialization, significantly as a result of top-down macroeconomic growth policies promoted by the national federal government through a centralized planned economy. The development policies of Malaysia have been geared toward moving the country from an agriculture-based nation to an industrialized nation (Howe and Kamaruddin 2016). Yet, despite significant economic growth, Malaysia's sociopolitical governance still leaves much to be desired, with distributive injustice meaning that the benefits of development have yet to be felt by some of the most vulnerable sections of society, and worse, big development projects causing irreparable harm to minorities as well as to the environment.

The building of Bakun Dam in the East Malaysian state of Sarawak displaced 10,000 indigenous people "forcibly" moved from the 70,000-hectare reservoir and catchment area to a 4,000-hectare sponsored resettlement site at Sungai Asap (Sovacool and Bulan 2011b). Bakun flooded about 700 square kilometers made up of farmland and forest. The people living in this area depended on the land around them for survival, with the majority relying on the forest for "their agriculture, hunting, and gathering of forest products" (Choy 2005). The problems of indigenous communities have been exacerbated by Malaysia's land policy that does not uphold native land tenure or native customary rights land unless the native communities "start cultivating crops, felling trees, and conducting rotational agriculture to claim ownership of the land" (Sovacool and Bulan 2011b). For most of the indigenous communities of Sarawak, their relationship to the land is more than just as it would be toward a farming place. Rather they feel bound to the preservation of the forest (Cooke 1997).

Choy (2004) has detailed four levels of land conceptualization among these communities: (i) *temuda,* the land area around the longhouses; (ii) *menoa,* forested land used for game hunting and gathering, which provides the people with all their daily subsistence needs; (iii) *dampor,* cultivated land which is located a distance away from their longhouse settlements; and (iv) *pulau,* or protected forest area, which provides the indigenous communities with traditional resources such as water catchments. Likewise, Cooke (1997) has noted how the Penan community, an indigenous hunter–gatherer tribe of Sarawak, view their relationship to the land as one not of ownership, but rather of trusteeship, wherein the rights of both the current and future generations are respected. Under Malaysia's national land-ownership regulations, however, land that is not cultivated is considered state land and thus the indigenous peoples' claims to these lands are not recognized.

A more extensive body of research and literature reviews by Sovacool and Bulan[1] show the extent of damage and hardship upon livelihood and survival of communities within the zones earmarked for "community relocation and resettlement." These negative impacts include the creation of "boom and bust towns" as well as inflation caused by the sudden increase and decrease of population due to the influx of construction workers during the development period (Sovacool and Bulan 2011b). The other two social problems identified by the authors include the impact on "navigation" capacities and "community livelihood" as well as "unfair compensation," which has affected the area and the people living near Bakun Dam. Hydroelectric dams also have negative impacts on the wider environment, which in turn adversely affect local communities. The Bakun Dam caused "irreversible destruction of 69,640 hectares of forest ecosystem" (Choy 2005). The overall environmental cost with the building of dams also manifests itself in the form of greenhouse gas emission "especially carbon dioxide and methane, arising from microbial decomposition of some 69,640 ha of submerged forest, vegetation, wildlife and soil" (ibid.). Dammed water is also prone to the proliferation of algae, which, if left unattended, leach oxygen from the water, bringing about a state of hypoxia (oxygen depletion). The Bakun reservoir is particularly prone to this phenomenon as the bottom holds massive quantities of vegetation (Howe and Kamaruddin 2016).

The state government has yet to release any environmental or social impact assessments despite the fact that the dams are promoted as part of the International Hydropower Association's Hydropower Sustainability Assessment Protocol. The International Association of Communication Activists criticized the fact that construction on Murum Dam began before the project's environmental and social impact assessments (ESIA) had even started and that the ESIA remains inaccessible to the public (Hurwitz and Herbertson 2013). The construction of the smaller Batang Ai Dam in the 1980s involved the relocation of the indigenous Iban (members of one of the largest indigenous groups of Sarawak, formerly animist, now majority Christian) from the Lubok Antu District. Here, too, the problems of insufficient compensation,

[1] Including *inter alia*: Sovacool BK, Bulan LC (2011a) Meeting targets, missing people: the energy security implications of the Sarawak Corridor of Renewable Energy (SCORE), Contemporary Southeast Asia: A Journal of International and Strategic Affairs 33(1): 56–82; Sovacool BK, Bulan LC (2011b) Behind an ambitious megaproject in Asia: The history and implications of the Bakun hydroelectric dam in Borneo, Energy Policy 39: 4842–4859; Sovacool BK, Bulan LC (2012) Energy security and hydropower development in Malaysia: The drivers and challenges facing the Sarawak Corridor of Renewable Energy (SCORE). Renewable Energy 40: 113–129.

lack of opportunities, and the difficulties of transition to a cash economy, including loan repayments, undermined the human security of the relocated groups (Osman 2000). If Bakun truly serves as a harbinger of things to come with the other planned Sarawakian dams—or a continuation of trends established with Batang Ai—then other indigenous communities must brace themselves for the sort of environmental damage and loss of human security endured by those already relocated (Howe and Kamarudin 2016).

10.4 Conclusion

As a result of their conflictual heritage, many Asian states have focused on state-centric security and development models. These have, however, not only proved insufficient to address human security and environmental considerations at the local, national, and transnational levels but have exacerbated them. They may even stimulate the generation of future interstate conflict as a result, for instance, of transborder pollution, refugee flows, and control or destruction of water resources.

The noninterventionary nature of international relations and organizations in Northeast and Southeast Asia in particular has made it difficult to address the life-threatening impact of transborder "yellow dust" and "haze," respectively. On the other hand, a number of human insecurity considerations have challenged the national or even international security perspective of neighboring states; as has occurred with refugees from Myanmar in Bangladesh, North Koreans in China, and Vietnamese in Hong Kong. Lack of human development can also lead to transborder migration, and transborder crime. Security concerns related to Asian transborder migration and refugee flows feature prominently on the traditional security radars of China (Vietnamese, North Koreans, and Burmese nationals), Thailand (Burmese and Lao nationals—particularly ethnic Hmong), Malaysia (Indonesians and Philippine nationals), and Australia (Chinese and Pacific Island nationals) (Howe 2013a). Environmental degradation can also pose national and international security challenges, with such conditions impacting on the human security and human development of vulnerable individuals, groups, and even nations, stimulating the movement of people just as surely as more traditional push factors. This was behind the 2007 controversial position of Australian Federal Police Commissioner Mick Keelty when he identified climate change in the Asia-Pacific region as the greatest security threat faced by Australia due to the relocation of peoples (Lauder 2007).

It is, therefore, in the enlightened self-interest of states and statesmen as well as the international community in East Asia and beyond, to pay attention to both environmental and human security concerns, recognizing the vicious cycle between national insecurity, human insecurity, and environmental insecurity. It becomes plausible then to understand that one way to address human insecurity is to help insecure states ameliorate their national security concerns, and vice versa, with the amelioration of human security concerns helping a target state feel less vulnerable. At the

Table 10.2 Circular insecurity causality

• National Development/Security Focus
• ***Causes***
• Threats to Human Security
• ***Which Cause***
• Unsustainable Development Practices
• ***Which Cause***
• National Development/Security Threats
• ***Which Cause***
• National Development/Security Focus

same time, environmental security is not just vital for one's own direct national security considerations, but also for regional and international considerations. Climate change and degradation of nature, relate quite literally, to the operating environment within which these mutually constitutive processes occur. There is a nonhierarchical, even circular causality operating in East Asia whereby (Table 10.2).

Thus, only holistic governance approaches can hope to deal with the complex web of causality and insecurity addressed in this chapter.

References

Acharya A (2002) Human security: what kind for the Asia Pacific? In: Dickens D (ed) The human face of security: Asia-Pacific perspectives. Australian National University Press, Canberra, pp 5–17

Acharya A (2003) Guns and butter: why do human security and traditional security co-exist in Asia? Glob Econ Rev Perspect East Asian Economies and Ind 32(3):1–21

Ahrens J, Rudolph PM (2006) The importance of governance in risk reduction and disaster management. J Contingencies Crisis Manage 14(4):207–220

Amnesty International. 2011. North Korea: images reveal scale of North Korean political prison camps, 10 May. http://www.amnestyusa.org/news/press-releases/north-korea-images-reveal-scale-of-political-prison-camps. Accessed 20 Feb 2018

Austin JE, Bruch CE (2000) The environmental consequences of war: Legal, economic, and environmental perspectives. Cambridge University Press, Cambridge

BBC (British Broadcasting Company). (2019) North Korea 'facing poor harvest' amid food shortages, 9 November. https://www.bbc.com/news/world-asia-50359435. Accessed 19 Sept 2020

Buzan B, Segal G (1998) Rethinking East Asian security. In: Klare MT, Chandrani Y (eds) World security: challenges for a new century. St. Martin's Press, New York, pp 96–112

Calder K, Ye M (2010) The making of Northeast Asia. Stanford University Press, Palo Alto

Cave R, Lawson A, Sherriff A (2006) Cluster munitions in Albania and Lao PDR: the humanitarian and socioeconomic impact. United Nations Institute for Disarmament Research, Geneva

CIA (Central Intelligence Agency). (2000) Global trends 2015: a dialogue about the future with nongovernmental experts. https://www.cia.gov/library/readingroom/document/0000516933. Accessed 23 Feb 2018

Chapman A, Van PDT (2018) Climate change is driving migration from Vietnam's Mekong Delta. Climate Change News, January 11. http://www.climatechangenews.com/2018/01/11/climate-change-driving-migration-vietnams-mekong-delta/. Accessed 20 Feb 2018
Choy YK (2004) Sustainable development and the social and cultural impact of a dam-induced development strategy—the Bakun experience. Pacific Aff 77(1):50–68
Choy YK (2005) Energy demand, economic growth, and energy efficiency—the Bakun dam-induced sustainable energy policy revisited. Energy Policy 33:679–689
Christie K, Roy D (2001) The politics of human rights in East Asia. Pluto Press, London
Collier P (2008) The bottom billion. Oxford University Press, Oxford and New York
Cook ADB (2017) Governance and insecurity in Myanmar. In: Howe B (ed) National security, state-centricity, and governance in East Asia. Palgrave Macmillan, London, pp 61–80
Cooke FM (1997) The politics of 'sustainability' in Sarawak. J Contemp Asia 27(2):217–241
Disaster Risk Reduction Working Group (2013) Strategic framework 2013–2018. (Myanmar). http://themimu.info/sites/themimu.info/files/documents/Core%20Doc_DRR_Strategic%20Framework%202013~18_DRRWG.pdf. Accessed 20 Feb 2018
Gerin R (2014) China gives new pledge on Lao Rail Project. Radio Free Asia, 06 October. http://www.rfa.org/english/news/laos/railway-project-10062014181543.html?searchterm:utf8:ustring=laos+Railway+. Accessed 20 Feb 2018
(GICHD) Geneva International Centre for Humanitarian Demining (2007) Lao PDR risk management and mitigation model. GICHD, Geneva
Global Security (2016) Myanmar—defense spending. http://www.globalsecurity.org/military/world/myanmar/budget.htm. Accessed 20 Feb 2018
Gramer R (2017) Infographic: Here's how the global GDP is divvied up. Foreign Policy, 24 February. http://foreignpolicy.com/2017/02/24/infographic-heres-how-the-global-gdp-is-divvied-up/. Accessed 20 Feb 2018
Haggard S, Noland M (2005) Hunger and human rights: the politics of famine in North Korea. US Committee for Human Rights in North Korea, Washington, DC
Han DH, Kim SA, Lee KC, Lee KS, Cho JA (2014) White paper on human rights in North Korea 2014. Center for North Korean Human Rights, Korea Institute for National Unification, Seoul
Handicap International (2004) Life after the bomb: a psychosocial study of child survivors of UXO accidents in Lao PDR. Handicap International, Vientiane
Heinimann A, Messerli P (2013) Coping with a land-grab world: Lessons from Laos. Global Change 80. Sweden: International Geosphere-Biosphere Programme, 12 June. http://www.igbp.net/news/features/features/copingwithalandgrabworldlessonsfromlaos.5.19895cff13e9f675e252ba.html. Accessed 23 Feb 2018
Holmes O (2017) Aung Sang Suu Kyi's office accuses aid workers of helping 'terrorists' in Myanmar. The Guardian, 28 August. https://www.theguardian.com/world/2017/aug/28/aung-sang-suu-kyis-office-accuses-aid-workers-of-helping-terrorists-in-myanmar. Accessed 20 Feb 2018
Howe B (2012) Governance in the interest of the most vulnerable. Public Adm Devel 32(4/5):345–356
Howe B (2013a) The protection and promotion of human security in East Asia. Palgrave Macmillan, Basingstoke
Howe B (2013b) Peace and Conflict Drivers. J Hiroshima Peace Institute 1(1):82–97
Howe B, Sims K (2011) Human security and development in the Lao PDR: freedom from fear and freedom from want. Asian Surv 51(2):333–355
Howe B, Park S (2015) Laos: the dangers of developmentalism? Southeast Asian affairs. ISEAS—Yusof Ishak Institute, Singapore, pp 165–186
Howe B, Kamaruddin N (2016) Good governance and human security in Malaysia: Sarawak's hydroelectric conundrum. Contemp Southeast Asia 38(1):81–105
Howe B, Bang G (2017) Nargis and Haiyan: the politics of natural disaster management in Myanmar and the Philippines. Asian Stud Rev 41(1):58–78

Hurwitz Z, Herbertson K (2013) Indigenous people confront industry on Sarawak dams. International Association of Communication Activists, 21 June. http://www.iacact.com/?q=node/316. Accessed 20 Feb 2018

International Rivers (2011) The Xayaburi Dam: A looming threat to the Mekong River. https://www.internationalrivers.org/resources/the-xayaburi-dam-2635. Accessed 23 Feb 2018

Kapucu N (2011) Collaborative governance in international disasters: Nargis cyclone in Myanmar and Sichuan earthquake in China cases. Int J Emerg Manage 8(1):1–25

KBS World (2020) FAO warns of acute food shortage in N. Korea due to COVID-19, 18 July. https://world.kbs.co.kr/service/news_view.htm?lang=e&Seq_Code=154937. Accessed 19 Sept 2020

Kingshill KP (1991) Present-day effects of United States bombing of Laos during the Vietnam War: can injured Laotians recover under the Federal Tort Claims Act. Loyola of Los Angeles. Int Comp Law J 13:133–178

Koga K (2009) The anatomy of North Korea's foreign policy formulation. North Kor Rev 5(2):21–33

Kreft S, Eckstein D, Melchior I (2017) Global climate risk index 2017: Who suffers most from extreme weather events? Weather-related loss events in 2015 and 1996 to 2015. https://germanwatch.org/de/download/16411.pdf. Accessed 20 Feb 2018

Landmine Monitor (2008) Landmine Monitor Report, Lao People's Democratic Republic. http://archives.the-monitor.org/index.php/publications/display?url=lm/2008/countries/lao.html. Accessed 23 Feb 2018

Lankov A (2015) The real North Korea, life and politics in the failed Stalinist utopia. Oxford University Press, New York

Lauder S (2007) Climate change, a huge security problem: Keelty. ABC News. 25 September. http://www.abc.net.au/news/2007-09-25/climate-change-a-huge-security-problem-keelty/680208. Accessed 20 Feb 2018

Leaning J (1993) War and the environment: Human health consequences of the environmental damage of war. In: Chivian E, McCally M, Hu H, Haines A (eds) Critical condition: human health and the environment. MIT Press, Cambridge, pp 123–137

Lwin HH, Maung KK (2011) A study of public awareness education training on natural disaster risk reduction in Myanmar. Univ. Res. J. 4(5):223–241

MacLean D (2014) Laos 'land grabs' drive subsistence farmers into deeper poverty. IRIN News, 22 May. http://www.irinnews.org/report/100116/laos-land-grabs-drive-subsistence-farmers-into-deeper-poverty. Accessed 20 Feb 2018

Mullany LC, Lee CI, Paw P, Shwe Oo EK, Maung C, Kuiper H, Masenior N, Beyrer C, Lee TJ (2008) The MOM Project: delivering maternal health services among internally displaced populations in eastern Burma. Reprod Health Matters 16(31):44–56

NCCR (Swiss National Centre of Competence in Research) (2008) Socio-economic atlas of the Lao PDR: an analysis based on the 2005 population and housing census. NCCR, Bern, Switzerland

ND-Gain (2017) University of Notre Dame Global Adaptation Index 2017. https://gain.nd.edu/our-work/country-index/rankings/. Accessed 20 Feb 2018

Nguyen HT (2017) Mekong Delta faces three big challenges. Vietnamnet, 4 October. http://english.vietnamnet.vn/fms/environment/187364/mekong-delta-faces-three-big-challenges–climate-change–hydropower.html. Accessed 20 Feb 2018

Nishikawa Y (2009) Human security in Southeast Asia: viable solution or empty slogan? Secur Dialogue 40(2):213–236

NRA (National Regulatory Authority for UXO/Mine Action Sector in Lao PDR) (2008) UXO Sector Annual Report, http://nra.gov.la/resources/Annual%20Reports/UXO%20Annual%20Report%202008_English.pdf. Accessed 20 Feb 2018

OHCHR (Office of the High Commissioner for Human Rights) (2014a) Report of the detailed findings of the commission of inquiry on human rights in the Democratic People's Republic of Korea–A/HRC/25/CRP.1 http://www.ohchr.org/EN/HRBodies/HRC/CoIDPRK/Pages/ReportoftheCommissionofInquiryDPRK.aspx. Accessed 20 Feb 2018

OHCHR (Office of the High Commissioner for Human Rights) (2014b) North Korea: UN Commission documents wide-ranging and ongoing crimes against humanity, urges referral to ICC, 17 February. http://www.ohchr.org/en/NewsEvents/Pages/DisplayNews.aspx?NewsID=14255&angID=E. Accessed 20 Feb 2018

OHCHR (Office of the High Commissioner for Human Rights) (2017) Report of OHCHR mission to Bangladesh. Interviews with Rohingyas fleeing from Myanmar since 9 October 2016. 3 February. http://www.ohchr.org/Documents/Countries/MM/FlashReport3Feb2017.pdf. Accessed 20 Feb 2018

Osman S (2000) Globalization and democratization: the response of the indigenous peoples of Sarawak. Third World Q. 21(6):977–988

Paul RA (1971) Laos: anatomy of an American involvement. Foreign Aff 49(3):533–547

Phoenix Clearance Limited (2001) UXO Lao Bombing Data. Vientiane, Phoenix

Pholsena V, Banomyong R (2006) Laos: From buffer state to crossroads?. Mekong Press, Bangkok

Redd (2013) Can REDD stop the land grabbing crisis in Laos and Cambodia? http://www.redd-monitor.org/2013/05/15/can-redd-stop-the-land-grabbing-crisis-in-laos-and-cambodia/. 15 May. Accessed 20 Feb 2018

Rigg J (1998) Uneven development and the (re-)engagement of Laos. In: Dixon C, Drakakis-Smith D (eds) Uneven development in South East Asia. Ashgate, Aldershot, pp 67–84

Ryan GE (2014) The Don Sahong Dam and the Mekong dolphin. WWF Science Brief. http://awsassets.panda.org/downloads/donsahong_dolphin_lr_feb2014.pdf. Accessed 20 Feb 2018

Selth A (2008) Even paranoids have enemies: Cyclone Nargis and Myanmar's fears of invasion. Contemp Southeast Asia 30(3):379–402

Selth A (2017) The fallen idol: Aung San Suu Kyi and the politics of personality. ABC Religion and Ethics. 12 September. http://www.abc.net.au/religion/articles/2017/09/12/4733319.htm. Accessed 20 Feb 2018

Shaw M (2000) Theory of the global state: globality as unfinished revolution. Cambridge University Press, Cambridge

Shaw R (2006) Community-based climate change adaptation in Vietnam: inter-linkages of environment, disaster, and human security. In: Sonak S (ed) Multiple dimensions of global environmental change. TERI Press, New Delhi, pp 521–547

Shikada M, Myint T, Gyi KK, Nakagawa Y, Shaw R (2012) Chapter 10: Reaching the unreachable: Myanmar experiences of community-based disaster risk reduction. In: Shaw R (ed) Community-based disaster risk reduction. Emerald Group Publishing Limited, Bingley, pp 185–203

Sisoulith T (2011) Statement at the General Debate of the 66th General Assembly of the United Nations, New York. 26 September 2011. https://gadebate.un.org/sites/default/files/gastatements/66/LA_en.pdf. Accessed 20 Feb 2018

Smith M (2007) State of strife: the dynamics of ethnic conflict in Burma. Policy Studies 36. Washington: East-West Center

Sovacool BK, Bulan LC (2011a) Meeting targets, missing people: the energy Security Implications of the Sarawak Corridor of Renewable Energy (SCORE). Contem Southeast Asia: J Int Strateg Aff 33(1):56–82

Sovacool BK, Bulan LC (2011b) Behind an ambitious megaproject in Asia: the history and implications of the Bakun hydroelectric dam in Borneo. Energy Policy 39:4842–4859

Sovacool BK, Bulan LC (2012) Energy security and hydropower development in Malaysia: the drivers and challenges facing the Sarawak Corridor of Renewable Energy (SCORE). Renewable Energy 40:113–129

Stuart-Fox M (2009) Laos the Chinese connection. Southeast Asian Affairs. Singapore, ISEAS—Yusof Ishak Institute, pp 141–169

Suh JJ (2013) Rethinking national and human security in North Korea. In: Park Kyung-Ae (ed) Non-traditional security issues in North Korea. University of Hawaii Press, Honolulu, pp 1–22

Tirman J (2015) The human costs of war and how to assess the damage. Foreign Affairs. 8 October. https://www.foreignaffairs.com/articles/middle-east/2015-10-08/human-cost-war. Accessed 20 Feb 2018

UNDP (United Nations Development Programme) (1994) Human development report: new dimensions of human security. Oxford University Press, New York

UNDP (United Nations Development Programme) (2001) National human development report Lao PDR 2001: Advancing rural development. http://hdr.undp.org/sites/default/files/lao_2001_en.pdf. Accessed 23 Feb 2018
UNDP (United Nations Development Programme) (2004) Reducing disaster risk—a challenge for development. http://www.undp.org/content/undp/en/home/librarypage/crisis-prevention-and-recovery/reducing-disaster-risk–a-challenge-for-development.html. Accessed 20 Feb 2018
UNDP (United Nations Development Programme) (2008) Hazardous ground: cluster munitions and UXO in the Lao PDR. UNDP, Vientiane
UNDP (United Nations Development Programme) (2015) What is human development? http://hdr.undp.org/en/content/what-human-development. Accessed 20 Feb 2018
UNDP (United Nations Development Programme) (2016) Issue Brief: Development and mine action. http://www.undp.org/content/undp/en/home/librarypage/poverty-reduction/development-and-mine-action-policy-and-programming.html Accessed 25 Feb 2018
UNEP (United Nations Environment Programme) (2009) Learning from Cyclone Nargis: investing in the environment for livelihoods and disaster risk reduction. https://wedocs.unep.org/bitstream/handle/20.500.11822/14116/myanmar_cyclonenargis_case_study.pdf?sequence=1&isAllowed=y. Accessed 20 Feb 2018
Unger DC (2012) The emergency state: America's pursuit of absolute security at all costs. The Penguin Press, New York
UXO Lao Bombing Data (2001) Map of UXO impact and bombing data 1965–1975. Phoenix Clearance Limited, Vientiane
Vandenbrink R (2013) Assistance for villagers resettled by Xayaburi Dam to last one year. Radio Free Asia. 18 June. http://www.rfa.org/english/news/laos/xayaburi-06182013164824.html. Accessed 20 Feb 2018
Vittal N (2001) Tribulation before trial in Cambodia: Confronting autogenocide. Econ Polit Weekly 36(3):199–203
Waever O (1995) Securitization and desecuritization. In: Lipschutz RD (ed) On security. Columbia University Press, New York, pp 46–86
Warner R (1996) Shooting at the moon: the story of America's clandestine war in Laos. Steerforth Press, South Royalton
Watson A (2004) An introduction to international political economy. Continuum, London
World Peace Foundation (2015a) The Korean War. https://sites.tufts.edu/atrocityendings/2015/08/07/korea-the-korean-war/. Accessed 20 Feb 2018
World Peace Foundation (2015b) The Vietnam War. https://sites.tufts.edu/atrocityendings/2015/08/07/vietnam-vietnam-war/. Accessed 20 Feb 2018
World Wildlife (2014) New dam on Mekong River could doom Irrawaddy dolphins. http://www.worldwildlife.org/stories/new-dam-on-mekong-river-could-doom-irrawaddy-dolphins. May 22. Accessed 20 Feb 2018
Zasloff JJ (1970) Laos: The forgotten war widens. Asian Surv 10(1):65–72

Dr. Brendan Howe (Ph.D., TCD) is Professor of International Relations at the Graduate School of International Studies, Ewha Womans University. He has held fellowships at the Freie Universität Berlin, University of Sydney, East–West Center, Georgetown University, UNIMAS, and Beijing Foreign Studies University. His research focuses on governance, human security, and post-conflict development in Asia. He has authored/edited more than 90 publications including *Regional Cooperation for Peace and Development: Japan and South Korea in Southeast Asia* (Routledge, 2018); *National Security, State-Centricity, and Governance in East Asia* (Palgrave, 2017); *Peacekeeping and the Asia-Pacific* (Brill, 2016); *Democratic Governance in Northeast Asia* (Palgrave, 2015); *Post-Conflict Development in East Asia* (Ashgate, 2014); and *The Protection and Promotion of Human Security in East Asia* (Palgrave, 2013).

Chapter 11
Mainstreaming CCA-DRVRM Using Probabilistic Multi-scenario Hazard Maps for Future Resilience in Haiyan-Affected Areas

Alfredo Mahar Francisco A. Lagmay, Joy Santiago, and Juan M. Pulhin

Abstract Resilience to future disasters means helping communities and ecosystems successfully adapt to changing climate. To do this, efforts must be made to mainstream climate change adaptation (CCA) and disaster risk and vulnerability reduction and management (DRVRM) into existing local development plans. This requires the use of probabilistic or multi-scenario hazard maps to contextualize adaptive measures to climate change. We describe the methods to prepare probabilistic hazard maps related to severe weather events (i.e., floods, storm surges, and landslides) with climate change projections modeled or simulated through computer simulations using high-resolution, 1:25,000- to 1:10,000-scale digital terrain models, rainfall/climatological data from at least 30 years of historical record, and other necessary input parameters such as soil and land cover type. We provide examples for multi-hazard maps prepared for several Haiyan-affected areas to mainstream CCA-DRVRM in comprehensive land use plans and local development plans of municipalities. This work aims to spread the same methods of preparing climate change-sensitive hazard and risk maps in the entire Philippines as a major input to local development planning to help advance human security in the country.

Keywords Climate change adaptation · Disaster risk and vulnerability reduction and management · Multi-scenario hazard maps · Resilience

---

A. M. F. A. Lagmay (✉) · J. Santiago
National Institute of Geological Sciences, University of the Philippines Diliman, Quezon City, Philippines
e-mail: mlagmay@noah.up.edu.ph

J. M. Pulhin
Department of Social Forestry and Forest Governance, College of Forestry and Natural Resources (CFNR), University of the Philippines Los Baños, Laguna, Philippines

J. M. Pulhin et al. (eds.), *Climate Change, Disaster Risks, and Human Security*, Disaster Risk Reduction, https://doi.org/10.1007/978-981-15-8852-5_11

11.1 Introduction

The Philippines experiences numerous types of hazards that could directly and adversely affect and threaten individuals and the necessities and values they possess. Ranking 13th on the Verisk Maplecroft 2016 climate change vulnerability index and 4th on the Germanwatch 2014 global climate risk index, it is one of the countries in Southeast Asia identified as a climate change hotspot (Yusuf and Francisco 2009).

To achieve resilience in facing future disasters, Philippine communities and ecosystems must cope with changing climate conditions. Efforts must be made to mainstream climate change adaptation (CCA) and disaster risk and vulnerability reduction and management (DRVRM) into existing local development plans such as the comprehensive land use plans (CLUP), comprehensive development plans (CDP), and local climate change action plans (LCCAP) at the city and municipal levels; at the provincial and regional levels, there are the provincial development physical framework plan and investment plan. Mainstreaming, integration, and convergence need to be done immediately in anticipation of the worsening impacts of climate change in the future.

In the development planning process of cities and municipalities, it is vital that future hazards, bigger than what is known, are captured during disaster risk assessment. It has been said that, "We cannot wholly rely on our knowledge of past events to anticipate future risk because some disasters that could happen have not yet happened" (UNISDR 2015). A probabilistic approach minimizes these limitations. It uses historical events, expert knowledge, and theory to simulate events that can physically occur but are not represented in the historical record (UNISDR 2015). Unlike deterministic risk assessment, which considers the impact of a single risk scenario (normally the worst case based on historical accounts), probabilistic risk assessment considers all possible scenarios, their likelihood, and associated impacts (UNISDR 2015).

While historical losses can explain the past, they do not necessarily provide a good guide to the future; most disasters that could happen have not happened yet (UNISDR 2015). Probabilistic risk assessment simulates those future disasters which, based on scientific evidence, are likely to occur. As a result, these risk assessments resolve the problem posed by the limits of historical data. Probabilistic models therefore complete historical records by reproducing the physics of the phenomena and recreating the intensity of a large number of synthetic events (UNISDR 2015).

As defined by the United Nations (UN), disaster risk is a function of hazard, exposure, and vulnerability, wherein hazards are described quantitatively by the likely frequency of occurrence of different intensities for different areas as determined from historical data or scientific analysis (UNISDR 2015). Reference to probability in the definition of hazard in the disaster risk equation is also articulated in Philippine Republic Act 10121 (Disaster Risk Reduction and Management Act). In that provision, risk assessments should include a review of the technical characteristics of hazards such as their location, intensity, frequency, and probability.

It is vital that probabilistic or multi-scenario hazard maps are used (e.g., 5-yr, 25-yr, 100-yr rain return floods, and climate change projections) for all disaster-prevention-and-mitigation-related efforts of the government. Probabilistic hazard maps related to severe weather events (i.e., flood, storm surge, and landslide hazard maps) with climate change projections modeled or simulated through computer simulations using high-resolution, 1:5,000- to 1:25,000-scale (Martini and Loat 2007) digital terrain models (DTMs), rainfall/climatological data from at least 30-year historical record (IPCC 2014), and other necessary input parameters such as soil and land cover type.

Disaster risk reduction management in the context of CCA therefore is not limited to the current weather and their potential hazard impacts but rather to climate-related hazard scenarios that can affect vulnerable communities and areas decades from today. As such, it is imperative that scientists and practitioners prepare multi-hazard maps of the future for early warning, incorporating possible scenarios predicted as a consequence of climate change. Its core aspect follows the triangle of survival: Paris Climate Agreement of the United Nations Framework Convention on Climate Change (UNFCCC), the Sendai Framework for Disaster Risk Reduction 2030 and the Agenda 2030, the Sustainable Development Goals (SDGs), which requires science- and risk-based approaches in their formulation and implementation.

This chapter describes the value of probabilistic multi-hazard maps that incorporate available climate change projections in resilience planning of communities to enhance human security. Examples are made for probabilistic multi-hazard maps prepared for several Haiyan-affected communities to mainstream CCA-DRVRM into CLUPs and CDPs of municipalities.

It is hoped that the same science-based methods be used to prepare climate change-sensitive hazard and risk maps for the entire Philippines based on learnings from the impacts of Typhoon Haiyan with the end view of advancing human security in the country. Called Yolanda, this was the worst typhoon ever to make landfall in recorded history, killing 6,300 Filipinos with scores missing and US$2 billion worth of damage to property in central Philippines (NDRRMC 2014).

11.2 Methodology

This section discusses briefly the techniques used to quantify climate change scenarios for municipal or city development planning and to generate climate-adjusted flood, landslide, and storm surge hazard maps in the Central Philippine region, areas that were affected by Typhoon Haiyan. The hazard methodologies were developed through the years by the Nationwide Operational Assessment of Hazards (NOAH) team starting in 2012 up to the present. A project funded by the Department of Science and Technology (DOST), NOAH is the DRR flagship program of the Philippines. The methods presented were enhanced under an ADB project entitled "Improved Resilience to Future Disasters" (JFPR-9175 Philippine emergency assistance and early recovery for poor municipalities affected by Typhoon

Yolanda) (CONCEP Inc. GEOS Inc., and ASSURE Inc. 2017), and United Nations Development Programme (UNDP) and Climate Change Commission (CCC) projects "Resilience Capacity Building for Cities and Municipalities to Reduce Disaster Risks from Climate Change and Natural Hazards (REBUILD)" (UP NIGS 2018), and "Resilience and Preparedness toward Inclusive Development (RAPID) Program" (UP NIGS 2017), and Enhancing Philippines Landslide Hazard Maps with LiDAR and High Resolution Imageries of Project NOAH under the Department of Science and Technology and University of the Philippines Diliman (Project NOAH 2015).

11.2.1 Climate Change Projections

The climate change scenarios that were used for a pilot project on "Improved Resilience to Future Disasters" is based on the Philippine Weather Bureau's Leyte and Samar climate change rainfall intensity-duration frequency (RIDF) baseline data and projections (PAGASA 2011). This is described to illustrate the concept of determining rainfall for a future rainfall scenario projection for climate change.

PAGASA's historical 100-year rain return period has an equivalent accumulated rainfall of 273 mm in 24 h. If translated to the Weather Bureau's projected 19.6% maximum seasonal increase in rainfall amount by the year 2050, this value of 273 mm in 24 h becomes a 24-year return period (Fig. 11.1). Extreme rainfall of 273 mm in 24 h that used to occur once every 100 year becomes more frequent in a climate

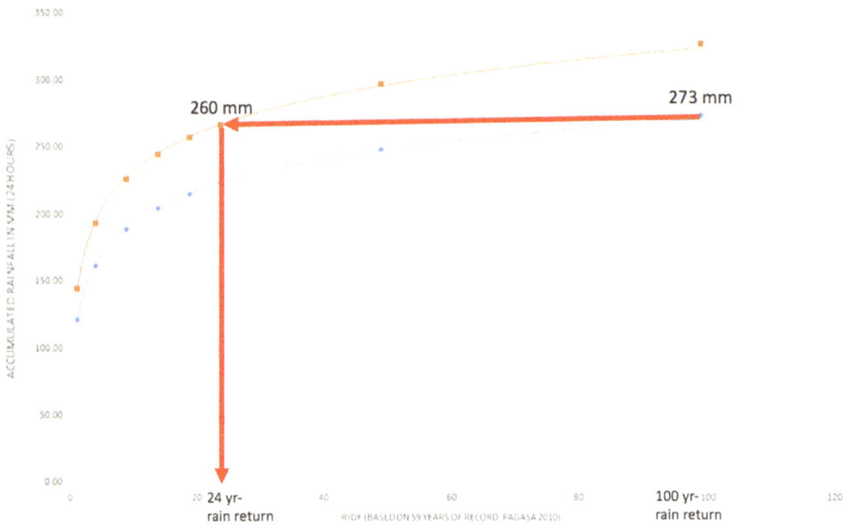

Fig. 11.1 Accumulated rainfall and climate change-adjusted rainfall for 24 h of different return periods for Leyte province

change scenario, which, by 2050, has a likelihood of occurrence almost three times in 100 year.

In the first quarter of 2017, PAGASA released climate change projections for precipitation and temperature (PAGASA 2017). Three datasets were available in their release: historical climate (1971–2000) and projected future climate; projections for 2020–2049 m, and projections for 2050–2079. According to their report, these statistically downscaled seasonal projections presented as ensemble means for seasons DJF (December, January, February), MAM (March, April, May), JJA (June, July, August), and SON (September, October, November) were derived at station level using six global climate models with two scenarios and the 20C3M/Historical (1971–2000) model run. Precipitation values from these model runs were translated into graphs of averaged 24-h rainfall vs RIDF choosing representative concentration pathways (RCPs) representing sustainable CCA strategies (Fig. 11.2). The RCPs chosen were RCP 4.5 (medium-range scenario) and RCP 8.5 (high-range scenario) as recommended by the national panel of technical experts (NPTE) of the CCC.

In the example provided showing climate change projections for the baseline (historical record), RCP 4.5 and RCP 8.5 for years 2049 and 2079, multiple scenarios are evident showing changes in frequency and probability of events and rainfall amount. A 100-yr rain return amounting to 280 mm of rainfall, when projected to the 100-yr return period for the RCP 4.5 projection for 2049, increases to 400 mm of rainfall. The same amount for an RCP 4.5 scenario becomes more frequent. Instead of an annual exceedance probability of 0.01, it changes to 0.08.

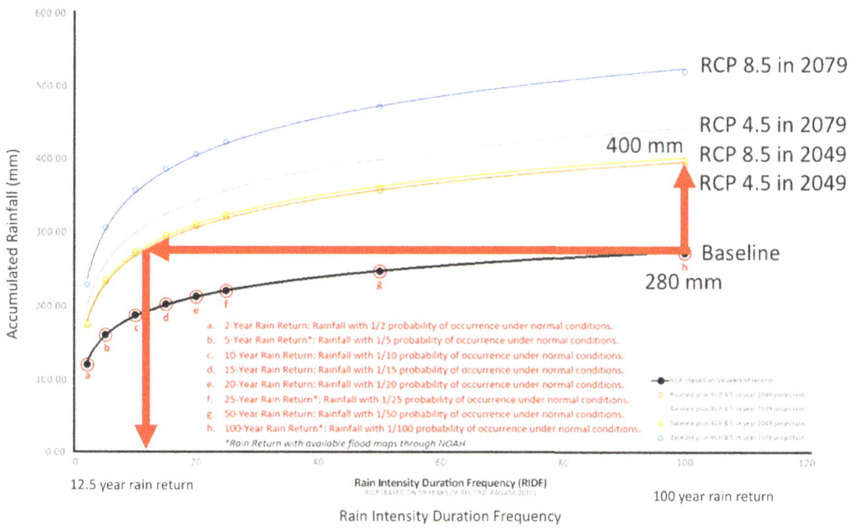

Fig. 11.2 Precipitation values for the historical and climate change scenarios using RCP 4.5 (medium-range scenario) and RCP 8.5 (high-range scenario). The example is for the province of Iloilo

The rainfall amounts are considered plausible as the range of values presented, which are from 280 to 520 mm in 24 h, was observed to happen in the Philippines. During the onslaught of Urduja in December 2017, total precipitation as recorded in Calubian, Leyte, was 574.50 mm, whereas during the Habagat events of 2012 and 2013, total precipitation was above 1000 mm in the span of 1.5 d, respectively. The change in frequency of the 100-yr rain return period to the 12.5-yr rain return period for a climate change scenario is also possible. In CDO, the rain events for both tropical storms Sendong in 2011 and Vinta in 2017 have a 100-yr recurrence interval based on baseline record but were repeated within a span of 6 yrs. Furthermore, the frequency change is consistent with predictions on climate change where the 100-yr return period flood is expected to become more frequent with 5- to 25-return periods in the twenty-first century in Asia and Southeast Asia for RCP 8.5 (Hirabayashi et al. 2013).

11.2.2 Hazard Maps

The changes in rainfall amount due to climate change as projected by PAGASA (i.e., RCP 4.5 and 8.5 for the years 2049 and 2079) are used as parameter input for multiple flood simulations. These future scenarios of floods are modeled over high-resolution topography (LiDAR or IFSAR) resampled to 10- × 10-m postings. Discharge flowing from the upper watershed into the model area was computed for each of the different scenarios with direct runoff values calculated using the hydrologic modeling system designed by the US Army Corps of Engineers Hydrologic Engineering Center. We used Flo-2D GDS PRO, a grid-based GIS-integrated software tool that utilizes the St. Venant equations to solve for continuity and momentum in computing for the velocity of water and inundation distribution.

The landslide hazard simulations generated climate-adjusted rain-induced shallow landslides and structurally controlled landslides. Modeling rain-induced shallow landslides to determine temporal and spatial distribution was conducted using the transient rainfall infiltration and grid-based regional slope-stability model (TRIGRS) software (Baum et al. 2008). This method employs transient pore pressure changes to indicate slope stability in terms of a factor of safety (Iverson 2000). The software presumes that pore water pressure is a function only of depth and time, with uniform soil physical properties based on an infinite slope model (Taylor 1948). An extension of TRIGRS called TRIGR-P uses a Monte Carlo probabilistic approach to account for variability and uncertainty in the parameters used in the simulation process (Raia et al. 2014). With this added module, critical aspects in slope stability analysis and rainfall intensity-duration conditions are included in the analysis (Alvioli et al. 2014). Several maps of the factor of safety were generated in multiple runs of TRIGRS-P, which were then compared with a rainfall-induced landslide inventory. This is to evaluate the geographic distribution of correct assignments and model errors (Guzzetti and Cardinali 2006). Using the probabilistic approach to modeling requires a longer processing time relative to the deterministic approach.

Only shallow landslide hazard maps are probabilistic as there are limitations in obtaining the probability of occurrence of structurally controlled landslides. In the identification of structurally controlled landslides, discontinuities are treated as the primary cause of slope failure. Once determined, the discontinuity sets are subjected to kinematic analysis using Matterocking software, which, in turn, is used to identify unstable slopes. Matterocking is an open-source software that was used in this study to compute the mechanisms for planar, wedge, and toppling failure. Results from structurally controlled simulation were used to determine the extent of the propagation of the landslides. Zones of landslide propagation were mapped using Conefall using a runout angle of 20 degrees.

Climate-adjusted storm surge hazard maps were simulated using the Japan Meterological Agency (JMA) storm surge model, a numerical model based on two-dimensional shallow water equations (Higaki et al. 2009), equation of motion, and the continuity equation with air pressure and wind fields as external forcing. The time-series data generated from the JMA model were then used as input parameter for the FLO-2D GDS PRO simulation. Projected increases in sea level (Rahmstorf and Coumou 2011) and typhoon intensity (Mei et al. 2015) were incorporated in the storm surge simulations.

11.3 Results

The output of the simulations made for the flood, landslide, and storm surge hazards was multi-scenario, including future hazards that incorporate the impacts of climate change. The number of scenarios depends on the number of models that are generated. In previous work conducted by the NOAH team (CONCEP Inc. GEOS Inc., and ASSURE Inc. 2017; UP NIGS 2018; UP NIGS 2017; Project NOAH 2015), 5-yr, 25-yr, and 100-yr hazard maps were created for the baseline, with the same number of simulations with return periods for RCP 4.5 and RCP 8.5 for the years 2049 and 2079. These RCPs were recommended by the NPTE of the CCC, government of the Philippines, and follow the guidelines of the Intergovernmental Panel on Climate Change (IPCC) Fifth Assessment Report in terms of RCPs. In total, 15 hazard models were generated for flood and landslide hazards. Only nine storm surge hazard maps were generated because only the RCP 4.5 projection was available. Example outputs are provided for each of these hazards.

11.3.1 Flood Models

An example of baseline and climate change-adjusted flood hazard maps is shown for the municipality of Zarraga, Iloilo (Fig. 11.3). The maps show that the extent of flooding increases for all return periods (i.e., 5-yr, 25-yr, and 100-yr return period) for all climate change scenarios (RCPs 4.5 and 8.5). The extent of flooding for a

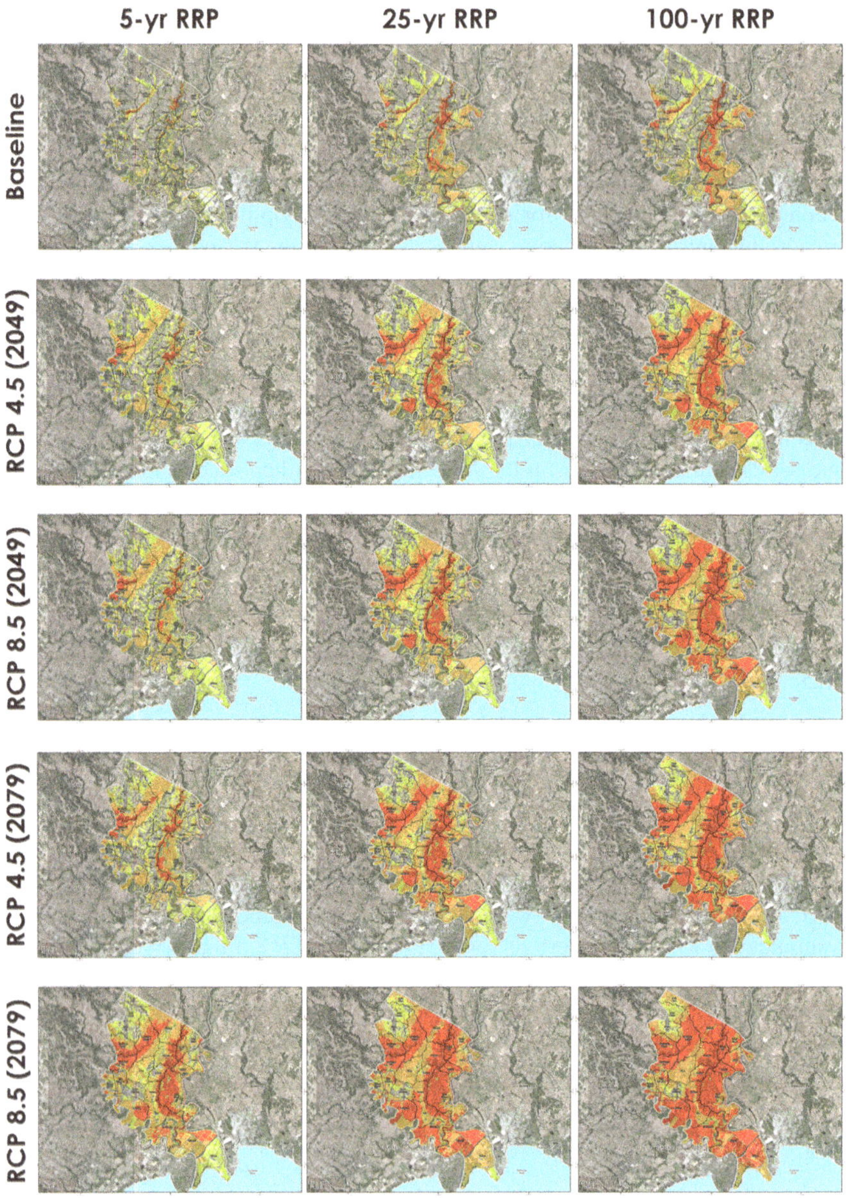

Fig. 11.3 Probabilistic (multi-scenario) flood hazard maps of the municipality of Zarraga, Iloilo, Philippines

100-yr return period baseline scenario is approximately 20,691,098 m^2. It increases by 49% for RCP 4.5 2049 and by 57% for RCP 8.5 2049. The same trend is apparent for 2079 scenarios; 63% for RCP 4.5 2017 and 73% for RCP 8.5 2079, respectively. The 5-yr and 25-yr scenarios also exhibit the same trend.

11.3.2 Landslide Models

Probabilistic multi-scenario rain-induced landslide hazard maps are shown for the municipality of Zarraga, Iloilo (Fig. 11.4). The factors affecting rain-induced landslides are ground conditions and amount of rainfall. The amount of rainfall used in the landslide modeling are baseline scenario and different climate change projections similar to those used in flood modeling. However, based on scientific research and hazard simulations conducted by the NOAH team, 90% of landslides in the inventory are triggered above a 100-mm threshold. This explains the extent of hazards shown in the set of maps in Fig. 11.4, which does not change significantly regardless of rain return period and scenario. In summary, the hazard extent for different landslide scenarios increase by no more than 20% compared with the baseline (Fig. 11.5).

11.3.3 Storm Surge

The probabilistic storm surge hazard maps show an increasing trend in the inland incursion of seawater. Unlike in the case of flood and landslide hazards, the available climate change projections for storm surge hazards are for RCP 4.5 year 2050 and 2100 based on the typhoon intensity projection of Mei et al. (2015). The extent of coastal flooding due to storm surges for the 5-yr tropical cyclone return period may increase by 22% in 2050 and by 76% in 2100 for RCP 4.5. An increase of 4% up to 17% is also evident for the 25-yr return period and an increase of 12% up to 19% for the 100-yr return period for RCP 4.5.

11.4 Discussion

The Philippines has a long history of suffering from meteorological hazards. Over the years, hazards have been enhanced by climate change impacts hitting hard on the more vulnerable and less developed communities, resulting in increased deaths and economic setbacks. Strengthening of DRR policies and practices is the initial step of action to establish more resilient communities. This includes comprehensive assessment of climate change vulnerability and disaster risks of hazardous areas. Assessments must be conducted using probabilistic (multi-scenario) hazard maps to produce scenario-based multi-hazard socioeconomic profiles, generate risk maps,

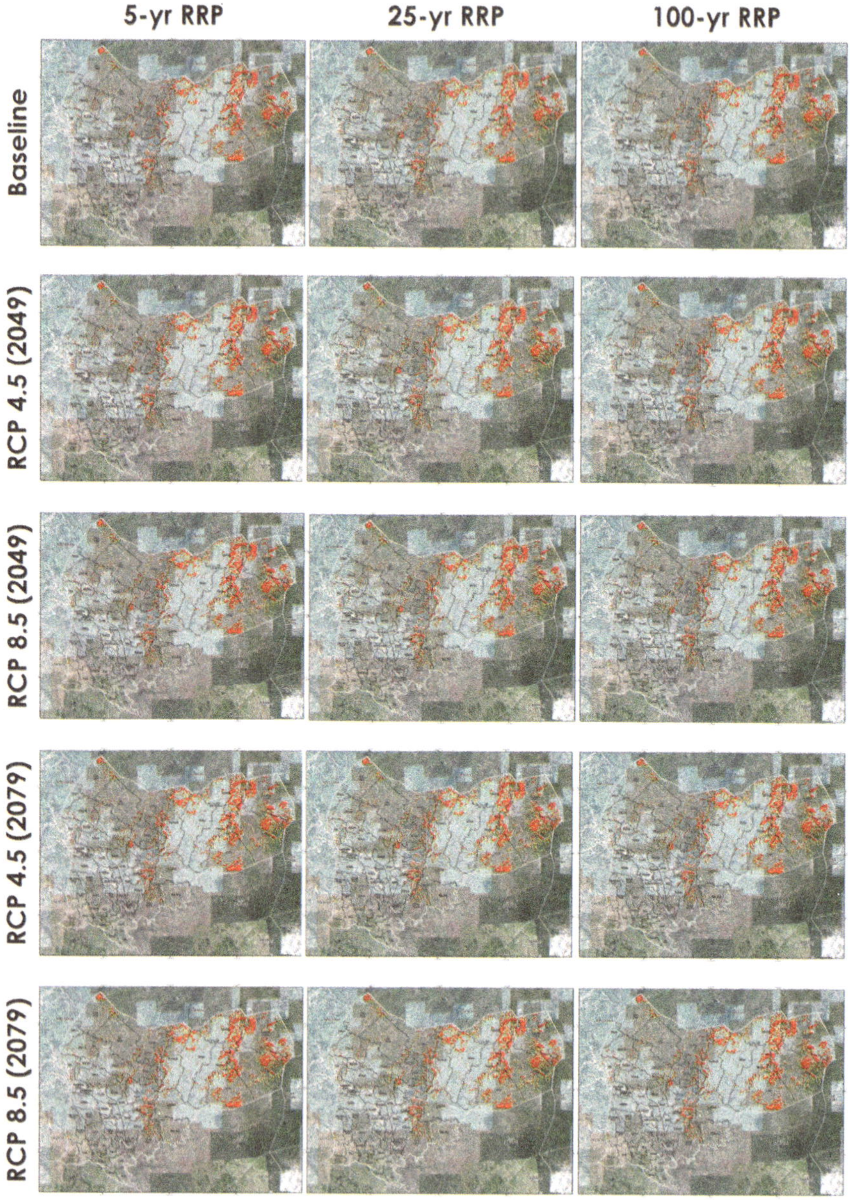

Fig. 11.4 Probabilistic multiscenario rain-induced landslide hazard maps of the city of Passi, Iloilo, Philippines

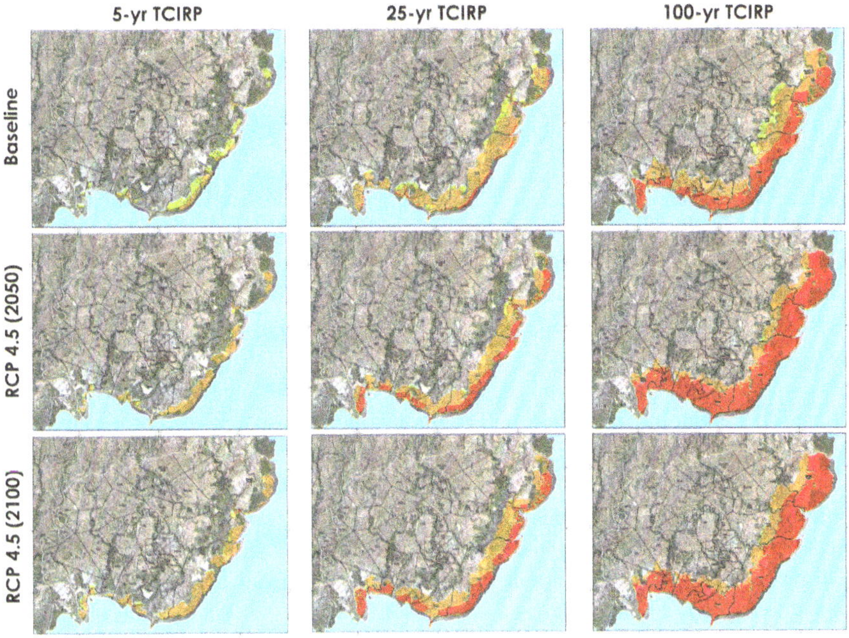

Fig. 11.5 Probabilistic multi-scenario storm surge hazard maps of the municipality of Dumagas, Iloilo, Philippines

and formulate vulnerability and adaptation reports as well as develop impact models to support decisions on adaptation measures.

Mainstreaming CCA-DRR in the Philippines is critical to the country's development. Improving the governing framework and advancing the capacity and competency of key local stakeholders in the cities and municipalities will enable key decision-makers to use vulnerability and risk assessment to formulate and implement measures for appropriate adaptation responses to ensure safety, to integrate CCA-DRR into local development planning and regulatory processes, to adopt proper CCA-DRR measures, and to develop climate-resilient livelihood and risk/sharing transfer models. Collectively, these efforts develop the ability of vulnerable communities to mitigate the effects of climate change.

11.4.1 Deterministic and Probabilistic Approaches to CCA-DRR in the Philippines

11.4.1.1 Deterministic Risk Assessment

In the Philippines, the practice of creating hazard maps is in general based on community accounts. This is called community-based disaster risk management (CBDRM). The key elements of CBDRM are (1) community ownership, (2) use of local knowledge about hazards, (3) communities as ultimate beneficiaries, (4) multi-stakeholder participation, (5) education and capacity building, (6) gender sensitivity, (7) cultural appropriateness, (8) sensitivity to local structure, (9) complementation of community-based and top-down approaches, (10) demonstrated potential for building economic resilience, (11) demonstrated transparency in procedures and processes, (12) commitment and accountability of stakeholders, (13) communication design, and (14) exit strategy and sustainability mechanisms (Torrente et al. 2008).

Item number 9 as identified by representatives from NGOs and national government agencies incorporates top-down approaches. After the Cherry Hills landslide disaster in 1999, the government embarked on a program to map out flood and landslide hazards through the issuance of DENR DAO 2000–33 on the conduct of an engineering geological geohazard assessment and the preparation of the corresponding report, with the Philippine Mines and Geosciences Bureau as the implementing agency, mapping of flood and landslide hazards became systemized on a nationwide scale by the year 2004 (Aurelio 2004).

The Department of Environment and Natural Resources Mines and Geosciences Bureau (DENR-MGB) has been implementing the National Geohazard Assessment and Mapping Program with the primary objective of identifying areas in the country that are susceptible or prone to various geological hazards and providing vital information to various stakeholders in order to lessen or mitigate the impacts of these events. There are various kinds of geohazards, but the DENR-MGB focused its efforts mainly on landslides and floods. The DENR-MGB has already completed the 1:50,000-scale assessment and mapping of 1,640 cities and municipalities nationwide in 2010; that at the 1:10,000-scale was done in 2014. The advisories are complemented by reports of geohazard assessment and GIS-based geohazard maps, which have been distributed to all local government units (LGUs) (MGB 2015). The methodology for the creation of the MGB susceptibility maps included (1) interpretation and analysis of remote sensing data (aerial photographs and satellite images) and other thematic maps such as geologic maps, drainage maps, slope maps, and vegetation maps; (2) ground truthing; (3) geomorphological and geological assessment; (4) ground-penetrating surveys; and (5) anecdotal accounts (Manzano 2014).

In June 2006, the Australian Government, through the Australian Agency for International Development (AUSAID), provided financial support to the Philippine government, through UNDP to strengthen the disaster preparedness capacities of communities in the Philippine eastern seaboard through the project "Hazards

Mapping and Assessment for Effective Community-based Disaster Risk Management," dubbed "READY" (PAGASA 2010). The project aimed to provide immediate, reliable information to communities at risk on the various geological and hydrometeorological hazards in their respective localities. It also meant to equip key stakeholder groups, particularly those in the target disaster-prone areas, with capacities to prepare for and cope with the impact of natural disasters, strengthen coordination processes and procedures for effective risk reduction, and start the process of mainstreaming risk reduction into local development planning (READY 2007).

The READY project undertook multi-hazard identification, specifically through mapping of hydrometeorological hazards (floods, storm surges, rain-induced landslides), and geologic hazards (ground rupture, ground shaking, liquefaction, earthquake-induced landslides, and tsunamis) and assessment of risks arising from these hazards; developed early warning systems and intensify information, education, and communication activities; and initiated mainstreaming of disaster risk management into local development planning. The project is implemented by the National Disaster Coordinating Council (NDCC) through the Office of Civil Defense of the Department of National Defense and the CSCAND Technical Working Group headed by the PHIVOCS, with PAGASA, Mines and Geosciences Bureau, and NAMRIA as members. Its implementation started in 2006 and was targeted to end by 2011 with the generation of multi-hazard maps for 27 provinces, including Leyte, Southern Leyte, Northern Samar, and Eastern Samar (READY 2007), those affected by Haiyan in 2013. The budget for the project amounted to USD 1.9 million, excluding work stations, salaries, and emoluments of government employees involved as these are the counterpart of the Philippine government (Alegre and Solidum 2009).

The approach in mapping exercises of the READY program involves table top analysis of the study area, including aerial photo and topographic map interpretation, remote sensing data analysis, mathematical modeling (especially for ground shaking, storm surge, and tsunami hazard mapping), and literature search. Preliminary hazard maps of these areas are then produced and subsequently verified in the field. Field verification involves interviews with local residents to gather local knowledge about the concerned hazards and checking out of the landforms and geologic features to verify initial table top interpretation (READY 2007).

In all of the CBDRM projects described above on DRR, the deterministic (single-scenario) approach for mapping hydrometeorological hazards was utilized. A single-scenario output for floods and landslides was generated by the MGB under the National Geohazard Assessment and Mapping Program for both 1:50,000 and 1:10,000 scale maps. Likewise, single-scenario storm surge hazard maps were also provided to the provinces of Leyte, Samar, and Eastern Samar by the READY Project (Fig. 11.6). Interesting to note are the storm surge hazard maps for Tacloban City and other municipalities provided by the READY Project in 2010 to the LGUs. According to the report entitled "Assessment of Early Warning Efforts in Leyte for Typhoon Haiyan/Yolanda," the official storm surge hazard map underestimated the inundation area of the storm surge. Furthermore, the official storm surge hazard map did not consider storm surges with a height of 7 m (the forecast height of PAGASA) and therefore it was not possible to relate the official map to the predicted surge

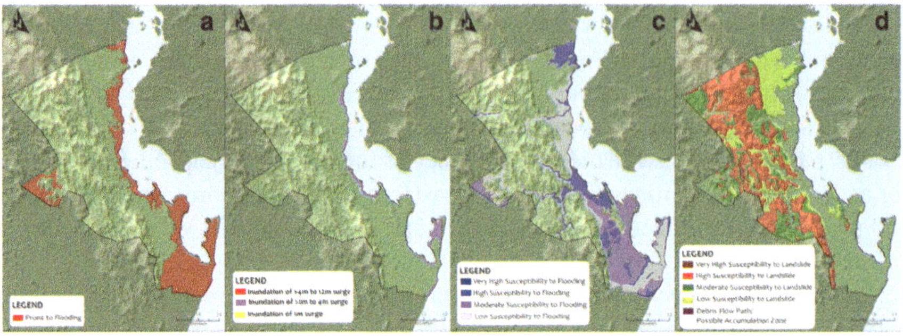

Fig. 11.6 Susceptibility maps generated by MGB under the National Geohazards Assessment and Mapping Program and the READY Project. **a** flood hazard; **b** storm surge hazard of Tacloban City, Leyte, at 1:50,000 from the READY Project (circa 2010). Coastal storm surge inundation as reflected in the READY map underestimated the actual extent of Haiyan; **c** flood susceptibility; and **d** landslide susceptibility of Tacloban City, Leyte, at 1:10,000 from MGB

(Neussner 2014). Several evacuation sites in Tacloban proved disastrous as 68% of the evacuation centers were overwhelmed by storm surges (Lagmay et al. 2015).

11.4.1.2 Probabilistic Risk Assessment

The Philippine government learned from shortcomings in the preparations for Typhoon Haiyan in 2013 (Lejano et al. 2015; Lagmay et al. 2015) and was able to limit the damage that last December's Typhoon Hagupit might otherwise have caused (Lagmay and Kerle 2015).

For example, storm surges during Haiyan extended inland by an unanticipated 2 km. Specific warnings of surges of up to 5.5 m were issued 2 d before Haiyan's landfall and were broadcast on primetime television by the country's president. Despite this, the warnings proved inadequate because the variability of coastal landscapes made it impossible to judge inundation extent on the basis of storm surge heights alone (Lagmay and Kerle 2015).

Probabilistic storm surge hazard maps, which represent scenarios bigger than the anecdotal accounts of the community, could have prepared the people of Tacloban and many other coastal areas in the Haiyan (Yolanda) corridor to anticipate and judge inundation extent (Fig. 11.7). Haiyan was the strongest typhoon to make landfall in recorded history and its unusual storm surge inundation extent could have been captured through numerical simulation of an event not represented in the historical record.

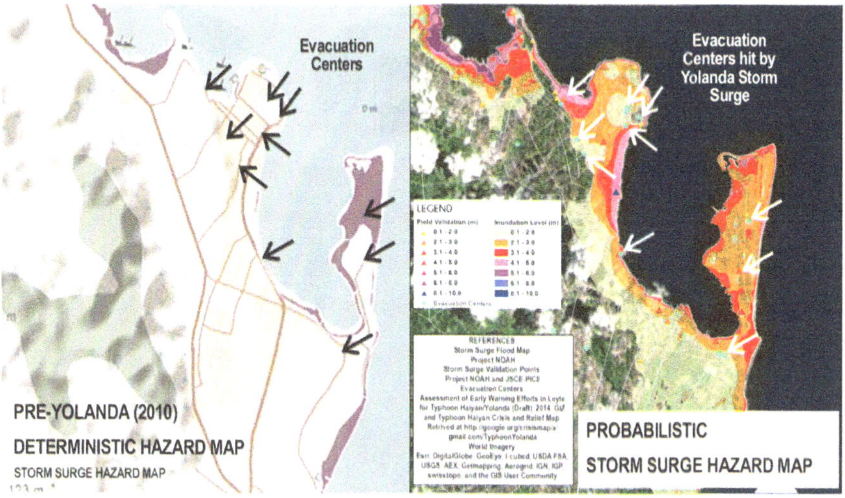

Fig. 11.7 Pre-Haiyan deterministic hazard map generated from anecdotal accounts of community members and expert opinion (left) and the hazard map for a Haiyan-type event (right). Arrows point to the location of evacuation centers

11.4.2 Mainstreaming Vulnerability and Risk Assessment into Local Development Planning

Vulnerability and risk assessment provides information for better understanding of impacts on communities of natural hazards and climate change. Assessments enable planners and decision-makers to come up with informed decisions during the CLUP formulation. Integration of vulnerability and risk results in the local plans can improve sectoral and subsectoral studies. It can be used to make adjustments in land demand projections, policy interventions for mitigating risks in identified high-risk/vulnerable areas, and other cross-cutting strategies for CCA-DRR (HLURB 2015).

Since the vulnerability and risk assessment exhibits the vulnerabilities and risks of the LGU, the results can be used to formulate a vision of resilience of the community. It may be a vision descriptor or it may act as a success indicator. Furthermore, the vulnerability and risk perspectives provide a better analysis of the planning environment and are also helpful in identifying issues, concerns, problems, and other challenges in development, which risk-sensitive policy interventions should address (HLURB 2015).

The probabilistic maps that capture climate change projections have been used in several municipalities in the Haiyan corridor. They have been used for siting evacuation centers that have been validated by people in the community during the mainstreaming of vulnerability and risk assessment into local development planning (Fig. 11.8). More importantly, the probabilistic hazard maps were used for

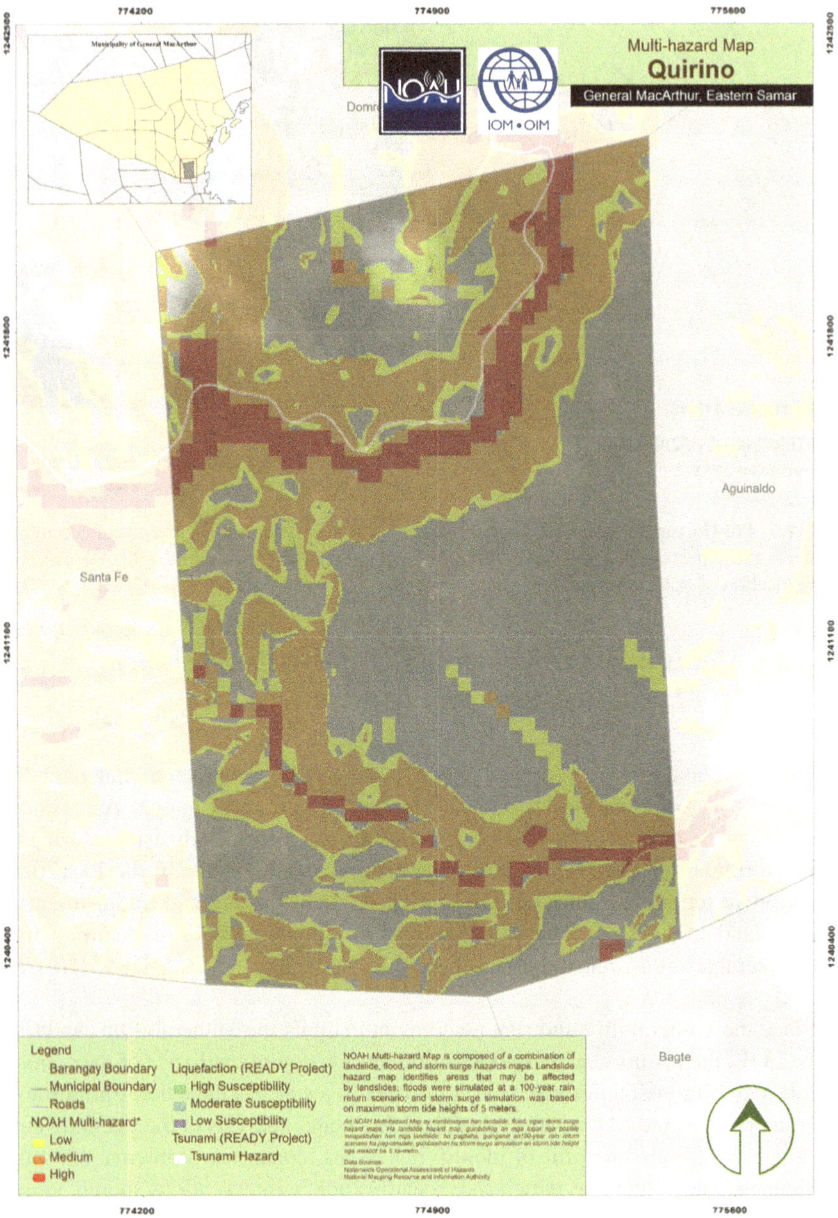

Fig. 11.8 Example of a multi-hazard map (flood, landslide, and storm surge hazards put together), where evacuation sites and routes at the barangay level are identified. Blue dots are evacuation centers, while yellow dots in the map are residential areas

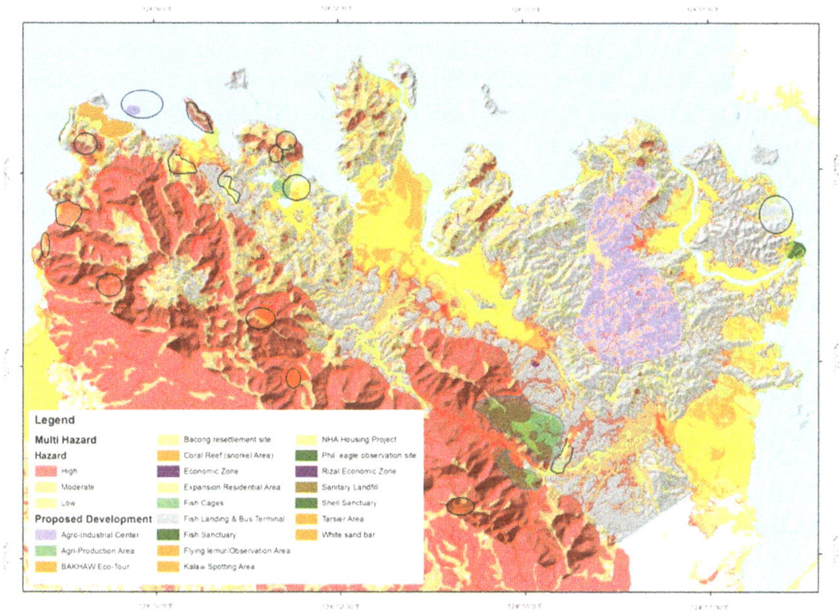

Fig. 11.9 Conceptual proposed land use map of Babatngon, Leyte, with a map overlay of flood, landslide, and storm surge hazards projected with the influence of climate change. Land use plans are easily visualized with this method

sectoral planning through integration of existing land maps and proposed development plans. An example is provided for the municipality of Babatngon, Leyte (Fig. 11.9) where future policy interventions across different sectors (i.e., coastal, health, agriculture, water, forestry, biodiversity, tourism, environment, energy, infrastructure, and mining) consider flood, landslide, and storm surge hazards with climate change projections. The multiple scenarios presented provided an opportunity to communicate different scenarios that the community can choose from in terms of level of risk they may want to prepare for, given their level of capacity.

It also reflects uncertainties, which are a characteristic of the "real world" and are essential in getting people to better understand and even accept the anticipated problems of climate change.

Other benefits of using probabilistic hazard maps with climate change projections have been recognized in the UNDP- and New Zealand-funded project entitled "Resilience Capacity Building for Cities and Municipalities to Reduce Disaster Risks from Climate Change and Natural Hazards (REBUILD)." Specific comments in the terminal evaluation (Arquiza 2018) of the project include the following: (1) Crafting of well-adjusted plans and investment programs; (2) Decision-making becomes more science-based and participatory arising from realistic feedbacks from LGU and communities; (3) Efficient budgeting and use of resources in prioritizing subprojects; (4) Correlation between hazards and potential loss, damage and disruption is established and understood; (4) Stronger relationships among stakeholders are built as ownership of plans is spread among them; (5) Development plans become climate-resilient (e.g., shelter; child protection); (6) Supportive river basin management councils; (7) Local partners (LGUs, academe, NGAs) highly involved and active; (8) Prospective investors guided on business locations, providing opportunities for barangays to become new preferred locations for new and expanding businesses; and (9) Recognition of the importance of climate-resilient infrastructure projects to prolong useful economic life through avoidance.

11.5 Summary and Conclusions

Disaster risk is an unresolved problem of development. By getting communities to plan better through local climate Change action plans, climate and disaster risk assessment, disaster risk management plans, and CLUPs, among others, genuine development will be ensured across all sectors. Disaster risk can significantly be reduced because the exposure and vulnerability of the community to hazards are decreased, thereby improving and expanding its capacity to mitigate the effects of disasters

Critical to smart and resilient development planning is the use of probabilistic hazard maps, which are multi-scenario and which depict future hazards that have not happened yet or not experienced by the community. According to United Nations International Strategy for Disaster Reduction (UNISDR), "We cannot wholly rely on our knowledge of past events to anticipate future risk, because some disasters that could happen have not yet happened." A probabilistic approach minimizes these limitations. The use of probabilistic hazard maps in risk assessment is not only embedded in Philippine RA 10121 of 2010 and the Philippine Development Plan (NEDA 2017); it is also the foundation for correct risk assessment and avoidance of maladapted communities.

A warmer climate would increase the risk of floods. So far, only a few studies have projected changes in floods on a global scale (Hirabayashi et al. 2013). This is perhaps true as well for studies on the effect of warmer climate on floods on a regional scale and for other hydrometeorological hazards such as rainfall-induced landslides and storm surges. On a local scale, an attempt was made to incorporate projected models of intensifying rainfall and wind and their consequent hazards were translated into maps to quantify the impacts of climate change on communities. Merging science-based information with local knowledge resulted in a promising and new perspective in resilience planning of Philippine communities. Further refinement of parameter inputs (i.e., climate change projections of rainfall amount and wind strength) to generate probabilistic hazards maps will further enhance our understanding of area-specific and hazard-specific climate change impacts. This, in turn, raises the ability of communities to anticipate and effectively plan for future hazards that are bigger than previously known, thereby giving them better options to pursue a more climate-resilient and human-secure future.

References

Alegre L Solidum R (2009) Hazard mapping and assessment for effective community-based disaster risk management or "READY" Project. Technical report. Office of Civil Defense and Philippine Institute of Volcanology and Seismology. http://drh.edm.bosai.go.jp/database/

Alvioli M, Guzzetti F, Rossi M (2014) Scaling properties of rainfall-induced landslides predicted by a physically based model. Geomorphology 213:38–47

Arquiza R (2018) Terminal evaluation report on Project ReBUILD (Resilience capacity building for cities and municipalities to reduce disaster risks from climate change and natural hazards). Unpublished presentation: Climate Change Commission

Aurelio MA (2004) Geological hazards mapping program of the mines and geosciences bureau: an institutional approach to hazard mitigation. Technical report. Quezon City, Philippines. Mines and Geoscience Bureau

Baum R, Savage W, Godt J (2008) TRIGRS—A Fortran program for transient rainfall infiltration and grid-based regional slope-stability analysis. Version 2.0. Technical report. USGS

CONCEP Inc. (Consultants for Comprehensive Environmental Planning Incorporated), GEOS Inc. (Geological and Environmental Consultants), and ASSURE Inc. (Alliance for Safe and Sustainable Reconstruction Incorporated) (2017). Final Report Vol 2 – Component C.1: Improved resilience to future disasters, JFPR-9175 Emergency assistance and early recovery for poor municipalities affected by Typhoon Yolanda. Quezon City. Asian Development Bank, Japan Fund for Poverty Reduction, & Department of the Interior and Local Government.

Guzzetti F, Cardinali M (2006) Landslide hazard assessment in the Collazzone area, Umbria, Central Italy. Nat Hazards Earth Syst Sci: 115–131. https://doi.org/10.5194/nhess-6-115-2006

Higaki M, Hayashibara H, Nozaki F (2009) Outline of the storm surge prediction model at the Japan meteorological agency, Tech. rep. Office of marine prediction, Japan Meteorological Agency, RSMC Tokyo-Typhoon Center Technical Review, Tokyo

Hirabayashi Y, Mahendran R, Koirala S, Konoshima L, Yamazaki D, Watanabe S, Kim H, Kanae S (2013) Global flood risk under climate change. Nat Clim Change 3:816–821

Housing and Land Use Regulatory Board (HLURB) (2015) CLUP guidebook: supplemental guidelines on mainstreaming climate change and disaster risks in the comprehensive land use plan. Housing and Land Use Regulatory Board (HLURB), Quezon City, Philippines

IPCC (Intergovernmental Panel on Climate Change) (2014) Climate change 2014. Synthesis Report. Fifth Assessment Report. World Meteorological Organization and United Nations Environment Programme, New York

Iverson R (2000) Landslide triggering by rain infiltration. Water Resour Res 36:1897–1910

Lagmay A, Kerle N (2015) Typhoons: storm-surge models helped for Hagupit. Nature: 414. https://doi.org/10.1038/519414b

Lagmay A, Agaton R, Bahala M, Cabacaba JBK, Caro C, Dasallas L, Gonzalo L, Ladiero C, Lapidez J, Muncal M, Puno J, Ramos M, Santiago J, Suarez J, Tablazon J (2015) Devastating storm surges of Typhoon Haiyan. Int J Disaster Risk Reduction 11:1–12. https://doi.org/10.1016/j.ijdrr.2014.10.006

Lejano R, Tan JM, Wilson M (2015) Communicating risk: learning from Typhoon Haiyan. Nature 518(7537): 35. https://doi.org/10.1038/518035d

Manzano L (2014) The MGB Geohazard assessment and mapping program. Technical report. Quezon City, Philippines. Mines and Geosciences Bureau

Martini F, Loat R (2007) Handbook on good practices for flood mapping in Europe. Eur Exch Circle Flood Mapp: 60. https://doi.org/10.1016/j.cageo.2011.11.027

Mei W, Xie S, Primeau F, McWilliams J, Pasquero C (2015) Northwestern Pacific typhoon intensity controlled by changes in ocean temperatures. Sci. Adv. 4:1–8

MGB (Mines and Geosciences Bureau) (2015) National Geohazard Assessment and Mapping Program. Unpublished manuscript. Quezon City, Philippines

NDRRMC (2014) Update on effects of Typhoon Yolanda (Haiyan) issued on 17 April 2014. Quezon City, Philippines. Available at: http://www.ndrrmc.gov.ph/attachments/article/1329/Update_on_Effects_Typhoon_YOLANDA_Haiyan_17APR2014.pdf. Accessed 21 Jan 2019
NEDA (2017) Philippine development plan 2017–2022. National Economic and Development Authority, Pasig City. Retrieved from http://pdp.neda.gov.ph/wp-content/uploads/2017/01/PDP-2017-2022-07-20-2017.pdfNOAH
Neussner O (2014) Assessment of early warning efforts in Leyte for Typhoon Haiyan/Yolanda. Manila, Philippines. Deutsche Gesellschaft Für Internationale Zusammenarbeit (GIZ) GmbH. https://doi.org/10.1016/J.JCIS.2008.03.007
PAGASA (Philippine Atmospheric Geophysical and Astronomical Services Administration) (2010) PAGASA Annual report 2010. Shaping the future though innovation. Manila: PAGASA, pp 1–56
PAGASA (Philippine Atmospheric Geophysical and Astronomical Services Administration) (2011) Climate change in the Philippines. http://web.pagasa.dost.gov.ph/index.php/climate-change-in-the-philippines
PAGASA (Philippine Atmospheric Geophysical and Astronomical Services Administration) (2017) Climate change projections for provinces. https://www1.pagasa.dost.gov.ph/index.php/93-cad1/472-climate-projectionsclimate-projections-for-provinces
Project NOAH (Nationwide Operational Assessment of Hazards) (2015) Landslide Hazards Map Atlas. Quezon City. Department of Science and Technology and University of the Philippines Diliman
Rahmstorf S, Coumou D (2011) Increase of extreme events in a warming world. Proc Natl Acad Sci: 55. https://doi.org/10.1073/pnas.1101766108
Raia S, Alvioli M, Rossi M, Baum R, Godt J, Guzzetti F (2014) Improving predictive power of physically based rainfall-induced shallow landslide models: a probabilistic approach. Geosci Model Dev 7:495–514. https://doi.org/10.5194/gmd-7-495-2014
READY (Hazards Mapping and Assessment for Effective Community-Based Disaster Risk Management) (2007) READY Project. Unpublished manuscript. United Nations Development Programme, Quezon City, Philippines
Taylor D (1948) Fundamentals of soil mechanics. Wiley, New York, p 700
Team (2017) Inception report. University of the Philippines Resilience Institute, Quezon City, Philippines
Torrente E, Zhang J, Le-Huu T (2008) CBDRM and poverty reduction. Partnerships for Disaster Reduction-South. http://www.adpc.net/v2007/Programs/CBDRM/Publications/Downloads/Publications/CountrySynthesisReport_CBDRMandPovertyReduction_Update.pdf
UNISDR (United Nations International Strategy for Disaster Risk Reduction) (2015) Making development sustainable: the future of disaster risk management. Global assessment report on disaster risk reduction. Geneva, Switzerland: United Nations Office for Disaster Risk Reduction (UNISDR)
UNISDR (United Nations International Strategy for Disaster Risk Reduction) (2017) Terminology on disaster risk reduction. United Nations Office for Disaster Risk Reduction (UNISDR), Geneva, Switzerland
UP NIGS (University of the Philippines National Institute of Geological Sciences) (2017) Inception Report – Project RAPID: Resilience and preparedness toward inclusive development program. Quezon City. Climate Change Commission, United Nations Development Programme, and Australian Aid
UP NIGS (University of the Philippines National Institute of Geological Sciences) (2018) Final Report – Project ReBUILD: Resilience capacity building for cities and municipalities to reduce disaster risks from climate change and natural hazards (Phase 1). Quezon City. Climate Change Commission, United Nations Development Programme, and New Zealand Aid Programme
Yusuf A, Francisco H (2009) Climate change vulnerability mapping for Southeast Asia: International Development Research Centre

Dr. Alfredo Mahar Francisco A. Lagmay is Academician of the National Academy of Science and Technology (NAST) and Professor at the National Institute of Geological Sciences, University of the Philippines (UP). He is Executive Director of the UP Resilience Institute and Director of the UP NOAH Center. He received his bachelor's and master's degrees from the University of the Philippines and holds a Ph.D. in Earth Sciences from the University of Cambridge. In 2013, he was awarded the Outstanding Professional of the Year Award in the field of Geology. Two years after, the European Geosciences Union (EGU) recognized his outstanding achievements in interdisciplinary natural-hazard research and natural-disaster engagement in the Philippines with a Plinius Medal.

Joy Santiago is a licensed environmental planner and a lecturer specializing in geographic information system (GIS), vulnerability and risk assessment, and land use planning. She was part of the Nationwide Operational Assessment of Hazards which efficiently mapped hydrometeorological hazards of the entire Philippines. She was awarded the 2016 outstanding student poster and PICO awards by the European Geosciences Union. She was also part of several projects which focuses on hazard, climate and disaster risk assessment, formulation of risk-sensitive comprehensive local development plans, and capacity building in collaboration with the Asian Development Bank, national and local government agencies, and non-government organizations. She also presented in numerous local and international conferences and co-authored internationally peer-reviewed scientific publications.

Dr. Juan M. Pulhin is full Professor and former Dean of the College of Forestry and Natural Resources, University of the Philippines Los Baños, Philippines. He was Lead Author of the 2007 Fourth Assessment Report of the Intergovernmental Panel on Climate Change (IPCC), Coordinating Lead Author of the Human Security Chapter of the 2014 IPCC Fifth Assessment Report and currently Lead Author of the Asia chapter of the IPCC Sixth Assessment Report. He has been a four-time a visiting professor at The University of Tokyo in 2007, 2010, 2016 and 2017 and a visiting scholar at Waseda University, Tokyo in 2018. He has written and edited 11 books and published more than 100 technical papers and book chapters on natural resources policy and governance, and on the human dimension of climate change and disaster risks. He is a fellow, invited expert, and council member of several international organizations and a member of the National Academy of Science and Technology (NAST), Philippines.

Chapter 12
Preventing International Assistance becoming a Threat to Human Security: Japan's Experience in the 1995 Kobe Earthquake and the 2011 Great East Japan Earthquake

Yosuke Okita

Abstract The Great East Japan Earthquake (GEJE) in 2011 was worth noting not only because of the disaster's massive scale but also because Japan received international assistance even as it is recognized as one of the countries most prepared to handle disasters. Due to globalization, more and more countries and organizations give assistance across state borders, with recipients not limited to only developing countries. Developed countries such as Japan, the United States, and New Zealand have also received assistance in recent years. While assistance itself is based on the goodwill of the international community, past examples have shown that receiving too much or unnecessary assistance can become a further burden to the affected countries, which can threaten human security. All countries now have to consider international assistance as part of their disaster response plan; relaying timely information from the affected country is the key to successful management of foreign assistance. This chapter looks at how the Japanese government managed the international assistance it received during the GEJE, focusing on information management and reception of international search and rescue teams. It also discusses how the government utilized the lessons learned from the 1995 Kobe earthquake. The government, getting support from the United Nations, successfully released timely information in English to the world. However, there were difficulties in coordinating activities to meet the needs of the affected areas. In requesting for foreign assistance, the government had to consider diplomacy and the feelings of families who lost their loved ones.

Keywords 2011 great east japan earthquake · 1995 kobe earthquake · UNDAC · Search and rescue teams · Disaster information management

---

Y. Okita (✉)
Graduate School of Media and Governance, Keio University, Kanagawa, Japan
e-mail: yosukeokita@mac.com

J. M. Pulhin et al. (eds.), *Climate Change, Disaster Risks, and Human Security*, Disaster Risk Reduction, https://doi.org/10.1007/978-981-15-8852-5_12

12.1 Introduction

Climate change presents a greater risk of large-scale disasters, posing a higher risk to human security in the affected areas. However, it is also a fact that globalization enables the international community to send more and more assistance across state borders. In December 2004, the Indian Ocean earthquake and tsunami killed more than 220,000 people in 14 countries in Southeast and South Asia. In response to this disaster, many countries and organizations sent assistance to the affected countries. The United Nations (UN), evaluating the response of the international humanitarian system, said that it "was able to provide massive relief to all tsunami-affected communities in the Indian Ocean, against all odds, in the course of a few weeks" (UNGA 2005). The development of information technology also allowed the timely spread of disaster information to the world when disasters strike, leading to further and quicker international assistance as seen in the 2015 Nepal earthquake.

On the other hand, there were instances in the past that demonstrated how receiving too much and unnecessary assistance can cause chaos in the affected areas, disrupting government efforts to assist disaster victims. For example, a temporary warehouse for relief goods was established in Indonesia at the time of the 2004 Indian Ocean earthquake and tsunami. An overwhelming amount of relief goods was sent from various countries without proper arrangements being made. The relief goods were kept in the warehouse for weeks, eventually damaged by mice, and were thrown away without being distributed. Receiving these international relief goods gave additional work to the Indonesian government, which was already overstretched in responding to the disaster. This is ironic as assistance from other countries in the spirit of goodwill becomes a burden to affected countries. As shown in this example, the international community sent, borrowing the word of the UN, "massive relief," but, in some countries, assistance was too much. It was even described as a "second tsunami" (Citraningtyas et al. 2010).

Another feature of recent international disaster relief efforts is developed countries also being at the receiving end. These countries also experienced problems in receiving assistance. As an example, the United States (US) received international assistance in response to Hurricane Katrina, which killed over 1,300 people in August 2005. The country later admitted that there was no mechanism in place to manage such a large quantity of foreign assistance. Some resources went unused, and many donor countries became frustrated (White House 2006).

The above examples show that, in the face of globalization, affected countries must consider incoming foreign assistance as part of their response plan. Foreign assistance can be on a massive scale, but, without proper management, it can be a burden for the affected governments, posing another risk to human security. The Great East Japan Earthquake (GEJE) on March 11, 2011 hit Japan in this context. Footage of areas hit by the tsunami was immediately shared with the world. Japan, one of the most disaster-prone and most prepared countries, also had to receive international assistance in the face of the GEJE.

This chapter analyzes how the Japanese government managed the incoming international assistance during the GEJE, focusing on the ways the government has improved the system based on lessons learned from the 1995 Hanshin-Awaji (Kobe) earthquake. Some reviews on the government response to the GEJE were conducted by officials in charge of receiving foreign assistance, but many of them were released only in Japanese (e.g., Asazuma 2012; Katayama 2013; Watabe and Murakami 2013). Furthermore, the reviews dealt with the government's appreciation of foreign assistance. It is rare to see a critical view of the international assistance received such as the ones conducted by the US and New Zealand governments after Hurricane Katrina and the 2011 Christchurch earthquake (e.g., White House 2006; McLean et al. 2012).

This chapter suggests possible improvements in dealing with international assistance for the future. The focus is on the international search and rescue (SAR) teams as Japan received SAR assistance during both disasters (Kobe and GEJE). It also discusses the importance of relaying timely information from affected countries so they can deal better with foreign assistance.

The author was part of the United Nations Disaster Assessment and Coordination (UNDAC) team that was deployed in Japan by the Office for the Coordination of Humanitarian Affairs (OCHA). OCHA, with inputs from the UNDAC team, issued situation reports (SitRep) regularly. This chapter relies on these SitReps as the main information source to review the GEJE response. SitReps are issued by OCHA, which is in a neutral position, and are widely used by international humanitarian workers as a reliable information source. As to the Kobe earthquake, reports issued by the Department of Humanitarian Affairs (DHA), later reorganized as OCHA, were reviewed. Also, the then governor of Hyogo prefecture (Kobe City is part of Hyogo prefecture) gives the view on international assistance from the perspective of local authorities. The Japanese government's reception of international assistance is evaluated from the operational point of view instead of the diplomatic one as studies on international disaster assistance have tended to focus on diplomacy (e.g., Kelman 2006).

First, the following section will discuss the importance (and difficulty) of relaying timely disaster information from affected countries using examples from Myanmar (Cyclone Nargis) and China (Sichuan earthquake). Second, it will introduce past examples of a "second tsunami," focusing on international SAR assistance: the Armenia earthquake in 1988 and the Christchurch earthquake in 2011. Third, it will analyze how the Japanese government responded to the GEJE in terms of information management and reception of international SAR teams, in the process showing how the government used the lessons learned from the Kobe earthquake. Lastly, it will provide some suggestions to improve the system of receiving international assistance so that it cannot be a threat to human security.

12.2 The Challenge of Relaying Timely Information in the Middle of a Disaster

In May 2008, two disasters hit Myanmar and China. Both countries received international assistance, but eventually, only the Myanmar government's response was criticized by the international community. Why? This section analyzes the difference in responses of the two countries and concludes that recognized capacity and timely information matter.

Category 3 Cyclone Nargis struck Myanmar on May 2 and 3, 2008, leaving nearly 140,000 dead and missing people. OCHA issued SitReps from 4 May based on information from the UN Resident Coordinator's Office in Myanmar, but, at this stage, details on the impact were not available (OCHA 2008a). In the first few days, it was extremely difficult for the international community to grasp what was going on in Myanmar. OCHA issued the third SitRep on 6 May and expressed concern about the unprecedented devastation, while humanitarian partners in the country just began collecting information about the destruction. The Myanmar government said that they would receive international aid. However, the issuance of visas of international responders was delayed. The UNDAC team assembled in Bangkok, but their visa applications were also pending (OCHA 2008b). Considering this situation, the French government urged the UN to invoke the concept of Responsibility to Protect so that the Myanmar government receives assistance (Parsons 2008).

About 10 days after Cyclone Nargis hit Myanmar, a magnitude-7.9 earthquake hit Sichuan province in China on 12 May, killing more than 80,000 people. The Chinese government, while putting their efforts on relief work, did not accept international assistance widely. For example, they did not request for an UNDAC team and did not ask other countries to deploy experts and professional personnel due to the limited local coordination capacity and transportation bottlenecks. Some international SAR teams were deployed on a bilateral basis, but timing was late. Consequently, the arrival of international SAR teams was also delayed. The first SAR teams arrived 3 days after the earthquake: The Russian SAR teams arrived on 15 May, while the Japanese SAR teams came on 16 May (OCHA 2008c). The probability of the trapped people surviving was already low when they arrived.

Although both Myanmar and China received only limited international assistance and the timing when they got it was off, only Myanmar was heavily criticized by the international community. One possible reason was the difference in the recognized capacity of the two countries to assist the affected and how much information they provided to the international community. In the response of the Myanmar government after Cyclone Nargis, information was scanty and the national efforts were described very briefly in the SitReps. In the first 3 days, the international community could only know that the government established the Emergency Operations Center, and that the police and military units were doing rescue operations (OCHA 2008b). In the Sichuan earthquake, the Chinese government response was described in detail (e.g., how many rescuers were deployed and how many tents were provided) (OCHA 2008c). Also, as of 19 May, eight international SAR teams have arrived in China and supported local authorities (OCHA 2008d). The deployed international SAR teams could also update their home countries on what was going on in affected areas. The international community got more information on the situation in China and was convinced that the government was trying to provide necessary support to the affected people.

From these two examples, one can see the need for the affected countries' governments to show that they satisfy their responsibility as states–i.e., that they provide assistance and security to the affected people. In case they cannot provide enough assistance, the affected countries need to request and receive international assistance. However, if the affected country does not release enough information and if the recognized capacity of the affected country is rather low, other countries may think that the security of the disaster victims is in danger. It can be said that releasing timely information on disasters to the international community has become an added responsibility of the affected countries. Although it is regarded as a "responsibility," it is also an advantage for them. If assistance is requested promptly, the affected country can receive assistance from the international community.

12.3 International SAR Assistance as a "Second Tsunami"—the Influx of International SAR Teams

In the face of globalization, many countries and organizations, including non-government organizations (NGOs) and private organizations, provide international disaster assistance across state borders. However, past experience has shown that this kind of assistance can sometimes become a burden if the affected counties receive too much or unnecessary assistance. The international SAR teams had a bitter experience on this about 30 years ago.

During the 1988 Armenia earthquake, which killed more than 25,000 people, many international SAR teams (mainly from European countries) rushed to the capital city, Yerevan. However, some of them were deployed without getting a request from or even the consent of the Armenian authorities; they conducted their SAR activities without coordinating with the local SAR teams. Several teams returned to their home countries without conducting any SAR activities, and the team members only stayed at hotels in Yerevan. This experience later led to the creation of the International Search and Rescue Advisory Group (INSARAG). The aim is to save lives through proper coordination of international SAR teams and the development of common operational procedures and standards such as the INSARAG guidelines. The Secretariat of INSARAG was located within the DHA, the current OCHA (DHA 1991; Nishikawa 1996).

The *INSARAG Guidelines* is a non-binding document, and INSARAG, as is the network of SAR teams, does not have any authority to stop unneeded international SAR teams. Who or which organization then can stop the unnecessary assistance? The UN General Assembly (GA) Resolution 46/182, *Strengthening of the Coordination of*

Humanitarian Assistance by the United Nations, confirms that the affected country has, first and foremost, the responsibility to take care of its affected people and stipulates that "humanitarian assistance should be provided with the consent of the affected country and in principle on the basis of an appeal by the affected country" (UNGA 1991). Although Stockton (2004) points out that the words "should" and "in principle" imply that, under certain conditions, neither consent nor appeal by the affected country is required, this Resolution clarifies that, in principle, only the affected countries can decide from which country and organization they receive international disaster assistance.

The affected countries thus have the right to decline the offer of unnecessary foreign assistance. But, in reality, it is difficult for them to decline the kind offer from other countries considering the diplomatic angle. As Kelman (2006) suggests, offering, sending, receiving, and declining international disaster assistance can positively or negatively affect the diplomatic relationship between the assisting and the affected countries. While declining offers of assistance might give the impression that the affected country can manage the disaster on its own, it also runs the risk of a deteriorating diplomatic relationship with countries that offered assistance and were refused, especially if assistance from some other countries has been obtained.

It becomes more complicated if foreign nationals are affected. One prominent example was the Christchurch earthquake in New Zealand in February 2011. Many foreign students, mostly Chinese and Japanese, were at the language school in the collapsed Canterbury Television (CTV) building. Considering the scale (the earthquake killed 185 people) and the robust response capability of the New Zealand SAR authorities, it seems that the government did not have to request for international SAR teams. However, eight international SAR teams, including the ones from China and Japan, were deployed to Christchurch, and these two teams operated at the CTV building. The New Zealand government, in their review report, described that declining the kind offers from other countries was the most difficult diplomatic task (McLean et al. 2012). Declining the offers of SAR assistance from countries whose nationals were affected should be much more difficult.

Timely information from the affected countries is the key to managing the incoming international disaster assistance, and it is difficult for the affected countries to simply say "no" to the kind offers from other countries even if it seems unnecessary. Also, receiving too much and unnecessary assistance can lead to a further burden for the affected countries. Keeping these in mind, the following sections will look at the Japanese government's response to the 1995 Kobe earthquake and the 2011 GEJE, focusing on their information management and reception of the SAR teams.

12.4 The 1995 Kobe Earthquake

The Kobe earthquake struck the Kansai area (western part of Japan) where Kobe City is and the surrounding area on January 17, 1995, killing more than 6000 people. The desk officer of DHA based in Geneva, Satoru Nishikawa (originally from Japan and in charge of earthquake response in DHA) later reported on the DHA response to this disaster (DHA 1995). He stated that, within a few hours after the earthquake, he received a call from the US asking for more information and if the Japanese government has requested for SAR assistance. Considering the magnitude of the earthquake and the fact that it occurred in a developed country, DHA only monitored the situation (while watching for an official request from the Japanese government).

According to the information that DHA received from Japan, the domestic SAR teams from other cities such as Tokyo, Nagoya, and Osaka were deployed to the affected areas, and there was no need for foreign SAR assistance. However, it was not very clear if the Japanese government actively conveyed this message and information on the impacts of the disaster to the international community. DHA kept on receiving a number of requests for information and offers of assistance. On the second day, the Western media started releasing footages from the affected areas. This then led to more offers of assistance to Japan. The footage from Kobe City might have given the impression that there was a shortage of SAR teams in the field, but the real bottleneck was the transport of SAR teams from other cities to Kobe and not the shortage of personnel. Local rescue services were not requesting for foreign SAR teams because it would entail looking for interpreters and more vehicles. They preferred to have domestic SAR teams that can immediately work under their command. The Japanese government, in response to offers of SAR assistance, sent this message: "We appreciate your offers of good will. We are mobilizing our teams in the country, and there is enough SAR personnel in Japan. At this stage, we do not foresee any need for international SAR teams" (DHA 1995).

The Swiss and French embassies, however, repeated the offers of sending their SAR teams. The local media reported on this and, finally, the Japanese government decided to receive the SAR teams from these two countries. Another NGO team from the United Kingdom (UK) was also deployed based on the strong recommendation from the UK Embassy. It is recorded that there was confusion in the expectations of both assisting and receiving sides. The Swiss team was asked to conduct their operations at the site where most wooden houses collapsed, while the French team was assigned at the site of a major landslide. The teams had no experience working on such sites. Moreover, here, the possibility of lives seemed to be low. The French team was thus reluctant to work there. The Swiss team was accepted by Japan because they had with them search dogs that the Japanese teams did not have. The people in the site must be kept away while the dogs are searching. However, it gave a wrong impression to the local people. Family members and nearby residents already knew where the missing people were sleeping and they expected the Swiss team to start rescue operations immediately. Thus, they could not understand the long wait. Finally, the Swiss team recovered nine bodies, while the French team recovered two bodies. There were no live rescues conducted by the international SAR teams (DHA 1995).

The local SAR authorities did not request for foreign SAR teams, but eventually, some teams were deployed to Kobe. According to the then governor of Hyogo Prefecture, Toshitami Kaihara, the Ministry of Foreign Affairs (MoFA) of Japan and the embassies became angry when local authorities declined the offers from other countries, putting the prefecture in a difficult position. The criticism for declining the offers came not only from the other countries but also from within Japan. The foreign SAR teams attracted public attention, and the media also covered them. Mr Kaihara later mentioned that foreign assistance should be deployed and received considering the situation in the affected areas (Kaihara 2009).

The next section will look at how the Japanese government dealt with foreign assistance in the GEJE and how the government utilized the lessons learned from the Kobe earthquake.

12.5 The 2011 Great East Japan Earthquake

The GEJE hit the northern part of Japan on March 11, 2011. The magnitude-9.0 earthquake also triggered a tsunami and the Fukushima nuclear accident. As in the case of the Kobe earthquake and because of the scale of the disaster, many offers of assistance were sent to Japan immediately. Eventually, the MoFA of Japan reported that the Japanese government received offers of assistance from 163 countries and regions and 43 organizations.[1].

This situation differed from Kobe in some points. First, more than 15 years have passed since the Kobe earthquake, and more countries and organizations were ready to send international assistance. Second, the main damage was caused by the tsunami and not by the earthquake. Thus, the need for international SAR assistance was relatively low. Tsunamis, by washing away collapsed buildings, usually do not leave voids, which are necessary for the survival of trapped people. Third, the Japanese government was required to release timely information on the disasters to the world because of the Fukushima accident. Many foreign citizens live in Japan, and the impact of the nuclear accident could affect other countries as well. The international community was carefully monitoring the situation in Japan.

To manage the incoming offers of assistance, the Japanese government released information to the international community in English with support from OCHA and the UNDAC team deployed in Japan. Despite the low possibility of live rescue, Japan received many foreign SAR teams. How and why did the Japanese government do that?

---

[1] https://www.mofa.go.jp/mofaj/gaiko/bluebook/2012/html/data/data1_01.html .

12.5.1 UNDAC and Information Sharing with the International Community

OCHA deployed an UNDAC team to Japan based on a request from the Japanese government, but the terms of reference (ToR) for the UNDAC team were relatively limited compared with those usually given to UNDAC missions. The UNDAC team arrived in Japan on 13 March and had an initial meeting with the MoFA of Japan on the same day. Agreement on the ToR was reached. During UNDAC missions, the team usually sets up a Reception/Departure Center (RDC) at the arrival area in the airport and an On-Site Operations Coordination Center (OSOCC) at the center of the affected area. These are in charge of registration and coordination of international assistance. However, the Japanese government did not request the UNDAC team to coordinate international assistance but instead asked them to focus on information management and giving advice to the government. They were requested to support the government by releasing correct and objective information on the disasters to the international community through the issuance of SitReps and by providing advice on receiving foreign assistance, including SAR teams and NGOs (Okita 2011; Watabe and Murakami 2013). The OSOCC was established in Tokyo instead of Tohoku area (the affected area), and the UNDAC team focused on collecting information and drafting the SitReps.

When the GEJE happened, the Japanese government's experience in Kobe enabled them to foresee the coming of the "second tsunami" of international assistance. The government wished to avoid receiving unnecessary or unqualified aid from other countries. However, it was difficult for the government to decline these offers because, in addition to diplomatic considerations, the government was still searching for missing people and trying to manage the nuclear accident. Also, although the Japanese government kept on releasing detailed information on the disasters and its response immediately after, it was mainly available in Japanese only. The UNDAC team and OCHA supported the government by releasing information in English from a neutral position and contributed to avoiding unnecessary assistance (Watabe and Murakami 2013).

As for relief goods, the Japanese government did not want countries and organizations to send materials without prior arrangements. This message was conveyed through the OCHA's SitReps on behalf of the Japanese government. The SitRep issued on 17 March cited the Japanese government as being engaged in identifying needs and establishing distribution mechanisms, and it was thus recommended that no relief goods would be sent without coordination with the government of Japan and the local governments (OCHA 2011b). The same SitRep mentioned some NGOs already assisting in Japan, but they were working through local partners. According to the SitRep, the Japanese government stated thus: "[the] [s]earch and [r]escue operation phase still continues in the affected areas at this moment, and the access to those areas is strictly limited to rescue workers. It is also reported that there is a temporary shortage of petrol in the affected area. International/foreign NGOs are recommended to wait until the situation improves so that those NGOs are able to conduct their activities in a self-sustainable way" (OCHA 2011b). Through messages issued by OCHA, the Japanese government clarified that incoming assistance groups must be self-sufficient and that they had to work with local partner organizations.

It is worth noting that the Japanese government did not clearly declare the end of receiving international assistance by themselves, and again, OCHA's SitRep conveyed this message on behalf of Japan. The UNDAC mission officially finished on 23 March, but some OCHA staff stayed in Japan to support the government (Okita 2011). The SitRep issued on 28 March, based on the OCHA staff's visit to the affected area, expressed the opinion that Japan had enough resources to respond and that the need for any further assistance was limited (OCHA 2011d). This message served as a declaration to end the process of receiving international assistance (Watabe and Murakami 2013).

In the GEJE response, the OCHA and the UNDAC teams supported the Japanese government by releasing timely information to the international community in English. They also conveyed some messages that were difficult for the government itself to express, such as declining unqualified assistance and ending the receiving-assistance phase. It seems that cooperation between the government and OCHA was smooth, but dealing with the nuclear accident was a sensitive issue for both parties. OCHA's SitRep no. 5 issued on 16 March described the situation thus: "[The] Fukushima Nuclear Power Plant crisis is still not under control with a fourth explosion" (OCHA 2011a). According to an UNDAC team member, the MoFA of Japan was not happy with the wording of the released SitRep and expressed this concern to the UNDAC team. The Japanese government received the UNDAC team to stop unnecessary assistance and to make the appeal that the Japanese government was making all efforts to contain the nuclear accident. In contrast, the phrase "not under control" could give a wrong impression to the world. It is difficult to judge whether the wording appropriately described the situation at that time. However, considering that UNDAC's first and foremost role was to support the affected country, the wording should have been carefully chosen. At the same time, OCHA must release information from a neutral position. If there is a situation where human security is in danger, OCHA and its SitRep must tell the world what is going on in the affected area.

12.5.2 Reception of International SAR Teams

Japan has received 20 international SAR teams, which represented 15 countries, by 20 March (OCHA 2011c). The Japanese government at first tried to receive only from four countries: Australia, New Zealand, Korea, and the US. That message was conveyed by the OCHA staff in Geneva on 12 March through their website called Virtual OSOCC, an OCHA's website used for exchanging information among disaster responders (Sakamoto 2013). It was not stated why these four countries were selected, but factors such as proximity and diplomatic relationship were considered. Receiving a SAR team from New Zealand can be explained from the diplomatic

Table 12.1 International SAR teams deployed to GEJE

Team name (country)	Number of personnel	Number of search dogs
Australia	72	2
China	15	
China (Taiwan)	28	
France	74	
Germany	43	3
Korea	107	2
Mexico	12	6
Mongolia	12	
New Zealand	45	
Russia (4 teams)	164	
Singapore	5 (dog handlers)	5
South Africa	49	
Switzerland	27	5
Turkey	33	
United Kingdom	63	2
United States (Fairfax)	72	6
United States (Los Angeles)	72	6

Note Adapted from OCHA (2011c)

angle. Japan's rescue team was just then deployed to Christchurch on 22 February, and when the GEJE happened, the third batch of Japan's rescue team was still in New Zealand. One possible reason for sending and receiving Japan's rescue team during the Christchurch earthquake was the existence of Japanese victims.

On the next day, however, the Japanese government changed its position and decided to receive international SAR teams from other countries as well. As in the case of the Kobe earthquake, the domestic rescue teams in the field did not request for any foreign SAR team, but the MoFA and the Cabinet Office argued that they should be useful if they were deployed in the field (Katayama 2013). As a result, as shown in Table 12.1, nearly 900 foreign rescue workers were deployed in Japan. The main damage was caused by the tsunami, not by the earthquake, so there were not so many sites where these international SAR teams could contribute to live rescue operations. Many teams helped in the recovery of dead bodies and the removal of debris, but the deployed international SAR teams could not rescue any live victims (Okita 2011).

Based on lessons learned from the Kobe earthquake in 1995, the Japanese government came up with procedures for receiving international disaster assistance. One remarkable development was the establishment of the so-called C7[2] Unit in 1998, which is in charge of receiving foreign assistance. At the time of GEJE, the C7 Unit had four staff members, including the ones from the Cabinet Office and the MoFA; this was later increased to 16 (Watabe and Murakami 2013). In receiving the SAR teams, the C7 Unit asked the local authorities if there was a need for foreign SAR teams. In many cases, receiving foreign SAR teams could be an added burden, but it was difficult for some local authorities to decline the offer from the central government (Sakamoto 2013).

The SitRep issued on 28 March said that the SAR phase was still going on (OCHA 2011d). Considering that the main damage was caused by a tsunami and that it was already 17 days after the earthquake, it is rare to see the SAR team still operating. Comparing the length of rescue operations in other countries, in Christchurch, New Zealand, SAR phase ended 9 days after the earthquake; Haiti earthquake in 2010, 11 days (Okita 2011); and Nepal earthquake, 8 days (Okita and Katsube 2016), it can be seen that the Japanese government was hesitant in declaring the end of the SAR phase. This is in consideration of the impact on the feelings of families and friends of missing people who believe that they are still alive. Some international SAR teams were puzzled because there were no clear demobilization instructions from the Japanese government (Asazuma 2012).

In terms of procedures with respect to receiving foreign SAR teams, there was some improvement from the time of the Kobe earthquake. For example, many international SAR teams brought their search dogs in the GEJE. Based on past experience (i.e., the search dogs of the Swiss team could not enter the country swiftly due to quarantine concerns), the Cabinet Office and other related agencies developed in advance quarantine guidelines in cases of emergency. The quarantine for search dogs did not become an issue in the GEJE response (Asazuma 2012).

12.6 Discussion

This paper has reviewed how the Japanese government managed international assistance in the GEJE in 2011, focusing on information management and international SAR teams. It shows how the government used the lessons of the Kobe earthquake to make further improvements. Below are some implications of the way the Japanese government responded to the GEJE. These can apply to other countries, especially developed ones with a high risk of natural disasters.

First, in the age of climate change and globalization, even developed countries can be recipients of international disaster assistance. The assistance emanates from the goodwill of other countries and organizations, but experience shows that too much

[2]When disasters on a massive scale happen, the government establishes Unit A (Review Department), Unit B (Information Department), and Unit C (Disaster Response Department) and they are based on the disaster management headquarters. Unit C has several groups (e.g., operations and transportation) and the 7th group is the overseas aid acceptance group, which is called C7 Unit.

or unnecessary assistance ("the second tsunami") can be a burden for the affected country, weakening the capacity of its government to assist the affected citizenry. Even Japan, one of the most prepared countries, also became a recipient of assistance. All countries should check the system and procedures of receiving international assistance to avoid the situation of a "second tsunami," which can threaten human security.

Second, releasing timely information after a disaster is important to avoid unnecessary assistance, and it is becoming an additional responsibility of the affected country. Japan already had a sophisticated disaster response system at the time of the GEJE, including information management in the country. However, the government had difficulty releasing timely information in English. The Japanese government requested OCHA to deploy the UNDAC team to support information management in English and dispatch the message of declining assistance. International disaster assistance is not limited only to, for example, SAR and medical assistance and cash contribution. Affected countries can also request international assistance in terms of coordination such as information management. Japan's case was an example of a successful support by the UNDAC team. Not only was unnecessary assistance avoided; release of information (such as the needs in the field) was effectively done and the necessary assistance was secured.

Lastly, the Kobe earthquake taught the Japanese government ways to organize procedures for receiving international disaster assistance, as in the case of the C7 Unit. While the technical side (such as the smooth quarantine process) has worked well in the GEJE, there was still room for improvement in making decisions about receiving international assistance. For example, communication between the affected local units and the central government remained difficult, and it was not clear if needs in the field were reflected in the final decision-making process. In both the Kobe earthquake and GEJE, the local SAR authorities did not ask for foreign SAR teams, but the government accepted them. Furthermore, the Japanese government did not declare the end of the SAR phase, and the foreign SAR teams found it hard to decide the timing of demobilization. It is understandable that, for affected governments, declining the kind offers from other countries and declaring the end of the SAR phase are difficult tasks, considering diplomacy, and assuaging the feelings of the affected family members. However, these decisions are crucial to controlling incoming international assistance.

The C7 Unit is composed of staff members from several ministries such as the Cabinet Office and the MoFA. While it is a good decision-making system, there may be problems in coordination as opinions of various ministries differ and making quick decisions becomes difficult. In Japan, there is no emergency management agency, not like the ones set up in other countries (e.g., Federal Emergency Management Agency (FEMA) of the US and the Ministry of Civil Defence and Emergency Management (MCDEM) of New Zealand). The establishment of an emergency management agency that can decide on foreign assistance issues is one possible option for future improvement.

12.7 Conclusion

The 2011 GEJE showed the world that even developed countries need international assistance. Climate change has increased the risk of large-scale disasters and globalization makes possible the assistance of more countries and organizations across state borders. However, receiving too much or unnecessary international assistance can be a burden for the affected countries, resulting in delayed efforts to assist the affected people. Ironically, assistance borne out of goodwill becomes a threat to human security.

To avoid this situation, all countries must recognize their vulnerability and look at the possibility that they, too, can be at the receiving end of international assistance and, in such a case, they must be able to develop appropriate protocols for this eventuality. In this connection, relaying timely and accurate information on the disaster is crucial so that affected countries only receive assistance that matches the needs in the field. Considering diplomacy and the victims' feelings, it is not easy to say "no" to offers of assistance from other countries. Governments are required to consider international assistance in their domestic disaster response plan. They should receive foreign support if needed, but they have to limit it if it is really not necessary. This prevents assistance from becoming a burden and a threat to human security.

References

Asazuma S (2012) *Higashinihondaishinsai ni okeru kaigai karano kinkyu enjo* (Foreign assistance in the Great East Japan Earthquake). *Kokusai Mondai* (International Affairs) 608:46–51

Citraningtyas T, Macdonald E, Herrman H (2010) A second tsunami?: The ethics of coming into communities following disaster. Asian Bioethics Rev. 2(2):108–123

DHA (Department of Humanitarian Affairs) (1991) INSARAG report of the inaugural meeting held at Beuggen, Germany, 11-13 December, 1991. DHA, Geneva

DHA (Department of Humanitarian Affairs) (1995) The Great Hanshin-Awaji (Kobe) Earthquake in Japan, 17 January 1995: the earthquake, on-site relief and international response. DHA, Geneva

Kaihara T (2009) *Hyogoken Chiji no Hanshin-Awaji Daishinsai 15 nen no Kiroku* (The Governor of Hyogo and the Great Hanshin-Awaji Earthquake: Record of the 15 years). Maruzen, Tokyo

Katayama Y (2013) *Higashinihondaishinsai ji no kokusai kinkyu shien ukeire to gaimusho* (Reception of international assistance and the Ministry of Foreign Affairs of Japan in the Great East Japan Earthquake). *Kokusai Kyoryoku Ronshu* (J. Int. Coop. Stud.) 20(2, 3): 45–70

Kelman I (2006) Acting on disaster diplomacy. J. Int. Affairs 59(2):215–240

McLean I, Oughton D, Ellis S, et al (2012) Review of the Civil Defence emergency management response to the 22 February christchurch earthquake. https://www.civildefence.govt.nz/assets/Uploads/publications/Review-CDEM-Response-22-February-Christchurch-Earthquake.pdf. Accessed on 19 Apr 2019

Nishikawa S (1996) *Hanshin-Awaji Daishinsai ni mirareta kokusai kyuen katsudo no misumatchi* (Mismatch of international relief activities in the Great Hanshin-Awaji Earthquake). *Chiiki Anzen Gakkai Ronbun Hokokushu* (Proceedings of the Annual Conference of the Institute of Social Safety Science) 6:261–268

OCHA (Office for the Coordination of Humanitarian Affairs) (2008a) Myanmar: Cyclone Nargis OCHA Situation Report No. 1 (4 May 2008)

OCHA (Office for the Coordination of Humanitarian Affairs) (2008b) Myanmar: Cyclone Nargis OCHA Situation Report No. 3 (6 May 2008)
OCHA (Office for the Coordination of Humanitarian Affairs) (2008c) Sichuan Province, China: Earthquake OCHA Situation Report No. 3 (15 May 2008)
OCHA (Office for the Coordination of Humanitarian Affairs) (2008d) Sichuan Province, China: Earthquake OCHA Situation Report No. 5 (19 May 2008)
OCHA (Office for the Coordination of Humanitarian Affairs) (2011a) Japan Earthquake and Tsunami Situation Report No.5 (16 March 2011)
OCHA (Office for the Coordination of Humanitarian Affairs) (2011b) Japan Earthquake and Tsunami Situation Report No.6 (17 March 2011)
OCHA (Office for the Coordination of Humanitarian Affairs) (2011c) Japan Earthquake and Tsunami Situation Report No.9 (20 March 2011)
OCHA (Office for the Coordination of Humanitarian Affairs) (2011d) Japan Earthquake and Tsunami Situation Report No.14 (28 March 2011)
Okita Y (2011) *Tohoku chiho taiheiyo oki jishin: kokuren saigai hyoka chosei chi-mu no katsudo ni tsuite* (The UNDAC team in response to the Tohoku-Pacific Ocean Earthquake). *Shizen Saigai Kagaku* (J. Jpn. Soc. Nat. Disaster Sci.) 30(2): 279–287
Okita Y, Katsube T (2016) Coordination of international urban search and rescue (USAR) teams in the 2015 Nepal Earthquake: Disaster literacy for international USAR. J. Jpn. Assoc. Earthquake Eng. 16(7):24–36
Parsons C (2008) France urges the UN council to act on Myanmar cyclone. Reuters. https://www.reuters.com/article/idUSL07810481. Accessed on 18 Dec 2018
Sakamoto M (2013) *Higashinihondaishinsai ni okeru kokusai kinkyu shien no ukeire chosei ni kansuru kenkyu: hisai chiho kokyo dantai ni yoru kokusai sosaku kyujo chi-mu no ukeire ni chumoku shite* (The coordination system to receive international assistance in the Great East Japan Earthquake: focusing on the coordination of disaster stricken municipalities for receiving international US&R team). *Chiiki Anzen Gakkai Ronshu* (J. Social Safety Sci.) 21: 199–207
Stockton N (2004) The changing nature of humanitarian crises. In: OCHA (ed.). The humanitarian decade, challenge for humanitarian assistance in the last decade and into the future. Vol. 2. UN, New York. pp 15–39
UNGA (United Nations General Assembly) (1991) UNGA Resolution 46/182. Strengthening of the coordination of humanitarian assistance by the United Nations. UNGA, New York
UNGA (United Nations General Assembly) (2005) In larger freedom: towards development, security and human rights for all. Report of the Secretary General. UNGA, New York
Watabe M, Murakami T (2013) *Kokuren jindo shien shisutemu no hatten to Higashinihondaishinsai* (International humanitarian system and the Great East Japan Earthquake). *Sekai Ho Nenpo* (Yearbook of World Law) 32:195–215
White House (2006) The federal response to Hurricane Katrina: lessons learned. White House, Washington, DC.

Yosuke Okita is pursuing his Ph.D. degree at the Graduate School of Media and Governance, Keio University in Japan while working for Japan International Cooperation Agency (JICA)'s Indonesia office in Jakarta. He has more than 10 years of experience working in international disaster and emergency management, including some emergency missions as a member of the Japan Disaster Relief team and the United Nations Disaster Assessment and Coordination (UNDAC) team. From 2014 to 2017, he worked at the Office for the Coordination of Humanitarian Affairs (OCHA) in Geneva as Asian regional focal point for International Search and Rescue Advisory Group (INSARAG) and UNDAC. He received an M.A. in International Relations and a Master of Diplomacy from The Australian National University.

Chapter 13
The 2009 Drought in the People's Republic of China: A Human Security Perspective

Ross Michael Pink

A journey of a thousand miles begins with a single step.—Lao Tzu

Abstract In spite of China's notable progress and policy innovation in socioeconomic development, the country confronts obstacles forestalling severe drought crises facing 1.4 billion people. Environmentally, China is threatened by the dual and paradoxical challenges of flooding in the south and drought in the north as well as extreme pollution levels. Although drought is historically an environmental phenomenon in China, it is evident that climate change will accelerate its impacts. In terms of the drought scenario for China, the impacts will be severe and will increase exponentially. Between January 2009 and April 2010, China experienced three severe drought episodes that were classified at the once-in-a-100-year level. This chapter argues that the human security perspective, first enunciated by the UNDP in a landmark 1994 report, offers a useful prism through which to examine the country's profound 2009–2010 drought crisis, the government response, and the continuing effect upon society nearly a decade later.

Keywords China · Drought · Climate change · Human security · Environment

13.1 Introduction: The Human Security Paradigm

Important features of human security emerged in the 1970s. Major developments included the creation of the Independent Commission on International Development Issues, chaired by former German Chancellor Willy Brandt, and the Independent Commission on Disarmament and Security Issues, chaired by Swedish Prime Minister, Olaf Palme, which raised important questions about common security. However, it was the landmark 1994 Human Development Report issued by the United Nations Development Programme (UNDP) that formally introduced the concept and

R. M. Pink (✉)
Political Science Department, Kwantlen Polytechnic University, British Columbia, Canada
e-mail: Ross.Pink@kpu.ca

J. M. Pulhin et al. (eds.), *Climate Change, Disaster Risks, and Human Security*, Disaster Risk Reduction, https://doi.org/10.1007/978-981-15-8852-5_13

profoundly advanced the human security ideal. The UNDP also added an important dimension of policy utility to the concept of human security. Many governments, including Canada, have adopted features of human security in their national and international policies. Since the paradigm of human security was first introduced, it has helped shape discourse and policy on security matters and new perspectives on human rights (UN Commission on Human Security 2003). The 1994 UNDP report outlined the need and urgency for global human rights policies that are "people-centered," noting seven specific threats to human security: economic, food-related, health; environmental, personal (including violence and abuse), community, and political (UN Commission on Human Security 2003).

The United Nations Commission on Human Security defines human security as follows: To protect the vital core of all human lives in ways that enhance human freedoms and human fulfillment. Human security means protecting fundamental freedoms—freedoms that are the essence of life. It means protecting people from critical (severe) and pervasive (widespread) threats and situations (UN Commission on Human Security 2003).

A UNDP-IUCN report noted cogently that the adoption of a human right places three distinct obligations upon the state. The first is the duty of a state to allow for the enjoyment of a human right. The second obligation relates to protection. The third obligation relates to a responsibility to fulfill. The emphasis is upon adopting policies that foster the realization of human rights and thus the provision of water rights must be secured through legislation and policies (Scanlon et al. 2004). The drought crisis in China clearly touches upon a profound human rights debate. Water security and environmental mitigation are parallel forces that will help combat drought. Moreover, the 2009–2010 drought crisis can be observed from multiple perspectives: social, health, economic, security, and indeed human rights.

The effects of climate change, including drought, flooding, and saltwater intrusion, will seriously impact an already dire water scarcity and pollution crisis facing many regions, towns, and cities in China. Five strategic priorities have been identified in China that pose profound barriers to economic growth and social development: risks of flooding, water scarcity, water pollution, aquatic ecosystem degradation, and under-improved management of water resources (China Country Water Resources Partnership Strategy 2013). Moreover, it is evident that severe drought episodes have far-reaching and long-term impacts upon society that are not limited to environmental, economic, or agricultural issues alone. Scientific data, along with future projections on impacts, reveal a scenario of continuing and worsening drought conditions in China and many regions of the world. An important analysis concluded that an increase in severe hazardous events will increase with climate change, thus undermining a host of human development basic needs related to health, food security, economic livelihood, and water security (Security and Climate Change 2001). Accordingly, the climate change phenomenon is intrinsically linked to the exacerbation of drought episodes in China in the twenty-first century.

13.2 Overview of the 2009 Drought

It is estimated that the drought in northern and southwestern China caused USD 3 billion in economic losses. The Chinese Office of State Flood Control and Drought Relief Headquarters and the Ministry of Civil Affairs reported that the drought-affected 25 million people and 18 million livestock by shortage of drinking water, as well as about 8 million ha of arable land nationwide (http://www.chinadaily.com.cn/china/2010drought/index.html). Chinese government authorities began issuing drought warnings in January 2009 and noted the following patterns: rainfall from October through December 2008 had been 50–80% below normal in most of northern China, a serious decline in rain and snow for several months, above-average temperatures in mid-January, and drought intensification in January and early February as soil moisture levels declined (USDA, Foreign Agriculture Service 2009). In a densely populated country, the impact of the drought was both pervasive and far-reaching. Rainfall levels were low and erratic, a symptom of climate change.

As reported by the China National Climate Center, the impacts were profound. Key environmental observations included a 2009 spring drought whereby rainfall in the affected regions declined by 50–80% compared with normal rainfall patterns, dramatically lower rainfall in Hebei Province that was the lowest level recorded since 1951, the third lowest rainfall in Henan Province, extreme low rainfall levels recorded in Liaoning, Jilin, Inner Mongolia, Hebei, and Shanxi that was 127.6 mm, also the lowest since 1951 (China National Climate Center 2009). The drought followed a long period of particularly dry weather that was heightened by dramatically reduced precipitation levels. This combination of weather phenomenon served to expedite the drought crisis in China.

An analysis by Barriopedro et al. noted the following main impacts: (1) accumulated precipitation from May 2009 to April 2010 that was lower by 25% compared with normal patterns, (2) the southwestern China drought from the summer of 2009 to the winter of 2010 was noted as the driest period since 1951, (3) declining reservoir levels caused a 20% decline in hydroelectricity production, and (4) crops in northern and eastern China were exposed to 8 months of steadily stressed vegetation between May 2009 and July 2010 (Barriopedro et al. 2012).

13.3 The 2009 Drought: A Human Security Perspective

The seven categories of threat noted in the human security paradigm are community, economic, environmental, food, health, personal, and political (UNDP 2004). All of these threats are outcomes of climate change and can also be correlated with drought scenarios. Community threats relate to threats to community life and the enjoyment of community standards, benefits, and opportunities. Economic threats are more direct and correspond to threats that limit or cease economic opportunities and livelihoods. The maintenance of basic income is a key feature of economic security that can be imperiled by a range of phenomena, including environmental disasters. Environmental threats are linked to damage or outright destruction of the natural world and habitat and may pose immediate dangers to local and national populations. Food threats correspond directly to food security and the provision or maintenance of adequate agricultural land, crops, cultivation, food supplies, and nutrition levels. Health threats relate to the general issues that limit or endanger the well-being of community, regional, and national populations. Health threats can be pervasive and multidimensional. Personal threats relate to individual security and thus are closely associated with climate change impacts that endanger security in multiple and destructive ways. Weather phenomena such as drought can have a direct and severe impact on safety and human life. Political threats relate to crises that impact not only the functions of government but also the legitimacy of a government that cannot meet the basic needs of its citizenry in normal life and in times of crisis.

The noted scholar, Amatya Sen, articulated the paradigm of capability deprivation that fosters a deeper understanding of multiple threats to individual development and security, apart from income poverty. Low income is therefore viewed as only one source of capability deprivation that may hinder individual opportunity and development. This paradigm cogently relates to environmental threats such as drought, which restrict and endanger individuals on multiple levels.

13.3.1 Community Threats

The immense and devastating drought had far-reaching implications for community life, well-being, and health. Community services and daily life were often affected with varying degrees of intensity, depending on the drought region. A potent example of community threats posed by the 2009 drought was recorded in five provinces. For example, Yunnan Province experienced desert conditions during the drought and the Yunnan Province Information Office reported that the drought continued until March 26, causing water shortages for 8.2 million people and 3.1 million ha of crops (Lu et al. 2011).

Agriculture was deeply affected and families and livelihoods were thus negatively impacted. Research by the Chinese Ministry of Civil Affairs Disaster Relief Division indicated that, due to this drought, agricultural crops in an area of 43486 km^2 were affected. The direct economic loss was of the order of 19.02 billion yuan, and the trend of the disaster was worsening. The fragile nature of the topography also renders agricultural land vulnerable to soil erosion and drought. These environmental burdens, which have been a persistent reality facing communities for many centuries, expose the populations to a myriad of socioeconomic problems and family hardships.

13.3.2 Economic Threats

The economic burden associated with the 2009 drought was severe. The Chinese government launched intense programs to combat drought. Overall, the government utilized 279,000 experts and technical staff; 19.7 million farmers were trained in drought-resisting technology, 31.7 million people were mobilized to offer various levels of assistance, 9.33 million ha of wheat fields were irrigated, and in Henan, China's leading wheat province, farmers irrigated approximately 95% of the wheat crop (USDA, Foreign Agriculture Service 2009). The economic costs associated with drought mitigation have been immense, extracting valuable human and financial resources away from other critical needs of society. Yet, the heavy financial obligations imposed upon the national and provincial governments were necessary and unavoidable due to the severity of the drought and the high civilian population affected. In the twenty-first century alone, the economic costs of drought mitigation, adaption, and relief will increase steadily as climate change impacts worsen.

A stark example of the economic dislocation caused by the drought was found in northeast China. It is clear that China is highly sensitive to drought because of massive crop areas and poor irrigation systems as affirmed in an analysis which noted "that 11% of the cities are in the extreme risk category; this category includes 26% of the cultivated land area and 11% of the total crop yields in the region. Twenty-three percent of the cities, accounting for 28.5% of the total cultivated land area and 26.4% of the crop yields of the study area, are in high-risk locations, and 77% of the cities with high and extreme risk levels are distributed in north and northeast China. Moreover, 64% of the cities in the monsoon region of east China have moderate- and low-risk levels. These cities are primarily located south of the Yangtze River (Xu et al. 2013).

It is incumbent upon Chinese authorities to aggressively pursue mitigation strategies, given the reality of persistent drought scenarios in the twenty-first century. Thus, managing drought conditions is the primary goal. Northern China is a region that is particularly impacted by the dual burdens of drought and water scarcity. An authoritative study on drought impacts assessed 65 cities in Anhui, Henan, Jiangsu, and Shandong provinces of China. The findings of the study were indicative of severe drought vulnerability noting four main conclusions: Anhui and Henan are more vulnerable to drought, inland regions have more drought vulnerability, southern cities are less vulnerable to drought compared with northern and central regions, and finally, extensive investment in drought mitigation and water efficiency methods are essential (Yuan et al. 2015).

13.3.3 Environmental Threats

The 2009 drought led to multiple environmental crises whose impacts are felt to this day. Moreover, the drought exacerbated an existing dire scenario related to water security and land fertility. According to China's State Environmental Protection Administration (SEPA), in 2006, 60% of the country's rivers were degraded so badly by pollution that they could not be used as drinking water sources. It is evident from the post-2009 drought scientific analysis that the environmental damages were significant. The drought exposed approximately 16 million people and 11 million livestock to water shortages, damaged 4 million ha of farmland, destroyed 25% of the harvest, and caused river recession ranging from 30 to 80% of normal flow and volume (Yu et al. 2014). Moreover, droughts also dried up some rivers, suspended electricity generation at some hydropower plants in southwestern China, and caused extensive disease and pest control problems in Yunnan Province and Inner Mongolia at a rate 70.7% higher compared with the previous year (EP-NDRCC 2010).

There are numerous adverse drought impacts on the environment. The damage to water systems, water security, and wetlands can be enormous and can persist for decades. An authoritative study reported multiple concerns: shrinking wetland areas which leads to a decrease in water conservation, wetland dewatering, wetland area shrinkage, reduced wetland water storage, and the health and environmental damage caused by 660 rivers in Yunnan Province that dried up (Cao et al. 2011).

13.3.4 Food Threats

Approximately 768 million people globally are living at or below the World Bank internationally defined poverty line of USD 1.90 USD per day. In China, approximately 11% of the population lived on less than USD 1.90 a day in 2010. Food security is intrinsically intertwined with climate change and drought. Although China continues to mark solid annual economic progress, the drought–climate change nexus that confronts the country will imperil food security for millions of citizens. Globally, scientists have predicted for years that global grain and general food prices will escalate sharply in the twenty-first century. At the peak of the drought, February 7, 2009, 10.866 million ha of cropland in northern China were impacted, 10 million ha of winter wheat in the leading eight wheat provinces were damaged with the worst conditions evident in Anhui, Henan, Shanxi, Gansu, Hebei, and Shandong provinces (USDA, FAS 2009).

China will experience massive health and socioeconomic hardship related to climate change, including drought, flooding, desertification, saltwater intrusion, rising food prices, and rising grain prices. Moreover, every year, about 13% of the irrigated area suffers from severe water scarcity as a result of dry weather events and drought. There are three pertinent points related to the north–south water nexus in

China: first, grain production in China has shifted from the humid south to the water-challenged north over the past 30 years; second, as desertification and dry conditions embattle the north, food production is increasingly reliant upon the overuse of groundwater; and third, inefficient irrigation infrastructure and inadequate water management practices have led to the severe depletion of groundwater aquifers and natural habitats. Moreover, farming income in numerous agricultural areas was dramatically affected by the drought. For example, cash crops such as tree fruits and perennial sugar cane, coffee, walnut, and medicinal and oil crops in the drought-affected provinces also suffered heavy losses (EP-NDRCC 2009). A statistical analysis covering the city of Chaoyang in Liaoning Province in September 2009 revealed that drought-induced grain shortages affected 1.38 million people; shortage of grains in 2009 was 80,000 t and 250,000 t in 2010 (Ye et al. 2012).

13.3.5 Health Threats

The effect of the 2009 drought on health was catastrophic. A range of human security "health threats" is linked to drought and water insecurity. Some 4.42 million people suffered from lack of drinking water and 2.2 million large domestic animals were likewise affected (USDA, Foreign Agriculture Service 2009).

Difficulty securing adequate water supplies is a barometer of drought severity. On the basis of nationally recognized standards, difficulty in accessing drinking water is measured by the distance one lives to a water source (supply point) that exceeds 1000 m (one way) or there is a 100-m difference in elevation or rural areas are without rainfall for more than 100 d. During the drought many citizens had extreme difficulty accessing drinking water because their regular water sources were unavailable (NDRC-MCA 2009).

13.3.6 Personal Threats

Personal threats in the human security paradigm refer to a broad range of difficulties and dangers that confront citizens. Personal threats can derive from violence and also environmental disasters. The 2009 drought confronted the affected Chinese population with clear and life-threatening dangers. Examples of personal threats include the following forms of deprivation and hardship: over 16 million people and 11 million livestock faced water shortages, devastated crops across more than 4 million ha of farmland and 25% yielded no harvest, rivers shrank to 30–80% of their normal volume (Yu et al. 2014). In the city of Chaoyang in Liaoning Province, the poverty-stricken population increased from 645,000 before the disaster to 1.54 million afterward. One assessment of the drought implication noted that both poverty and food security were dramatically affected. Many farmers, unaware of the looming drought crisis, invested a considerable amount of their capital stock in crop production so that when the drought hit hard, they had limited resources, which exacerbated income stress and livelihoods (Ye et al. 2012).

13.3.7 Political Threats

Politically, there have been impacts from the 2009 drought that will take years to resolve. Questions have been raised about the political response in terms of preparation and mitigation. Although the government received high marks for responding to the human crisis and engaging in emergency efforts, questions have been raised about the overall government preparation plan and mitigation strategies. Moreover, a severe drought in China is not a rare occurrence but rather a frequent environmental reality that experts believe will intensify in the twenty-first century. Clearly, the hardship, human suffering, and protests have been felt at the political level both provincially and nationally. In China, although political protest is tolerated by the authorities with more restraint, it is also well documented that activists, even famous ones, have been arrested and imprisoned.

It is evident once the human toll of the drought was enumerated, that millions of people and livestock had difficulty accessing drinking water, tens of millions of hectares of cropland were damaged or placed at risk, and long-term devastation was imposed upon the economy, society, and ecological environment (Ye et al. 2012).

By March 22, 2010, about 51 million people faced water shortages in a number of provinces (Asianews.it. 03-22 2010). Commodities including sugar cane, flowers, tea, fruit, potatoes, rapeseed, medicinal ingredients, tobacco, wheat, rubber, and coffee have been severely affected, with output reduced by as much as 50% (*Chang G, April 9*, 2010). Authorities began to fear unrest due to soaring food prices and sent more than 10,000 armed police to the affected regions to ensure stability and help with water supplies (Asianews.it. 03-22 2010; *Chang G, April 9*, 2010). More than 20 million people are left without adequate drinking water in Yunnan, Guangxi, Guizhou, Sichuan, and Chongqing, and many wells in Yunnan have gone dry (*China, NTDTV, April 15*, 2010). These severe socioeconomic and environmental crises also carry profound political consequences.

One of the primary goals of the Chinese state is to maintain stability and social order. Yet, the escalating water and drought crisis in China threatens the government in terms of stability, social order, and political legitimacy. The number of major environmental protests in China grew by 120% from 2010 to 2011, according to Yang Chaofei, Vice-Chairman of the Chinese Society for Environmental Sciences. Moreover, according to research by the Pew Center, 74% cite air pollution as a big problem and 66% so named water pollution and 80% of the Chinese think protecting the environment should be made a priority (Pew Research Center, July 22, 2008). Although climate change and drought are largely beyond the capacity of any government to control, the political fallout from devastated populations and communities can be difficult for political leaders at all levels. Drought complicates an already severe and worsening water crisis facing the nation. As these problems become exacerbated in the twenty-first century, political protest will rise considerably and with it, the political legitimacy of the state.

The United Nations Intergovernmental Panel on Climate Change 2014 Report expressed concerns that China will be affected dramatically by climate change. Moreover, future scenarios of water scarcity and growing climate change-induced threats, particularly drought in northern China, are severe and demand emergency planning by Chinese officials, as noted in the 12th Five-Year Plan released by the Ministry of Environmental Protection. The country has approximately 20% of global population but only 7% of global freshwater. Moreover, China experiences marked regional diversity in water supply, characterized by flooding in the south and drought in the north. The North China Plain, which comprises 42% of the population, has only 8% of the country's water resources. In 2014, an estimated 108 million Chinese citizens had no access to improved drinking water sources (UNICEF press release, March 21, 2014). Water insecurity and scarcity were exacerbated by the 2009 drought and will continue to challenge the population as these problems persist.

13.4 Conclusion

Drought episodes in China will escalate according to scientific data and projections. A recent study based on drought modeling reported evidence that repeated extreme droughts began in the recent two decades in northern China with drying trends in "west northwest China at a rate of 4.90%/10 years. Persistent multiyear droughts dominated in the last two decades in west northwest with percentage area over 40% in 5 consecutive years in the 2000s. The percentage area of drought condition even reached almost 90% during 2008, and nearly 70% both in 1997 and 2006. Obvious increases in dry areas were also seen in northeast China and east northwest China, with a rate of 3.80% and 3.96% per decade, respectively" (Yu et al. 2014).

The Chinese government responded to the drought crisis of 2009 with impressive organization, resources, and financial support. The difficult reality that policymakers are addressing is the need for more aggressive adaption and mitigation policies to forestall the worst effects of drought. Human security is a vital and relevant dimension through which to comprehend and respond to the enormous challenges of drought. Accordingly, water scarcity and drought are two monumental environmental challenges that will test the resilience and resourcefulness of the Chinese government and people in the twenty-first century.

References

Asianews.it. (2010) Worst drought in a century wipes out harvests in southwestern China
Barriopedro D, Gouveia CM, Trigo RM, Wang L (2012) The 2009/10 drought in China: possible causes and impacts on vegetation. J Hydrometeorology 13:1251–1267. https://doi.org/10.1175/JHM-D-11-074.1
Cao CX, Zhao J, Gong P, Ma GR, Bao DM, Tian K, Tian R, Niu ZG, Zhang H, Xu M, Gao MX, Zheng S, Chen W, He QS, Li XW (2011) Wetland changes and droughts in southwestern China. Geom Nat Hazards Risk 3(1):79–95. https://doi.org/10.1080/19475705.2011.588253
Chang GG (2010) Will China run out of water? *Forbes.com*
China National Climate Center (2009) China drought and flood climate bulletin 242 2009b (in Chinese)
EP-NDRCC (Expert Panel of National Disaster Reduction Committee of China) (2009) Internal report (in Chinese)
EP-NDRCC (Expert Panel of National Disaster Reduction Committee of China) (2010) Yunnan drought disaster survey report (internal report, in Chinese)
Lu E, Luo Y, Zhang R, Wu Q, Liu L (2011) Regional atmosphere anomalies responsible for the 2009–2010 severe drought in China. J Geophysics Res 7(4)
NDRC-MCA (National Disaster Reduction Center, Ministry of Civil Affairs) (2009) *Nationwide Disaster Monthly* (in Chinese). 2012 China Disaster Reduction 4:61–63
PRC (Pew Research Center) (2008) The Chinese celebrate their roaring economy as they struggle with its Cost. Washington, DC: PRC
Scanlon J, Cassar A, Nemes N (2004) Water as a human right. IUCN Environmental Policy and Law Paper No. 51. Cambridge: UNDP-IUCN
Security and Climate Change (2001) Jon Barnett Macmillan Brown Centre for Pacific Studies, University of Canterbury, New Zealand. Tyndall Centre Working Paper No. 7
UN Commission on Human Security, 2003
UNDP (United Nations Development Programme) (2004) Human development report. Oxford University Press, Oxford, New York
UNICEF press release, March 21, 2014
USDA (United States Department of Agriculture) (2009) Foreign agriculture service, commodity intelligence report. Washington, DC: USDA
World Bank (2013) China country water resources partnership strategy. Washington, DC
Xu X, Ge Q, Zheng J et al. (2013) Agricultural drought risk analysis based on three main crops in prefecture-level cities in the monsoon region of east China. Nat Hazards 66:1257. https://doi.org/10.1007/s11069-012-0549-y
Ye T, Shi P, Wang J et al. (2012) China's drought disaster risk management: perspective of severe droughts in 2009–2010. Int J Disaster Risk Sci 3:84. https://doi.org/10.1007/s13753-012-0009-z
Yu M, Li Q, Hayes MJ, Svoboda MD, Heim RR (2014) Are droughts becoming more frequent or severe in China based on the Standardized Precipitation Evapotranspiration Index: 1951–2010? Int J Climatology 34:545–558. https://doi.org/10.1002/joc.3701
Yuan XC, Wang Q, Wang K et al. (2015) China's regional vulnerability to drought and its mitigation strategies under climate change: data envelopment analysis and analytic hierarchy process integrated approach. Mitigation and Adaptation Strategies for Global Change 20(3). Netherlands: Springer. https://doi.org/10.1007/s11027-013-9494-7

Dr. Ross Michael Pink teaches Sustainability, Politics, and International Law at Kwantlen Polytechnic University in Canada. Dr. Pink is the author of two recent book, "Water Rights in Southeast Asia and India" and "The Climate Change Crisis: Solutions and Adaption for a Planet in Peril," both published by Palgrave Macmillan. Dr. Pink is also the Co-Founder of Global Water Rights (www.globalwaterrights.com).

Chapter 14
After the 2004 Bangladesh Flood: Integrated Management for Flood Disaster Risk Reduction in Four Different Areas

Kumiko Fujita, Rajib Shaw, and Gulsan Ara Parvin

Abstract Flood disaster is a critical issue in Bangladesh, with the most severe cases of flooding occurring in July and August. Regular river floods affect 20% of the country, increasing up to 68% in extreme years. After such catastrophic floods, the policy, action, strategy, and plan for flood disaster risk reduction have been developed. The floods of 1987, 1988, 1998, 2004, and 2007 were particularly catastrophic, resulting in large-scale destruction and loss of lives. After the 1988 flood, a strategy of "living with the floods" was developed to achieve sustainable agricultural development and this concept was integrated into the country's Flood Action Plan. In 1998, 68% of Bangladesh was inundated by floods. The earthen embankments were unable to give protection against the floods in 1998, even against some medium floods in the 1990s. Catastrophic disasters in the 1990s affected institutional development, giving rise to an integrated water resource management program. In 2004, Bangladesh experienced devastating floods from July to September, with 56,000 km^2 (about 38% of the country) inundated. Consequently, a more comprehensive risk reduction program was established to effectively address the complexity of water management. This paper looks at the flood history of Bangladesh, the different flood phenomena and approaches used after 2004, and the ways climate and social change affected flood disaster mitigation and people's livelihood. It then discusses an integrated program for disaster risk reduction and resource management.

Keywords Bangladesh · Flood disaster risk reduction · Climate change · Social change · Local people's participation

---

K. Fujita (✉)
International Consortium on Landslides, Kyoto, Japan
e-mail: kumiko.fujita@gmail.com

R. Shaw
Graduate School of Media and Governance, Keio University, Kanagawa, Japan

G. A. Parvin
Graduate School of Global Environmental Studies, Kyoto University, Kyoto, Japan

J. M. Pulhin et al. (eds.), *Climate Change, Disaster Risks, and Human Security*, Disaster Risk Reduction, https://doi.org/10.1007/978-981-15-8852-5_14

14.1 Introduction

Because of weather and topography, floods occur almost every year in Bangladesh. Heavy rainfall and the warm and tropical climate trigger flooding. Annual rainfall is at least 2000 mm in most parts of the country, except in the relatively dry western region where about 1600 mm of rainfall occurs. There are four prominent seasons: winter (December to February), pre-monsoon (March to May), monsoon (June to early October, major rainy season), and post-monsoon (late October to November). In addition, rainfall in India, Nepal, China, and Bhutan drains into Bangladesh through three mighty rivers–the Ganges, the Brahmaputra, and the Meghna (commonly known as the GBM river system, one of the largest river systems in the world) and their tributaries. Bangladesh is located at the lower part of the GBM river basin (Fig. 14.1). Almost half of the area is within the 10-m-above-mean-sea-level range and floodplains make up 80% of the country's landscape. Thus, the monsoon and the flat topography are regarded as major causes of floods.

Flooding brings fertile soil from upstream. As rice is the staple and as traditional rice-based agriculture is dependent on the monsoon rain and flood-conveyed fertile soil, Bangladeshi farmers welcome the regular floods during the monsoon season. However, the increased volume of rainfall caused by climate change has intensified the flood problem (Brouwer et al. 2007). Catastrophic disasters provide the turning points of policy, action, strategy, and plan for flood disaster risk reduction. In addition, social changes also affect these risk reduction interventions. In this research paper, the flood history of Bangladesh is reviewed with the focus on the 2004 catastrophic

Fig. 14.1 The Ganges, Brahmaputra, and Meghna basins (Mirza 2002)

flood, highlighting the need for integrated management for flood disaster risk reduction. Possible integrated management approaches to reduce flood disaster risks are discussed.

14.2 Flood History

Catastrophic floods occurred in Bangladesh in 1988, 1998, and 2004. If the extent of inundated area and the number of people affected are more than what the annual floods bring about, the floods are considered extreme. Twenty percent of the country is inundated by regular floods, with the area affected increasing up to 68% in years of extreme flooding. Approximately 37, 43, 52, and 68% of the area is inundated by floods with return periods of 10, 20, 50, and 100 years (DMB 2010). Floods inundating more than 30% of the country occurred in 1974, 1987, 1988, 1998, 2004, and 2007 (Fig. 14.2). The largest recorded flood in terms of depth and duration was in 1998, where 68% of the country was under water for several months (FFWC 2005; Nishat et al. 2000). The flood in 1988 also brought extreme damage to Bangladesh as inundation area reached 61%. Both floods in 1998 and 1988 were caused by a combination of heavy rainfall within and outside Bangladesh and the peak flows of the major rivers. As Fig. 14.2 shows, the inundation rate in 1988 and 1998 stood out as there was large-scale destruction and loss of lives.

In addition to area, duration of inundation also affects the extent of damage. In the 1998 flood, 53 districts (nearly 100,000 km^2 area) were inundated for 65 days (Shaw et al. 2013). Figure 14.3 shows the duration of flooding, defined as the number of consecutive days at or above the danger level registered for the water level station in Dhaka from 1909 to 2009 (Thiele-Eich et al. 2015). For Dhaka, 32 events were noted between 1909 and 2009, with an average duration of 16.3 days (Thiele-Eich et al. 2015).

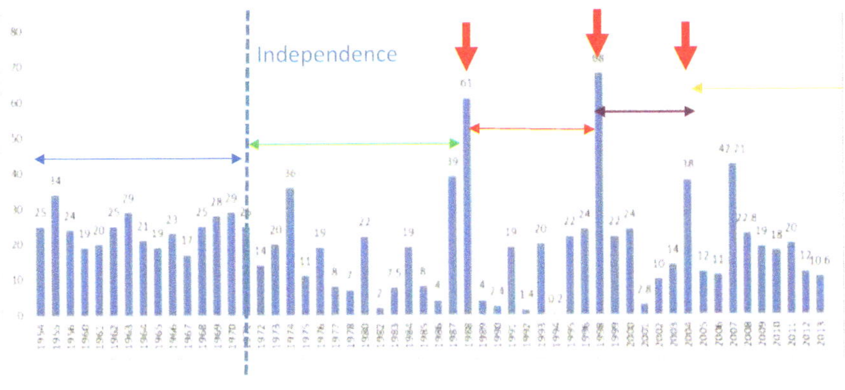

Fig. 14.2 Percentage of inundated area over the time period 1954–2013 (*Source* Department of Disaster Management 2014)

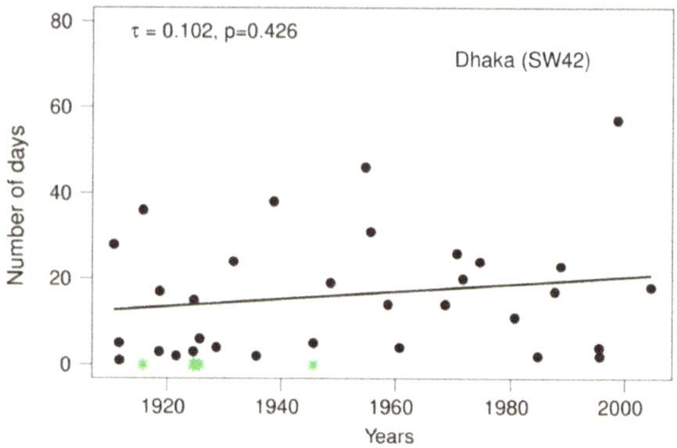

Fig. 14.3 Duration of flood events in Dhaka (If two flood events are separated by only 6 days recording below-danger level, these are considered one continuous event marked by green stars adapted from Thiele-Eich et al. 2015: 1205)

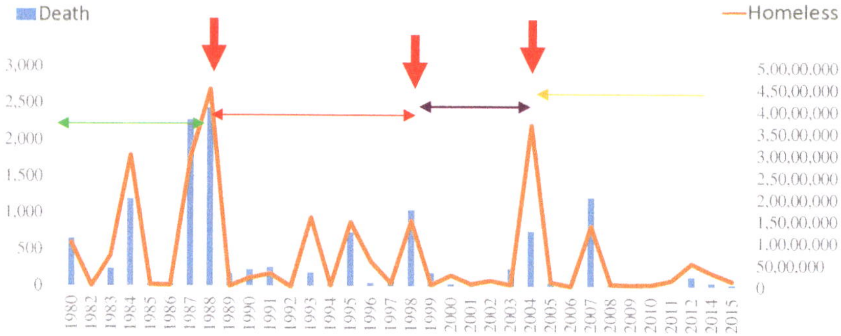

Fig. 14.4 Flood-affected people (population in 2016: 163,000,000) (*Source* CRED 2016)

Figure 14.4 shows the number of deaths and people rendered homeless by the floods since 1980. Because of the gentle slope and water velocity of the rivers, the floods were not fast enough to deprive people's lives directly. However, the inundation damaged the farmland, resulting in death due to starvation. In addition, lack of access to safe drinking water contributed to disease epidemics.

More than 2,000 people died in 1987 and 1988 and more than 35 million people were affected by the floods in 1988 and 2004. Table 14.1 is prepared by selecting the number of deaths that exceed 750 from Fig. 14.4 and adding the area of inundation from Fig. 14.2.

Population density also relates to high rates of mortality and morbidity (Cash et al. 2013). Bangladesh has 166 million people (in 2014) and has a high population density, 1,277.4 persons/km^2 (CIA 2014). Because of population pressure and land scarcity, many poor communities are forced to live in flood-prone areas. Seventy percent of the population live in regions at risk of floods; 26% reside in regions at risk of cyclones (DMB 2010). Almost 74% of the country's agricultural output comes from utilizing the fertile basin of the GBM rivers (Brammer and Khan 1991). In addition, the increasing population (Fig. 14.5) makes it difficult for people to

Table 14.1 Number of deaths (more than 750) since 1980

Year	Deaths (no.)	Homeless (no.)	Inundated area	
1984	1,200	30,000,000	19%	About 28,000 km^2
1987	2,280	29,700,000	39%	About 57,000 km^2
1988	2,440	45,000,000	61%	About 90,000 km^2
1998	1,050	15,000,050	68%	Nearly 100,000 km^2
2004	761	36,871,700	38%	About 56,000 km^2
2007	1,230	13,851,440	42%	About 62,000 km^2

Source Based on Department of Disaster Management (2014), CRED (2016)

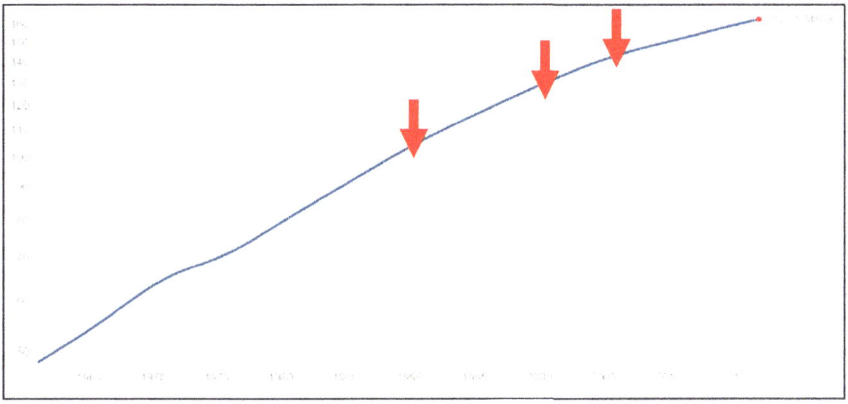

Fig. 14.5 Bangladesh population (162.95 million, December 2016) (*Source* World Bank 2016, http://www.multpl.com/bangladesh-population)

survive the food shortage during the floods.

Climate change is expected to change the frequency, intensity, duration, and magnitude of flooding (IPCC 2007). It is also likely to create food supply problems because Bangladesh is highly dependent on monsoon farming (Shukla 2003). Food production is being disrupted by floods more frequently and more severely than before (Douglas 2009). As population increases, more and more people have no recourse but to live in flood-prone areas, thereby increasing the number of flood victims.

The floods in 1988, 1998, and 2004 were the turning point. A flood disaster risk reduction program was crafted in this country that gained independence in 1971. The phases are divided into four.

14.2.1 Phase I (1955–1971): Post-disaster Recovery to Protect Agricultural Land

Official flood management in Bangladesh started in the 1960s. During this phase, protecting the agricultural land was the main focus because agriculture was the main industry. Development of infrastructure was given high priority. The Coastal Embankment Project (CEP) in 1960–1970 has been the largest structural intervention in the coastal areas. Flood control, drainage, and/or irrigation (FCD/I) projects have been implemented by the Bangladesh government since the mid-1960s. Embankments, drainage channels, and sluice gates were constructed, with a total coverage area of 5.37 million ha (Rahman and Salehin 2013). The Bangladesh Water Development Board (BWDB) built more than 100 polders in the coastal area in the early sixties to protect designated areas from tidal floods, salinity intrusion, and sedimentation. Though the projects gave positive impacts (i.e., increase in agricultural production and economic activities; reduced damage to infrastructure inside the protected areas) (Rahman and Salehin 2013), the scale was too large and needed a large amount of time and money. In spite of existing projects, even moderate floods caused lots of damage to the people and agricultural land. Although the embankments were able to prevent the intrusion of salinity water, disturbances in the river flow proved inconvenient for the local people and they could not fully utilize the water resource. The local people destroyed some of the embankments for their own convenience. As a result, these structures were not able to function as envisioned. Moreover, the poor operation and maintenance of the embankments decreased fish stocks and caused deposition of silt in the riverbed as well as on agricultural land. Consequently, non-structural measures were also considered as a means for mitigating flood damage. Established in 1972, the Flood Forecasting and Warning Center of BWDB was responsible for making flood forecasts and flood warning during the flood season (Rahman and Salehin 2013).

14.2.2 Phase II (1971–1988): Shift to Small-Scale Projects

In the 1970s–1980s, large-scale polderization was promoted to meet the increasing demand of agricultural production. However, this has resulted in severe drainage congestion and waterlogging (Fig. 14.6). Around the same time, demand for shrimp has increased in the international market, jacking up its price. As polders are suitable for intensive shrimp farming, salt water was allowed to enter under polder shrimp culture. Land previously used for agriculture and mangroves were used for shrimp farming (Islam 2006). The government focused on water resource development and management in addition to protecting agriculture. The National Water Plan (NWP) was formulated in 1982 and finalized in 1986. In addition to the large-scale structural solutions, the government adopted small and medium-scale FCD projects to provide early benefits.

Fig. 14.6 Poor polder drainage (left) and sedimentation resulted in inundations marked by levels of discoloration in buildings (middle) (Kapotakho in Khulna District, 2014)

14.2.3 Phase III (1988–1998): Living with the Floods

This phase is characterized by the shift from flood control to flood management and the change in focus from structural solutions to a combination of structural and non-structural measures. There were particularly catastrophic floods in 1987 and 1988, resulting in large-scale destruction and loss of lives. Awareness of the need for a more integrated, multisector approach to surface water management became evident in the late 1980s. As a result, various flood assessment studies were conducted and there was recognition that although floods are a hazard, they can also be beneficial in many aspects. The strategy of "living with the floods" was applied, which is the more traditional way of dealing with these floods (Rahman and Salehin 2013). This concept of "living with the floods" for sustainable agricultural development was integrated into the Flood Action Plan (FAP) and has become widespread from 1990 to 1996. The FAP was initiated to save the country from devastating floods such as the two consecutive floods that occurred in 1987 and 1988 (Haque 1993). Gradually, FAP regional studies paid attention to the interdisciplinary nature of flood management. What came out was an integrated water management plan that covered issues such as flood, drainage, irrigation, navigation, the environment, and the economy.

14.2.4 Phase IV (1998–2004): Integrated Water Resource Management

The earthen embankments were unable to give protection against severe floods in 1998 and even against some medium floods in 1991, 1993, and 1995 (Salehin et al. 2007). The disasters in the 1990s affected institutional development. A real start toward "integrated" management took place in the realms of policy and practice. The National Water Policy (NWP) was prepared in 1999 and subsequently the National Water Management Plan (NWMP) was finalized in 2001 (Rahman and Salehin 2013). The NWP makes clear the government's intention to implement the policy of integrated water resource management (IWRM) and to pursue all necessary measures to manage the country's water resources in a comprehensive, integrated, equitable, and environmentally sustainable manner. The NWP also focuses on stakeholder participation. The NWMP is a framework under which the ministries define the strategies; agencies, departments, and local bodies that will be involved in the projects.

14.2.5 Phase V (2004 to Date): Shift from Response and Relief Practice to More Comprehensive Risk Reduction Approach

In 2004, Bangladesh experienced one of the most devastating floods in its nearly 50 years of existence. About 38% of the country went under water (DMB 2010). After this catastrophe, the government resolved to improve disaster response and preparedness at the local level, with provision of immediate rescue resources, emergency funding mechanisms, better information management, and contingency planning (DER 2004). A paradigm shift in disaster management from conventional response and relief practice to a more comprehensive risk reduction system slowly took place (Zimmermann et al. 2009).

Based on climate change and social and economic conditions, the measures and plans for flood disaster risk reduction have shifted as follows:

- From "response and recovery" to "preparedness" (Fig. 14.7)
- From "control" to "management" to "comprehensive risk reduction"
- From "structural measures" to "comprehensive measures," including non-structural ones
- From "government" to "multi-stakeholder participation" (Fig. 14.8).

The clear trend is toward integrated management for disaster risk reduction.

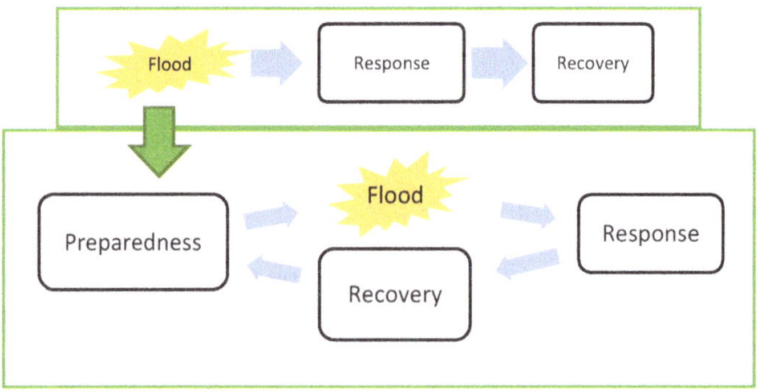

Fig. 14.7 Shift in focus from "response and recovery" to "all stages"

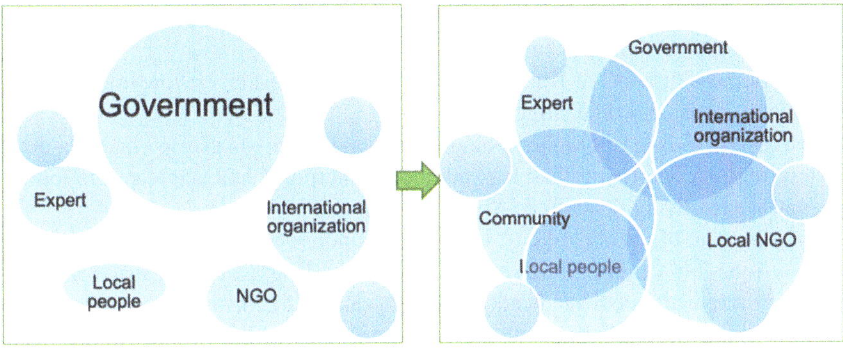

Fig. 14.8 Shift from "government-led approach" to "multi-stakeholder participation"

14.3 Review of the 2004 Flood

There are several turning points in the development of an approach to flood disaster risk reduction. Since the flood in 2004 was relatively recent and since it is pivotal in the shift to a more comprehensive and integrated approach toward flood risk reduction, the details of this particular disaster are reviewed.

The 2004 flood lasted for 3 months, from July to September. The government reported massive damage in 39 out of 64 districts; 56,000 km^2, about 38% of the country's total area, were inundated (WARPO 2005), including 8,000 km^2 of agricultural land. About 2,500 km of embankments were damaged and about 74 primary school buildings were washed away (GoB 2010). More than 1,200 primary schools were destroyed or severely damaged and another 2,000, though not inundated, were damaged through their use as official flood shelters. Thus, nearly 18,000 primary schools were affected (LCG 2004). In total, impacts of the flood were felt by 24,000 educational institutions, including primary schools, high schools, colleges, and madrasa (LCG 2004).

By mid-September, about 800 people have died of various diseases, the result of having no access to clean water. Thirty million people lost their homes. Economic losses reached about 2,200 million US dollars (GoB 2010). Food scarcity became a common phenomenon in flood-affected areas during and after the flood. Crop failure has significant negative effects on household income (GoB 2010).

Flooding conditions vary in different areas because of topography and prevailing weather. Annually, the northern, central, and northeastern districts are subjected to 2–3-m flood depths for 4–5 months during the monsoon (July–September) (Ali 2007). These areas were also inundated in the 2004 flood. Several eastern districts (Noakhali, Comilla, and Sylhet) remained under deep floodwaters until the end of September.

The western districts of Rajshahi, Chapai Nawabganj, Natrore, and Pabna are in the Ganges river floodplain and the northwestern districts of Dinajpur and Rangpur regions are in the Teesta-Karotoa river floodplains. These regions experienced moderate to high flooding in July.

The southern districts of Khulna, Barisal, and Patuakhali have low flood risk and experience occasional tidal surges.

The eastern hills in Chittagong, Rangamati, and Bandarban districts and the south-western region comprising Jessore, Satkhira, Meherpur, Chuadanga, and Joypurhat districts are relatively flood-free (Ali 2007).

Forty percent of the capital city, Dhaka, was under water in July 2004. The flood had a serious impact on children's education in the Dhaka area because 7% of the school children dropped out from school (Afrin 2015). In the urban areas, slum dwellers living on poorly drained land suffered the most. Since pure drinking water was not available during the flood, 100,000 people in Dhaka suffered from diarrhea (Afrin 2015).

In the rural area, around 52% of the people are engaged in agriculture, mostly as labor. As a result, landless laborers and small farmers were the most severely affected (GoB 2010). The rice crop was devastated and so were important cash crops such as jute and sugarcane (Fig. 14.9).

14.4 Four Types of Flood-Prone Areas: Different Phenomena in Different Areas

Since flooding phenomena and damage are different in various areas, the flood disaster risk reduction approach applied in each area also differed. In this section, the different floods, their causes, and the measures implemented are reviewed in order to come up with a more comprehensive and integrated management system.

Bangladesh experiences several types of flood, generally categorized into four (see Fig. 14.10 for the location of these flood types):

i. River floods (also called river bank erosion or monsoon floods) along major rivers during the monsoon season (June–September).
ii. Flash floods caused by overflowing of hilly rivers in eastern and northern Bangladesh (April–May and September–November).
iii. Urban floods (also called normal floods, pluvial floods, or rainfed floods) caused by drainage congestion during heavy rains.
(Although it is included in the river flood area, it is separately discussed in this paper inasmuch as the phenomena and measures along the major rivers and those in Dhaka are different.)
iv. Coastal floods caused by storm surges (April–May and October–November) and tidal surges (April–May and October–November in addition to June–September during high tide).

(i) River floods occur along major rivers such as the Brahmaputra and Ganges during the monsoon season from June to September every year. They are

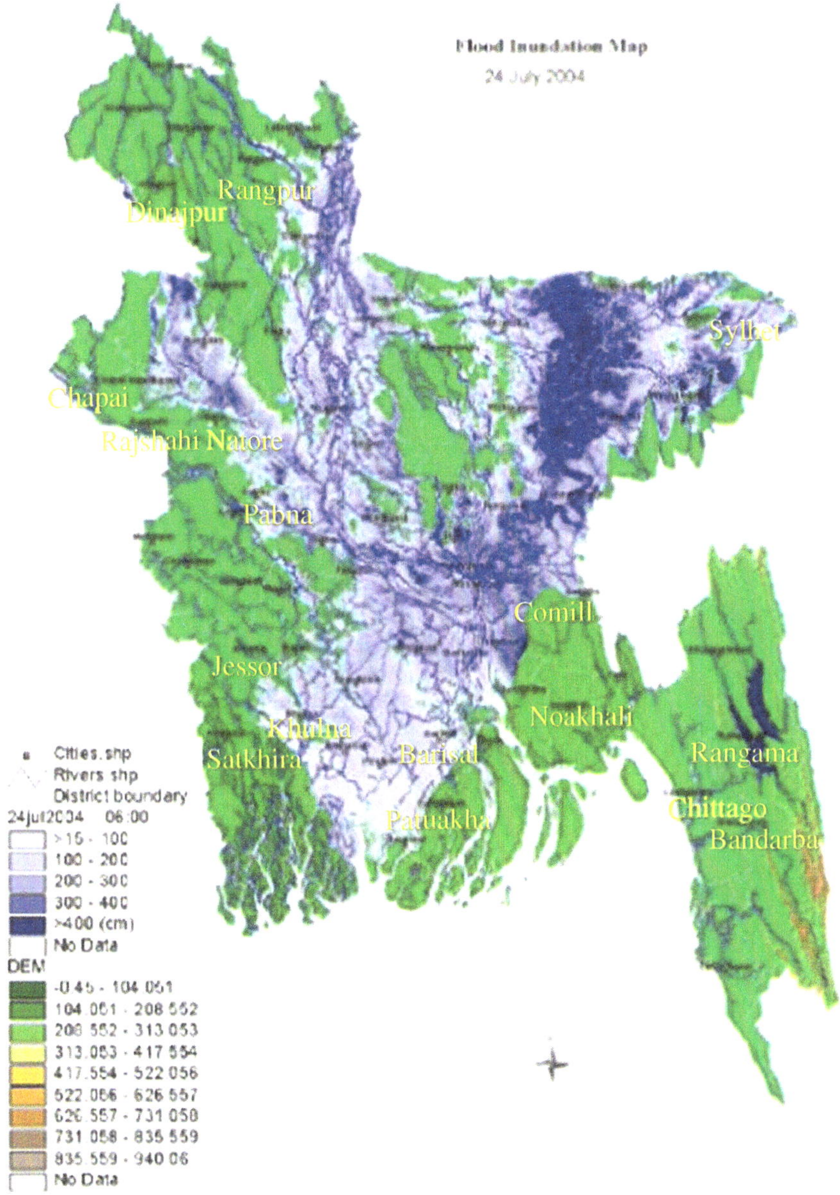

Fig. 14.9 Flood depth and inundation map, 24 July 2004 (*Source* Based on OCHA/ReliefWeb 2004)

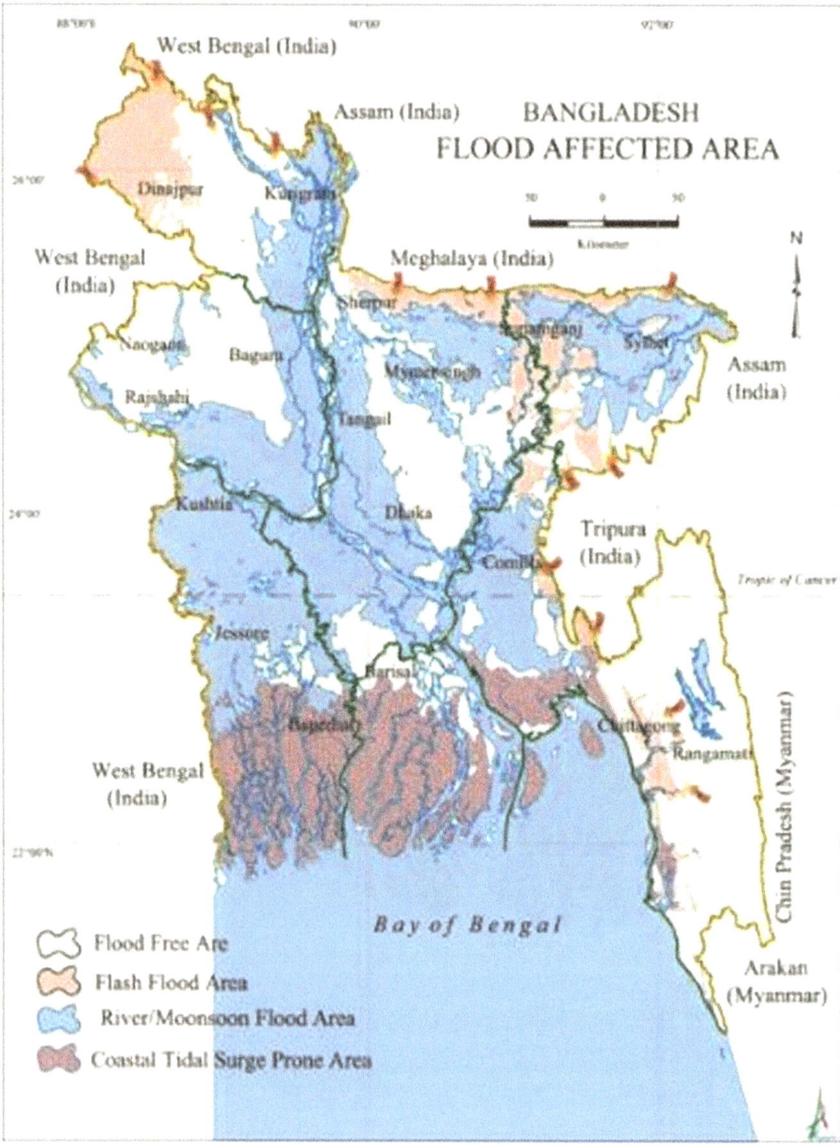

Fig. 14.10 Different flood phenomena in different areas of Bangladesh (*Source* http://en.banglapedia.org/images/thumb/0/06/FloodAffectedArea.jpg/457px-FloodAffectedArea.jpg)

caused by erosion and bank overflow. According to BWDB, about 1,200 km of river banks are actively eroding and about 100,000 people live on these sites (Zimmermann et al 2009). About 80% of the country is vulnerable to extreme river floods. The most destructive river ones occur when peak discharges of both the Brahmaputra (June/July) and the Ganges (August/September) coincide. Structural and non-structural measures have been developed against river floods. Indigenous ways such as raising/elevating their houses with plinth (Fig. 14.11) and placing sand bags to prevent erosion on the river bank (Fig. 14.12) are common and effective in mitigating the annual river floods. Migration is the major non-structural measure implemented in river flood areas. Flood is one of the main reasons of migration in Bangladesh (Khatun 2013). Loss of life directly due to a river flood is a rare occurrence; however, erosion and inundation deprive farmers of their farmland and livelihood, and the low yield of crops and the ensuing food shortage put their families on the brink of starvation. Seasonal migration to find jobs and food is needed to avoid

Fig. 14.11 Houses on plinth. A woman shows the height of the recent flood (Sundarganj District, Gaibandha Division, 2014)

Fig. 14.12 Sand bags and concrete blocks are placed on the embankment to prevent erosion (Fulcchari District, Gaibandha Division, 2014)

starvation. Usually, the villagers spend a few weeks to a month to prepare for migration. This seasonal/circular migration is an important strategy. In addition to seasonal migration, families move several times when they lose their land. Riverbank erosion displaces 50,000–200,000 people in Bangladesh every year (Mehedi 2010). The number of internal migrants to urban cities is also increasing (Planning Commission 2010) because of increasing population and extreme river flooding caused by climate change. Food shortages are already serious issues in the rural areas, particularly in the eroded areas. The poor people migrate to Dhaka City and generally settle in the slums (Asfar 1999).

(ii) Flash floods and the associated landslides are the major disasters in the hilly regions of Bangladesh because of its physiology, hydrology, and climatology (Sarker and Rashid 2013). Flash floods occur in the north and in the east, particularly in the Meghna River basin, and in the southeastern hilly areas in Chittagong. The northeast districts, such as Sylhet, Moulvibazar, Sunamganj, Habiganj, Kishoreganj, Netrokona, Rangamati, Bandarban, and Khagrachari are the main flash flood-prone districts. Flash floods in the foothills and Haor basin damage agriculture and, to a lesser degree, infrastructure. Localized heavy rainfall, as well as water flow from the upstream Meghalaya hills in India through the hilly rivers, results in a sudden rise in river flows within a short time, causing flash floods in April–May and September–November. Since flow velocity is fast, a flash flood can occur within a few hours and can be over within 5–6 h. It recedes rapidly, within minutes to a few hours, and often dissipates quickly. However, flash floods with high hydraulic force and erosive power frequently damage crops, property, and fish stocks in wetland areas. In the northeast and southeast regions, flash floods often affect the winter rice crop at harvesting stage during April–May. With rainfall forecast and timely warning, the local people are able to receive flash flood information through public announcement systems (Fig. 14.13) and neighbors in advance. They have enough time to bring their property to higher ground and go to a safe place (Fig. 14.13).

In 2002, 2004, 2007, 2009, and 2010, flash floods devastated the rice crop of northeast Haor region (MoDMR 2014). This was attributed to the changing

Fig. 14.13 Farmers are given advance information about the floods through a public announcement system

timing of flash floods. Recently, as flash floods come earlier than usual, farmers do not get to harvest their rice on time. Therefore, fast-growing rice varieties become crucial as farmers need enough time to harvest their crop after flood warnings have been issued. Moreover, land use changes, deforestation, hill-side cutting, and unplanned infrastructural development influence the kind of mitigation measures applied to lessen the impact of flash floods and landslides (Sarker and Rashid 2013).

(iii) Urban floods, also called rainfed floods, occur in many parts of the country but mainly in Dhaka.
With intense rainfall, the natural drainage system cannot carry the runoff and temporary inundation results. This phenomenon is increasingly being observed in the urban areas (BFSC 2014; Rahman et al. 2014). The drainage systems have been disturbed by various human interference such as construction of roads, river courses, water bodies, etc. The culprit is unplanned and rapid urbanization.

(iv) The coastal area is highly vulnerable to coastal flooding, which is caused by storm surges and astronomically high tide. The southern part of Bangladesh is mainly coastal, with a coastline of about 800 km. About 32% of the land in Bangladesh fall in the coastal zone. This area is found in the Bay of Bengal where shallow continental shelves extend to about 20–50 km. The eastern coastline is conical and funnel-shaped. Because of the topography, the height of storm surges generated by cyclonic storms in this area is higher than those reported in other coastal areas in the world. The highest recorded storm surge caused by super cyclones was 10–15 m, which caused flooding in the entire coastal belt. The coastal area was continuously hit by severe cyclones in the 1990s and 2000s. The flood that caused a huge loss of human lives and property occurred on April 29, 1991. This catastrophic cyclone affected a large construction project, including 21 coastal polders. The CERP was implemented from June 1996 to June 2003. The overall project objective was to secure life, property, crops, and livestock along the cyclone-prone coastal areas. Cyclone Sidr (November 15, 2007), a 10-year-return-period cyclone, and Cyclone Aila (May 25, 2009) provided recent examples of devastating storm surges. Both cyclones posed extensive loss and damage in lives and livelihood and caused long-term inundation and salinity intrusion in vast areas.

In addition to the storm surge floods, tidal floods also occur in coastal areas. Tidal floods are caused by the water-level rise due to the combination of storm surge and the astronomically high tide. The sea-level rise of normal high tide is caused by the southwest monsoon wind during the June to September monsoon season. This rise in water level can cause extreme flooding in coastal areas, particularly when a storm surge coincides with the normal high tide.

Floods caused by erosion also occur along rivers in the coastal areas. It is not destructive in the short term. However, long-term inundation can destroy farm land, which may result in food shortage. Since the duration of inundation is long, the local people live there as long as they can and then migrate to other areas instead of going to evacuation shelters. Increased river bank erosion and saline water intrusion in the

Fig. 14.14 A house with the floor raised higher (left) and the original structure with traditional plinth (right), Goainghat, Sylhet, April 2017

coastal areas are likely to displace hundreds of thousands of people. They will be forced to migrate, often to slum areas in Dhaka and other big cities (Ministry of Environment and Forests 2009a).

The frequency of "above-normal or above-common" floods has shown an increasing trend since 1950 (Hofer and Messerli 2006). Both sea-level rise and other climate-induced changes could submerge one-fifth of Bangladesh (Stern 2007). Sea-level rise has already caused land erosion and increased salinity in the coastal areas, affecting biodiversity, food production, and fisheries (Khatun 2013). Climate change impact on natural disasters and agricultural system is critical, since these two aspects are closely related to life, livelihood, and food security. Frequent hydro-meteorological disasters such as floods, cyclones, and storm surges disrupt agricultural systems and cause shortage of food (Parvin et al. 2015).

As mentioned in the section on flood history, the government of Bangladesh has paid great attention to coastal planning and land use management since the 1950s because of two major reasons: preservation of its rich resources and mitigating the devastating cyclone and storm disasters. The coastal zones have been considered areas of enormous economic potential because of the country's biodiversity and natural resource endowment. Therefore, these areas have been used intensively (Islam et al. 2011) despite the threat of disasters. After a number of cyclones hit the coastal zones in the 1960s, protective measures were started. Not much attention was given if the disaster was deemed not destructive in the short term (e.g., river floods). Since coastal land use management is considered one of the key elements for national development (Parvin et al. 2017), many institutional, structural, and non-structural measures have been developed to protect the coastal areas from devastating disasters

Fig. 14.15 A cyclone shelter houses both livestock (first floor) and people (second floor). A warning system with lights that signify levels of danger has been installed. Sarankhola, Bagerhat District, Khulna Division, August 2014

Fig. 14.16 A loudspeaker from a mosque is used to warn people about changes in weather conditions. Gabura Union, Satkhira, Khulna Division, August 2014

such as cyclones and storms. Structural measures, such as building cyclone shelters and raising houses with plinth were implemented. The cyclone shelter is useful during floods caused by storm surges because flood duration lasts for several days (Figs. 14.14 and 14.15). As to non-structural measures, a warning system was put in place and assistance from domestic and international organization, NGOs, and the government was secured. Some of the measures developed to lessen cyclone damage are likewise used for flood disaster risk reduction (Fig. 14.16).

The main features of the four types of flood and the corresponding mitigation measures are summarized in Table 14.2.

Table 14.2 Four types of flood and corresponding mitigation measures

	(i) River flood	(ii) Flash flood	(iii) Urban flood	(iv) Coastal flood	
Area	Along rivers, especially Brahmaputra and Ganges rivers	Haor and foothills in northeastern areas, hills in southeastern areas	Flood plain, southwestern area, area with poor natural drainage system	Coastal areas (coastline of about 800 km along the southern part)	
Season	Monsoon (Jun.–Sept.) • Brahmaputra: Jun.–Jul. • Ganges: Aug.–Sept.	Apr.–May. and Sept.–Nov.	Rainy season (Apr.–Sept., after every heavy rain)	**Storm surge flood**: Apr.–May and Oct.–Nov.	**Tidal flood**: combination of storm surge (Apr.–May and Oct.–Nov.) and astronomically high tide in monsoon season (June–Sept.)
Cause	Erosion and bank overflow	Sudden rise in river flows by short-term localized intensive rainstorms	Poor drainage: disturbance in natural drainage system by unplanned urbanization and road construction	**Storm surge flood**: tropical cyclone with high storm surges	**Tidal flood**: water-level rise due to the combination of storm surge and astronomically high tide
Principal vulnerable/affected livelihood	Farmers	Farmers	Informal sectors, vendor, petty traders	Farmers, fisherman	
Structural measures By public support	• Embankment		• Land use planning, zoning • Embankment • Drainage system improvement: box-culvert, water pumping station, etc.	**Storm surge flood**: • Embankment • Polder • Shelter	**Tidal flood**: • Embankment • Polder • Shelter • Tidal basin management
Individually	• Sand bag • Plinth for houses	• Plinth for houses		• Sand bag • Plinth for houses	

(continued)

Table 14.2 (continued)

	(i) River flood	(ii) Flash flood	(iii) Urban flood	(iv) Coastal flood
Non-structural measures By public support		Warning system	Better management of solid waste by behavior improvement and motivation	Warning system
Individually	Migration • seasonal migration • long distance migration	Evacuation		Migration Evacuation
Availability of warning	Water-level information for major rivers available on the website of the Flood Forecasting and Warning Center, Bangladesh Water Development Board	1. About a week before for harvesting of crops 2. A day before	No specific urban flood warning, general weather forecast available in media	Yes, for cyclone and storm surge No, for flood
Existing approach for warning local people	Mostly online and through media	Microphones used	Mostly online and media, weather and rainfall forecast	Flag, microphone, volunteers under the cyclone preparedness program
Necessary preparation time for evacuation/migration	A few weeks to a month to find available places to go to	1. A week for harvesting crops 2. 5–6 h to keep necessities in higher places	No evacuation generally	A few hours (3–4) for evacuating people and livestock
Means of evacuation/migration	On boats (vehicle on land)	On foot, vehicle	No evacuation generally	On foot, vehicle (rickshaw, van)

(continued)

Table 14.2 (continued)

	(i) River flood	(ii) Flash flood	(iii) Urban flood	(iv) Coastal flood
Necessary things to bring for evacuation/migration	As much as possible, all property, including livestock	Belongings which cannot be wet	No evacuation generally	Important documents (national ID, land ownership record), basic clothes, livestock, seeds, food grain
How long are they affected by flood?	A few months (2–3 months)	5–6 h in the morning	Between 1–2 h and 2–3 days in different parts of the cities/towns	Several days to several weeks
External help	Local NGOs and local government during and after floods	Local government for warning	Government, NGOs personal/individuals	Government, local and international NGOs, private support

Source Flood Forecasting and Warning Center http://www.ffwc.gov.bd/index.php/about-us/

14.5 Need for an Integrated Management Approach for Flood Disaster Risk Reduction

The flood in 2004 highlighted the need for an integrated management approach for flood disaster risk reduction. As agriculture is a major industry in Bangladesh, annual flooding is not considered a disaster and is welcomed by farmers and fisherfolk. However, it has harmful effects and threatens human security because of changes in the social and climate scenarios. An increasing population has to be considered a social issue in flood-prone areas. In Bangladesh, the most densely populated country in the world, it is very difficult to relocate people from flood-prone areas to places with less exposure to flooding. They are even forced to live in locations that are more flood-prone. Because of overpopulation, some farmers give up living in agricultural areas near the river; they go to urban areas and become part of the labor force. Many of these migrants work in the textile industry, which has recently become more popular. The population is still increasing. Though one major non-structural measure for mitigating annual/common floods is migration or evacuation, this is not suitable in cases of extreme flooding. The increasing incidence of extreme floods caused by climate change must be considered. In 2004, the number of homeless families was high despite 38% inundation in the area; this was much less than figures recorded during other devastating inundation events like those in 1988 (61%) and 1998 (68%) (Fig. 14.4 and Table 14.1).

Though integrated flood management for disaster risk reduction and resource management are critical, mitigation measures should be discussed separately for the different areas inasmuch as the flood phenomena and the social background vary greatly.

14.5.1 River Flood Area

As in coastal flood, the duration of inundation is a typical feature of the river flood. Since inundation period is long (i.e., a few months), there is some discontinuation in farming and education activities. Compared with cyclone and flash flood duration, the period of flooding caused by erosion in the river flood area is longer. Farmers have difficulty cultivating the land, and children are not able to go to school for a few months. The long inundation period affects the economy and disrupts the education system. Even short-term inundation makes it difficult for children to continue their education.

Before the overpopulation problem became critical, people were able to live near the flood area, commuting to the fertile land to farm after the floods. People were pleased with the fertile soil carried by the flood. They enjoy this benefit and manage farming by themselves. However, as the population increased, people were forced to live in the flood-prone area, farming the same way as before. Traditional farming, relying only on flood-conveyed fertile soil, did not produce enough food. Farmers turned to using chemical fertilizers, which affect their health.

Though seasonal and long-term migration to urban cities is considered an adaptive measure, frequent migration makes it difficult for people to accumulate property and continue their children's education. In this area, in addition to individual efforts, help from external sources is expected.

14.5.2 Flash Flood Area

Flash floods are a recurrent and hazardous phenomenon. It occurs in the northeastern part of Bangladesh. After a flash flood, paddy fields in the hoar area remain under water for half of the year. Farmers can grow paddy only in the dry season. Early flash floods destroy the crop, especially boro rice. In addition to developing short-duration rice varieties, putting a warning system in place is considered effective.

14.5.3 Urban Flood Area

Unplanned urbanization and road construction caused disturbances in the natural drainage system in the urban areas. The result is poor drainage, especially in the capital city, Dhaka. This area gets priority in terms of investment in infrastructure, which can address the flood issues to some extent. As people rely heavily on public works, there is a need to raise individual awareness toward averting flood disasters.

14.5.4 Coastal Flood Area

As in the case of urban areas, flood protection activities often become a flood-causing factor in coastal areas. Embankment and polderization cause waterlogging and erosion, then induce flood inundation. In these locations, the natural environment has to be considered when structural measures (embankments, dikes, polders) are taken. A proper environmental impact assessment must be done to protect the coastal ecosystem and resources.

The coastal area got a lot of attention from international and domestic organizations, NGOs, and governments, which provided external help. The focus was mainly recovery and reconstruction from short-term disasters such as cyclones and storm surges. Recent efforts now include comprehensive projects that involve flood disaster risk reduction, disaster preparation, all stakeholder participation, etc. In many cases, the agencies work independently, thus some programs/projects are biased, overlapping, or few. It will be more effective if there is coordination among these organizations, working together and allocating specific roles to each to achieve the target goals.

Since Bangladesh is a deltaic country, the floods have typical features. They are not destructive in the short term, but are rather profitable for farming. However, long-term inundation affects education and economic development. Sometimes, long-term inundation causes starvation indirectly because of food shortages caused by the disruption of farming activities. Maximizing the use of each locality's characteristics in the four different areas can bring about an effective integrated flood disaster risk reduction and resource management scheme.

14.6 Conclusion

Bangladesh is a flood-prone country because of its topography and climate. Floods in Bangladesh do not threaten human security directly. As almost 50% of the area is within 10 m above mean sea level and the floodplains make up 80% of the landscape, river velocity, and therefore flood velocity, is not fast. The people are able to utilize the flat topography, rainwater, and flood-conveyed fertile soil for farming, and flood is rather considered a benefit. Seventy percent of the population of Bangladesh thus depends on agriculture. However, recent changes in the social milieu and climate begin to affect human security. Social change such as rapid population forced more people to live in flood-prone areas. In addition, climate change has critical impact on flood inundation duration and area coverage. Long-term inundation disturbs agriculture and causes food shortage in both rural and urban areas, and people die of starvation. Thus, floods in Bangladesh do not directly affect people's lives; however there is that big possibility to threaten human security indirectly.

The catastrophic floods resulted in the development of the country's policy, action, strategy, and plan for flood disaster risk reduction. The years 1987, 1988, 1998, 2004, and 2007 saw large-scale destruction and loss of lives. The 2004 flood became the turning point to shift to a more comprehensive and integrated approach for flood risk reduction. In the various regions, different flood phenomena such as river floods, flash floods, urban floods, and coastal floods were seen, necessitating specific measures in each region. In the river flood area, duration of inundation affected people's livelihood. External help in this area is not much, considering that this type of flood is not as destructive as a flash flood or that the resulting economic loss is not as big as that observed in an urban flood area or a coastal area. In the flash flood area, timing of floods is a serious concern since flooding occurs during the harvest season. Damage is less if farmers are able to harvest their crops before the water comes. In the urban area, poor drainage system caused by unplanned urbanization makes the situation worse. However, since Dhaka is the capital city, it gets priority in getting mitigation support. The coastal area is also prioritized because of the need to protect the country's rich ecosystem from salinity intrusion and destructive cyclones and storm surges. There is thus a long history of these protection efforts in Bangladesh. Analyzing the different flood phenomena and assessing mitigation measures in the different regions is effective in understanding the complexity of water management and how climate and social changes affect flood disaster interventions and people's livelihood.

Acknowledgment The authors wish to thank the Japan International Cooperation Agency (JICA) and the Japan Science and Technology Agency (JST) for their support of the SATREPS Project "Research project on disaster prevention/mitigation measures against floods and storm surges in Bangladesh."

References

Afrin S (2015) Livelihood pattern of flooded families in Bangladesh: a study in two villages. Dhaka University Institutional Repository, Institute of Social Welfare and Research, University of Dhaka, Dhaka (Bangladesh)

Asfar R (1999) Is migration transferring rural poverty to urban areas? An analysis of longitudinal survey data of Dhaka city. Paper presented at the workshop 'Changes and Determinants of Urban Poverty.' Grameen Trust, Grameen Bank, Dhaka (Bangladesh)

Ali A Md S (2007) September 2004 flood event in southwestern Bangladesh: a study of its nature, causes, and human perception and adjustments to a new hazard. Nat Haz 40(1):89–111

BFSC (Bangladesh Food Security Cluster) (2014) Contingency plan for flooding. Food Security Cluster, Bangladesh. https://fscluster.org/sites/default/files/documents/FSC%20Bangladesh%20Contingency%20Plan%20for%20Flooding_12.08.2014.pdf (accessed 4 Feb 2019)

Brammer H, Khan HR (1991) Bangladesh country study. In Asian Development Bank (ed) Disaster mitigation: in Asia and the Pacific. Asian Development Bank, Manila (Philippines)

Brouwer R, Akter S, Brander L, Haque E (2007) Socioeconomic vulnerability and adaptation to environmental risk: A case study of climate change and flooding in Bangladesh. Risk Anal 27(2):313–326. https://doi.org/10.1111/j.4539-6924.2007.00884.x

Cash RA, Halder SR, Husain M, Islam Md S, Mallick FH, May MA, Rahman M, Rahman MA (2013) Reducing the health effect of natural hazards in Bangladesh. In Bangladesh: Innovation for universal health Coverage 5, vol 382. www.thelancet.com

CIA (Central Intelligence Agency) (2014) World fact book. https://www.cia.gov/library/publications/the-world-factbook/geos/bg.html. Accessed 23 Oct 2014

CRED (Centre of Research on Epidemiology of Disasters) (2016) EM-DAT The international disaster database. http://www.emdat.be/advanced_search/index.html. Accessed 23 Aug 2016

DDM (Department of Disaster Management) (2014) Flood response preparedness plan of Bangladesh, June 2014. Ministry of Disaster Management and Relief, DDM, Dhaka (Bangladesh)

DER (Disaster and Emergency Response) Sub-Group (2004) Post-flood needs assessment. Summary report, Monsoon floods 2004. DER, Dhaka (Bangladesh), 30 Sept 04

DMB (Disaster Management Bureau) (2010) National plan for disaster management 2010–2015. DMB, Dhaka (Bangladesh)

Douglas I (2009) Climate change, flooding and food security in south Asia. Food Secur 1:127–136

FFWC (Flood Forecasting and Warning System) (2005) Consolidation and strengthening of flood forecasting and warning services, Final report. Volume II—Monitoring and evaluation. Bangladesh Water Development Board, Dhaka (Bangladesh)

GoB (Government of the People's Republic of Bangladesh) (2010) National plan for disaster management 2010–2015. Disaster Management Bureau, Disaster Management and Relief Division, Dhaka (Bangladesh)

Haque CE (1993) Flood prevention and mitigation actions in Bangladesh: the 'sustainable floodplain development' approach. Impact Assessment 11(4):367–390
Hofer T, Messerli B (2006) Floods in Bangladesh. History, dynamics and rethinking the role of the Himalayas. United Nations University Press, Tokyo
IPCC (Intergovernmental Panel on Climate Change) (2007) Climate change 2007: Synthesis report. The Fourth Assessment Report, IPCC. http://www.un-documents.net/ipcc-ar4/syr.pdf. Accessed 4 Feb 2019
Islam MR (2006) Managing diverse land uses in coastal Bangladesh: Institutional approaches. In: Environment and livelihoods in tropical coastal zones: CAB international, pp 237–248. http://www.pg-du.com/cru/wp-content/uploads/2013/12/Hoanh_1845931076-Chapter18.pdf. Accessed 4 Feb 2019
Islam M, Sakakibara H, Karim M, Sekine M, Mahmud Z (2011) Bacteriological assessment of drinking water supply options in coastal areas of Bangladesh. J Water Health 9(2):415–428
Khatun M (2013) Climate change and migration in Bangladesh: golden Bengal to land of disasters. Bangladesh e-J Sociol 10(2):64–79
LCG (Local Consultative Group) (2004) Monsoon floods 2004: post-flood needs assessment. Summary report. Disaster and Emergency Response Sub-Group, LCG, Dhaka (Bangladesh)
OCHA/ReliefWeb (2004) Bangladesh: Monsoon floods 2004—Draft assessment report, 28 July 2004. http://reliefweb.int/report/bangladesh/bangladesh-floods-situation-report-19-jul-2004
Mehedi H (2010) Climate-induced displacement: case study of cyclone Aila in the southwest coastal region of Bangladesh. HumanityWatch, Khulna
Mirza Q (2002) Global warming and changes in the probability of occurrence of floods in Bangladesh and implications. Global Environ Change 12(2):127–138
MoDMR (Ministry of Disaster Management and Relief) (2014) Flood response preparedness plan of Bangladesh, June 2014. Department of Disaster Management, MoDMR, Dhaka (Bangladesh)
MoEF (Ministry of Environment and Forests) (2009a) Bangladesh climate change strategy and action plan 2009, government of the People's Republic of Bangladesh. MoEF, Dhaka (Bangladesh)
MoEF (Ministry of Environment and Forests) (2009b) Bangladesh climate change strategy and action plan. http://www.moef.gov.bd/climate_change_strategy2009.pdf
Nishat A et al (2000) The 1998 flood: impact on environment of Dhaka City. Ministry of Environment and Forest and IUCN (The International Union for Conservation of Nature), Dhaka (Bangladesh)
Parvin GA, Fujita K, Matsuyama A, Shaw R, Sakamoto M (2015) Climate change, flood, food security and human health: Cross-cutting issues in Bangladesh. In: Habiba U, Hassan A, Abedin M, Shaw R (eds) Food security and risk reduction in Bangladesh. Disaster risk reduction (methods, approaches and practices). Springer, Tokyo (Japan)
Parvin GA, Ali MH, Fujita K, Abedin MA, Habiba U, Shaw R (2017) Land use change in southwestern coastal Bangladesh: consequence to food and water supply. In: Banba M, Shaw R (eds) Land use management in disaster risk reduction. Disaster risk reduction (methods, approaches and practices). Springer, Tokyo (Japan), pp 381–401
Planning Commission (2010) Outline perspective plan of Bangladesh 2010–2021 (vision 2021). Planning Commission, Dhaka (Bangladesh)
Rahman R, Salehin M (2013) Flood risk and reduction approaches in Bangladesh. In: Shaw R, Mallick F, Islam A (eds) Disaster risk reduction approaches in Bangladesh. Springer, Tokyo/New York, pp 65–90
Rahman MM, Hossain MA, Bhattacharya AK (2014) An analytical study of flood management in Bangladesh. IOSR J Eng 4(1):1–6
Salehin M, Haque A, Rahman MR, Khan MSA, Bala SK (2007) Hydrological aspects of 2004 floods in Bangladesh. J Hydrol Meteorol Nepal 4(1):33–44
Sarker AA, Rashid AKMM (2013) Landslide and flashflood in Bangladesh. In: Shaw R, Mallick F, Islam A (eds) Disaster risk reduction approaches in Bangladesh. Disaster risk reduction (methods, approaches and practices). Springer, Tokyo (Japan)

Shaw R, Aminul I, Fuad M (2013) Chapter 3: National perspectives of disaster risk reduction in Bangladesh. In: Disaster risk reduction approaches in Bangladesh. Springer, Tokyo (Japan). https://books.google.co.jp/books?isbn=4431542523
Shukla PR (2003) Climate change and India. University Press India, Hyderabad (India)
Stern N (2007) The economics of climate change: The Stern review. Cambridge University Press, Cambridge
Thiele-Eich I, Burkart K, Simmer C (2015) Trends in water level and flooding in Dhaka, Bangladesh and their impact on mortality. Int J Environ Res Public Health 12(2):1196–1215
WARPO (Water Resources Planning Organization) (2005) National Adaptation Program of Action (NAPA): water, coastal areas, natural disaster and health sector. WARPO, Dhaka (Bangladesh)
WB (World Bank) (2016) http://www.multpl.com/bangladesh-population. Accessed 1 Dec 2016
Zimmermann M, Glombitza K-F, Rothenberger B (2009) Disaster risk reduction programme for Bangladesh 2010–2012. Swiss Agency for Development and Cooperation, Bern, Switzerland

Kumiko Fujita is a researcher at the International Consortium on Landslides. She has more than 20 years of experience in education, research, and development for natural disaster risk reduction. Her expertise is in applying Japanese technologies of mitigating and preventing natural disasters, such as landslides, debris flow, erosion, and flood to Asian countries. She had coordinated training courses for foreign engineers provided by the Japan International Cooperation Agency (JICA). She also coordinated science programs at the United Nations Educational Scientific and Cultural Organization (UNESCO) in Indonesia. Recently, she works as a researcher for a project titled "Development of Early Warning Technology of Rain-Induced Rapid and Long-Travelling Landslides." She has a doctorate degree in Global Environmental Studies from Kyoto University.

Dr. Rajib Shaw is Professor at the Keio University in Japan and former Executive Director of the Integrated Research on Disaster Risk (IRDR) programme. He is also Senior Fellow of Institute of Global Environmental Strategies (IGES) Japan, and Chairperson of SEEDS Asia, and CWS Japan, two Japanese NGOs. Previously, he was Professor at the Graduate School of Global Environmental Studies, Kyoto University, Japan. His expertise includes community-based disaster risk management, climate change adaptation, urban risk management, and disaster and environmental education. He is the editor of a book series on disaster risk reduction, published by Springer, Editor-in-Chief of the journal Progress in Disaster Science. He severs as Chair of United Nations Science Technology Advisory Group for disaster risk reduction, and is Co-chair of the UNDRR's Asia-Pacific Science Technology Academic Advisory Group (ASTAAG). Prof. Shaw has published more than 45 books and more than 300 academic papers and book chapters.

Gulsan Ara Parvin is a researcher of Laboratory of Rural Planning and Sustainable Rural Development, Graduate School of Global Environmental Studies, Kyoto University, Japan. She was Associate Professor of Urban and Regional Planning Department of Bangladesh University of Engineering and Technology. She earned her Ph.D. from the Urban Engineering Department of the University of Tokyo. Her expertise is in disaster risk reduction and community resilience. She has also conducted different researches related to women empowerment and microfinance in Bangladesh. She has published about 50 research papers in different journals and books. She has research and working experiences in Bangladesh, Japan, Malaysia, Thailand, and USA.

Chapter 15
The 2004 Indian Ocean Earthquake and Tsunami: Resettlement and Demographic Challenges

Syamsidik, Agus Nugroho, and Mirza Fahmi

Abstract The 2004 Indian Ocean tsunami has set a number of changes in affected communities and in the world of disaster management policy. In modern history, this disaster is acknowledged as one of the deadliest, taking around 230,000 lives from 21 nations around the Indian Ocean region. For tsunami survivors, the 13-year recovery period is a very long process for them to accept the dramatic changes brought about by these disasters. One of them is a place to live. Resettlement programs in tsunami-affected regions are mostly completed. However, this relocation task has not been easy. In Aceh, Indonesia and in Sri Lanka, the relocation program gave rise to another type of risk. Although the relocation is deemed complete, an increasing coastal population in tsunami-affected zones has been observed. This chapter aims to elaborate on the return of the tsunami risk from the aspects of resettlement and increasing coastal population. The return migration also puts disaster managers and authorities to task: protection of lives becomes part of the human security development agenda. Besides the focus on demographic challenges, this chapter also reviews the best practices instituted to increase the community's preparedness toward handling similar disasters in the future. It offers a multi-aspect review of the 2004 Indian Ocean tsunami and its 13-year recovery process.

Keywords Tsunami · Risk · Re-population · Coastal zone · Return migration · Recovery

15.1 Introduction

The 2004 Indian Ocean tsunami affected about 21 countries around the basin of the Indian Ocean, causing almost 230,000 deaths. Around 130,000 casualties alone came from Aceh, Indonesia. The second largest number was recorded in Sri Lanka where 38,000 people died (Pickrell 2005). Considered one of the deadliest disasters in recent decades, the tsunami not only caused a staggering number of human casualties but

Syamsidik (✉) · A. Nugroho · M. Fahmi
Tsunami and Disaster Mitigation Research Center (TDMRC)/Civil Engineering Department, Universitas Syiah Kuala, Banda Aceh, Indonesia
e-mail: syamsidik@tdmrc.org

J. M. Pulhin et al. (eds.), *Climate Change, Disaster Risks, and Human Security*, Disaster Risk Reduction, https://doi.org/10.1007/978-981-15-8852-5_15

also resulted in a dramatic change of world perspective on disaster management. The concepts of prevention, mitigation, and risk reduction became the center of discussions among disaster managers, practitioners, and scientists. In Indonesia, the tsunami made the country take a long hard look at its disaster management approach. The result is not only the establishment of a new national agency that specifically addresses disaster management but also the enactment of a new law on disaster management. Law No. 24 year 2007 on Disaster Management has been a major factor in creating massive changes. Although the law does not explicitly state the concept of human security, it adopts the concepts of eliminating threats produced by natural and non-natural process and reducing vulnerabilities of a region or community to disasters. These two adopted concepts have been made part of the main approaches followed to achieve human security (Busumtwi-sam 2008).

Protecting lives is part of the concept of human security. In a 1994 Human Development Report released by the United Nations Development Program (UNDP), a contemporary definition of human security was introduced. Seven dimensions of human security were identified: economic security, food security, health security, environmental security, personal security, community security, and political security. Safety from disasters is part of the environmental security dimension (Fukuda-Parr and Messineo 2012). Although attaining human security adopts multidimensional approaches, disasters could put environmental and food security dimensions as the most critical. These two dimensions of human security could overpower other commonly feared security aspects, such as health security (Bambals 2015). A mega-disaster, as in the case of the 2004 Indian Ocean tsunami, proves that the lives of a million people could dramatically change in minutes.

Coastal communities around the region were severely affected in the first decade after the disaster struck. A number of coastal villages were demolished immediately after the tsunami or temporarily relocated farther inland. The demographic changes, followed by land-use changes, are seen in several places during the recovery process. The relocation of affected communities to a safer area met difficulties in Aceh and Sri Lanka (Løvholt et al. 2014; Cheng 2007). Return migration was also observed in a number of affected regions. In Aceh, Indonesia, the community resettlement process did not fully follow the initial plan. A master plan for tsunami reconstruction and rehabilitation was released by the Indonesian Development and Planning Agency (BAPPENAS) as stipulated in Presidential Decree no. 30 in 2005 (BAPPENAS 2005; Matsumaru et al. 2012). It was intended to make the coastal area within 500 m from the coastline free from any houses or buildings; it would be replaced by coastal vegetation as a tsunami mitigation measure (referred to as a green belt area). However, the affected coastal community challenged the plan and people requested that they be allowed to settle in their original area. The Tsunami Rehabilitation and Reconstruction Agency for Aceh and Nias (BRR Aceh Nias), which was established in April 2005 by the government of Indonesia, was not able to fully implement the master plan. In early 2015, another study revealed that the return migration in Banda Aceh has given rise to tsunami risks in the city due to lack of consideration of the hazards faced by the coastal community when deciding where to live (Syamsidik et al. 2017).

In Sri Lanka, similar problems were observed because of the lack of proper land-use planning with tsunami mitigation measures integrated into it.

One of the objectives of introducing the concept of human security is to guarantee human quality of life. As in the case of the 2004 Indian Ocean tsunami-affected area, the challenge lies on ensuring that the tsunami recovery process could prevent the return of tsunami risks. Compared with the 2011 Great East Japan Earthquake and Tsunami recovery process,.the approach taken in developing countries was different. The basic livelihood needs of the survivors dominate the process, especially in the early stages. Often, the fulfillment of the people's needs got more attention from reconstruction agencies/institutions and preventing future tsunami risks became secondary.

In the context of the 2004 Indian Ocean tsunami, there are more complications in Aceh and Sri Lanka as the two regions also face security problems brought about by internal conflicts. In Aceh, the Free Aceh Movement (GAM) has been a major security problem to the Indonesian Government as it started to rise up in arms in 1976. On the other hand, the Liberation Tigers of Tamil Eelam Movement (LTTE), which operated in the eastern part of Sri Lanka, also posed the same security concerns. Interestingly, the LTTE was also founded in 1976, the same year the GAM was established in Aceh.

Figure 15.1 and Fig. 15.2 show the casualty distribution in Aceh and Sri Lanka,

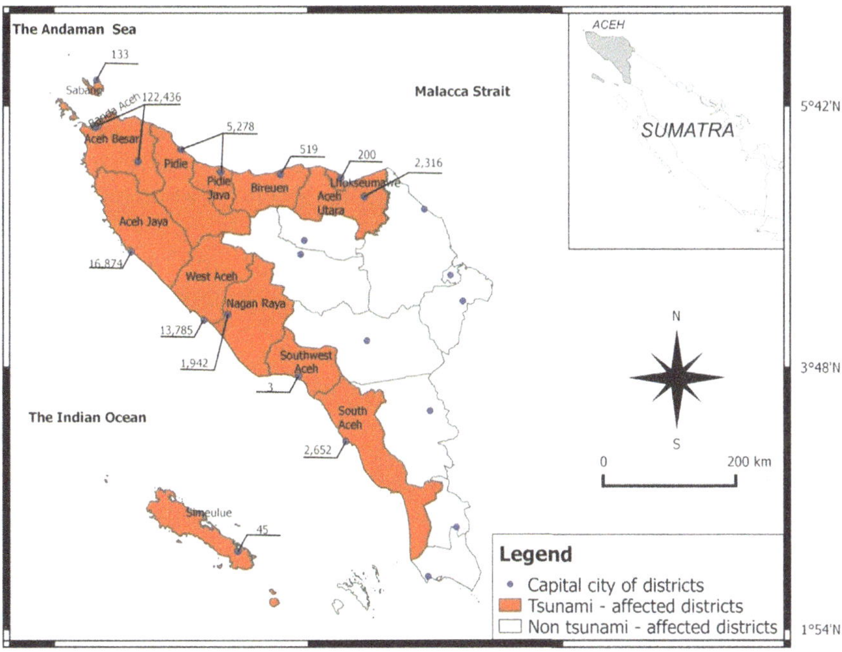

Fig. 15.1 Distribution of 2004 tsunami victims in Aceh, Indonesia. Numbers indicate casualties reported in the district (modified from BRR Aceh-Nias 2009a)

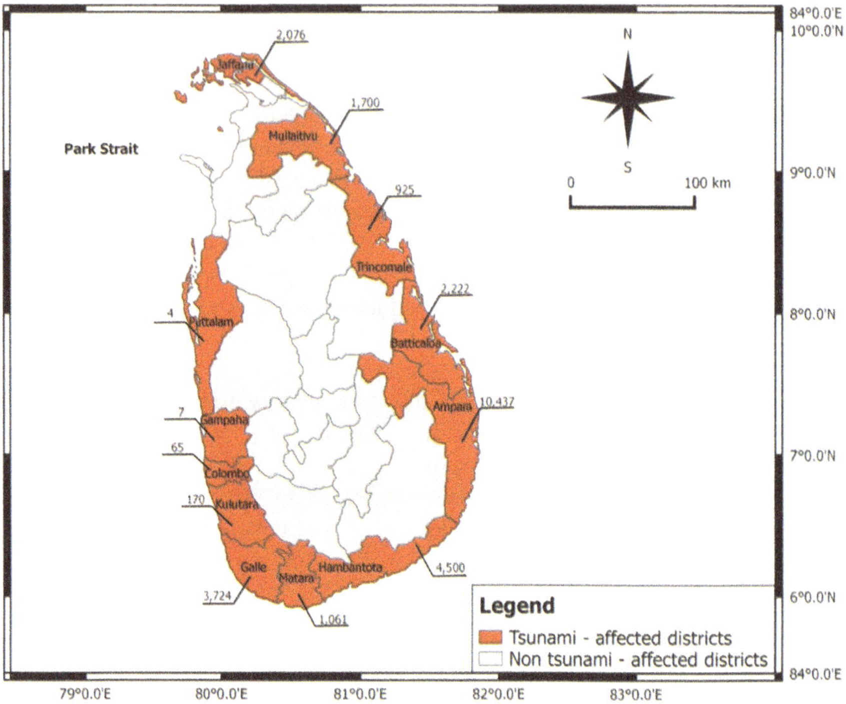

Fig. 15.2 Distribution of casualties in the wake of the 2004 Indian Ocean tsunami in Sri Lanka (re-digitized from UNHCR 2004)

respectively. In Aceh, human casualties were mostly found in Banda Aceh, which is the capital city of Aceh Province. In Sri Lanka, most of the dead were reported to have come from the southern part of its main island.

This chapter discusses the tsunami recovery process as part of human security development in tsunami-affected regions. Two main questions are addressed: (1) have disaster managers in tsunami-affected areas identified ways to reduce vulnerability? and (2) are there any mitigation measures instituted in the area to protect lives from future tsunami occurrences? Selected as study areas were the Aceh Province of Indonesia and Hambantota District of Sri Lanka. The two countries offer interesting aspects from the perspective of human security. First, the countries face challenges in responding to the tsunami survivors' demands and in providing sufficient tsunami mitigation measures. Second, the two countries have problems with security due to internal conflicts. Third, in recent years, there is return migration in tsunami-affected regions that should be fully managed. The lack of proper tsunami mitigation measures may expose the region once again to a catastrophic situation. This chapter could offer useful insights to other tsunami-prone countries.

15.2 The 2004 Indian Ocean Tsunami and Its Recovery Context

Aceh is situated in the northern part of Sumatra Island. In 2017, it has a population of about 4.2 million people. This region is prone to a number of geological hazards such as tsunamis and earthquakes. Two main actively tectonic sources are situated across Sumatra Island (the Sumatran fault system) and the Indo-Australian subduction zone. These frequently generate earthquakes. In the recent century, earthquakes in the subduction zone had generated several tsunami events such as those in 2004 and 1907. The two tsunamis were classified as near-field tsunami, where arrival time of the first tsunami wave was less than 1 h after the earthquake. The short arrival time gave little time to the coastal community to properly evacuate from the estimated tsunami inundation area (Syamsidik et al. 2015).

Meanwhile, Sri Lanka is situated rather far from the collision zones of the earth's major plates. Small inland seismic activities were recorded in Sri Lanka, making the country relatively free from major earthquake events (Dissanayake 2005).

In Sri Lanka, the tsunami rehabilitation and reconstruction process was led by TAFREN-CNO (Task Force to Rebuild the Nation-Center for National Operation). It was one of three task forces established by the government of Sri Lanka to specifically find solutions to problems brought about by the 2004 tsunami. The other two were TAFRER (Task Force for Rescue and Relief) and TAFLOL (Task Force for Logistic and Law and Order). In November 2005, RADA (Reconstruction and Development Agency) was created. The mission of RADA is to accelerate and coordinate reconstruction and development in tsunami-affected areas. It also aims to ensure a sustainable recovery process to mitigate the effects of similar disasters in the future. Considering the mission of the agency, RADA is seen as a permanent organization. In the case of Indonesia, a similar organization at the national level only came about in 2007 with the creation of the Indonesian National Disaster Management Agency or Badan Nasional Penanggulangan Bencana (BNPB).

15.3 Coastal Demographic Changes

Changes in demographic data in the tsunami-affected regions could reflect the progress of the recovery process. The repopulation or migration process was driven by the way resettlement efforts were accomplished. In this regard, locations of reconstructed houses play a significant role in influencing the demographic change. As explained earlier, the resettlement process did not follow the prescribed scenario under the reconstruction process proposed by both governments of Indonesia and Sri Lanka.

In Indonesia, reconstructed houses for tsunami survivors numbered 140,304 (BRR 2009a). The number of relocated coastal villages was smaller compared with that of non-relocated settlements. Similar conditions were also seen in Sri Lanka.

Following the reconstruction process, the government of Sri Lanka created a buffer zone, not allowing any house to be built within 100 m of the southern coast of Sri Lanka and 200 m from the coastline east and northeast of the country (Ingram et al. 2006). However, this plan failed because of lack of community consultation (Shanmugaratnam 2005). This section presents two cases (Banda Aceh and Hambantota) to give a more detailed picture of demographic change in tsunami-affected cities.

15.3.1 Banda Aceh

Banda Aceh is divided into nine *kecamatan* (subdistricts), four of which are categorized as coastal. The major fishery and marine-driven economy sectors are situated around Meuraxa, Kutaraja, Kuta Alam, and Syiah Kuala. The 2004 tsunami resulted in a large number of human casualties from these kecamatans (BRR 2009b). In the last 12 years, massive reconstruction projects have been implemented. With newly reconstructed houses, new harbor facilities, and other types of buildings, people were attracted to reside in the four kecamatans.

Figure 15.3 shows the changes in demographic data of the four kecamatans. Meuraxa and Kutaraja were the most severely damaged.

A drastic decrease of the coastal population may be seen in kecamatans Meuraxa and Kutaraja. Although kecamatans Kuta Alam and Syiah Kuala are situated near the coastal area, most of the population resided farther inland. The population in Meuraxa and Kutaraja has increased significantly since 2008 as house reconstruction was completed.

The percentage of the coastal population in Banda Aceh can be seen in Fig. 15.4.

The 2004 tsunami caused a drop in coastal population to below 40%. However, in recent years, the percentage has been stable to around 47%. It is lower than the population before the 2004 tsunami occurred, which was around 54%.

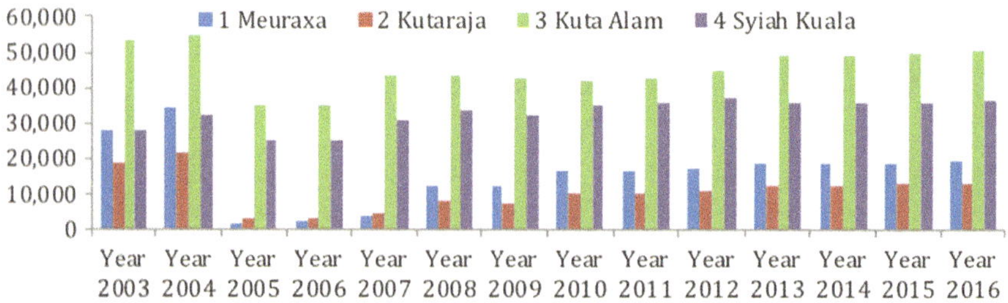

Fig. 15.3 Demographic changes in the coastal subdistricts of Banda Aceh from 2003 to 2016

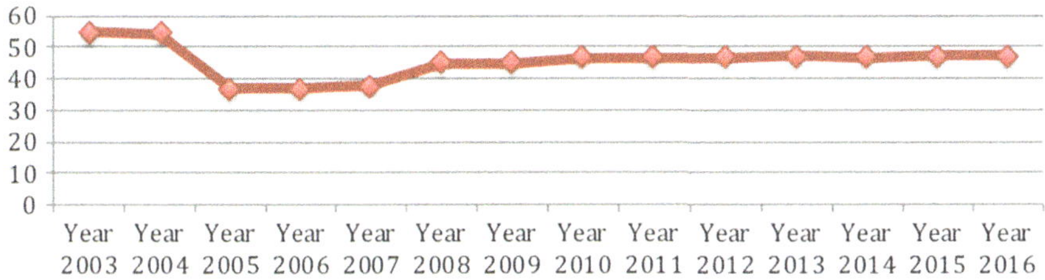

Fig. 15.4 Percentage of coastal population in Banda Aceh from 2003 to 2016

15.3.2 Hambantota—Sri Lanka

The population of Hambantota District was about 525,000 in 2004 before the tsunami. The number of casualties from this city was 1,507 (around 17%); 17,723 people were displaced; and about 20% of its coastal population was affected by the tsunami (ADB 2005).

The Hambantota Divisional Secretariat (DS), which is the capital city of the district, had a population of 46,757 people in 2001. Four of the DS were affected by the 2004 tsunami: Hambantota DS, Tangalle DS, Ambalantota DS, and Tissamaharama DS. About 54% of the population decided to seek shelters within Hambantota DS, while the other 28.9% evacuated to other areas outside Hambantota.

Figure 15.5 to Fig. 15.7 show the settlement profiles in Hambantota before the tsunami (2003), 3 months after the tsunami (March 2005), and in 2017 (Fig. 15.6).

These images were those of Panawala Village in Hambantota DS. The village has a lagoon (Mawalla Lagoon) that was used as a fishing ground for communities around the area. The figures illustrate the impact of the 2004 tsunami where most of the houses between the lagoon and the beach area were demolished. Later, in 2017, houses dominate the coastal area, although the area around the inlet of the lagoon remains deserted.

15.4 Livelihood Recovery Process

Livelihood recovery in tsunami-affected areas became one of the most serious challenges after the 2004 Indian Ocean tsunami. In Aceh, recovery efforts were led by BRR Aceh-Nias, although direct interventions were made by many parties, including the Indonesian government. Agriculture, fishery, and small-medium enterprises became the focus of the livelihood recovery process. One immediate measure done in Aceh was to give some financial assistance to tsunami survivors (*Jatah Hidup/Jadup*). A short-term program was also implemented in Aceh through the Cash-For-Work (CFW) Program. Under this program, affected community members were given small honoraria (between IDR 50,000 and IDR 100,000) for cleaning up their own houses or the area around their houses. The program ran for 3–6 months in

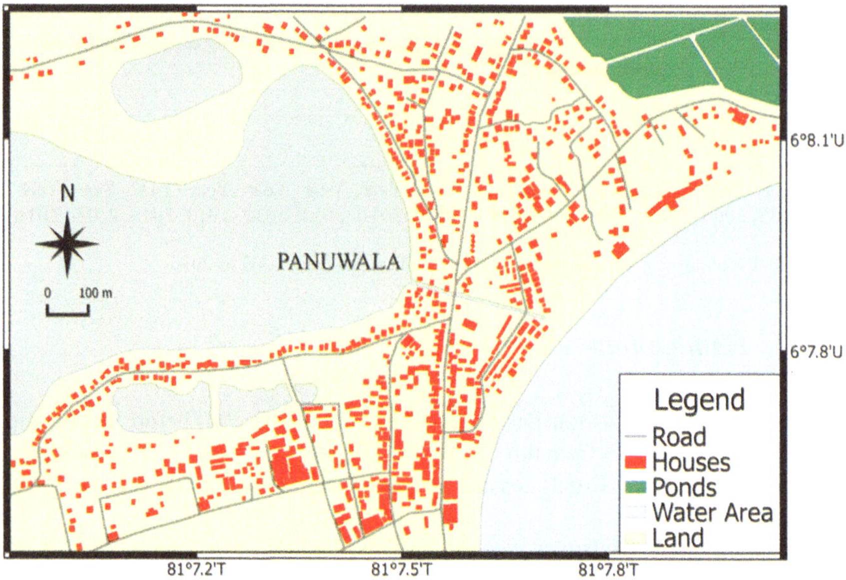

Fig. 15.5 Settlement condition in Hambantota DS before the 2004 Indian Ocean tsunami (digitized based on 2003 image)

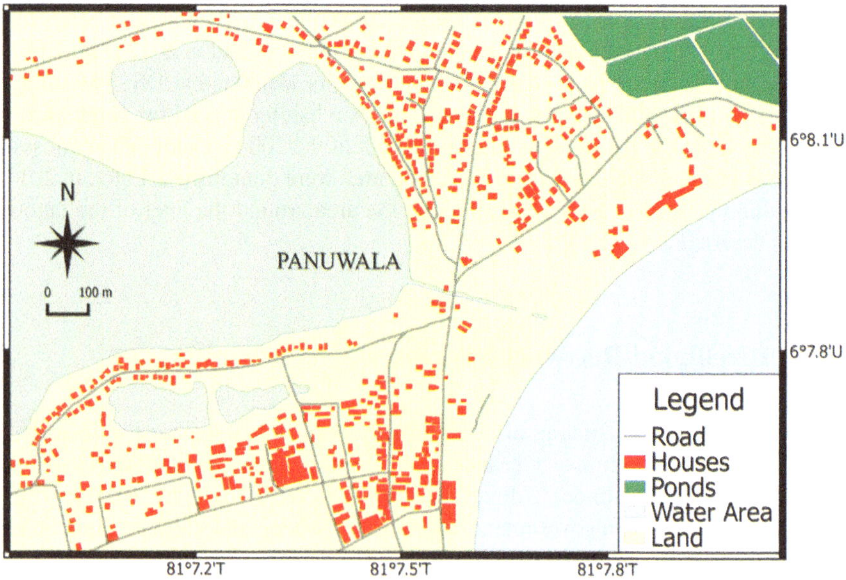

Fig. 15.6 Settlement condition in Hambantota DS after the 2004 Indian Ocean tsunami (digitized based on 2005 image)

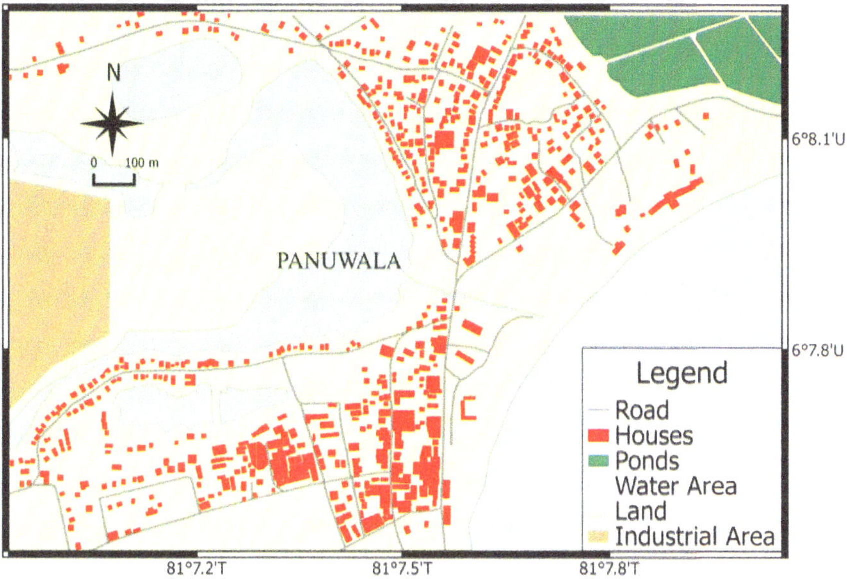

Fig. 15.7 Settlement condition in Hambantota DS in 2017, 13 years after the 2004 Indian Ocean tsunami

the first year of the recovery process in 2005. The CFW Program was also extensively used to reactivate ponds and rehabilitate small public facilities in the affected region (Adam 2006, Anand and Newport 2005).

In Sri Lanka, a similar recovery process was also done, coordinated by TAFREN. The International Labor Organization (ILO) played a significant role in Sri Lanka's livelihood recovery process. Initially, TAFREN ran the Rapid Income Recovery Program. Later, the program name was changed to Income Recovery Program (IRP). Figure 15.8 shows this scheme as proposed by ILO to TAFREN.

In the first stage of the recovery process, the government of Sri Lanka focused on food relief and cash grant to the affected communities. This is similar to what was done in Aceh.

However, criticisms of the program surfaced in both countries for the following reasons:

a. Inadequate targeting of marginalized communities as in the case of Sri Lanka (Parakrama et al. 2006);
b. Erosion of social capital in affected communities as in the case of Aceh (Syukrizal et al. 2009);
c. Creation of dependency among recipients (Kutty 2018; Doocy et al. 2006); and
d. Drastic cost inflation at the local level (Thornburn 2009; Alexander et al. 2006).

Conversely, the livelihood recovery process after the 2004 tsunami gained some positive impacts:

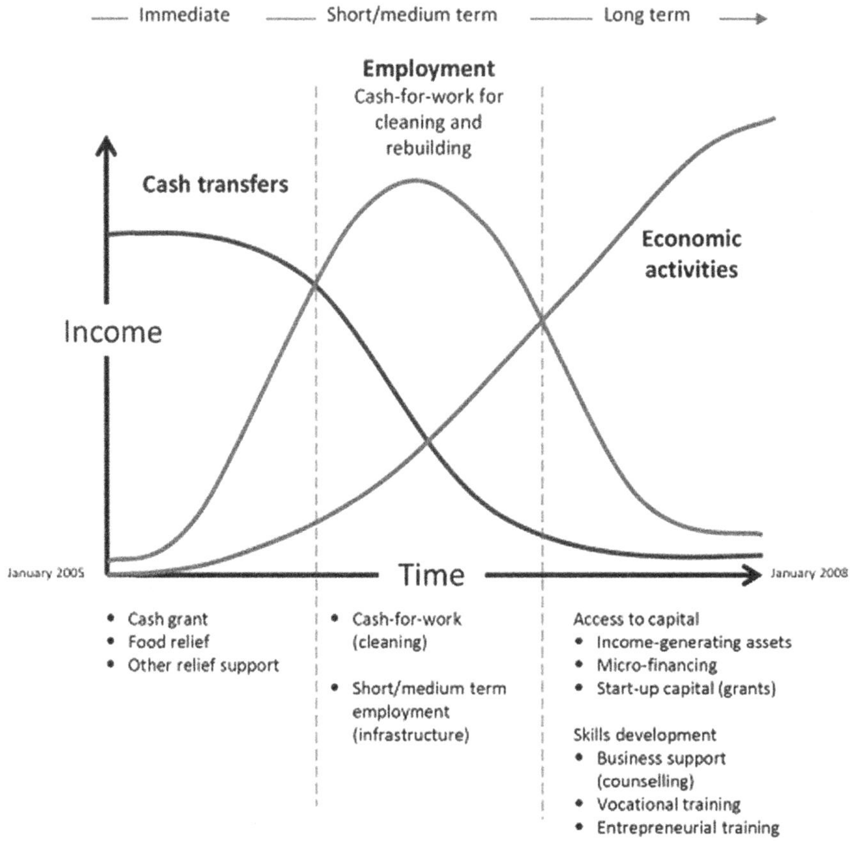

Fig. 15.8 Schematic process of the Income Recovery Program done in Sri Lanka (Kapadia 2014)

a. Providing early income source for affected communities after the tsunami (Doocy et al. 2006; Thornburn 2009);
b. Enabling communities to recover fast by creating economic security of survivors (SCF 2005; Kendrick 2005);
c. Supporting efforts to mobilize local human resources by cleaning up their homes and villages (Samuels 2010).

As the recovery process involved multisectoral development, it was difficult to comprehensively interrelate the sectors. In Sri Lanka, community recovery programs were conducted without considering the integration of cross-sectoral and cross-level issues. Furthermore, community participation in the process was found to be minimum (Wiek et al. 2010). In Aceh, the tsunami recovery process has missed the opportunity to include a proper mitigation approach in land-use planning. Interestingly, the community economic revitalization was found to be successful after more than a decade of the recovery process (Meilianda et al. 2017). In this regard, Aceh has demonstrated some positive aspects in the framework of human security, particularly in ensuring economic security after the impact of the tsunami. However, similar achievements were not found in the case of Sri Lanka.

15.5 Tsunami Risk Reduction

Due to the complexity of the recovery process in Aceh and Sri Lanka, both governments should provide alternative ways to reduce tsunami risk. Relocating the affected communities farther inland may not be the best solution in both areas. Most of the families in the affected coastal area were those of fishermen and their income was strongly related to coastal resources (Griffin et al. 2013). Moving them away to the hinterland could force them to change their job, not an easy task for most people (Syamsidik et al. 2018). Therefore, alternative solutions must be provided.

In recent years, tsunami risk reduction efforts in Indonesia gave some positive outcomes. The establishment of BNPB has resulted in significant resource mobilization (in terms of financing, human resources, regulations, and facilities). At the provincial level, the impacts can also be seen–Aceh established its provincial disaster management agency, the Aceh Disaster Management Agency (BPBA), in 2010. A set of regulations was set up. The Aceh governor ordered all districts to allocate at least 1% of their annual budget to activities geared toward disaster risk reduction. A number of "escape" buildings were also constructed in Banda Aceh. Until 2017, there are six formally assigned escape buildings in Banda Aceh and one each in districts Aceh Besar, Aceh Jaya, West Aceh, and Bireuen. As these cannot accommodate all people in the tsunami-prone areas, alternative private and public buildings were set for the same purpose. In addition, tsunami evacuation signs have been sufficiently constructed to mark evacuation routes in the area. At the national level, in 2013, the government of Indonesia formulated the National Tsunami Mitigation Master Plan. It aims to develop a comprehensive system, the infrastructure, and the country's capacity to mitigate the impacts of future tsunami events. The plan was the consequence of the April 2012 twin earthquakes that revealed the readiness level of the country in anticipating tsunami occurrences. The proposed activities include strengthening the tsunami early warning system.

In Sri Lanka, tsunami risk reduction is mainly done by the Disaster Management Center (DMC), which was established in 2005. DMC later created the National Council for Disaster Management (NCDM), the main organization responsible for tsunami mitigation (Ratnasooriya et al. 2007). As previously mentioned, Sri Lanka does not have any near-field tsunami source. Therefore, the country depends on a tsunami early warning system that is managed for far-field tsunami sources such as those around the Andaman Sea. In this regard, communication with the Indian Ocean Tsunami Early Warning System (InaTEWS) is important. In several major earthquake cases, the country relied on the information provided by the government of Indonesia through its Geophysics, Climate, and Meteorological Agency (BMKG). To increase community awareness about tsunamis, formal, non-formal, and informal education activities are suggested (Oktari et al. 2015; Morin et al 2008). In the formal education system, disaster-related topics have been included in existing subjects at the elementary and secondary levels in Aceh. Furthermore, a compulsory subject on disaster management has been introduced in Syiah Kuala University, which is the largest university in Aceh. The informal education activities included disseminating tsunami awareness materials such as those displayed in some places in Banda Aceh as tsunami heritage (Miller and Bunnell 2011). Though Sri Lanka does not preserve as much tsunami heritage, activities related to disaster education and creation of a safety culture are in progress (Suppasri et al. 2015).

15.6 Conclusions

This chapter specifically addressed issues in the process of recovery from the 2004 Indian Ocean tsunami in Aceh-Indonesia and Sri Lanka from the perspectives of human security. Both areas showed similarities in terms of maintaining a balance between giving into the demands of the affected communities and protecting them from future tsunami disasters. In addition, both faced lengthy internal conflicts that required specific attention as the tsunami recovery process is conducted. The resettlement process that involved relocating the community to the hinterland faced difficulties as survivors were not able to adapt to new jobs and the new area. This resulted in return migration and the increase in coastal population several years after the tsunami. To mitigate the impacts of future tsunami events, the government must provide facilities such as escape buildings and escape routes. In addition, disaster education must be pursued and a safety culture within the communities be established. In general, the recovery process in both affected areas has not sufficiently incorporated the human security aspects, particularly environmental security. The increasing coastal population has put back the tsunami risks of the area to the same degree as before the 2004 Indian Ocean tsunami. The governments of Indonesia and Sri Lanka face the serious task of coming up with a disaster management plan that would effectively prevent a repeat of the 2004 disaster.

Acknowledgements The authors would like to thank the Partnership Enhanced Engagement in Research (PEER) for financial support (USAID Grant No. 5-395) of the project titled "Incorporating Climate Change Induced Sea Level Rise Information into Coastal Cities' Preparedness towards Coastal Hazards" under Federal Award Identification Number (FAIN): AID-OAA-A- 11-00012 (subaward no. 2000007546).

References

Adams L (2006) Learning from cash response to the tsunami. Humanitarian Policy Group, London, England

ADB (Asian Development Bank) (2005) Asian tsunami impact: preliminary damage and needs assessment for Sri Lanka. ADB, Manila

Alexander B, Chan-Halbrendt C, Salim W (2006) Sustainable livelihood considerations for disaster risk management: implications for implementation of the Government of Indonesia tsunami recovery plan. Disaster Prev Manage 15(1):31–50

Anand KTS, Newport J (2005) Role of microfinance in disaster mitigation. Disaster Prev Manage 14(2):176–182

Bambals R (2015) Human security: an analytical tool for disaster perception research. Disaster Prev Manage 24(2):150–165

BAPPENAS (Indonesia National Development Planning Agency) (2005) Master plan of Aceh-Nias rehabilitation and reconstruction: the main book. BAPPENAS, Jakarta, Indonesia

BRR (Aceh-Nias Tsunami Reconstruction and Rehabilitation Agency) (2009a) Housing: roofing the pillars of hope. BRR book series. BRR Aceh-Nias, Banda Aceh-Indonesia

BRR (Aceh-Nias Tsunami Reconstruction and Rehabilitation Agency) (2009b) A map of tsunami victims due to earthquake and tsunami in Aceh and Nias in Bahasa Indonesia. http://atdr.unsyiah.ac.id:8080/jspui/bitstream/123456789/7233/1/20090100_Map_Korban_Meninggal_Tsunami_Gempa_Bumi_NAD_Nias.pdf. Accessed 3 Feb 2018

Busumtwi-sam J (2008) Contextualizing human security: a 'deprivation-vulnerability' approach. Policy Soc 27(1):15–28

Cheng MH (2007) Health and housing after the Indian Ocean tsunami. Spec Rep Lancet 369(9579):2066–2068. https://doi.org/10.1016/S0140-6736(07)60961-1

Dissanayake CB (2005) A new plate boundary near Sri Lanka: implications for future geohazards. J Nat Sci Foundation Sri Lanka 33(1):5–8

Doocy S, Gabriel M, Collins S, Robinson C, Stevenson P (2006) Implementing cash for work programmes in post-tsunami Aceh: experiences and lessons learned. Disasters 30(3):277–296

Fukuda-Parr S, Messineo C (2012) Human security: a critical review of literature. CRPD Working Paper No. 11. Katholieke Universiteit Leuven, Leuven

Griffin C, Ellis D, Beavis S, Zoleta-Nantes D (2013) Coastal resources, livelihoods and the 2004 Indian Ocean tsunami in Aceh. Indonesia Ocean Coast Manage 71:176–186

Ingram JC, Franco G, Rio CR, Khazai B (2006) Post disaster recovery dilemmas: challenges in balancing short-term and long-term needs for vulnerability reduction. Environ Sci Policy 9(7–8):607–613

Kapadia K (2014) Sri Lankan livelihoods after the tsunami: searching for entrepreneurs, unveiling relation to power. Disasters 39(I):23–50

Kendrick A (2005) Restoring coastal livelihoods in tsunami-affected areas of Aceh: a needs assessment on Aceh's eastern coast. Save the Children, Toronto, Canada

Kutty NFM (2018) The role of NGOs in post-tsunami reconstruction in Aceh, Indonesia. Doctoral thesis (draft), University of New South Wales, Australia

Løvholt F, Setiadi NJ, Birkman J et al (2014) Tsunami risk reduction—are we better prepared today than in 2004? Int J Disaster Risk Reduct 10:127–142. https://doi.org/10.1016/j.ijdrr.2014.07.008

Matsumaru R, Nagami K, Takeya K (2012) Reconstruction of the Aceh region following the 2004 Indian Ocean tsunami disaster: a transportation perspective. IATSS Res 36:11–19

Meilianda E, Safrida Direzkia Y (2017) Linking global context of sustainable recovery with a local context post-tsunami recovery at Banda Aceh city. Indonesia Int J Disaster Manage 1(1):20–34

Miller MA, Bunnell T (2011) Post-disaster urban renewal: memories of trauma and transformation in an Indonesian City. Asian Research Institute, Working Paper Series No. 154. National University of Singapore, Singapore

Morin J, Coster BD, Paris R, Lavigne F, Flohic F, Floch DL (2008) Tsunami-resilient communities' development in Indonesia through educative actions: lessons from the 26 December 2004 tsunami. Disaster Prevention Manage 17(3):430–446
Oktari RS, Shiwaku K, Munadi K, Syamsidik Shaw R (2015) A conceptual model of a school-community collaborative network in enhancing coastal community resilience in Banda Aceh. Indonesia Int J Disaster Risk Reduct 12:300–310
Parakrama A, Scheper E, Gunawardena S (2006) Impact of the tsunami response on local and national capacities: Sri Lanka country report. Tsunami Evaluation Coalition, London
Pickrell J (2005) Facts and figures: Asian tsunami disaster. https://www.newscientist.com/article/dn9931-facts-and-figures-asian-tsunami-disaster/. Accessed 10 Jan 2018
Ratnasooriya HAR, Samarawickrama SP, Imamura F (2007) Post-tsunami recovery process in Sri Lanka. J Nat Disaster Sci 29(1):21–28
Samuels A (2010) Remaking neighbourhoods in Banda Aceh: post-tsunami reconstruction of every day life. A disertation. Leiden University, Leiden-Netherlands
SCF (Save the Children Fund) (2005) Livelihood assessment in NE Aceh, Aceh Province. SCF, Jakarta
Shanmugaratnam N (2005) Tsunami victim's perceptions of the proposed buffer zone and its implications in Eastern Sri Lanka. Eldis in affiliation with the Department of International Environment and Development Studies, Norwegian University of Life Science. http://sacw.insaf.net/free/Sri LankaTsunami_Reflections.pdf. Accessed 12 June 2017
Suppasri A, Goto K, Muhari A, Ranasinghe P, Riyaz M, Affan M, Mas E, Yasuda M, Imamura F (2015) A decade after the 2004 Indian Ocean tsunami: the progress in disaster preparedness and future challenges in Indonesia, Sri Lanka, Thailand, and Maldives. Pure appl Geophys 172:3313–3341
Syamsidik Fahmi M, Fatimah E, Fitrayansya A (2018) Coastal land use changes around the Ulee Lheue Bay of Aceh during the 10-year 2004 Indian Ocean tsunami recovery process. Int J Disaster Risk Reduct 29:24–36. https://doi.org/10.1016/j.ijdrr.2017.07.014
Syamsidik Oktari RS, Munadi K, Arief S, Fajri IZ (2017) Changes in coastal land use and reasons for selecting places to live in Banda Aceh 10 years after the 2004 Indian Ocean Tsunami: Nat. Haz 88(3):1503–1521
Syamsidik Rasyif TM, Kato S (2015) Development of accurate tsunami estimated times of arrival for tsunami-prone cities in Aceh. Indonesia Int J Disaster Risk Reduct 14(4):403–410
Syukrizal A, Hafidz W, Sauter G (2009) Reconstructing live after the tsunami: the work of Uplink Banda Aceh in Indonesia. IIED, London
Thornburn C (2009) Livelihood recovery in the wake of tsunami in Aceh. J Bull? Indonesian Econ Stud 45(1):85–105
UNHCR (United Nations High Commissioner for Refugees) (2004) Divisions Affected by Tsunamis. https://reliefweb.int/sites/reliefweb.int/files/resources/875D7DA12DAFBEE2C1256F80005B7642-unhcr_srilanka_tsunami_affected301204.pdf. Accessed 1 Dec 2017
Wiek A, Ries R, Thabrew L, Brundiers K, Wickramasinghe A (2010) Challenges of sustainable recovery processes in tsunami-affected communities. Disaster Prevention Manage 19(4):423–437

Dr. Syamsidik is a lecturer at the Civil Engineering Department, Universitas Syiah Kuala in Banda Aceh, Indonesia. He is also Coordinator for Tsunami Research Cluster at Tsunami and Disaster Mitigation Research Center (TDMRC) at Universitas Syiah Kuala. He obtained his doctoral degree from Toyohashi University of Technology in Japan in March 2009, majoring in Coastal Engineering. Since then, he has been developing his research interests on tsunami mitigation, tsunami numerical simulations, disaster preparedness, and sustainable disaster recovery. He was a visiting scholar at the International Research Institute of Disaster Science (IRIDeS) in Tohoku University, Japan in 2018 and in 2019. He also served as a visiting scholar at the University of Pittsburgh and at Georgia Technology Institute in the USA in 2017 and 2018, respectively.

Dr. Agus Nugroho is a lecturer at the Agribusiness Department, Universitas Syiah Kuala in Banda Aceh, Indonesia. He is also Coordinator for Socio Economic of Disaster Research Cluster at Tsunami and Disaster Mitigation Research Center (TDMRC) at Universitas Syiah Kuala. He obtained his doctoral degree from Kyoto University, Japan in May 2016, majoring in Natural Resource Economics. His research interests are on social and economic aspects of disaster, natural resource risk management, and disaster risk financing and insurance (DRFI). He was also involved as a panelist in developing of DFRI Strategies in Indonesia established by the Ministry of Finance, Republic of Indonesia.

Mirza Fahmi is a lecturer at Civil Engineering Department of Politeknik Negeri Lhokseumawe, Indonesia. He obtained his master's degree from Universitas Syiah Kuala in 2017, majoring in Water Resource Management. Recently, he has been developing his research interests on tsunami mitigation, spatial analysis, and coastal zone management. Besides his academic activities, he is also actively involved in promoting coastal ecology and conservation through community-based activities.

Chapter 16
Regional Policies and Initiatives on Climate Change and Disaster Risks: How Can Peacebuilding Assistance and Climate Change Adaptation Be Integrated?

Mikio Ishiwatari

Abstract Climate-fragility risks are regarded as "threat multipliers" that undermine the security of humans and the state. This chapter considers various development assistance approaches in managing climate-fragility risks. While development assistance agencies are supporting developing countries to strengthen efforts in climate change adaptation and peacebuilding, existing approaches in these two areas are not connected enough. The chapter examines gaps in the assistance approaches of managing climate-fragility risks: risk assessment, program time scale, and integration of approaches. Climate risk information is currently not used for peacebuilding programs, while risk assessment with respect to water availability and disasters under a changing climate would be useful in preparing projects under unstable regional situations. Peacebuilding programs focus on immediate needs such as rehabilitating the infrastructure destroyed, improving the livelihoods of local communities, and creating government capacities. Climate change adaptation programs aim at resolving issues from a long-term perspective, such as flood risk management and capacity building. Incorporating a long-term perspective into peacebuilding programs is a challenge. The chapter proposes practical approaches to fill these gaps in order to merge assistance approaches of peacebuilding with climate change adaptation.

Keywords Climate-fragility risks · Climate change adaptation · Peacebuilding · Human security · Development assistance

16.1 Introduction

Climate-fragility risks are regarded as "threat multipliers" that catalyze water and food scarcity, pandemics, and displacement, which undermine human and state security (EU 2016). The 2015 G7 Foreign Ministers' Meeting, which was held in Hiroshima, recognized that climate change poses serious threats to the stability of

M. Ishiwatari (✉)
Graduate School of Frontier Sciences, the University of Tokyo, Tokyo, Japan
e-mail: ishiwatari.mikio@jica.go.jp

J. M. Pulhin et al. (eds.), *Climate Change, Disaster Risks, and Human Security*, Disaster Risk Reduction, https://doi.org/10.1007/978-981-15-8852-5_16

states and societies and has significant implications on international security (MFA 2017b). The G7 countries commissioned an independent report, which warned that states and local communities could be seriously impacted by climate-fragility risks (Rüttinger et al. 2015).

The Asia Pacific region, where the most dynamic economies in the world are currently found, needs to manage climate-fragility risks to ensure continued economic growth and poverty reduction. The region has experienced some 8% economic growth annually from the 1980s to the mid-2000s, except in 1998 with the economic crisis, and has achieved 6% growth in 2015 (Kuroda 2017). The number of people living in extreme poverty decreased by more than a billion from 1990 to 2012 (ADB 2016).

The region is facing serious water issues that would exacerbate with climate change. Nearly 90% of all water-related disasters in the world happen in this region (ADRC 2015). Annual disaster damage is estimated at USD 53 billion and the annual death toll reaches 16,000 (ADB 2016). Countries in the region need to invest to mitigate damages. Annual flood protection expenses are estimated at USD 14.76 billion. In addition, climate change adaptation costs are calculated at USD 1.74–3.21 billion in East and South Asia and Pacific regions (Ward et al. 2010). Fant et al. (2016) predict that 1 billion additional people will live in Asia under water stress by 2050 because of socioeconomic growth and climate change.

The Asia Pacific region that hosted some 3.5 million refugees, 2.7 million internally displaced persons, and 1.6 million stateless people needs to cope with human security issues. Most displaced people were refugees from Afghanistan and Myanmar (UNHCR 2016). The Ministry of Foreign Affairs of Japan (2017a) stresses that climate change policies should include urbanization and socioeconomic issues, as well as ecosystem conservation in the region to mitigate human security risks caused by climate change.

Development assistance agencies do not fully recognize the complexity of the risks. The agencies support developing countries to strengthen efforts in climate change adaptation and peacebuilding, but existing approaches of these agencies to these areas are not adequately connected. Currently, concerned organizations are assessing climate and fragility risks separately using different approaches.

This chapter proposes several ways of giving development assistance to manage climate-fragility risks. It examines gaps and tackles risk assessment, program time scale, and integration of approaches.

16.2 Relationship Between Climate and Fragility Risks

Climate change poses risks to human security through various processes, such as reducing access to natural resources, undermining livelihoods, compromising culture and identity, and increasing migration (Adger et al. 2014; Barnett and Adger 2007; Nett and Ruttinger 2016). Damaged livelihoods and increasing inequality among ethnic and religious groups are sensitive climate change issues and increase the risk of violent conflicts. Climate change and conflicts create complex risks that lead to the emergence of non-state armed groups, such as the Islamic State and Boko Haram (Nett and Ruttinger 2016).

Fragile and conflict-affected countries are vulnerable to natural disasters and have limited capacity to manage disasters. More than 750,000 people were displaced by natural disasters in 23 fragile and conflict-affected countries in 2014 (IDMC 2015). Persons already affected by conflicts are often exposed to displacement because of disasters and the changing climate. The majority of the 59.5 million people of concern to the United Nations High Commissioner for Refugees (UNHCR)-refugees, internally displaced persons, and affected others-are situated in climate change hotspots (UNHCR 2015).

Food insecurity in poverty areas increases the likelihood and intensity of armed conflicts and refugee outflows. For example, resource scarcity exacerbated by the worsening weather conditions is regarded as a multiplier of conflicts in the East and Horn of Africa, where armed conflicts have continued. Changing rainfall patterns worsened access to food, water, and other resources. This led to a massive influx of Somalis into neighboring countries in 2011 and 2012 (Afifi et al. 2012; UNHCR 2017).

Floods have induced migrations from Bangladesh to India, and population displacement has become a thorny issue in bilateral relations (Martin et al. 2017; Quencez 2011). Migration can be a coping strategy for households affected by high tide and flooding in Bangladesh (Foresight 2011). More than 4.7 million people were newly displaced by water-related disasters between 2008 and 2014 (IDMC 2015). Eight to nine percent of total displaced people are estimated to be dislodged across international borders (Displacement Solutions 2012).

Bangladesh is a densely populated country and it frequently experiences natural disasters such as cyclones, high tides, and floods. Climate change would exacerbate the situation. An extra 9.6 million people are likely to migrate from areas affected by floods, storms, and riverbank erosion from 2011 to 2050 (Kniventon et al. 2013). The prime minister of Bangladesh states that the affected people would become "climate migrants" (UN 2014).

Development assistance agencies have not developed practical tools to assess climate-fragility risks. Mobjörk et al. (2016) suggest developing analytical tools for assessing climate-related security risks to prevent conflicts. Japanese experts propose the development of hazard maps as analytical tools for policymaking to manage climate-fragility risks (MFA 2017c). The maps integrate both climate change predictions and factors of conflicts over local resources as well as ethnic, economic, and social conflicts. Ishiwatari (2017) argues that Japanese organizations should combine the scientific-engineering approach for climate change programs and the socioeconomic approach for peacebuilding endeavors to assess climate risk and peacebuilding impacts in an integrated manner.

The Brookings Institution, Georgetown University, the Institute for the Study of International Migration, and the UNHCR formulated the "Guidance on Planned Relocations" to plan and implement relocation programs. These aim to protect people from disasters and environmental change. Furthermore, this document is useful in formulating relocation laws, policies, and plans (Brookings Institution, Georgetown University, UNHCR 2015).

16.3 Gaps Between Climate Change Adaptation and Peacebuilding Programs

This section reviews the gaps in operation approaches between climate change adaptation and peacebuilding programs by examining JICA's operations. The existing approaches of the agency in these areas are not adequately connected.

16.3.1 JICA's Approach and Operation in Climate Change and Peacebuilding

JICA recognizes that climate change poses a substantial threat to human security, stability, and prosperity across societies (JICA 2016). Consequently, the following activities and practices are given priority:

(i) promoting low-carbon, climate-resilient urban development, and infrastructure investment: JICA is providing an ODA loan to construct a mass transport system in Bangkok to facilitate a modal shift from road to rail.
(ii) enhancing comprehensive climate risk management: JICA promotes comprehensive risk management covering disasters, food, and water scarcity.
(iii) supporting climate policy and institutional development: JICA conducts a study on low-carbon technology to reduce greenhouse gas emissions in Vietnam.
(iv) enhancing conservation and management of forests and other ecosystems. JICA is supporting Indonesia to promote wildfire management and carbon management in peat-forest by introducing satellite monitoring and prediction models.

The climate change office focuses on the relationship between climate change adaptation and mitigation for projects in all sectors and recommends incorporating climate risk assessment and risk management measures as necessary.

The priority activities of JICA's operations in peacebuilding include (i) social capital, (ii) economic recovery, (iii) governance functions, and (iv) security enhancement. The operations are expected to promote reconciliation among conflict groups and to involve socially vulnerable groups (JICA 2011b). JICA gives support to capacity enhancement to achieve reconciliation and development at the community level in Afghanistan, Myanmar, the Philippines, and Nepal.

16.3.2 Gaps in JICA Operation

The major approaches of JICA operations in climate change adaptation and peacebuilding can be summarized thus:

	Climate change	Peacebuilding
Risk assessment	Physical impacts	Socioeconomic knowledge
Project activities	Technical solutions	Software measures
Time scale	Long term	Immediate response

Gaps in risk assessment. Different assessment methodologies are used in climate change adaptation and peacebuilding. JICA focuses on the physical impacts of climate change in its climate risk assessments. For example, a JICA handbook prescribes climate vulnerability assessment, which starts with hazard analysis. It recommends using the information on climate impacts predicted by the general circulation models in the IPCC assessment reports to project the occurrence of extreme events of floods and droughts (JICA 2010a). Impacts on precipitation and heavy rains due to climate change can be predicted, while the uncertainty of prediction needs to be dealt with. Based on the prediction, experts can assess risks of floods and droughts, including economic damage and frequency through the use of scientific and engineering approaches. Various simulation models of river flooding are available. Risk information is currently not used for peacebuilding programs. On the other hand, risk assessment on water availability and disasters under a changing climate would be useful in preparing for regional and local conflicts.

JICA has developed peacebuilding need and impact assessment (PNA) as a tool to manage peacebuilding projects at the planning, implementation, monitoring, and evaluation stages (Seto 2015). The agency recognizes the fact that conflict-affected countries and areas have common issues: (a) instability of security, political, and social conditions; (b) conflicts among stakeholders and communities; (c) malfunctioning of government and mistrust from communities; and (d) remaining or emerging factors that cause conflicts (JICA 2017). PNA covers the analysis of a wide range of sectors such as politics, governance, security, socioeconomics, as well as factors that stabilize and destabilize societies.

Temporal gaps. While development assistance agencies regard climate change events as having long-term impacts, they tend to focus on immediate problems when it comes to peacebuilding programs. People in conflict-affected areas face difficulties in accessing basic public services and finding jobs. Thus, peacebuilding programs involve rehabilitation of infrastructure, enhancing livelihood, or improving security; at times, they cannot afford to pay attention to climate change events that underline these problems (MFA 2017c).

Gaps in project activities. JICA programs focus on technical solutions to manage risks that could increase because of climate change. For example, flood management projects support dike construction and early warning systems. Infrastructure development needs engineering knowledge. On the other hand, peacebuilding efforts go to social infrastructure, economic recovery, and security enhancement. Peacebuilding programs often include socioeconomic activities at the community level, such as rehabilitating livelihood and building community infrastructure. Policies related to climate change and peacebuilding both include capacity building. Integrating the scientific approaches of climate change adaptation into these community activities is another challenge.

16.3.3 Integrating Assessments of Climate Risk and Peacebuilding Impacts

This subsection compares climate change vulnerability analysis (CCVA) and PNA (JICA 2011a, 2017). Table 16.1 shows that the two assessments use different approaches and cover different areas. CCVA does not typically cover security risks caused by water shortages or floods in conflict-affected areas, such as livelihood insecurity or other economic impacts. Neither does PNA cover fragility risks posed by climate change. The common areas covered by the two assessments are limited: socioeconomic situations and institutional capacity or stakeholder analysis.

Table 16.1 Comparison between climate change vulnerability analysis (CCVA) and peacebuilding needs and impact assessment (PNA)

CCVA	PNA
(i) Risk and sensitivity assessment (a) Assessing current climate risks	• Backgrounds and current situation of conflicts • Politics and security • Progress of peace process, national building, and peacebuilding • Characteristics of communities: politics, administration, security, effects of conflicts, socioeconomics • Stabilizing or destabilizing factors
Analyzing socioeconomic situations	
(b) Predicting change in climate (c) Assessing sensitivity to climate change	
Predicting changes and areas to focus on	
(ii) Capacity assessment: Assessing adaptation capacity in institutions, infrastructure, information, management, etc.	• Stakeholder analysis (a) Characteristics of stakeholders (b) Relationship between stakeholders
(iii) Vulnerability assessment: Examining local vulnerability by combining results of risk and sensitivity assessment and capacity assessment	

New integrative assessment methodologies can be developed by considering climate change impacts and fragility and conflict risks. All areas analyzed must be integrated to assess climate-fragility risks.

16.4 The Case of Mindanao in the Philippines

This section examines the process of integrating assessments of climate-fragility risks by looking at JICA assistance in Mindanao, Philippines. Natural disasters, which will be more serious with climate change, will increase fragility and fuel social unrest. Historical data show that displacement caused by large floods fueled existing conflicts, particularly in developing countries (Ghimire et al. 2015).

16.4.1 Conflict in Mindanao

The conflict between the Government of the Philippines and Islamic rebel groups has continued for more than 40 years in the Mindanao region in the southern Philippines. The recurring conflicts produced casualties of more than 100,000 people and have driven poverty rates in Mindanao to levels well above the national average. Government poverty statistics in 2015 show that the Autonomous Region in Muslim Mindanao remains mired in poverty, at 53.4% poverty rate, higher than the national average of 21.1%. Also, the conflict has deprived this region of opportunities to access social services and quality investments and jobs.

JICA has promoted development assistance in balance with the peace process in Mindanao (Ishikawa 2014). JICA specialists joined an international monitoring team to bridge peacebuilding and development efforts in areas of conflict (Ishikawa 2017). JICA launched the Japan-Bangsamoro Initiatives for Reconstruction and Development Program to support reconstruction and development activities in 2006. The Japanese Government provided 15.1 billion yen as official development assistance to the region, supporting projects to improve infrastructure, livelihood, and other public services.

16.4.2 Typhoon Disaster and Climate Change Adaptation

Recently, typhoons have affected Mindanao, a region previously considered as outside the typhoon belt. Typhoon Bopha hit conflict areas in Mindanao in December 2012, and the disaster left some 2000 people dead or missing. These areas were severely affected by a 40-year conflict, and people have been displaced by the conflict. Some 6.2 million people were affected by the disaster, and more than 1 million people lost their houses (Bamforth 2015). The typhoon drastically reduced income from the agriculture sector by 23%, and 12% of families in the area completely lost their income (WFP 2013).

JICA assessed the risks of climate change and estimated economic damage in the Tagaloan River Basin in Mindanao Island (JICA 2010b). Considering climate change impacts, the agency supported a flood management project in the river basin. Climate change effects on rainfall intensity in the Philippines were calculated by a statistical downscaling method. JICA experts analyzed a subset of the general circulation models used in the IPCC assessment reports (Ishiwatari 2011). Based on the relationship between local average temperature rise and incremental rate of storm rainfall in the Philippines, the future increments of a probable 2-day storm rainfall were estimated. Flood discharge was predicted to increase by 11–20% in 2050 and by 14–29% in 2100. The targeted safety level of 25 years in the current situation is likely to decrease to 11–17 years in 2050 and to 8–15 years in 2100. The economic damage of once-in-100-year floods was estimated at PHP 2.3 billion, including damage to buildings of PHP 870 million and damage to industry of PHP 820 million.

16.4.3 Assessment of Climate-Fragility Risks from the JICA Programs

JICA has identified various factors that cause conflicts in peacebuilding programs. Factors related to disasters are economic inequality among ethnic and religious groups and the limited capacity of the government. The poverty ratio in the conflict area is the highest in the country. Infrastructure investment and access to resources are limited because of the prolonged conflicts. The limited water access and inadequate investment in agricultural facilities made productivity of agriculture, the region's main industry, lower than that in other regions. Since the government has limited capacity, support to local communities in disaster management is not enough.

The extent of flood damage among ethnic groups and religious groups needs to be examined to evaluate the impact on society's stability. Vulnerable groups suffer the most from natural disasters, leading to instability. Often living in buildings with a weak structure in at-risk areas, people cannot spend money on preparedness activities and they cannot avail of safety net benefits provided by the government. Impacts on individual groups are not clear in the JICA climate change adaptation (CCA) program as economic damage was estimated as aggregate data.

CCA and peacebuilding programs usually include capacity assessment, but these assessments focus on different aspects. The assessments should be combined. CCA programs cover the capacity of dealing with disasters and using scientific knowledge to assess a changing climate and to respond to disasters. Peacebuilding programs cover the capacities of institutions and societies in security and community development.

16.5 Conclusion

This study examines the gaps in approaches between CCA and peacebuilding programs. It was found that such gaps need to be filled in various areas. Climate change impacts, such as increased damage from natural disasters, should be considered in peacebuilding programs. Also, fragility risks should be incorporated into CCA programs.

While the link between climate change and conflicts is recognized, practical methods of assessing climate-fragility risks are not available for practitioners of development assistance. Scientific-engineering knowledge used for CCA and the socioeconomic approach for peacebuilding should be combined to manage climate-fragility risks. Taking the case of Mindanao, Philippines, the study proposed practical methods of assessing climate-fragility risks. Development assistance agencies are expected to develop assessment methods of climate-fragility risks.

References

ADB (Asian Development Bank) (2016) Asian water development outlook 2016: strengthening water security in Asia and the Pacific. Mandaluyong City (Philippines): ADB

Adger WN, Pulhin JM, Barnett J, Dabelko GD, Hovelsrud G K, Levy M, Oswald Spring Ú, Vogel CH (2014) Human security. In: Field CB, Barros VR, Dokken DJ, Mach KJ, Mastrandrea MD, Bilir TE, Chatterjee M, Ebi KL, Estrada YO, Genova RC, Girma ES, Kissel AN, Levy S, MacCracken PR, Mastrandrea LL, White B (eds.) Climate change: Impacts, adaptation, and vulnerability. Part A: Global and sectoral aspects. Contribution of Working Group II to the Fifth Assessment Report of the Intergovernmental Panel on Climate Change. Cambridge and New York: Cambridge University Press, pp 755–791

ADRC (Asian Disaster Reduction Center) (2015) Natural disaster data book 2015: an analytical overview. Kobe (Japan): ADRC

Afifi T, Govil R, Sakdapolrak P, Warner K (2012) Climate change, vulnerability and human mobility: perspectives of refugees from the East and Horn of Africa. Report No. 1. Partnership between UNU and UNHCR. Bonn: United Nations University Institute for Environment and Human Security

Bamforth T (2015) Social impact of Typhoon Bopha on indigenous communities, livelihoods, and conflict in Mindanao. In: Companion M (ed) Disaster's impact on livelihood and cultural survival losses, opportunities, and mitigation. CRC Press, Boca Raton, pp 199–210

Barnett J, Adger WN (2007) Climate change, human security and violent conflict. Polit Geogr 26:639–655

Brookings Institution, Georgetown University, UNHCR (2015) Guidance on protecting people from disasters and environmental change through planned relocations. https://georgetown.app.box.com/s/qwx6dcvl9762fv9itnqn98ogx1h3sjzz. Accessed 9 March 2018

Displacement Solutions (2012) Climate displacement in Bangladesh: the need for urgent housing, land and property rights solutions. Geneva: Displacement Solutions

EU (European Union) (2016) Shared vision, common action: a stronger Europe, a global strategy for the European Union's foreign and security policy. https://eeas.europa.eu/archives/docs/top_stories/pdf/eugs_review_web.pdf. Accessed 9 March 2018

Fant C, Schlosser CA, Gao X, Strzepek K, Reilly J (2016) Projections of water stress based on an ensemble of socioeconomic growth and climate change scenarios: a case study in Asia. PLoS ONE 11(3):e0150633. https://doi.org/10.1371/journal.pone.0150633. Accessed 9 March 2018

Foresight (2011) Migration and global environmental change: future challenges and opportunities. Final project report. London: The Government Office for Science

Ghimire R, Ferreira S, Dorfman JH (2015) Flood-induced displacement and civil conflict. World Dev 66:614–628

IDMC (Internal Displacement Monitoring Centre) (2015) Global estimates 2015: people displaced by disasters. Geneva: IDMC

Ishikawa S (2014) The role of a development agency in peacebuilding: track one-and-a-half mediation in Mindanao. Asian J. Peacebuilding 2(1):79–95

Ishikawa S (2017) A new perspective on conflict resolution in Asia: integration of peace and development for the Philippines, JICA RI working paper. Tokyo: JICA Research Institute

Ishiwatari M (2011) Climate change adaptation in flood risk management through utilizing datasets produced by supercomputer. In: Proceedings of an international symposium on disaster simulation and structural safety in the next generation. Kobe

Ishiwatari M (2017) Climate and fragility risks in Japanese development cooperation: implications of adaptation and peacebuilding experiences. Adelphi, Berlin

JICA (Japan International Cooperation Agency) (2010a) Handbook on climate change adaptation in the water sector: a resilient approach that integrates water management and community development. Tokyo: JICA

JICA (Japan International Cooperation Agency) (2010b) Preparatory study for sector loan on disaster risk management in the Republic of the Philippines. Final report. Tokyo: JICA

JICA (Japan International Cooperation Agency) (2011a) Climate finance impact tool for adaptation. Tokyo: JICA

JICA (Japan International Cooperation Agency) (2011b) Thematic guidelines on peacebuilding. Tokyo: JICA

JICA (Japan International Cooperation Agency) (2016) JICA climate change cooperation strategy. JICA, Tokyo

JICA (Japan International Cooperation Agency) (2017) Handbook on practices for peacebuilding needs and impact assessment (in Japanese). JICA, Tokyo

Kniventon D, Rowhani P, Martin M (2013) Future migration in the context of climate change. Climate change-related migration in Bangladesh. Briefing paper 3. Brighton: Sussex Centre for Migration Research and Refugee and Migratory Movements Research Unit, Dhaka

Kuroda H (2017) Next growth model of Asia economy: beyond world factory. Bank of Japan, Tokyo

Martin M, Kang YH, Billah M, Siddiqui T, Black R, Kniveton D (2017) Climate-influenced migration in Bangladesh: the need for a policy realignment. Dev Policy Rev 35:O357–O379

MFA (Ministry of Foreign Affairs) (2017a) Analysis and proposal of foreign policies regarding the impact of climate change on fragility in the Asia-Pacific Region with focus on natural disasters in the region. http://www.mofa.go.jp/files/000287344.pdf. Accessed 9 March 2018

MFA (Ministry of Foreign Affairs) (2017b) The follow-up review meeting on the roundtable seminar on climate change and fragility implications on international security: summary. http://www.mofa.go.jp/files/000252482.pdf

MFA (Ministry of Foreign Affairs) (2017c) Roundtable seminar on climate change and fragility implications on international security. http://www.mofa.go.jp/files/000226918.pdf

Mobjörk M, Gustafsson MT, Sonnsjö H, van Baalen S, Dellmuth LM, Bremberg N (2016) Climate-related security risks: towards an integrated approach. Stockholm International Peace Research Institute, Slona

Nett K, Ruttinger L (2016) Insurgency, terrorism and organised crime in a warming climate: analysing the links between climate change and non-state armed groups. Berlin: Climate Diplomacy and Adelphi

Quencez M (2011) Floods in Bangladesh and migration to India. The state of environmental migration. In Gemenne F, Brücker P, Ionesco D (eds.) The state of environmental migration. Paris: IDDRI

Rüttinger L, Smith DF, Stang G, Tänzler D, Vivekananda J (2015) A new climate for peace: taking action on climate and fragility risks. Adelphi, Berlin

Seto N (2015) Fragility and conflict assessment measure of the Japan International Cooperation Agency. In: Proceedings of experts' roundtable meeting on assessing fragility for a differentiated approach in dealing with fragile and conflict-affected situations, 2015 Bangkok, Thailand. Mandaluyong City (Philippines): ADB

UN (United Nations) (2014) Unparalleled global pressures tempered in General Assembly debate by appeals to forge bonds with neighbours. Mitigate man-made, natural disasters: Sixty-ninth General Assembly. http://www.un.org/press/en/2014/ga11564.doc.htm

UNHCR (United Nations High Commissioner for Refugees) (2015) The environment and climate change. UNHCR, Geneva

UNHCR (United Nations High Commissioner for Refugees) (2016) Global report 2016. UNHCR, Geneva

UNHCR (United Nations High Commissioner for Refugees) (2017) Climate change and disaster displacement: an overview of UNHCR's role. UNHCR, Geneva

Ward PJ, Strzepek K, Pauw WP, Brander LM, Aerts JCJM (2010) Costs of adaptation related to industrial and municipal water supply and riverine flood protection. Climate change discussion paper no. 6. Washington, DC: World Bank

WFP (World Food Programme) (2013) Emergency food security, nutrition, and livelihoods assessment for Typhoon Bopha (Pablo). https://www.humanitarianresponse.info/sites/www.humanitarianresponse.info/files/assessments/emergency_food_security_nutrition_and_livelihood_assessment_for_typhoon_bopha.pdf. Accessed 9 March 2018

Mikio Ishiwatari is Senior Advisor on Disaster Management and Water Resources Management at Japan International Cooperation Agency, and Visiting Professor at the Graduate School of Frontier Sciences, The University of Tokyo. He has been engaged in projects and research works on disaster risk reduction (DRR), climate change adaptation, and water. He led the formulation of the Japanese assistance policies of climate change adaptation and community-based disaster management. He worked at the World Bank as Senior Disaster Risk Management Specialist, and produced the "Learning from Megadisaster: Lessons from the Great East Japan Earthquake." He worked at various positions at the Ministry of Land, Infrastructure, and Transport, Japan for 17 years.

Chapter 17
National Policies and Programs on Climate Change and Disaster Risks that Address Human Security

Mark Anthony M. Ramirez, Linda Anne Stevenson, Juan M. Pulhin, and Makoto Inoue

Abstract The IPCC 5th Assessment Report (AR5) provides strong evidence that human security will be progressively threatened as climate changes for the worse. Climate change poses a great challenge for states to provide a supportive backdrop, infrastructure, and protection for its people under present and future climate regimes. At the core of addressing this challenge is the formulation and effective implementation of appropriate national policies and programs that advance human security. This chapter reviews key national policies and programs that are implemented to improve human security in selected Asian counties. The review highlighted three important observations. First, national policies and programs across Asia are currently at various stages of development, initiating innovative solutions in response to country-specific problems and threats associated with climate change and disaster risk. Second, in spite of positive developments, key challenges remain on how these national policies are translated and effectively implemented at local levels to achieve human security. Lastly, with the anticipated increasing risk associated with climate change and other forms of hazards in the future, national policies and programs need to evolve in a dynamic manner to incorporate advancement in science and technology and integrate local and indigenous knowledge and experience in order to be more responsive to the changing dynamics of risk and resilience and to prioritize accordingly. The chapter points out that regular monitoring and evaluation of national policies and programs, especially at the ground level, is crucial to achieve human security.

M. A. M. Ramirez (✉)
Resources, Environment and Economics Center for Studies (REECS), Quezon City, Philippines
e-mail: mark.ramirez@reecs.co

L. A. Stevenson
Asia Pacific Network for Global Change Research (APN), Kobe, Japan

J. M. Pulhin
Department of Social Forestry and Forest Governance, College of Forestry and Natural Resources (CFNR), University of the Philippines Los Baños, Laguna, Philippines

M. Inoue
Faculty of Human Sciences, Waseda University, Saitama, Japan

J. M. Pulhin et al. (eds.), *Climate Change, Disaster Risks, and Human Security*, Disaster Risk Reduction, https://doi.org/10.1007/978-981-15-8852-5_17

Keywords Climate change · Disaster risks · Human security · National policies · Asia

17.1 Introduction

In the last decade, there have been numerous reports that address climate change adaptation (CCA) and disaster risk reduction (DRR) as these are intrinsically linked, particularly as extreme weather events brought about by climate change are on the rise. In 2013, the Asia-Pacific Network for Global Change Research (APN) launched a special call for proposals for focused activities that link CCA, DRR, and Loss and Damage (APN 2013). Prior to this, a workshop on linking these three elements was conducted; it focused on the linkages, priorities and limitations, and the challenges and opportunities involved. Clear was the manifestation of extreme and slow-onset events in the face of disasters, and disaster risk management (DRM) and climate adaptation practices are now being planned and incorporated at the local, national, subregional, and regional levels in Asia and the Pacific. A number of challenges were presented at the workshop: the so-called disaster amnesia, cultural practices and traditional knowledge, lack of financial capacity, lack of information/data, lack of consultation in project design (leading to maladaptation), and insufficient education to deal with adverse impacts of weather events, among others. This clearly links to the aspect of human security in climate change. Thirteen projects funded by APN are under way in this area and information can be accessed via the network's electronic library.

Loss and damage is now a fast emerging issue in research agenda as well as in national policies and programs in the Asia-Pacific region. For example, the International Institute for Environment and Development (IIED) and the International Centre for Climate Change and Development (ICCCAD) have undertaken research to better understand loss and damage in Bangladesh. Discussions with policymakers at both the national level and throughout the Asia-Pacific region have revealed an urgent need to understand more about potential future losses and damages and how they can be addressed. Researchers are now beginning to respond to this need. However, research on loss and damage is complicated by both the uncertainty of future climate change and the linkages with a range of disciplines, including DRR, CCA, and development and insurance. By creating an internet forum to disseminate knowledge on loss and damage, the forum profiles related research and sheds light on the challenges and potential solutions to address loss and damage and, hence, human security. In addition, the Asia-Pacific Forum on Loss and Damage allows researchers to connect and learn from others conducting research in the region (APN 2017).

According to a report published by UNISDR (2010) in Asia and the Pacific, policy aspects for CCA and DRR operate in a silo manner and in isolation in most countries, allowing for limited interaction. The report notes that, for historical reasons,

DRR and CCA are different due to political history and departmental and institutional development. As a result, environment authorities tend to be responsible for CCA, and authorities for DRR tend to be covered by departments such as disaster management, civil defense, or home affairs. "The real limitation to adaptation is the political dimension associated with issues of compensation, forcing policymakers to isolate climate risks attributable to climate change" (Berse et al. 2011). Importantly, the report also notes that impact assessments both on health and the environment are effective entry points for interactions between DRR and CCA. Both are high policy priorities, and activities that improve human security can provide useful entry points for DRR and CCA interaction due to their sensitivity to climate change and disasters. Extreme weather as a result of climate change is increasing and will likely cause more disasters. Such disasters, especially those linked to drought, can be the most important cause of impoverishment, impeding progress on poverty reduction (Shepard et al. 2013). This raises wide concern for human security in the face of disaster risk associated with climate change. O'Brien et al. (2008) note that human security is closely linked to the development of human capabilities in the face of change and uncertainty. Individuals and communities faced with rapid change and increasing uncertainty are challenged to respond in new ways that protect their social, environmental, and human rights. Considering human security as a rationale for DRR and CCA in the face of climate change emphasizes both equity issues and the growing connections among people and places.

This chapter reviews key national policies and programs relevant to climate change and disaster risk that are implemented to improve human security in selected Asian counties. The scope of the present chapter covers two countries in each of the three main subregions of Asia. These are Nepal and Bangladesh in South Asia; the Philippines and Lao People's Democratic Republic (PDR) in Southeast Asia; and Japan and the Republic of Korea in East Asia (see Fig. 17.1). The choice of these countries is primarily based on the availability of information and their vulnerability to climate and other forms of hazards.

17.2 South Asia

At the time of writing, South Asia is in its 2017 monsoon season. This year, the monsoon has brought unprecedented, devastating, and relentless flooding to Bangladesh, India, and Nepal. As of August 25, 2017, deaths recorded in these three countries topped 1,200. Rescue workers are struggling to provide much-needed aid to the millions of people impacted and displaced by the flooding, which is the worst the Asian subregion has seen for years (The Strait Times 2017).

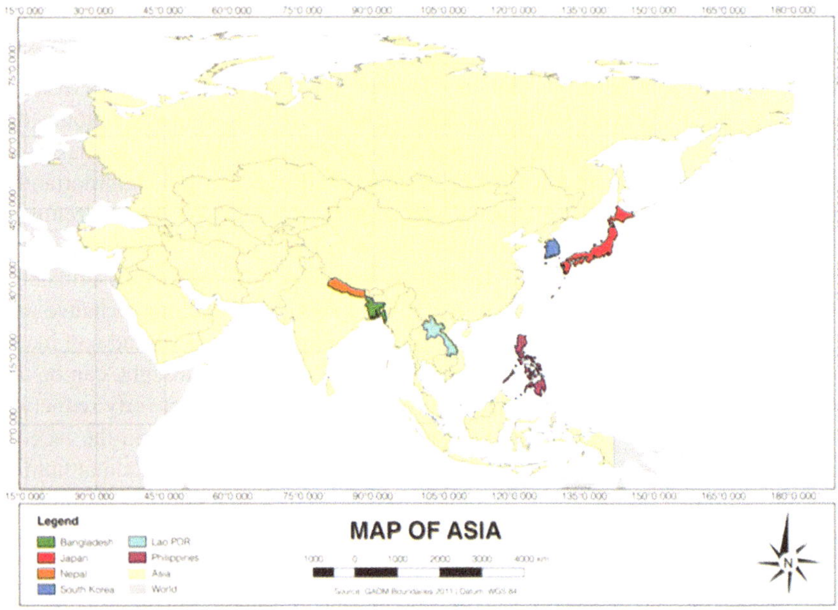

Fig. 17.1 Location map of country studies

17.2.1 Nepal

Nepal, officially known as the Federal Democratic Republic of Nepal, is a landlocked central Himalayan country with an area of 147,200 km^2. Its largest city is Kathmandu. Nepal's neighboring countries are India, China, and Tibet. Nepal's ecological zones run east to west about 800 km along its Himalayan axis, 150–250 km north to south, and is vertically intersected by river systems. Its longest river is river Karnali. The country can be divided into three main geographical regions: Himalayan region, mid-hill region, and Terai region. The highest point in the country is Mount Everest, which stands at 8,848 m (29,029 ft). The lowest point is in the Terai plains of Kechana Kalan in Jhapa (60 m). The population of Nepal is approximately 29,000,000 (2016 est.), with an urban population of about 18.6% (2015). This represents a striking 81.4% living in rural areas in Nepal. The rate of urbanization of Nepal has increased by about 4% in the period 2010–2015. According to 2016 estimations, the population growth of Nepal is 1.24% (Index Mundi 2017a).

17.2.1.1 Disaster-Prone Areas and Disaster Management

Nepal is a disaster hotspot and among the most climate-vulnerable countries in the world (MoHA and DPNet 2015). The country is exposed to many natural and human-induced disasters, including floods, landslides, avalanches, and glacial lake outburst floods (GLOFs). This affects livelihood, property, and infrastructure and has an enormous impact on human security. Nepal's vulnerability is not only due to its steep, rugged, and fragile terrain but also to anthropogenic factors such as unsustainable land-use practices and low adaptive capacity, all of which are compounded by poverty. Lack of resources, social inequities, power imbalance, and limited opportunities mean that poor and marginalized communities, women, and children are among the most vulnerable to disasters and the hardest hit when they happen (Raut and Meyer 2017). According to CARE Nepal, CCA and DRR stand on a shared agenda of resilience building through vulnerability reduction and sustainable development, following similar processes, tools and institutions, and led by local governments. Despite a few issues and challenges, there are ample opportunities to build resilience using different context-specific options for integration and mainstreaming CCA and DRR so as to enhance efficiency, cost effectiveness, and synergies.

17.2.2 National Policies and Programs to Improve Human Security in Nepal

Upreti et al. (2013) describes human security as a state that is achieved when and where individuals and communities have the options necessary to mitigate or adapt to threats to their human, environmental, and social rights. The Government of Nepal initiated the Nepal Climate Change Support Programme (NCCSP), which aims to enhance the capacity of the Government of Nepal and non-government institutions to implement climate change policies and adaptation actions, mainstream climate change in key development sectors, and implement CCA through public–private partnerships (Raut and Meyer 2017). The NCCSP is the first significant government-led initiative to implement Nepal's National Adaptation Programme of Action (NAPA), with a focus on priority area 1–promoting community-based adaptation through integrated management of agriculture, water, forest, and biodiversity sectors. The Government of Nepal facilitated the preparation and implementation of the NAPA and endorsed the National Framework on Local Adaptation Plans for Action (LAPA). The aim of the program, essentially, is to integrate CCA activities into local and national development planning processes to ensure climate-resilient development. The program is aimed at mainstreaming climate change into Nepal's national development agenda and contributing to poverty reduction, livelihood diversification, and community resilience. By introducing LAPA, the most urgent and immediate adaptation actions could be implemented so that the poorest and the most vulnerable communities in Nepal could adapt to the effects of climate change.

The LAPA framework was adopted as a national framework in 2011 by the Government of Nepal. In response to this framework, NCCSP facilitated the preparation of 70 LAPAs in 14 program districts of the mid- and far west regions of Nepal. Across these districts, more than 5,000 LAPA actions have been identified, out of which over 3,000 are urgent and need immediate action plans. This work formed the basis for the NCCSP implementation plan. The LAPAs represent local priorities around agriculture, livestock and food security, forest management and biodiversity, water resource and energy, climatic hazards and physical infrastructure, human resources and livelihoods, and human health. The key results of this program are expected to be 100 LAPAs developed in 14 districts of the mid- and far-western regions of Nepal.

Nepal has recorded a significant number of natural disasters over the past 46 years. An average of 900 lives are lost each year from floods and landslides, rendering families homeless and inflicting heavy damage to properties. The year 2017 saw the worst flood in South Asia happening in Nepal, killing 160 people and destroying 43,000 houses. Raut and Meyer (2017) estimated the losses in agriculture and other livelihoods at more than NPR 8 billion. Prior to this, Nepal was hit by earthquakes in 2015 that killed 9,000 people mainly due to weak infrastructure that collapsed under heavy shaking. This prompted the government to push for a legislation to address risks of natural disasters. The Parliament finally passed the "Disaster Risk Reduction and Management Act, 2074" that defines the roles of federal, provincial, and local governments and calls for a more proactive approach in dealing with disasters. The Ministry of Home Affairs is currently leading the effort in preparing the National Policy and Strategic Action Plan for Disaster Risk Reduction that will set targets and identify actions to reduce casualties in times of disasters. If completed, Nepal will be only the second country in South Asia next to India to have a national plan in line with the Sendai Framework. The plan will take into consideration the impacts of climate change, which have been increasingly felt in Nepal in the past few years. This will require active cooperation from different sectors of society and is expected to result in increased resilience and enhanced adaptive capacity of the country (Fig. 17.2).

17.2.3 Bangladesh

The geography of Bangladesh is varied and has an area characterized by two distinctive features: a broad deltaic plain subject to frequent flooding and a small hilly region crossed by swiftly flowing rivers. The country has an area of 147,610 km^2 and extends 820 km^2 north to south and 600 km^2 east to west. Bangladesh is bordered on the west, north, and east by a 4,095-km land frontier with India and, in the southeast, by a short land and water frontier (193 km) with Myanmar. A highly irregular deltaic coastline of about 580 km lies to the south, with many rivers and streams flowing into the Bay of Bengal. The population of Bangladesh is approximately 156,186,882 with an annual population growth of 1.04% as of 2016 (Index Mundi 2017b). Its rate of urbanization has increased by about 3.5% in the period 2010–2015. Urbanization

Fig. 17.2 Women from a rural farming community at Kusadevi VDC in Kavre, Nepal, performing a tradition dance in thanks to a climate adaptation project that is sustaining their agriculture

is growing fast and it is estimated that only 30% of the population entering the labor force in the future will be absorbed in agriculture, although many will likely find other kinds of work in rural areas. The areas around Dhaka and Comilla are the most densely populated. The Sundarbans, an area of coastal tropical jungle in the southwest, and the Chittagong Hill Tracts on the southeastern border with Myanmar and India are the least densely populated with an urban population of about 34.3% (2015).

17.2.3.1 Disaster-Prone Areas and Disaster Management

The country is very densely populated (1,125 km^2) and many Bangladeshis live on land that is exposed to floods, riverbank erosion, or cyclones. The groundwater used for drinking water and irrigation is widely contaminated with naturally occurring arsenic in some floodplain areas. Water-borne diseases are prevalent with surface water widely polluted by industrial, agricultural, and urban effluents, affecting domestic supplies and inland fisheries. Soil is degraded due to intensive cropping, depletion of organic matter, and imbalance in the use of fertilizers. Deforestation and soil erosion are common in hilly areas (MoEF, n.d.).

Salt water intrusion is another issue that besets Bangladesh due to over-abstraction of water. It affects the quality of drinking water, especially in urban areas near the coasts. This is expected to worsen due to the impacts of climate change and population growth. Water constraints will also affect food production as there will be less irrigation water in times of drought. These problems will exacerbate the already precarious situation of the people in terms of food security and livelihood, which has now resulted in inland migration of some Bangladeshis (Johnson 2017).

Disaster management is a problem that can only be properly addressed if action is taken at all levels. This is not only an environmental concern but also a development concern for Bangladesh. For this reason, human security in Bangladesh needs to be rethought in a holistic way and not split into different dimensions that could be treated separately (Yeasmin and Sattar 2015). The floodplain area of Bangladesh is about 80% and it has an extensive coastline, rendering the nation at great risk of widespread damage. While more permanent defenses, strengthened with reinforced concrete are being built, many embankments are composed purely of soil and turf and made by local farmers (Shrestha et al. 2017).

17.2.3.2 National Policies and Programs to Improve Human Security in Bangladesh

Bangladesh's vulnerability to climate change has pushed the government to respond through policy formulation and creation of innovative ways on how to deal with its impacts. It has prioritized and strengthened policy development both for disaster management and CCA. The country has created the Ministry of Food and Disaster Management (MoFDM) and the Ministry of Environment and Forestry (MoEF), which are primarily responsible for DRR and CCA, respectively (Hasan et al. 2013).

Bangladesh's National Adaptation Program of Action (NAPA) was developed in 2005 after consultation with communities across the country, professional groups, and members of civil society following the guidelines of the United Nations Framework Convention on Climate Change (UNFCCC). Among the policies the country enacted are the National Plan for 2009 and the Bangladesh Climate Change Strategy and Disaster Management (NPDM) for 2010–2015. An action plan, the Bangladesh Climate Change Strategy and Action Plan (BCCSAP) was also developed in line with the Hyogo Framework of Action in 2010 through partnership among various national and international institutes and agencies (Hasan et al. 2013).

Bangladesh is said to be the pioneer in prioritizing DRR in national fiscal planning among the least developed countries. Over the last 35 years, it has invested more than USD 10 billion to make the country resilient in coping with natural disasters (MoEF n.d.) and created the Climate Change Trust Fund (CCTF) from the government's coffers. For the past 7 years, the government has allocated more than Tk 30 billion to undertake 440 climate-related projects (Rahman 2017). Another fund that is supporting Bangladesh in its climate change initiatives is the international donor-supported Bangladesh Climate Change Resilience Fund (BCCRF). The BCCRF was allocated USD 189 million for afforestation projects, cyclone shelter construction, and improvements to embankments and dikes identified by BCCSAP. So far, more than 5,000 km of coastal and river embankments have been rebuilt using the funds

(Johnson 2017). Still, more funds are needed for CCA, which is estimated to cost around USD 40 billion until 2030 (Rahman 2017).

Due to disappearing opportunities in the coastal areas brought about by salt water intrusion and sea level rise, more than 400,000 people are moving from Dhaka each year to migrate inland (Johnson 2017). This is expected to continue in the future, compromising the livelihood of these people (Rahman 2017). To address this situation, government and development organizations are advocating for initiatives such as the refortification of houses so that communities do not have to relocate as often. This process can be as simple as raising the foundation of homes so that they are more resistant to increasing tides and floods (Johnson 2017).

These and similar initiatives are being promoted by the government through its comprehensive Seventh Five-Year Plan: Accelerating Growth, Empowering Citizens for FY2016–FY2020 in late 2015 (GED 2015). The 736-page plan is centered on three themes, one of which places more emphasis to a sustainable pathway that is resilient to disasters and climate change, ensuring the sustainable use of natural resources and the empowerment of every citizen to participate in the development process. It addresses numerous issues related to adaptation, resilience, and human security to ensure that the nation can withstand future impacts of climate change and facilitate the process of CCA. To do this, the government plans to implement a number of strategies over the course of the seventh phase that include the following:

- Revision of the BCCSAP, which expired in 2008, and the completion of the National Action Plan (NAP);
- Integration of CCA into all development projects;
- Institutionalization of a coordination mechanism between and among various ministries in implementing the BCCSAP to make it more functional and more pro-active;
- Development of a mechanism on how institutions can integrate community-based adaptation (CBA) through participatory planning;
- Establishment of a climate fiscal cell in the Finance Division of the Ministry of Finance that can generate reports on climate change-related expenditure and disseminate the information to policymakers;
- Mainstreaming gender in climate change action as envisioned in BCCSAP and NAPA.

The Seventh Five-Year Plan also seeks to achieve "transformative adaptation" rather than the traditional "climate resilience," which aims to empower vulnerable communities to carve their own actions on climate change. This will require them to understand their vulnerabilities with the help of local authorities, technical experts, and civil society using available participatory tools. Protection measures will be formulated to address the needs of displaced people in accordance with relevant national and international policies. Knowledge gaps will also be addressed by the government, especially in the areas of food security and health, through research and development (i.e., climate-resilient crops, diseases that would likely spread due to climate change) (GED 2015).

17.3 Southeast Asia

Perhaps the most affected region of the world in terms of impacts of climate change, Southeast Asia has been witness to the devastating effects of super typhoons, which claimed thousands of lives and inflicted heavy economic losses. The region also has the highest sea level rise in the world that has threatened the struggling economies of Southeast Asia. This is where both climate change and human security need to be addressed simultaneously to reduce the vulnerabilities of communities. The combined impacts of sea level rise, typhoons, and landslides are expected to worsen inequalities, increase food insecurity, and exacerbate poor living conditions (ADB 2017).

17.3.1 Philippines

Strategically located near the equator, the Philippines is one of the countries of the Southeast Asian region blessed with an abundance of natural resources. It is bounded by the Pacific Ocean on the east, the Celebes (Sulawesi) Sea on the south, and the South China Sea on the west and north. It has a total land area of about 300,000 km^2 and is composed of 7,150 islands (Pulhin et al. 2007). With a tropical climate, the Philippines has two distinct seasons—a wet season (June–October) and a dry season (November–May)—conducive to agriculture and a thriving biodiversity. The country has three major island groups: Luzon is the largest, followed by Mindanao, and the Visayas. As of 2017 estimates, the Philippines has a population of 104,256,076 with a growth rate of 1.57% (Index Mundi 2017c). Around 44% of the population live in urban areas, while 55% dwell in rural areas (Doroteo 2015). A large proportion of its growing population resides along the country's coastlines, where about 60% of its cities and municipalities are situated (Long and Giri 2011). Despite its vulnerability to climate-induced disasters, the Philippines is considered as one of the fastest growing economies in the region.

17.3.1.1 Disaster-Prone Areas and Disaster Management

The Philippines ranked fifth among countries most severely affected by climate change over a 20-year period according to Global Climate Risk Index 2018 (Eckstein et al. 2017). The country's vulnerability is mainly due to its location and topography. It is an archipelago, which is exposed to weather-related disasters, including seasonal typhoons, flooding, and landslides, as well as natural disasters such as earthquakes and volcanic eruptions since it is also situated in the Ring of Fire. In recent decades, the Philippines has experienced rising temperatures; extreme rainfall is also more frequent and intense, resulting in flashfloods in urban areas and landslides in rural areas. Several coastal areas are also at risk from sea level rise and storm surges during typhoons. An average of 21 typhoons with increasing intensity, rainfall volume, and wind speed pass by the country every year (PAGASA 2011; Doroteo 2015). These often result in loss of lives and severe damage to property and infrastructure.

As seen in the devastation brought about by typhoons Yolanda in 2013, Glenda in 2014, and Lando in 2015, millions of Filipinos have been affected and communities incurred costly damages and were forced to rebuild. The effects of climate change were accurately depicted when Typhoon Haiyan (also known as super typhoon Yolanda) hit the Philippines in the morning of November 8, 2013. With sustained winds of about 315 kph, Haiyan is considered one of the strongest tropical cyclones to hit land in recorded history. In its aftermath, 6,300 persons perished and 1,061 persons were considered missing. Total damage to infrastructure, agriculture, and property was pegged at PhP 571 billion. In anticipation of stronger typhoons hitting the country, CCA and mitigation become vital to the development and preparedness of local government units (LGUs), the people, and other sectors of society in order to address human security issues (CCC, n.d.). Several interventions are now in place in the Philippines to anticipate the impacts of climate change through the adoption of appropriate policies and programs as well as the production of scientific information needed by communities for disaster preparedness.

17.3.1.2 National Policies and Programs to Improve Human Security in the Philippines

As the Philippines is prone to natural disasters, policies on disaster management have been put in place since 1954 through the passage of the Civil Defense Act that created the National Civil Defense Administration (Florano 2013). This was amended by Presidential Decree No. 1566 (issued in 1978), which created the National Disaster Coordinating Council, which has counterparts in the provincial and local levels. Under this scheme, local governments are mandated to allocate funds for the purchase of equipment and training of teams to enable them to prepare and respond to the needs of communities during disasters (Rey 2015). During this period, DRR focused mainly on disaster preparedness and response and not on identifying hazard-prone areas and other parameters that increase people's exposure to disasters, thereby endangering human security. This has been addressed by Republic Act 10121 (National Disaster Risk Reduction and Management [NDRRM] Law) enacted in 2010, which ushered in a paradigm shift from a reactive to a more proactive approach in dealing with disasters. It focuses on risk reduction through preparedness, mitigation, and prevention. Part of the provisions of this law is the creation of the National Disaster Risk Reduction Management Council (NDRRMC) and the formulation of a National DRRM Plan 2011–2028 (NDRRMP) that outlines the expected outcome, indicators, key activities, and implementing partners in operationalizing the goals of the DRRM. The local governments are also directed to produce its local DRRM plan (LDRRMP) to minimize the effects of disasters in their respective localities.

With the heavy casualties brought by Typhoon Haiyan, the government mobilized the scientific community to produce up-to-date information that will help the communities and the government prepare for disasters and prevent similar catastrophes from happening again. The government funded its umbrella project NOAH (Nationwide Operational Assessment of Hazards), which provides a more accurate, integrated, and responsive disaster prevention and mitigation system, especially in high-risk areas throughout the Philippines. It has a web-based system that is easily accessible to Filipinos in times of disaster, providing early warning to communities and information that will help local governments pre-position resources to cover high-risk areas. Under NOAH is the Disaster Risk and Exposure Assessment for Mitigation (DREAM) project, which has two complementary programs, the PHIL-LiDAR 1 for 3-D flood and hazard mapping using LiDAR for the Philippine river systems and PHIL-LiDAR 2 for detailed resource mapping of different sectors such as agriculture, renewable energy sources, forestry, aquaculture, and irrigation assessment. Both programs produced high-resolution flood hazard maps that are important inputs in development planning, addressing issues related to settlements, DRR, and land use zoning (Blanco 2014).

To respond to the impacts of climate change, the Climate Change Act or Republic Act 9729 was enacted in 2009, 2 years ahead of the NDRRM Law. This law created and mandated the Climate Change Commission (CCC) to mainstream climate change, in synergy with DRR, into national, sectoral, and local development plans and programs. Its framework strategies are centered both toward adaptation and mitigation and recognize the important role of local governments in the formulation and implementation of climate change action plans through their Local Climate Change Action Plan (LCCAP) consistent with the National Climate Change Action Plan (NCCAP). The NCCAP addresses the immediate needs of the country in terms of the adverse effects of climate change through key strategic priorities: food security, water sufficiency, ecosystem and environment stability, human security, climate-smart industries, sustainable energy, and knowledge and capacity building (CCC, n.d.). As of 2016, however, only 160 out of the 1,700 local governments have produced their LCCAPs. The small number is primarily attributed to the limited technical capacity in the local governments to identify vulnerabilities and integrate climate risk adaptation in their development plans (Serafica 2016). The lack of capacity is rooted on limited manpower and financial resources as well as lack of access to technology and inadequate technical knowledge and understanding of government staff.

The country also enacted Republic Act 10174 that created the People Survival Fund (PSF), a special fund in the national treasury for financing adaptation programs and projects. It has an initial appropriation of PHP 1 billion that can be accessed by local governments by submitting proposals to the national government. The proposals will be prioritized in terms of poverty incidence, exposure to climate risk, and presence of identified key biodiversity areas in the municipality. These laws call for every social development program aimed at poverty reduction to take into account the vulnerabilities of the poor communities to both climate and disaster risks to effectively formulate interventions that minimize risks and increase their adaptive capacities.

Aside from CCA measures, the government is also pursuing activities that are aimed for mitigation. While the Philippines is considered a net carbon sink (absorbing more greenhouse gases [GHGs] than they produce), the country aims to reduce GHG emissions by around 70% by 2030, relative to its business as usual scenario of 2000–2030, conditional on receipt of external finance and technical assistance as part of its intended nationally determined contribution (RECOFTC and AWG-SF 2017). The government has recognized that social forestry can contribute to both adaptation and mitigation as forests under community management have the potential to store and sequester around 90 Mtc of carbon (Lasco et al. 2010). At the center of the government's strategy is the National Greening Program (NGP), initiated in 2011, that aims to arrest the continuing decline of forest cover. It aims to rehabilitate 7.2 million ha of forest land to contribute to poverty reduction, food security, biodiversity conservation, and climate change mitigation and adaptation through various modes and partnerships. As of 2014, a total of 592 million seedlings have been planted in degraded lands, covering an area of more than 1 million ha (FMB 2013). Despite several issues in its implementation such as poor planning, lack of site-species matching, and lack of suitable areas, the program has been extended up to 2028.

17.3.2 Lao PDR

Lao PDR is a landlocked and mountainous country located in Southeast Asia. The country is bounded to the north by China, to the northwest by Myanmar, Thailand to the west, Cambodia to the south, and Vietnam to the east. It has three agroclimatic zones composed of the mountainous north, hilly and mountainous areas in the central and the southern portions, and the alluvial plains along the Mekong River. It has a total land area of 230,800 km^2 with around 58% forest cover as of 2010 (RECOFTC and AWG-SF 2017). It has a population of about 6.25 million living in its 17 provinces in 2014 where 67% still reside in the rural areas. It has an urbanization rate of 4.9% per year (NIDM 2018).

17.3.2.1 Disaster-Prone Areas and Disaster Management

Lao PDR is one of the countries most vulnerable to the impacts of climate change in the Southeast Asian region as it is a low-income country with an agriculture-based economy. The low-lying areas along the Mekong River are considered the most vulnerable in the country. Lao PDR regularly experiences natural hazards, the most dominant of which are floods, droughts, and typhoons. Natural disasters heavily affect all development sectors, with agriculture, transport, and housing being the hardest hit. It recorded around 30 major flood events over the last 40 years and typhoons cause an economic average annual loss of USD 17.6 million, about USD 8.3 million from floods, and around USD 4.7 million from droughts, which could otherwise be used to tackle poverty alleviation. Due to the high poverty incidence in Lao PDR, even low-intensity disasters can have heavy negative impacts on rural households. The country is believed to have a low capacity to adapt to climate change, owing to its poor socioeconomic development. Climate change is projected to heavily affect the food security of the nation due to shifts in rainfall patterns. In developing cities such as the capital Vientiane, urban centers are expanding rapidly without adequate land-use planning, creating new vulnerabilities and human security issues (NIDM 2018).

17.3.2.2 National Policies and Programs to Improve Human Security in Lao PDR

Lao PDR is the 42nd country in the world that is most vulnerable to the impacts of climate change and the 36th country least ready to adapt, based on the Global Adaptation Institute's Index of 2013 (RECOFTC 2014). Its land use change and forestry sector produces more than 60% of total emissions in the country (CAIT Climate Data Explorer 2015).

Based on a study conducted by RECOFTC in 2014, climate change is a relatively new concept to the citizens of Lao PDR and detailed understanding of its consequences and impacts is limited to a few institutions and individuals. Hence, long-term scientific studies to establish baseline data at the national level for adaptation and mitigation planning are needed. This supports an earlier finding of UNDP in 2008 that Lao PDR requires massive funding and technologies to make itself ready for climate change (Tong 2009).

Lao PDR farmers usually turn to the forest for basic food and livelihood by collecting non-timber forest products (NTFPs) during times of agricultural failure. This means that the negative effects of climate change will increase the reliance of rural communities on forest resources for survival, particularly for the poorest households. Women will be hardest hit since they are the ones responsible for ensuring food security for the families through the collection of food crops from the forest. The World Food Programme estimates that around 157,000 people in the country will become food-insecure if there is less hunting, gathering, and fishing in the immediate future. Maintaining the integrity of forest ecosystems is therefore critical in addressing climate change in Lao PDR (RECOFTC 2014).

Villagers extract many NTFPs, which include food, fiber, condiments and medicinal products, inputs for chemical and perfume industries, bamboo poles, rattan, and fuelwood. Around 40% of the annual household income is from the sale of NTFPs (RECOFTC 2014). Wild foods from the forest are estimated to contribute between 61% and 79% of non-rice food consumption by weight, providing an average of 4% of energy intake, 40% of calcium and vitamins A and C, and 25% of iron. Hence, these are very important resources during times of emergency and seasonal food shortages as a result of climate change (RECOFTC 2014). Aside from NTFPs, sustainable timber harvesting can also be a source of income for villages with mature secondary forests and areas that have scenic views or nature trails can be improved for the ecotourism industry.

Two of the most relevant pieces of government documents on climate change are the Strategy on Climate Change of the Lao PDR (2010) and the National Adaptation Program of Action to Climate Change (NAPA) that was submitted to UNFCCC in 2009. Under the NAPA, forests are defined as an area of focus, along with agriculture, water, transport, urban development, and public health. It includes projects on strengthening the capacity of village foresters in forest management. NAPA also envisions training village forestry volunteers to conduct vulnerability assessments and adaptation planning. As of May 2013, Lao PDR has implemented projects that will increase the resilience of forest production systems and strengthen technical capacity in managing forests for climate change mitigation and adaptation under the Least Developed Country Fund (LDCF) (RECOFTC 2014).

The government has taken important steps toward building resilience to natural hazards in recent years. It has formulated the National Disaster Management Plan for 2001–2020, which aims to identify focal points in all sectors of government, develop an early warning and information system in all of its 142 districts, and conduct training on disaster management for all sectors. In addition, the Seventh National Socio-Economic Development Plan (7th NESDP 2011–2015) seeks to mainstream DRM into development planning and to create legal frameworks and strategies for DRM. Still, existing decrees and regulations on DRM can still be improved as these are limited in defining mandates and functions, resulting in difficulties in implementation.

17.4 East Asia

East Asia is an export-oriented powerhouse that accounts for 30% of the world's energy-related emissions. This means that addressing the impact of climate change in this region has huge consequences that can affect the world in general. This also suggests that initiatives to mitigate climate change must include this region. While attaining economic growth, the region is also vulnerable to the impacts of climate change. Japan and the Republic of Korea are among the urban areas in the world severely exposed to sea level rise that can significantly affect the lives of coastal communities. Moreover, climate change, if not addressed, may reduce yields of key crops in the region, thereby endangering food security (Westphal et al. 2013).

17.4.1 Japan

Japan, an archipelago situated in East Asia, is composed of 6,852 islands covering a total area of 377,930 km^2. It is located in the volcanic zone of the Pacific Rim of Fire where seismic and volcanic activities regularly occur. Due to its geographical and climatological conditions, Japan has the highest natural disaster risk in the developed world as it is vulnerable to almost all kinds of natural disasters such as typhoons, torrential rains, and heavy snow, not to mention volcanic eruptions, earthquakes, and tsunamis. These do not prevent the country, however, from achieving a high economic status as it is considered the third largest economy in the world. The country has four main islands: Honshu, Hokkaido, Kyushu, and Shikoku. Tokyo and many other cities in Japan can be found in Honshu, including Yokahama, Osaka, Nagoya, Kobe, Kyoto, Kawasaki, Saitama, Hiroshima, and Sendai. Seventy-three percent of Japan is forested, mountainous, and unsuitable for agriculture or residential use. This is why most of the coastal areas have extremely high population densities. As of 2014, its population stood at 127 million where around 79% are found in urban areas. Japan has a population density of 341 persons km^2 (NIDM 2017a).

17.4.1.1 Disaster-Prone Areas and Disaster Management

With the constant threat of natural disasters, the entire country of Japan is said to be the riskiest region on earth. The threat of earthquakes and tsunamis is experienced daily by the Japanese people. The Great East Japan Earthquake (GEJE) or the Tohoku Earthquake that happened on March 11, 2011 was the most powerful earthquake to ever hit Japan. Regarded as the world's fifth most powerful earthquake in modern times, it unleashed a tsunami of 40.5 m in the city of Miyako and claimed the lives of 15,894 individuals. It also severely damaged the Fukushima Daiichi Nuclear Power Plant that made policymakers in Japan rethink its climate mitigation commitment, especially in harnessing nuclear power for electricity. Aside from earthquakes, frequent typhoons, river flooding, and landslides are regular occurrences in the country, affecting people's lives, livelihood, and property since ancient times. Climate change is also affecting weather patterns in Japan, and the number of extremely hot days and cold nights is increasing. Frequency of flooding is predicted to increase from 1.8 to 4.4 times in the future. Coastal areas will be increasingly exposed to the risk of rising sea level resulting in inundation (MOE 2016). While Japan is the country most at risk, it is also the most prepared when dealing with natural and anthropogenic disasters. The government has designed "disaster countermeasures" that are regarded as investments rather than costs in achieving safe and secure living (Cabinet Office Japan 2015). These countermeasures are reflected in the policies and programs of the government that call for modern approaches in dealing with disasters, using technologies that mitigate the effects of climate change and strategies based on lessons learned in the past.

17.4.1.2 National Policies and Programs to Improve Human Security in Japan

Japan's primary response to climate change is on how to attain its commitment under its intended nationally determined contribution. It has pledged to reduce its GHG emissions by 26% from 2013 levels by 2030 (Kiko Network and NRDC 2017). Many analysts describe this as unambitious, but the reduction in its pledge was mainly driven by its goal to shift to other sources of energy since the Fukushima nuclear plant disaster. Debates on climate change have been reframed by policymakers as debates on nuclear power, hence the government is now focusing its attention to renewable energy and coal power plants (Kameyama 2015). Prior to the earthquake, nuclear power provided around 25–30% of the country's electric supply, but today, all nuclear plants have ceased to operate and natural gas and coal power plants are now the main sources of power (Kameyama 2015). Electric power has a significant implication on the economic growth of Japan since most of its industries depend on electricity for production. Without these industries, employment will suffer, resulting in loss of livelihood and rendering the Japanese people more vulnerable to disasters.

Japan's contribution to global emission reduction is largely attributed to the diffusion of its cutting-edge technology combined with market incentive and a responsive tax system. For instance, techniques have been developed and applied in its iron and steel mills that have contributed to a reduction of 50 million tons of carbon dioxide annually (Mochizuki 2017). Another strategy is to shift toward a less carbon-intensive economy through the imposition of a carbon tax that will change people's consumption behavior (Kameyama 2015). With this, many Japanese home appliances have improved their energy efficiencies and more people are buying hybrid cars to save on costs for gasoline. The government has implemented its Best Available Technologies guidelines, which provide a list of technologies and their efficiencies. These guidelines are updated annually (Kuramochi 2014). Japan also introduced the Feed-in Tariff Scheme in 2011 that promotes the installation of solar photovoltaics (PVs) and obligated power companies to buy all renewable energy being generated by implementing the scheme. This has seen the growth of renewable energy in the country.

In terms of DRM, Japan has a highly respected disaster response system and is considered a world leader in this area, given its vulnerability to natural disasters. Japan initiated the establishment of a planned and comprehensive disaster prevention system in 1961 through the enactment of the Disaster Countermeasures Basic Act (MoFA 2018). The Act calls for measures in prevention, emergency relief, and recovery that are linked interactively. Prior to this, the Forest Conservation and Flood Control Urgent Measures Law was passed in March 1960, followed by the planned implementation of flood control measures and other projects designed to enhance Japan's resiliency in the face of disasters. Countermeasures against disasters have the following components: (1) research into the scientific and technical aspects of disaster prevention; (2) reinforcement of the disaster prevention system, its facilities, and equipment, and other preventive measures; (3) construction projects designed to enhance the country's ability to defend against disasters; (4) emergency measures and recovery operations; and (5) improvement of information and communication systems.

Since the 1960s, the country has celebrated "Disaster Prevention Day" every year on September 1 as a means of disseminating disaster prevention knowledge and raising public awareness (MoFA 2018). Over the years, the government has invested in structural and non-structural measures to prevent disasters and reduce their impacts. Structural measures include the establishment of coastal dikes and breakwaters in areas affected by the GEJE and deployment of early warning systems, while nonstructural measures involve community-based disaster risk management, DRM education, and a business continuity plan (Toyama and Sagara 2013). The country is continuously improving its DRM systems by promoting national land conservation, improving weather forecasting technologies, and upgrading disaster information communication systems. This has resulted in a decline in disaster-related casualties. It has a well-established risk residential earthquake insurance program that helped victims recover from the GEJE earthquake (Mahul and White 2013). With their long experience in disaster management, Japanese experts are usually deployed to assist other nations in responding to disasters.

Although mitigation is the prominent strategy on climate change in Japan, the government published the *Approaches to Climate Change Adaptation* in 2010, a document produced for relevant departments in national and local governments to show how to implement adaptation measures in their respective jurisdiction. Owing to the fact that the impacts of climate change differ by region in Japan due to its diverse geographical and climatic features, each prefecture and municipality is encouraged to prepare its own adaptation strategies. A study conducted by Baba et al. (2017) found that very few local governments have completed the preparation of adaptation strategies. They attributed this to the challenges that are being faced by the local governments, such as lack of expert knowledge and experience in climate change adaptation and compartmentalization of government bureaus. Climate change adaptation is a relatively new issue for local governments and expertise in the area is still developing. Formulation of adaptation strategies also requires the collaboration of multiple bureaus, such as agriculture, public health, and disaster prevention, but collaboration is difficult to achieve due to the nature of bureaucratic government.

17.4.2 Republic of Korea

The Republic of Korea (ROK), also known as South Korea, is a capitalist country in East Asia. It is classified as a newly developed nation, owing to the impressive economic growth of its industries in recent years. The Republic of Korea is surrounded by the Yellow Sea in the west and the Sea of Japan in the east. It has a total area of 99,000 km^2 where 70% is mountainous, which is mainly found in the northern and the eastern fronts. The western and southern parts are covered with plains. The coast is surrounded by approximately 3,400 islands. The country has a total population of 49 million as of 2014 where 84% are living in urban areas. Due to the political tension with North Korea, there is a constant threat of war, hence, South Koreans are always prepared for any eventuality, which is also the case in its preparedness and response to disasters.

17.4.2.1 Disaster-Prone Areas and Disaster Management

South Korea has four distinctive seasons: spring, summer, autumn, and winter with hazardous weather patterns throughout the year. Typhoons usually hit the Korean Peninsula in the second half of the year with torrential rains resulting in flooding. Floods brought about by heavy rainfall have led to loss of lives and damage to property. Typhoons originating from the Philippines frequently hit South Korea, often triggering landslides and flooding. Hazards in the country that have major impacts are due to both natural and anthropogenic disasters. Natural hazards include drought, heavy snow, earthquake, and the yellow-dust phenomenon, while acts of terrorism, oil spills, domestic violence, and economic crisis are considered man-made. The country is also susceptible to earthquakes and tsunamis, though to a lesser extent, as it is geographically close to Japan (NIDM 2017b). Climate change is further aggravating this situation as the average temperature in the Korean Peninsula has increased by 1.7 °C and rainfall has increased by 19% from 1912 to 2008. This has resulted in an increase in average rainfall intensity and strengthened the average minimum pressure of typhoons (Dolcemascolo et al. 2011).

17.4.2.2 National Policies and Programs to Improve Human Security in the Republic of Korea

South Korea has completely embraced the concept of green growth to support its industries and, at the same time, contribute to climate change mitigation. This, in effect, allows its economy to grow while addressing the impacts of climate change and human security of its population. While other nations have just started to explore the concept of green growth, South Korea has adopted it as a national vision embedded in its governance system (GGGI 2015). South Korea's green growth strategy underscores the need to continue growing economically by reducing GHG emissions through the promotion of green technology, green industries, and green jobs. It is based on the notion that contributing to climate change mitigation need not hamper economic growth.

This vision is embodied in the Framework Act on Low Carbon Green Growth, enacted in 2010, that mobilized various government ministries to formulate comprehensive green growth plans at the sectoral, national, and local levels in honoring South Korea's commitment to reduce its GHG emissions by 30% in 2020 (GGGI 2015). The Act integrates economic, environmental, and social objectives in one framework. This was followed by the introduction of the First National Basic Energy Plan, the country's strategy for the energy sector that calls for interventions to reduce energy demand of households and industries by improving energy efficiency and deploying renewable energy to conserve the environment. Strategies to achieve green growth include green procurement, carbon labeling, green education, and pay-as-you-throw waste management programs and greening the value chain by promoting high-tech industries through innovative arrangements such as public–private partnerships.

Given its vulnerability to disasters, the government enacted several policies on how to respond to and manage disasters. The main framework for rescue activities was established though the Framework Act on Civil Defense in 1975, which was later strengthened by the Disaster Control Act in 1995. This was further improved by the Emergency and Safety Management Basic Act in 2004. The National Emergency Management Agency was formed as the first disaster management organization of the nation. With the presence of a government department in charge of disasters, the basic concept of disaster prevention policy shifted from recovery to prevention and field response system. Part of South Korea's readiness exercises are actual evacuation drills conducted three times a year to anticipate heavy rains and typhoons, as well as the devil air defense evacuation drills in the event of a war with North Korea. Before any large-scale developments projects are carried out, prior examination system of disaster impact is required. This includes the prediction and analysis of factors that can possibly cause disasters and the establishment of structural and nonstructural countermeasures (Yoon et al. 2014).

Apart from these, the government regularly improves its land-use planning guidelines to mitigate the impacts of natural disasters in floodplains and on steep slopes; implements an efficient system of water resource management; upgrades building standards for schools, hospitals, and other critical infrastructure; and enhances the protection of its forests and wetlands. All of these rely on accurate climate information provided by the Korean Meteorological Administration. Each city is also required to come up with a Comprehensive Flood Mitigation Plan, including the construction of drainage systems that can absorb the anticipated volume of floodwater as forecast from climate projections and scenarios. In addition, they are also required to conduct community-based preparedness activities, a key feature of building resilience in Korean culture, thus bringing climate information directly to people's homes (Dolcemascolo et al. 2011).

Actions in terms of climate change adaptation are included in the National Climate Change Adaptation Master Plan (NCCAMP) covering the years 2009–2030, published by the South Korean government in 2008. This was followed by the formulation of the National Strategic Plan for Climate Change Adaptation 2011–2015 in 2010 and the establishment of the National Government Adaptation Committee to implement the plan. The strategic plan has 87 major projects in 10 sectors: public health, disaster management and infrastructure, agriculture, forestry, marine and fisheries, water, ecosystem, climate change monitoring and projection, adaptation business and industry, and publication, education and international cooperation. It also has provisions for local action planning, which is critical in adopting the plan and involves the coordination and cooperation of 13 government agencies (Nachmany et al. 2015).

17.5 Discussion and Conclusion

There is no denying that the impacts of climate change have affected and will affect the lives of a greater number of people as human insecurities are compounded by poverty, inequality, and unequal access to resources, especially in the least developed countries in Asia. Nations are forced to rethink their development agenda to include the effects of climate change in their economic growth projections. Guided by international agreements, national policies and plans dealing with climate change and disaster risk and reduction management are being formulated to address the externalities brought about by sea level rise, flooding, droughts, and landslides. However, human security concerns may or may not be directly stated in these policies and plans but are assumed to be part of its goals as these have similar aims of reducing vulnerabilities and risks being faced by the marginalized sectors of society.

As the region most prone to natural disasters, given its geographic location and climatic conditions, most of the countries in Asia covered in this chapter started framing their policies toward addressing the risks being faced by their population from natural disasters. This is fairly evident in early policies on disaster management in all the Asian countries. It is only recently that the impacts of climate change have been recognized as a necessary part of the disaster risk paradigm, placing more emphasis on more proactive ways to handle disasters. Climate change is now perceived as an important factor in exacerbating disasters and the focus is on how these disasters create and enhance vulnerabilities. Individual countries began to formulate their own policies and plans on how to adapt and mitigate the effects of climate change, this time, recognizing the inherent relationship between CCA and DRR. Both the concepts of DRR and CCA are focused on reducing climate risks to attain a certain level of sustainable development.

Since there are significant differences among the Asian countries discussed in this chapter in terms of spatial, social, economic, demographic, cultural, and current political realities, these countries are in various stages of policy formulation to address the impacts of climate change, where developed countries are clearly ahead in terms of policy and plan implementation. These differences also mean that there can be no prescriptive approach in policymaking and plan formulation. Policies and plans must be supported by a robust information system at the country level for them to be responsive to climate change. In the Asian context, policies on climate change are often dictated by the current economic priorities of the government. For instance, Japan and South Korea are taking the path to climate change mitigation in pursuing green growth, while the lesser developed countries of Nepal, Bangladesh, and Lao PDR are strengthening their capacities to respond to disaster by prioritizing CCA while aiming for sustained economic growth. It is only the Philippines that gives equal attention to both adaptation and mitigation as dictated by policies and priorities of the government.

Increasing the resiliency of human and ecological systems is also the focus in developing countries such as the Philippines and Lao PDR, which is reflected in the need for sustainable management of natural resources, particularly that of forests.

This is based on the assumption that increasing the natural resource base will provide the needed livelihood of people and protect them from natural disasters, thereby keeping losses and damage at bare minimum. Efforts in these countries, together with Nepal and Bangladesh, are now increasing to develop local plans on CCA using a participatory and inclusive approach in crafting strategies and activities to counter the negative effects of climate change. The usual entry point for this planning process is to relate the effects of the changing climate to existing livelihoods of the local people, food security, and capacity to access natural resources—areas covered by a human security perspective and areas that people can easily relate to. It is based on the principle that when local people are able to address these concerns, their vulnerabilities to climate change are reduced and their adaptive capacities are enhanced. Mitigation projects are also being implemented in these countries, especially in pioneering REDD + projects funded by international donor agencies and bilateral agreements.

Bangladesh has already invested in fortifying areas prone to disasters. It has also established finance mechanisms on its own and with multilateral agencies that can be tapped to initiate adaptation projects at the local level. This is expected to increase the country's adaptive capacity in the long run, but efforts to think of innovative approaches should continue. Nepal has just recently passed its comprehensive law on disaster risk reduction and management that is anchored on the Sendai Framework and is seen by many to improve its response to disasters. In Southeast Asia, the Philippines has the most advanced policies and systems in place to address both the impacts of climate change and natural disasters. Development planning at the local level now includes climate proofing using current scientific information to anticipate future risks. The government has also opened up a financing mechanism that can be tapped by local governments in implementing adaptation projects, ensuring that communities are protected against climate-induced hazards. Efforts to fast-track the formulation of local climate change action plans, however, are needed, especially in terms of building the technical capacities of the local governments in using technologies and tools to produce local vulnerability assessments.

Developed countries, on the other hand, have given attention to changing behavior and consumption patterns by imposing taxes and giving incentives to companies toward reducing their GHG emission. While industries are largely being blamed for much of the GHG emissions, efficient technologies are now in place to mitigate the impacts of climate change without sacrificing production as shown by the cases of Japan and South Korea. In the wake of the GEJE, Japan, however, has reduced its original commitment to cut its GHG emission as it shifts to renewable energy and coal from the use of nuclear power for electricity, a move described by some sectors as unambitious, but, given its own catastrophic experience, can be viewed as justifiable. Both Japan and South Korea have introduced structural (i.e., improved drainage system, climate proofing in infrastructure, climate-enhanced land-use planning) and nonstructural (risk insurance, climate-enhanced land-use planning, and green technologies) countermeasures that address the impacts of climate change. Being considered as developed countries would also mean a high adaptive capacity to respond to the effects of climate change for both countries, with a wide range of resources at their disposal.

Amidst these positive developments in the Asian region, the challenge remains on how these national policies and plans are translated and effectively implemented at the local level to achieve the goal of human security. When dealing with impacts of climate change and disasters, the main battleground is at the community level. Poor communities in urban and rural areas often have limited capacity at the local level to address policy adjustments and this is an area where efforts and resources should now be shifted by governments. The possibility of cooperation by family and relatives who migrated from the community should also be taken into consideration in development planning. Critical in this regard is the presence of institutions with clear authority and mandate, budget, and technical personnel, factors necessary to translate policies into practices, which, unfortunately, are still lacking in least developed countries. Governments should follow the footsteps of Japan that considers climate change and disaster management as investments rather than costs, which can accrue benefits to the greater society.

More often than not, policies and plans that are not properly disseminated to the communities lead to maladaptation. Also, inappropriate policies that are not applicable to the local condition may accelerate and amplify human security and conflict. These policies can increase the chances of having adverse reaction to climate change such as migration from rural to urban areas. Migrants are more at risk in destination areas in the absence of a comprehensive plan from government on how to address different aspects of their daily lives. The best chance of increasing their resiliency is to craft adaptation strategies in the area of origin as they have already established relationships and social capital within their communities. It is therefore imperative for any government to harness social capital in addition to the political capital as represented by these policies, as adaptation and mitigation strategies that are imposed on communities are more likely to negatively impact these communities. In this case, the main role of the national government is to provide up-to-date scientific climate data such as climate change projections and technologies as shown in the case of the Philippines, the only time that a top-down approach is necessary, in aiding localized adaptation planning. Increasing resilience would not only involve the government but should also tap the private sector and civil society to form a triangle of cooperation (Elliot 2014). In most cases, especially in the least developed countries of Asia, government resources are limited and support from other sectors is needed to address human security issues.

Climate is dynamic and, to be able to choose appropriate adaptation strategies, there is a need to update existing knowledge. Therefore, policies should evolve from time to time, considering new scientific information and cultural practices. Revisiting policies for amendment or passing a completely new policy is part and parcel of the feedback process necessary in attaining the goals of human security. However, it is imperative to assess first the effectiveness of a certain policy before formulating a new one as passage of unnecessary policies can only result in overlapping mandates and eventually maladaptation.

Transforming our world, the 2030 Agenda for Sustainable Development states its vision to build the resilience of the poor and those in vulnerable situations and reduce their exposure and vulnerability to climate-related extreme events and other economic, social, and environmental shocks and disasters (UN General Assembly 2015). This can only be attained by having responsive policies at different governance scales—international, national, and local—prioritizing both social and environmental vulnerabilities at the grassroots level where the impacts of climate change are more pronounced and are being experienced on a more regular basis. The main role of national policies is to provide the mandate and guidance as well as to allocate adequate resources and technical expertise to assist communities in crafting their own adaptation actions in order for them to move from reactive to anticipatory resilience.

As depicted in this chapter, national policies and programs across the Asian region are currently at various stages of development, initiating innovative solutions in response to country-specific problems and threats associated with climate change and disaster risk. In spite of these positive developments, however, key challenges remain on how these national policies are translated and effectively implemented at the local level to achieve human security. With increasing risks associated with climate and other forms of hazards in the future, national policies and programs need to evolve in a dynamic manner to incorporate advancements in science and technology and to integrate local and indigenous knowledge and experience to enable nations to be more responsive to rapidly changing times and priorities. This necessitates regular monitoring and evaluation of national policies and programs, especially at the ground level.

References

ADB (Asian Development Bank) (2017) A region at risk: the human dimensions of climate change in Asia and the Pacific. ADB, Manila, Philippines

APN (Asia-Pacific Network for Global Change Research) (2013) Workshop on climate change adaptation, disaster risk reduction, and loss and damage. http://www.apn-gcr.org/2013/09/10/workshop-on-climate-change-adaptation-disaster-risk-reduction-and-loss-and-damage/. Accessed 12 Oct 2017

APN (Asia-Pacific Network for Global Change Research) (2017) Asia Pacific forum on loss and damage. http://www.apn-gcr.org/resources/items/show/1981. Accessed 30 Aug 2017.

Baba K, Matsuura M, Kudo T, Watanabe S, Kawakubo S, Chujo A, Tanaka H, Tanaka M (2017) Climate change adaptation strategies of local governments in Japan. https://doi.org/10.1093/acrefore/9780190228620.013.597. Accessed 13 Oct 2017

Berse K, Tran P, Velasquez J (2011) At the crossroads: climate change adaptation and disaster risk reduction in Asia and the Pacific. UNISDR Asia and Pacific Secretariat, Bangkok, Thailand

Blanco A (2014) DRRM in the Philippines: DRRM projects, geoportals and socio-economic integration. Paper presented at the regional workshop on Geo-referenced information system for improving disaster risk preparedness and management in Asia and the Pacific, UNCC, Bangkok, Thailand, 26–27 June 2014

Cabinet Office Japan (2015) White paper: disaster management in Japan. https://www.bousai.go.jp/kaigirep/hakusho/pdf/WPDM2015_Summary.pdf. Accessed 23 December 2017

CAIT Climate Data Explorer (2015) Climate watch. World Resources Institute, Washington, DC http://cait.wri.org. Accessed 13 Sept 2017
CCC (Climate Change Commission) (nd) Climate-smart planning and management: training program modules. CCC, Philippines. (unpubl.)
Dolcemascolo G, Kim Y, Mu TL (2011) Reducing disaster risk in cities—the Republic of Korea's experience. https://public.wmo.int/en/bulletin/reducing-disaster-risk-cities-—-republic-korea's-experience. Accessed 15 Jan 2018
Doroteo H (2015) Philippines: disaster risk profile and disaster risk reduction (DRR) framework: natural calamities. University of Oviedo—Department of Medicine Unit for Research and Emergency and Disaster, Asturias, Spain
Eckstein D, Künzel V, Schäfer L (2017) Briefing paper: global climate risk index 2018: who suffers most from extreme weather events? Weather-related loss events in 2016 and from 1997 to 2016. Germanwatch, Berlin, Germany
Elliot L (2014) Human security, climate change and social resilience. In: Elliot L, Caballero-Anthony M (eds) Human security and climate change in Southeast Asia. Routledge, Taylor and Francis Group, London and New York
Florano E (2013) Mainstreaming integrated climate change adaptation and disaster risk reduction in local development plans in the Philippines. In: Leal Filho W (ed) Handbook of climate change adaptation. Springer, Berlin, Heidelberg
Forest Management Bureau (FMB) (2013) Philippine forest facts and figures. FMB, Manila, Philippines
GED (General Economics Division) (2015) Planning commission, government of the people's Republic of Bangladesh seventh five year plan FY2016-FY2017: accelerating growth, empowering citizens. https://www.lged.gov.bd/UploadedDocument/UnitPublication/1/361/7th_FYP_18_02_2016.pdf. Accessed 15 Nov 2017
GGGI (Global Green Growth Institute) (2015) Korea's greengrowth experience: process, outcomes and lessons learned. GGGI, South Korea
Hasan Z, Akhter S, Ahmed S, Kabir A (2013) Challenges of integrating disaster risk management and climate change adaptation policies at the national level: Bangladesh as a case. Global J Human Soc Sci Geogr Geo-Sci Environ Disaster Manage 13(4):55–65
Index Mundi (2017a) Nepal demographics profile 2017. https://www.indexmundi.com/nepal/demographics_profile.html. Accessed 22 Nov 2017
Index Mundi (2017b) Bangladesh demographics profile 2017. https://www.indexmundi.com/bangladesh/demographics_profile.html. Accessed 22 Nov 2017
Index Mundi (2017c) Philippines demographics profile 2017. https://www.indexmundi.com/philippines/population.html. Accessed 22 Nov 2017
Johnson A (2017) Commentary: Bangladesh's climate change challenge, The National Bureau of Asian Research. https://www.nbr.org/research/activity.aspx?id=806. Accessed 3 Jan 2018
Kameyama Y (2015) Working paper: recent developments in climate change policy in Japan and a discussion of energy efficiency. University of Texas at Austin, USA
Kiko Network, NRDC (2017) The road from Paris: Japan's progress toward its climate pledge. https://www.nrdc.org/sites/default/files/paris-climate-agreement-progress-2017-japan-ib.pdf. Accessed 12 Jan 2018
Kuramochi T (2014) Working paper: GHG mitigation in Japan: an overview of the current policy landscape. World Resources Institute (WRI) and Institute for Global Environmental Strategies (IGES), Japan
Lasco RD, Evangelista RS, Pulhin FB, Lopez R (2010) The potential of community-based forest management to mitigate climate change in the Philippines. Small-scale For 9(4):429–443
Long J, Giri C (2011) Mapping the Philippines' mangrove forests using Landsat imagery. Sens 11(3):2972–2981. https://doi.org/10.3390/s110302972
Mahul O, White, E (2013) Knowledge note 6–2. Cluster 6. The economics of disaster risk, risk management, and risk financing: Earthquake risk insurance. World Bank, Washington DC

Mochizuki J (2017) How the Tohoku Disaster is changing Japan. Risk and resilience program. https://blog.iiasa.ac.at/2017/02/08/how-the-tohoku-disaster-is-changing-japan/. Accessed 15 Jan 2018

MoE (Ministry of the Environment) (2016) Ecosystem-based disaster risk reduction in Japan: a handbook for practitioners. MoE, Tokyo (Japan)

MoEF (Ministry of Environment and Forests), Government of the People's Republic of Bangladesh (n.d.) Climate change and disaster management in Bangladesh: Information brief. https://cmsdata.iucn.org/downloads/disaster_1.pdf. Accessed 20 Jan 2018

MoFA (Ministry of Foreign Affairs of Japan) (2018) Disaster prevention. www.mofa.go.jp/policy/disaster/21st/2.html. Accessed 15 Jan 2018

MoHA, DPNet (Ministry of Home Affairs and Disaster Preparedness Network) (2015) Nepal disaster report 2015. MoHA, DPNet, Kathmandu, Nepal

Nachmany M, Fankhauser S, Davidová J, Kingsmill N, Landesman T, Roppongi H, Schleifer P, Setzer J, Sharman A, Singleton CS, Sundaresan J, Townshend T (2015) Climate change legislation in South Korea: an excerpt from the 2015 global climate legislation study "A review of climate change legislation in 99 countries." Grantham Research Institute on Climate Change and the Environment, The Global Legislator Organization and Inter-Parliamentary Union, United Kingdom

NIDM (National Institute of Disaster Management) (2017a). Japan. https://www.nidm.gov.in/easindia2014/err/pdf/country_profile/Japan.pdf. Accessed 17 Dec 2017

NIDM (National Institute of Disaster Management) (2017b). South Korea. https://www.nidm.gov.in/easindia2014/err/pdf/country_profile/south_korea.pdf. Accessed 17 Dec 2017

NIDM (National Institute of Disaster Management) (2018) Lao People's Democratic Republic. https://www.nidm.gov.in/easindia2014/err/pdf/country_profile/Laos.pdf. Accessed 6 Jan 2018

O'Brien K, Sygna L, Leichenko R, Adger WN, Barnett J, Mitchell T, Schipper L, Tanner T, Vogel C, Mortreux C (2008) Disaster risk reduction, climate change adaptation and human security. Report prepared for the Royal Norwegian Ministry of Foreign Affairs, Global Environmental Change and Human Security (GECHS) Project, GECHS Report 2008:3

PAGASA (Philippine Atmospheric, Geophysical and Astronomical Services Administration) (2011) Climate change in the Philippines. PAGASA, Manila, Philippines

Pulhin JM, Ramirez MA, Pulhin P (2007) The evolution of community-based forest management. In: Decentralisation and state-sponsored community forestry In Asia: seven country studies of transition in forest governance, contemporary forest management and the prospects for communities to contribute to and benefit from sustainable forest management. Institute for Global Environmental Strategies (IGES), Japan

Rahman R (2017) Climate change policy is crucial for Bangladesh. The Opinion Pages. bdnews24.com. https://opinion.bdnews24.com/2017/12/26/climate-change-policy-is-crucial-for-bangladesh/. Accessed 13 Nov 2017

Raut K, Meyer R (2017) Disaster risk reduction: critical for Nepal. The Himalayan Times. https://thehimalayantimes.com/opinion/disaster-risk-reduction-critical-nepal/. Accessed 13 Dec 2017

Rey A (2015) Timeline: PH policies on climate change and disaster management. https://www.rappler.com/move-ph/issues/disasters/111307-climate-change-disaster-management-policies-philippines. Accessed 11 Mar 2018

Serafica R (2016) Only 1 in 10 LGUs has plan on how to respond to natural disasters. https://www.rappler.com/nation/138836-lgus-climate-change-action-plan. Accessed 11 Mar 2018

Shepard A, Mitchell T, Lewis K, Lenhardt A, Jones L, Scott L, Muir-Wood R (2013) The geography of poverty, disasters and climate extremes in 2030. https://www.odi.org/publications/7491-geography-poverty-disasters-climate-change-2030. Accessed 11 Nov 2017

Shrestha M, Panthi J, Sarker M, Hashmi Z, Suneetha P, Fox J, Aziz A (2017) Runoff scenario and water-based adaptation strategies in South Asia. APN E-Lib. http://www.apn-gcr.org/resources/items/show/1891. Accessed 4 Sept 2017

The Center for People and Forest (RECOFTC) (2014) Community forestry adaptation roadmap to 2020 for Lao PDR. RECOFTC, Bangkok, Thailand

The Center for People and Forest (RECOFTC), ASEAN Working Group on Social Forestry (AWG-SF) (2017) Social forestry and climate change in the ASEAN region: situational analysis 2016. RECOFTC—The Center for People and Forests, Bangkok, Thailand
The Strait Times (2017) Death toll from floods in India, Bangladesh, and Nepal crosses 1,200. https://www.straitstimes.com/asia/south-asia/death-toll-from-floods-in-india-bangladesh-and-nepal-crosses-1200. Accessed 17 Sept 2017
Tong PS (2009) Lao People's Democratic Republic Forestry outlook study: Asia-Pacific Forestry sector Outlook Study II Working Paper Series. Food and Agriculture Organization, Bangkok, Thailand
Toyama M, Sagara J (2013) Knowledge Note 6–1: Cluster 6. The economics of disaster risk, risk management, and risk financing: measuring the cost-effectiveness of various DRM measures. World Bank and CTI Engineering, Washington DC
UN General Assembly (2015) Transforming our world: the 2030 agenda for sustainable development. http://www.refworld.org/docid/57b6e3e44.html. Accessed 12 Sept 2017
United Nations International Strategy for Disaster Reduction (UNISDR) (2010) Synthesis report on 10 ASEAN countries' disaster risk assessment: ASEAN disaster risk management initiative. http://www.unisdr.org/files/18872_asean.pdf. Accessed 12 Sept 2017
Upreti BR, Bhattarai R, Wagle GS (2013) Human security in Nepal: concepts, issues and challenges. Nepal Institute for Policy Studies (NIPS) and South Asia Regional Coordination Office of NCCR (North-South). NIPS, Kathmandu, Nepal
Westphal MI, Hughes GA, Brömmelhörster J (2013) Economics of climate change in East Asia. Asian Development Bank, Mandaluyong City, Philippines
Yeasmin S, Sattar A (2015) Human security and clmate change-induced disaster: Bangladesh perspective. NU J Humanities Soc Sci Business Studies 2 (1): inclusive pages?
Yoon J, Choi J, Kim D, Kim S, Seo J (2014) National report for Habitat III. Republic of Korea. Ministry of Land, Infrastructure and Transport. https://www.habitat3.org/wp-content/uploads/National-Report-Asia-Pacific-Republic-of-Korea-Final-in-English.pdf. Accessed 17 Jan 2018

Mr. Mark Anthony M. Ramirez is the Executive Director of the Resources, Environment and Economics Center for Studies (REECS) based in the Philippines. He has more than 20 years of experience in the field of watershed management, community forestry, natural resources management, forest land use planning, climate change and DRR and has co-written articles and publications in these fields. A licensed forester, he has a master's degree in Social Development from the Ateneo de Manila University.

Dr. Linda Anne Stevenson is Head of Communication and Scientific Affairs at the Asia-Pacific Network for Global Change Research (APN), which is an intergovernmental organization with 22 member governments. With 17 years of experience in APN, she oversees over 70 national and regional projects annually across a broad range of disciplines, and works closely with researchers and decision-makers in Asia and the Pacific. She has significant experience in scientific capacity building that includes developing capacity of early-career researchers; plans dialogues that cross the science-policy; and engages with UNFCCC, IPCC, and IPBES. She has over eight years of lecturing experience in Japanese universities on "Global Environmental Systems." She holds a doctorate degree in Applied Chemistry from the University of Strathclyde, UK.

Dr. Juan M. Pulhin is full professor and former dean of the College of Forestry and Natural Resources, University of the Philippines Los Baños, Philippines. He was lead author of the 2007 Fourth Assessment Report of the Intergovernmental Panel on Climate Change (IPCC), Coordinating Lead Author of the Human Security Chapter of the 2014 IPCC Fifth Assessment Report and currently Lead Author of the Asia chapter of the IPCC Sixth Assessment Report. He has been a four-time visiting professor at The University of Tokyo in 2007, 2010, 2016 and 2017 and a visiting scholar at Waseda University, Tokyo in 2018. He has written and edited 11 books and published more than 100 technical papers and book chapters on natural resources policy and governance, and on the human dimension of climate change and disaster risks. He is a fellow, invited expert, and council member of several international organizations and a member of the National Academy of Science and Technology (NAST), Philippines.

Dr. Makoto Inoue is professor at the Faculty of Human Sciences, Waseda University, Japan and former professor at The University of Tokyo. He serves as dean of Graduate School of Human Sciences, Waseda University since September 2020. His specialty is environmental sociology, forest governance, and Southeast Asian studies. He has written and edited/co-edited 34 books, and published more than 140 original articles in peer-reviewed international and national journals on the social aspects of forestry and the environment. He serves as President of The Japanese Association for Environmental Sociology (June 2019–present), Vice President of Japanese Society for Current Anthropology (January 2020–present), Vice President of Society for Environmental Economics and Policy Studies (April 2018–March 2020), Councilor of The International Association for the Study of the Commons (January 2019-present), a board member of various academic societies in Japan, and an editorial board member of several international journals.

Chapter 18
Addressing Climate Change and the Risks of Disaster to Human Security: The Role and Initiatives of Civil Societies in Asia

Takako Izumi

Abstract Natural disasters and climate change affect human security as they undermine access to livelihood, clean water, food, property, homes, health care, and education. There are strong links among climate change adaptation (CCA), disaster risk reduction (DRR), and human security. Therefore, creating communities resilient to climate change and disasters enhances human security, and engagement by multiple stakeholders is crucial. In particular, involvement of civil society organizations (CSOs) is necessary to meet the present challenges to both CCA and DRR. Depending on the type of CSO, support and contributions vary. Nongovernment organizations (NGOs) customarily have knowledge, experience, and project funds. International NGOs (INGOs) are particularly significant in countries that cannot prioritize CCA and DRR because they confront other intractable issues. National and local NGOs understand the communities' challenges, but they have smaller budgets and less experience than INGOs. Adequately funded national and local NGOs could be ideal partners for communities, but when resources are inadequate, CCA and DRR initiatives require support from INGOs and international organizations. CSO networks can assist national and local CSOs in developing capacity, strengthening knowledge, supplying tools and guidance, and providing opportunity to meet donors. This chapter review challenges to implementing CCA and DRR projects by CSOs and CSO networks and shares case studies of ongoing projects in Indonesia, Afghanistan, the Philippines. It discusses how CSOs have contributed to CCA and DRR, and their implementation.

Keywords Civil society organizations · Climate change · Disaster risk reduction · Asia

T. Izumi (✉)
International Research Institute of Disaster Science, Tohoku University, Tohoku, Japan
e-mail: izumi@irides.tohoku.ac.jp

J. M. Pulhin et al. (eds.), *Climate Change, Disaster Risks, and Human Security*, Disaster Risk Reduction, https://doi.org/10.1007/978-981-15-8852-5_18

18.1 Introduction

Climate change is increasingly evident in the ascending sea levels, eroding coastlines, warming temperatures, declining supply of fresh water, and proliferation of endemic diseases. More frequent and extreme climate-related events escalate threats to life and property. They are associated with greater numbers of displaced people and property damage. Intensified seasonal weather patterns affect livelihood and agriculture. The globe's changing climatological risks demand improved disaster risk reduction (DRR) strategies (Solecki et al. 2011; Mercer 2010; Van Aalst 2006).

DRR strategies were developed by analyzing how historical hazards and socioeconomic and biogeophysical trends boost exposure to environmental perils. Given the global concern over climate-related disasters, it is urgent to determine whether the extant DRR strategies address the potential risks from climate change. Those strategies normally entail two elements. The first is upscaling policies to mitigate structural and nonstructural hazards as models update the information about the changing exposure and vulnerabilities to climate change (Solecki et al. 2011). The second element is the establishment of relationships among climate change adaptation-(CCA) and DRR-related institutes to share information and to work together.

There is considerable potential for synergy between CCA and DRR (O'Brien et al. 2006) in assessing risks, vulnerability to livelihoods and wellbeing, and institutional capacity. However, gaps between CCA and DRR establishments, inadequate communication between them, and their inherently different focus hinder joint efforts to tackle future challenges. Unless CCA and DRR enhance their synergies, human security remains at risk.

Birkmann and von Teichman (2010) identify scale, knowledge, and norms as impediments to synergy between institutions charged with undertaking CCA and DRR. CCA is analyzed on a global scale, whereas DRR focuses on local vulnerabilities and risks. CCA falls under ministries of the environment or meteorology, whereas ministries of the interior, land-management, defense, or development are responsible for DRR. CCA was viewed initially as an environmental problem for environmentalists and natural scientists who advocated reducing greenhouse gases as the best response. Only after the dangers of climate change were recognized was the need for CCA raised. Furthermore, scientific knowledge about climate change has not been communicated successfully to the public, and there are discontinuities in local and scientific data/tradition. However, combining different fields of knowledge could ensure the necessary linkages and effective management of CCA and DRR. Joint efforts are key to minimizing the risks of climate change and environmental disaster for human security.

Human security is "a condition that exists when the vital core of human lives is protected and when people have the freedom and capacity to live with dignity" (Adger et al. 2014). The Human Development Report by the United Nations Development Programme (UNDP) describes seven categories of human security: economic, food, health, environment, personal, community, and political (UNDP 1994). Human security is affected as natural disasters and climate change undermine access to livelihoods, clean water, food, property, homes, health care, and education (Adger et al. 2014; Ruttinger et al. 2015). Izumi and Shaw (2014) document strong links among CCA, DRR, and human security,

Creating communities resilient to climate change and disasters enhances human security, and engagement by multiple stakeholders is crucial (Jimba et al. 2012). Civil society organizations (CSOs) and the CCA and DRR establishments are active in Asia. CSOs range from small, informal, community-based organizations to large, high-profile international non-government organizations (INGOs) working through local partners across the developing world (UNDP 2013; Shaw and Izumi 2014). CSOs undertake several major roles (Izumi and Shaw 2011):

a. Provide support following small-scale climate-related disasters that occur yearly
b. Raise awareness of and increase the capacity of local governments and other CSOs to address climate change, its impact, and DRR initiatives
c. Partner with local governments to shape and implement their strategic plan for climate-related disasters
d. Mobilize additional support needed to manage climate-related disasters and expand collaborations with stakeholders into wider partnerships.

A significant characteristic of CSOs is that they are close to communities and better understand local risks, challenges, and capacities. Therefore, CSOs often lead community-based initiatives in collaboration with local governments. This chapter reviews challenges in implementing CCA and DRR projects by CSOs and shares case studies of ongoing projects in Indonesia, Afghanistan, the Philippines, and by CSO networks. It discusses how CSOs have contributed to CCA and DRR and their implementation.

18.2 Challenges to CCA

Adger (2003) emphasizes that many climatological risks require unilateral planning and intervention by the state; yet, CCA strategies require individuals and communities to act collectively. CSOs can empower the capacities of local actors to address these matters and to participate in setting local and national policy (Vignola et al. 2009).

However, traditional CCA efforts invoke top-down approaches because adapting to climate change was characterized originally as a problem of global pollution (Karim and Thiel 2017; Van Aalst 2006). The growing dissatisfaction with top-down, scenario-driven approaches prompted a search for locally relevant CCA methods developed from the bottom up (Van Aalst 2006). The impact of climate change is first visible in harm to residents' health and food and water shortages after droughts and floods. Therefore, community-based CCA and DRR involve community stakeholders in assessing current situations and challenges, discussing how to reduce vulnerability to climate change, and strengthening adaptive capacities (Allen 2006; Van Aalst 2006; Mansanet-Bataller 2010). However, communities often know little about climate change and adaptation strategies, perhaps reflecting the deficiencies of long-term, top-down approaches to climate change. Thus, CSOs, the private sector, and local governments become instrumental in raising awareness and developing communities' capacity to adapt to climate change (Thomalla et al. 2006; Ampaire et al 2017). Many CSOs engage in those activities, and CCA projects under DRR initiatives are increasing.

18.3 Challenges to DRR

The Hyogo Framework for Action (HFA), adopted at the 2005 UN World Conference on Disaster Reduction in Kobe, Japan, encouraged the involvement of multiple stakeholders in DRR (UNISDR 2005). The Sendai Framework for Disaster Risk Reduction (SFDRR), an international DRR blueprint drawn at the 2015 UN World Conference on Disaster Risk Reduction in Sendai, anticipates active participation by government, academia, the private sector, CSOs, and communities committed to collaboration with other stakeholders.

In particular, the SFDRR encourages the use of science-based data for setting environmental policy and making decisions via collaboration with academia and scientists (UNISDR 2015). CCA initiatives require the same approach. The SFDRR includes as a new element the seven global targets for monitoring and evaluation. Individual governments are expected to monitor and evaluate the implementation of the SFDRR based on the targets. Inadequate data concerning disaster damage and losses render it difficult to measure that progress, identify core issues and challenges, assess the economic impacts of disasters, and develop sound national policies. Unfortunately, many countries lack systems and mechanisms to collect data, and establishing them is challenging (Groeve et al. 2014). Thus, many countries neither incorporate science and technology into DRR nor link it effectively to the populace. For instance, science and technology groups enjoy little stature in developing local evacuation plans, and science-based early warnings and forecasts often are not disseminated publicly (Shaw et al. 2016). These deficiencies highlight the disparity between science and practice and the need to strengthen the links between science and the populace. Collaboration among scientists, scholars, and practitioners is indispensable to scale up current DRR measures.

In sum, the primary challenge to CCA is to strengthen collection of local data, community-based projects, and collaboration with CSOs. DRR initiatives need more robust policy based on science and technology, data for monitoring, links between science and the populace, and collaboration with practitioners such as CSOs.

Involvement of CSOs is crucial for ameliorating the present challenges to both CCA and DRR. Concerning CCA, involvement of CSOs can help communities understand climate change and its impact. CSOs can initiate and lead CCA projects with local governments and communities based on their knowledge and experience. CCA is often considered a scientific and environmental matter for which a top-down approach works well; however, the involvement of CSOs and local stakeholders in community-based, bottom-up approaches is absent in the current situation. Conversely, community-based approaches and involvement of local stakeholders have been the norm in DRR for many years. However, DRR lacks the scientific approach. It has become obvious that CCA and DRR require both local inclusivity and the application of science and technology. That endorses strongly the need for both establishments to work together and to learn from each other.

18.4 CCA and DRR Case Studies

This section presents case studies of CSO-based CCA and DRR projects in Indonesia, the Philippines, Afghanistan, and by two CSO networks to showcase various types of CSO involvement in CCA and DRR. Questionnaire surveys and interviews were conducted with CSOs in Indonesia, the Philippines, and those based in Japan that are working for Afghanistan as well as regional and international CSO networks.

a. Indonesia: case study of a CCA project by a CSO.
b. Philippines: case study of a DRR project by a CSO of journalists that incorporates CCA.
c. Afghanistan: case study of a DRR project by an international NGO that incorporates CCA.
d. CSO networks: a case study of CSO networks that develop the capacity of CSOs to meet climate-related disasters.

18.4.1 Indonesia

Established in 2001, the YAKKUM Emergency Unit (YEU) focuses on emergency response initiatives and reduction of risks from environmental disaster. Given increases in climate-related disasters in Indonesia, YEU is concerned with how climate change affects communities, particularly their food security and livelihoods (farming). YEU initiated a CCA project: climate-adaptive farming in drought-prone areas. YEU introduced climate-adaptive agriculture to four villages in the Gunungkidul District that are highly vulnerable to water and food shortages. It sought to strengthen crop resilience in Yogyakarta, a city in Java, with a grant from the Indonesian Climate Trust Fund. Problems communities face there include the following:

a. Each year, farmers suffer crop losses from landslides, volcanic ash, and attacks by long-tailed macaques. Climate change has led to droughts, long dry seasons, unpredictable rainfall, and long rainy seasons. Its secondary effects include pest incursions and environmental degradation.

b. Direct farming losses attributable to climate change include reduced yields, crop failures, scarcity of green grass for livestock, and lake water drying up. Alongside the economic losses farmers incurred are expenses of buying water and green grass during the long dry seasons. Farmers commonly sell their livestock to buy water and green grass.
c. The price of seedlings is high, and farmers cannot produce their own seedlings.

Eventually, farmers identified the following adaptation plan:

a. Climate-adaptive farming that combines organic farming, biological pest control, and drought-resistant seedlings from local varieties.
b. Conservation of an artificial lake as source of water during dry seasons.
c. Fermenting livestock feed to reduce costs of buying green grass during dry seasons (Fig. 18.2). YEU conducted capacity development on how to make fermented livestock feed from straw, grass, and leaves available during dry seasons (Fig. 18.1).

Farmers initially resisted shifting from traditional farming to climate-adaptive organic farming without guarantees of bigger harvests. In such cases, national NGOs that know local cultures, histories, and traditions were important in explaining new ideas and innovative approaches to farmers. After understanding the evidence and the results of organic farming, farmers were more motivated toward its practices.

Fig. 18.1 Farmers learn how to reproduce the biopesticide *Paenibacillus polymyxa* (photo provided by YEU)

Fig. 18.2 Making livestock feed in Giripurwo with an extension officer (photo provided by YEU)

18.4.2 Philippines (an NGO of Journalists)

Information and data concerning climate change are often complex, making it difficult to frame issues and identify their consequences. Mass media are essential in disseminating information understandably and issues widely. The Center for Community Journalism and Development (CCJD) is a non-profit platform for exchange between local news media and the populace on issues that have an impact on communities in the Philippines. The CCJD builds capacities for news reporting and develops mechanisms for media-citizen engagement. Alongside a broader engagement with human rights, governance and development, peace and conflict, and gender, the CCJD addresses the environment, reducing disaster risks, and adapting to climate change. It seeks for journalists to facilitate discussion among local governments, CSOs, and communities about local challenges and to identify solutions jointly. Climate change generally is considered a scientific matter; therefore, the media has a role to inform how it affects daily lives, how to reduce its risks, and how to prepare for its consequences. "Public journalism" helps communities understand climate change as a local problem.

The CCJD received funding in 2008 from the UN Democracy Fund for a 2-year project—Media, Democracy, and Development—to develop free, responsible, independent, professional news. It is a member and former lead convener of the DRR Network Philippines, a coalition of organizations and individuals advocating community-based DRR. In concert with this network, DRR projects implemented by the CCJD have included

a. DRR and media dialog in partnership with Oxfam GB and Christian Aid

b. A flashflood public awareness campaign via community media on Camiguin Island in partnership with the Center for Disaster Preparedness and JICA
c. Documenting lessons from early recovery efforts after Typhoons Ondoy and Pepeng in partnership with the Crisis Prevention Unit of UNDP Philippines
d. Reporting disaster risk, climate change, and water issues in central Mindanao in partnership with Oxfam

New DRR initiatives have been developed by journalists who participated in the CCJD training–a weekly DRR radio program and a training program for local journalists. Although the CCJD is not the lead organization in monitoring the implementation of DRR-related law in the Philippines, it works with CSOs and provincial media partners to monitor local compliance. Its media-citizen initiatives enhance transparency and accountability in local governance.

18.4.3 Afghanistan (CWS)

Afghanistan is beset by long-term conflicts, refugees, terrorism, human rights issues, and food shortages. These ongoing difficulties have redirected attention from DRR efforts to mitigate natural disasters, a serious threat there. The climate change component is often combined with DRR initiatives to raise awareness of root causes of frequent and devastating natural disasters. Despite the impact of climate change (Fig. 18.3), the urgency of other issues compels many countries to neglect the risks and influence of climate change. They are unable to reflect its importance in policy, and do not share information with the public. Those countries must depend on support by NGOs and international organizations. However, not all CSOs know enough about climate change to conduct CCA. Therefore, involvement and support by experienced INGOs are crucial.

Founded as an international humanitarian organization in 1947, the Christian World Service (CWS) is an INGO that operates in Japan, Myanmar, Thailand, Cambodia, Indonesia, Vietnam, Pakistan, and Afghanistan. CWS Japan is leading a DRR project in Afghanistan to enact legal frameworks to address floods, landslides, and earthquakes and to develop a hazard map in collaboration with government, the private sector, and communities. Hazard maps enable communities to understand disaster risks and evacuation routes and destinations. The project includes promoting broadcasters' role in providing early warnings of disaster, explaining climate change, and encouraging communities to undertake CCA projects guided by DRR approaches (Fig. 18.4).

Fig. 18.3 Riverside community in Bihsud District in Nangahar Province. Recurrent floods affect the community and erosion of river bank poses a significant threat (photo provided by CWS Japan)

Fig. 18.4 Meeting with local authority to communicate specific hazard risks in Bihsud District in Nangahar Province (photo provided by CWS Japan)

18.4.4 CSO Networks: GNDR and ADRRN

Support from international NGOs knowledgeable about CCA and DRR is indispensable in countries where government is occupied with other problematic issues. However, national and local NGOs focus on problem solving in their respective countries and have few opportunities for outside knowledge, new technology, and tools. Their ability to develop capacity, raise awareness, and share information is enhanced by CSO networks that provide opportunity for CSOs to learn from each other and to build contacts with other stakeholders. Following are instances of regional and international CSO networks.

18.4.4.1 Asian Disaster Reduction and Response Network (ADRRN)

The ADRRN is a network established in 2002 to promote coordination, information-sharing, and collaboration among CSOs and other stakeholders for effective disaster reduction and response in the Asia-Pacific. Asia is a large and varied landmass with multiple river basins and floodplains and high population densities in disaster-prone regions; the region cannot avoid the impact of climate change. The ADRRN is raising CSOs' concerns globally, promoting best practices in disaster reduction and response, and developing members' capacities for DRR and CCA (Izumi and Shaw 2012).

For instance, the ADRRN organizes yearly regional workshops for all members to discuss common concerns and urgent issues. This year's workshop dealt with localization and weather-related disasters in Asia—i.e., why disasters occur and how to tackle them. It emphasized understanding the risks of climate change, preparing for weather-related disasters, and developing DRR measures. Presentations by speakers from different sectors and group exercises enlarged awareness of climate change, weather-related disasters, and the need to disseminate information and cope with future impacts of climate change.

Networks that assemble diverse stakeholders allow CSOs to identify partners, increase knowledge, exchange information, and identify funding opportunities. For instance, the UK-based Humanitarian Innovation Fund (HIF) funded YEU's project in Indonesia. YEU and HIF met through the ADRRN. Staff exchanges and collaborations are ongoing.

18.4.4.2 Global Network of Civil Society Organizations for Disaster Reduction (GNDR)

The GNDR was established in 2007 to connect civil societies globally. It raises a collective voice for action to reduce risks for and increase the resilience of the most vulnerable. It envisions a world of resilient communities wherein the vulnerable can prepare for, mitigate, recover from, and adapt to hazards of a changing climate. CCA is its major focus. Its membership exceeds 850 organizations and 1,400 individuals in all regions.

This network is supported by donor agencies such as the U.S. Agency for International Development, the Federal Department of Foreign Affairs/Swiss Confederation, World Bank GFDRR, and the Department for International Development. Support and funding allowed the GNDR to establish a secretariat with 10 staff members in the UK and to open access to funding DRR and CCA projects.

The GNDR publishes guides, toolkits, case studies, and reports such as *Views from the frontline*, a global initiative of 500 organizations across 69 countries to monitor and measure progress of the UN and governments in strengthening community resilience. It highlights the gaps between national policies and local practices, identifies needs for collaboration to implement local projects, and brings voices from the frontline to higher levels.

18.4.5 Value of CSO Participation in DRR and CCA Projects

The Indonesia case study highlighted the importance of ensuring that CSOs at the national level have working experience with communities and understand the local risks of climate change. Due to their prior experience with DRR projects, YEU recognized how to work with communities and how important it is to understand their situation and maximize their existing capacity. CSOs at national and local levels have the ability to understand local needs and risks through long-term experience with community-based projects and prioritize community needs, capacities, and vulnerabilities.

What is unique about the Philippine case study is that the CSO is composed of a group of professional journalists. One of the challenges of CCA that was pointed out in the previous section is that communities often know little about climate change and adaptation strategies. Journalists in this CSO recognize that gap and possess the best tools for disseminating information via various media formats (newspapers, radio, TV, etc.). Through their work as journalists, they can have a tremendous impact on communities by raising awareness and sharing information.

In cases of particularly challenging environments such as the current situation in Afghanistan, the contribution and involvement of INGOs is crucial for DRR and CCA because national and local CSOs are often fully occupied with providing emergency assistance. However, without any disaster preparedness and risk reduction measures in place before an emergency occurs, immediate damages and long-term impacts will be beyond the response capacity that they own and expect. Therefore, it is crucial to have support of INGOs that normally have more funds, capacity for and experience with raising awareness, and capacity for development. Even if it is not a fully dedicated project for CCA or DRR as CWS is, it is most useful to attach DRR and CCA elements to existing projects in order to save costs and human resources.

The role of CSO networks is often neglected because they do not directly conduct a project; rather, their primary role is to support capacity development and advocacy efforts of member organizations that are required for CSOs. In addition, without the support of CSO networks, there are insufficient means of bringing the concerns and important messages raised by CSOs at national and local levels to regional and international levels. All levels of CSOs and CSO networks have different roles to play. As CCA and DRR are such diverse issues, collaboration and involvement of CSOs at all levels are indispensable.

18.5 Conclusion

O'Brien et al. (2008) argue that human security entails responding to climate change and risks of disaster in ways that reduce vulnerability and conflict and create a more equitable, resilient, and sustainable future. Climate change and the potential for environmental disasters raise concerns about long-term human security (Barnett and Adger, 2007). The term "human security" encompasses all the effects that climate change exerts on food and water shortages, insufficient housing and education, livelihoods, and health. It extends to the climatological effects on social welfare and rights, gender, age, class, culture, traditions, and living conditions (O'Brien et al. 2008). It is crucial to acknowledge that CCA and DRR are imperative to human security.

CSOs are crucial for implementing CCA and DRR initiatives (Allen 2006; Van Aalst et al. 2007; Vignola et al. 2009) and for reducing the gaps in CCA and DRR identified in this chapter. One major challenge is to give communities sufficient knowledge about climate change and risks of disaster. Depending on the type of CSO, support and contributions vary. NGOs customarily have knowledge, experience, and project funds. They engender opportunities for countries with few or inactive national and local NGOs with lesser knowledge, experience, and budgets. INGOs are particularly significant in countries such as Afghanistan that cannot prioritize CCA and DRR because they confront other intractable issues. CSOs at national and local levels understand communities' challenges, but they have lesser budgets and experience than INGOs. Adequately funded CSOs at national and local levels could be ideal partners for communities, but when resources are inadequate, CCA and DRR initiatives require support from INGOs and international organizations is needed, as shown in Indonesia and the Philippines. CSO networks can assist national and local organizations in developing capacity, strengthen their knowledge, supply tools and guidance, and provide opportunity to meet donors.

CCA and DRR address issues that threaten the quality of life and human life itself. It is crucial that CSOs inform communities about connections between climate change and environmental disasters and the connections among CCA, DRR, and human security. Doing so enables the populace to understand potential risks and the importance of approaches that transcend the traditional.

Acknowledgements The author's heartfelt appreciation goes to Ms Hepi Rahmawati (YEU), Mr. Takeshi Komino (CWS Japan), and Mr Red Batario (CCJD) whose comments and information were of inestimable value for this study.

References

Adger WN (2003) Social capital, collective action and adaptation to climate change. Econ Geogr 79(4):387–404

Adger WN, Pulhin JM, Barnett J, Gabelko GD, Hovelsrud GK, Levy M, Oswald, Sprging U, Vogel CH (2014) Human security. In: Climate change 2014: impacts, adaptation, and vulnerability. Part A: global and sectoral aspects. Contribution of Working Group II to the Fifth Assessment Report of the Intergovernmental Panel on Climate Change. Cambridge University Press, Cambridge, United Kingdom and New York, NY, USA, pp 755–791

Allen KM (2006) Community-based disaster preparedness and climate adaptation: local capacity-building in the Philippines. Disasters 30(1):81–101

Ampaire EK, Gassogne L, Providence H, Acosta M, Twyman J, Winowiecki L, Van Aten O (2017) Institutional challenges to climate change adaptation: a case study on policy action gaps in Uganda. Environ Sci Policy 75:81–90

Barnett J, Adger EK (2007) Climate change, human security, and violent conflict. Polit Geogr 26(6):639–655

Birkmann J, von Teichman K (2010) Integrating disaster risk reduction and climate change adaptation: key challenges—scales, knowledge and norms. Sustainable Sci 5:171–184

Groeve TD, Poljansek K, Ehrlic D, Corbane C (2014) Current status and best practices for disaster loss data recording in EU member states: a comprehensive overview of current practice in the EU member states. Publication Office of the European Union

Izumi T, Shaw R (2011) Roles of civil society in climate and disaster resilience of cities and local governments. Community Environ Disaster Risk Manage 6:261–280

Izumi T, Shaw R (2012) Effectiveness and challenges of an Asian NGO network for disaster reduction and response. Risk, hazards and crisis in public policy 3(2):1–15

Izumi T, Shaw R (2014) Civil society and cross-cutting issues for risk reduction: food security, health, human security, environment and microfinance. In: Shaw R, Izumi T (eds) Civil society organizations and disaster risk reduction: The Asian dilemma. Springer, pp 159–176

Jimba M, Hubbard S, Sase E, Suzuki T, Otsuka K (2012) Human security approaches for disaster recovery and resilience. Int Med Community 54(5):338–341

Karim MR, Thiel A (2017) Role of community-based local institutions for climate change adaptation in the Teesta riverine area of Bangladesh. Clim Risk Manage 17:92–103

Mansanet-Bataller M (2010) The challenges of adapting to climate change Climate report no. 21. CDC Climate Research

Mercer J (2010) Disaster risk reduction or climate change adaptation: are we reinventing the wheel? J Int Dev 22:247–264

O'Brien G, O'Keefe P, Rose J, Wisner B (2006) Climate change and disaster management. Disasters 30(1):64–80

O'Brien K, Sygna L, Leischenko R, Adger WM, Barnett J, Mitchell T, Schipper L, Tanner T, Vogel C, Mortreux C (2008) Disaster risk reduction, climate change adaptation and human security: A recommended report for the Norwegian Ministry for Foreign Affairs. University of Oslo, Norway

Ruttinger L, Smith D, Stang G, Tanzler D, Vivekananda J (2015) A new climate for peace: taking action on climate and fragility risks. An independent report commissioned by the G7 members. G7 Germany

Shaw R, Izumi T (2014) Civil society and disaster risk reduction: an Asian overview. In: Shaw R, Izumi T (eds). Civil society organizations and disaster risk reduction: The Asian dilemma. Springer, pp 1–13

Shaw R, Izumi T, Shi P (2016) Perspectives of science and technology in disaster risk reduction of Asia. Inter J Disaster Risk Sci 7:329–342

Solecki W, Leichenko R, O'Brien K (2011) Climate change adaptation strategies and disaster risk reduction in cities: connections, contentions, and synergies. Curr Opi Environ Sustain 3:135–141

Thomalla F, Downing T, Spanger-Siegfried E, Han G, Rockstrom J (2006) Reducing hazard vulnerability: towards a common approach between disaster risk reduction and climate adaptation. Disasters 30(1):39–48
UNDP (United Nations Development Programme) (1994) Human development report. UNDP
UNDP (United Nations Development Programme) (2013) Working with civil society in foreign aid: possibilities for South-South operation? UNDP
UNISDR (United Nations International Strategy for Disaster Reduction) (2005) Hyogo framework for action 2005–2015? United Nations
UNISDR (United Nations International Strategy for Disaster Reduction) (2015) Sendai framework for disaster risk reduction 2015–2030. United Nations
Van Aalst MK (2006) The impacts of climate change on the risk of natural disasters. Disasters 30(1):5–18
Van Aalst MK, Cannon T, Burton I (2007) Community-level adaptation to climate change: the potential role of participatory community risk assessment. Glob Environ Change 18:165–179
Vignola R, Locatelli B, Martinez C, Imbach P (2009) Ecosystem-based adaptation to climate change: what role for policymakers, society, and scientists? Mitig Adapt Strat Glob Change 14:691–696

Takako Izumi is Associate Professor at the International Research Institute of Disaster Science (IRIDeS), Tohoku University, Japan. She also serves as Program Director of the Multi Hazards Program under the Association of Pacific Rim Universities (APRU), which comprises 56 universities and academic institutes in the Pacific Rim. Her research interests include international and regional frameworks/strategies for disaster risk reduction, international humanitarian assistance, and DRR initiatives at the local and community levels. In May 2015, she has been appointed as a member of the UNISDR Asian Academia, Scientific and Research Advisory Group (ASTAAG) to support the effort of governments and stakeholders in the implementation of the Sendai Framework for DRR. She holds a Ph.D. in Global Environmental Study from Kyoto University, Japan.

Chapter 19
Response and Initiatives from the Private Sector to Address the Challenges Posed by Climate Change and Disaster Risks on Human Security

Perlyn Pulhin-Yoshida, Rodel D. Lasco, and Ana Veronica S. Gabriel

Abstract The private sector, encompassing a range of actors, is increasingly exposed to climate change and related disaster risks which are disrupting business operations. These risks are also posing a threat to human security at large, given the key role of the private sector in economic growth, job generation, and poverty reduction. While the role of the private sector in climate change adaptation (CCA) and disaster risk reduction and management (DRRM) is becoming more prominent, there are few literatures on how their initiatives help confront the challenges posed by climate change and disaster risks on human security. This chapter highlights two cases from the Philippines which describes private sector initiatives, entry points for private sector engagement, and actual practices on the ground. Key informant interviews and small group discussions were conducted to validate the gathered information from secondary data. The essential requirements common to these two private sector experiences in order to successfully carry out CCA and DRRM-related programs that contribute to human security are awareness and understanding of the climate change risks to business operations and value chains; recognition of the capacities to enable solutions and sustain action; and strong commitment to actively cooperate with the government, community, and other actors in collectively sharing resources to achieve mutual interests and goals. The stories suggest lessons on how the sector can maximize its resources and enhance its role as a key player in recognizing and adapting to climate change and making unique contribution to ensure a human-secured, progressive, sustainable, and resilient future.

Keywords Private sector · Climate change · Disaster risks · Human security · Resilience

---

P. Pulhin-Yoshida (✉)
Freelance Researcher-Writer, Osaka, Japan
e-mail: perlynyoshida@gmail.com

R. D. Lasco
The Oscar M. Lopez Center for Climate Change Adaptation and Disaster Risk Management Foundation, Inc. (OML Center), Pasig City, Philippines

A. V. S. Gabriel
Forest Foundation Philippines, Makati City, Philippines

J. M. Pulhin et al. (eds.), *Climate Change, Disaster Risks, and Human Security*, Disaster Risk Reduction, https://doi.org/10.1007/978-981-15-8852-5_19

19.1 Introduction

Climate change and disaster risk pose a threat not only to individuals and communities but also to the private sector, which encompasses a range of actors from private households, smallholding farmers, small and medium enterprises, corporate conglomerates, and global multinational companies. Direct exposure of core business operations to climate change impacts and indirect exposure through other market changes are challenging the private sector—whether they are big firms or small enterprises. Extreme weather, for instance, will increasingly cause business interruption and damage to physical assets. The Global Environment Facility (GEF) noted that the number of enterprises already affected by extreme weather events is growing, highlighting examples of droughts in wheat-growing regions leading to dramatic price rises and food riots, and floods in Thailand that are severely disrupting global manufacturing and posing a threat to human security (DCED 2016).

While these climate-related risks are becoming increasingly serious, some leading firms have started to take into account climate change in their business strategies and are assessing both threats and opportunities. Governments, communities, and non-government organizations (NGOs) are beginning to look for opportunities to build partnerships with businesses in taking climate actions. In the same way, the private sector is seeking partnerships with government and civil society to expand its participation in climate-resilience building. While the role of the private sector in climate change adaptation (CCA) and climate mitigation is becoming more prominent, literature is scant on how private sector initiatives help confront the challenges posed by climate change and disaster risks on human security. This chapter highlights case stories of these initiatives, the entry points for private sector engagement, and actual practices on the ground. The chapter also assesses the challenges the private sector is facing in optimizing its role toward human-secured and climate-resilient development and how these challenges can be addressed.

Human security, in the context of climate change, is defined by Adger et al. (2014) as "a condition that exists when the vital core of human lives is protected, and when people have the freedom and capacity to live with dignity." A number of factors including discrimination of many kinds, poverty, technological disasters, and climate-related impacts or extreme natural disasters undermine human security. Since the publication of the Intergovernmental Panel on Climate Change Fourth Assessment Report (IPCC AR4), many studies have demonstrated the linkages between climate change and human security through collaborative research initiatives and programs (Gleditsch 2012; Oswald Spring 2012). Systematic examination of these linkages is reported in several chapters of the IPCC Fifth Assessment Report (AR5), citing the different dimensions of human security such as public health, food security, livelihoods, culture, migration, conflict, and well-being, among others.

Findings from Adger et al. (2014) revealed the negative impacts of climate change on the following human security dimensions: livelihoods (including access to food and prices, housing, clean water, employment, and the avoidance of direct risks to health), culture and individual identity, migration, and conflict. Other studies show the adverse impacts of climate change on human capital and health that can lead to eroded human capability (UNDP 2007). Climate change can also undermine the ability of states and nations to provide the conditions necessary for human security. Some dimensions of human security are less sensitive to climate change and affected by other economic and social factors. While human security is at risk at varying degrees for vulnerable communities everywhere, many scientific studies suggest that the impacts of climate change on human security will be felt most in developing countries (Berrang-Ford et al. 2011), particularly by those populations that are already socially marginalized, have limited capital and financial assets, and are natural resource-dependent.

The private sector has resources and particular competencies that can significantly contribute to addressing disaster risks associated with climate change and its implications to human security. The sector can take the lead in introducing innovative technologies, designing resilient infrastructure, developing and implementing improved information systems, and managing major projects as part of the CCA strategies (Villamejor-Mendoza 2014). They are also in a unique position to place pressure on national governments to adopt more ambitious policies and cooperate actively in international negotiations. With better means, the private sector can implement its own initiatives to reduce its carbon footprint and ensure that it respects human rights across all of its activities (UNEP 2015).

As climate change impacts continue to pose a serious threat to the private sector's role as a key contributor to economic growth, job generation, and poverty reduction, it soon realized that its interventions must go beyond its own business agenda. Under the realm of their corporate social responsibility (CSR) and corporate social investment, many private sector organizations began engaging in activities to help rehabilitate and build resilient communities. Other business opportunities emerged that aim to help others in reducing their climate risks such as generating new finance to fill the huge deficit in available funds for adaptation; providing insurance and other risk management tools; and designing, fabricating, manufacturing, and distributing or offering products and services that can reduce the vulnerability of individuals and communities to climate change and disaster risk (Atteridge 2010; Villamejor-Mendoza 2014).

Two case stories from the Philippines of private sector response and initiatives toward human security amidst climate change and disaster risks are featured in this chapter. Key informant interviews and small group discussions were conducted to validate the information that was generated from the available secondary data. The case stories suggest lessons on how the sector can maximize its resources and enhance its role as a key player in recognizing and adapting to climate change and making unique contribution to ensure a human-secured, progressive, sustainable, and resilient future. The chapter benefits from a number of technical reports and publications that tackled and documented the impacts of climate change on the private sector and their completed, ongoing, and planned initiatives to address them. The synthesis of the reviewed literature and the analysis of the lessons gained from the case stories hope to fill some gaps and provide insights to better understand the motivation in adopting responses or practices at different circumstances to address the challenges posed by climate change and disaster risks on human security as gleaned from the private sector experience and perspective.

19.2 Impacts, Uncertainties, and Risks Posed by Climate Change and Disaster to Private Sector

19.2.1 Direct and Indirect Threats

As noted earlier, climate change poses direct and indirect threats to the private sector, which is understood in this chapter as any privately owned enterprise composing of a broad range of actors, including private households, smallholding farmers, small and medium enterprises, corporate conglomerates, and global multinational companies. Direct threat refers to direct exposure of the core business operation to climate change impacts. Besides the more obvious business interruption and damage to physical assets brought by extreme weather events, this also covers the impact on staff health and crop and livestock productivity (in the case of agribusiness) due to temperature changes. Water scarcity due to drought, for instance, will pose problems for hydroelectricity, river transport, and industrial cooling, to name a few. Agriculture, infrastructure operators, and tourism businesses are the sectors that will most likely be affected (DCED 2016).

Indirect threat pertains to indirect exposure such as disruption in the supply chain, increased competition for some resources, and changes in market demand as clients respond to climate change impacts. Other market changes will affect many enterprises, including food retail, finance, and insurance (PWC 2010).

19.2.2 Potential Climate Change Risks

Agrawala et al. (2011) categorized the private sector into three subsectors: goods, services, and joint goods and services. The goods subsector develops physical products such as merchandise, commodities, and minerals. The services subsector offers non-physical products like education, banking, and accounting. The joint good and services subsector provides both goods and services or needs goods to provide services.

The business components (raw materials, assets, supply chain, market, regulations, logistics, customers/clients, competitors, intermediate goods) of each of these subsectors are affected by climate change, either through direct damages and losses or indirect losses and disruption (Agrawala et al. 2011). Table 19.1 shows the potential climate change risks for the subsectors and for the sample industries under each subsector.

Table 19.1 Potential private subsectoral climate change risks (lifted from Agrawala et al. 2011)

Private sub-sector	Sample industries	Potential climate change risks
Good-producing sectors	Manufacturers	• *Physical risks*—disruption in operations due to extreme weather events; damage to infrastructure; restrictions in production due to rising temperature, variations in water quality and water availability
	Agriculture and mining businesses	• *Physical risks*—extreme weather events increase physical risks to business operations; risk of overflow of storage due to increased rainfall; resource extraction could be limited by sea level and water availability • *Supply chain and raw material risks*—water scarcity affects production • *Product demand risks*—changes in quality, quantity, and type of agricultural products • *Logistics risks*—risks to the transport corridors and transport hubs from where raw materials are processed and exported
Service-providing sectors	Retailers and distributors	• *Physical risks*—damage to products during transport due to extreme events • *Supply chain and raw material risks*—interruption, inefficiency, or delays in supply chain; difficulties with water scarcity and increased fuel prices • *Reputational risks*—decrease in product quality affecting reputation and consumer satisfaction
	Transportation	• *Physical risks*—extreme weather events causing delays, supply disruptions, and losses of goods; access to transport routes affected by flooding, permafrost thawing, and mass movements, subsidence due to drought

(continued)

Table 19.1 (continued)

Private sub-sector	Sample industries	Potential climate change risks
	Utilities	• *Physical risks*—disruptions of supply due to flooding or extreme events; business interruption due to extreme weather • *Supply chain and raw material risks*—reduced output due to water scarcity impacting hydropower and power plants using a thermal plant cooling system • *Product demand risks*—demand effects due to temperature changes • *Regulatory risks*—increasing pressure to conserve water in water-scarce areas
Good- and service-providing sectors	Financial businesses	• *Financial risks*—risks in investment portfolio where investments are made in areas with climate vulnerabilities; increased risk of customer default
	Information businesses	• *Physical risks*—disruptions of operations due to extreme weather events; difficulties in transportation
	Real estate businesses	• *Physical risks*— delays and disruptions in construction projects; damage to buildings and drainage problems; additional costs due to temperature changes increasing cooling loads • *Regulatory risks*—changes in building and design requirements • *Financial risks*—loss of value due to climate change impacts
	Other service businesses	• *Product demand risks*—tourism industry affected in its infrastructure and by changes in tourism demand caused by different climatic conditions

In terms of the private sector's view on climate change, two general tracks emerged–climate change as a risk but also as an opportunity (Davies and Klingel 2015). However, only a few private entities have yet taken climate change impacts into consideration on their existing activities and the possible market opportunities that will come out domestically, and even globally, in their planning horizon. Business climate change initiative is perceived to be either irrelevant or just an extension of their CSR (Villamejor-Mendoza 2014). The case studies done by Agrawala et al. (2011) reveal that only a few companies have carried out risk assessments and even fewer have assessed their adaptation options despite being aware of the climate risks they are faced with.

19.2.3 Impacts and Uncertainties

A single extreme weather event (e.g., flood in Thailand, December 2011) can displace workers by eliminating tens of thousands of job opportunities, disrupt global supply chains for manufacturing products from cars to computers, and reduce a country's gross domestic product by several percent (Biagini and Miller 2013). The indirect effect of extreme weather incidents can also severely impact business, for instance, as ports may suffer from disruptions in the transport of goods due to floods. A subtle but even more dramatic impact is how commercial business is linked to extended droughts in several wheat-growing regions, resulting in shortages, rapid rise in prices, food riots, and civil unrest, which are considered human security dimensions (Pauw 2015).

A review of *Physical Risks from Climate Change: A Guide for Companies and Investors on Disclosure and Management of Climate Impacts* indicated losses from extreme weather incidents that are progressively becoming a factor in business balance sheets (Ceres 2012). Following are the main reported losses:

The 2011 floods harmed more than 160 companies in Thailand's textile industry. A quarter of the country's garment production stopped.

Driven by droughts in 2010, agribusiness and food company Bunge reported a USD56 million quarterly loss in its sugar and bioenergy segments in its main growing areas.

The 2011 heat wave in Texas reduced the quarterly earnings of Constellation Energy (an electric power company) to about USD 0.16 per share, which forced it to buy incremental power at peak prices.

In terms of uncertainties, businesses encounter these in many forms (e.g., changes in customer choices, shifts in government policies, and competitors' actions). Business planning has long considered climate variability as an important uncertainty since biblical times, when farmers have had to prepare for droughts and floods. Utilities based their critical planning on the hottest summer demands. Given the uncertainties, it has always been a challenge to anticipate and quantify or even qualify the associated risks. A better knowledge of the risk will allow for more targeted and specific prevention measures and interventions. Unfortunately, business planning is normally defined by financial timeline and spatial scale much smaller than what current climate science and models can provide (Biagini and Miller 2013).

19.3 Private Sector Role in Addressing the Impacts of Climate Change and Disaster Risks on Human Security

For businesses to thrive and become productive, political stability and civil harmony should be maintained and so human security is crucial in business growth and sustainability. Business operations are equally important to human security as they generate jobs and can have lasting impact on human and environmental health, especially on the communities that are within or near their scope of operation. As the main driver of major economic activities in a community, the private sector plays a role in building and maintaining peace. Instability and conflict may arise or persist if the private sector will (a) worsen "tension through hiring local employees along divisionary/divisive ethnic lines," (b) employ "violent security forces," (c) exercise "poor governance," and (d) allow "bribery and other corrupt practices that reinforce inequity and contribute to unrest" (Appiah and Jackson 2015). On the other hand, if a company has initiatives toward human security, the adaptive capacity of the community to overcome climate-induced disasters may increase.

As climate change will continuously impact the private sector, measures to address it must be incorporated in the businesses' strategic management processes (Haigh and Griffiths 2009). The private sector's role in confronting the challenges posed by climate change and disaster risks on human security should also be strengthened. A number of factors can drive the private sector to engage: operational or strategic, commitment to CSR, response or compliance to regulation, and broader stakeholder relations (PWC 2010). As the private sector gets more involved, new goods/services may become in demand though other existing business approaches may also become unviable. This may motivate or discourage other businesses to engage and strengthen their involvement or disengage.

The following sections will highlight various private sector responses and capacities to combat climate and disaster risk, private sector initiatives in building resilience, some challenges, and entry points to improve private sector engagement in CCA and disaster risk management (DRM) efforts.

19.4 Private Sector Response

Public funding alone will not be enough to address the challenges posed by climate change. Given the currently constrained public budget to fully support climate action, policymakers advocate the active involvement of the private sector (Pauw 2015).

While there is a growing private sector engagement in trying to contribute to CCA initiatives and programs (Davies and Klingel 2015), there are very few available data that document private sector contributions to adaptation (Brown et al. 2015). Buchner et al. (2015) recorded that the private sector contributed around USD 242 billion in 2014 (project developers, 38.1%; corporate actors, 24%; households, 17.8%; commercial financial institutions, 19%; private equity/venture capital/infrastructure funds, 0.7%; institutional investors, 0.9%) toward low-carbon and climate-resilient growth. Project developers comprised the most responsive class and half of their investments (USD 45 billion) targeted onshore wind projects, which originated from and were invested in East Asia and the Pacific region (46%), western Europe (25%), and the Americas (15%). In a response to a Carbon Disclosure Project by member countries of The Business of a Better World (BSR),[1] it appeared that 75% of the members are addressing climate change in some form. However, their activities remain limited mostly to helping reduce greenhouse gas emissions (Avory et al. 2015).

A recent study found that a total of USD 270 million private sector adaptation financing by multilateral development banks (MDBs) was invested in 2013-2014. This amount made a total of USD 1.5 billion of MDB investment more climate-resilient. Since the financing mechanism is highly leveraged, the total project value of the investments made with adaptation components was only USD 5.5 billion. The investment portfolio was dominated by infrastructure, water, and agribusiness projects and more than 80% was deployed in middle-income countries. This mechanism shows how MDBs can enhance their effectiveness by continuing to mainstream climate change across banking teams and monitor adaptation finance (DCED 2016).

Another step taken by the private sector to respond to climate change is to prepare for and manage their exposure. Climate proofing of public sector investments emerged as a usual and sound response. Design and construction of workplaces and industrial zones and factories, among others, took into account the environmental and safety standards to protect business from sudden climate-related catastrophes. Other areas that were covered by the private sector, based on a survey of existing climate change-related programs across the world, are capacity building, education, and training; finance and insurance; food, agriculture, forestry, and fisheries; technology and information and communication technology; water resources; science, assessment, monitoring, and early warning; business; and human health (Villamejor-Mendoza 2014).

[1]BSR is a global nonprofit organization that works with its network of more than 250 member companies to build a just and sustainable world.

19.4.1 Private Sector Capacities

While the private sector is known for its ability to fund climate change initiatives as an immediate response, particularly on disaster relief-related activities, it also has the resources, knowledge, skills, and competencies that can be harnessed to elicit valuable contribution in managing risks and addressing impacts. "Most of the private sector skills and knowledge are based on efforts to manage risks from severe weather events such as insurance, modeling, and reconstruction; but there remains limited evidence of input on slow-onset events" (Surminski and Eldridge 2015). At the local setting, the private sector has the capacity to steer development and help alleviate poverty toward becoming more adaptive in responding to climate-related disasters. Some of the private sector initiatives and programs, however, may not be tagged as responses to climate change but as part of their business risk management or resiliency plan (Agrawala et al. 2011).

In terms of contribution to human security, private sector operations can inadvertently create livelihood opportunities and increase living standards through provision of goods and services. Its CSR also allows provision of land and shelter, access to education and health, improvement of infrastructure, and increased economic status. In times of climate-related disasters, the private sector provides financial aid and manpower, therefore leveraging the efforts of the government. Appiah and Jacksons (2015) summarized private sector contributions to human security through "advocacy, direct development assistance, generation of employment opportunities, community-based multi-stakeholder dialogues, philanthropy, and direct involvement in peace negotiations."

The private sector also has the capacity to reach poorer countries and people as noted in the case studies done by the United Nations Framework Convention on Climate Change (UNFCCC) Private Sector Initiative. For instance in the agricultural sector, café´Direct trains thousands of coffee and tea farmers to adapt, while Unilever Tea responds to deforestation and changing rainfall patterns by investing in efficient irrigation equipment, drought-tolerant tea varieties, and reforestation (UNFCCC 2014). Both examples demonstrate how the companies' investments benefit the supply chains that can drip down to smallholder farmers, eventually increasing their agricultural production.

19.4.2 Private Sector Initiatives in Building Resilience

This chapter adopts the IPCC definition of resilience, which is "the ability of a system and its component parts to anticipate, absorb, accommodate, or recover from the effects of a hazardous event in a timely and efficient manner, including through ensuring the preservation, restoration, or improvement of its essential basic structures and functions" (IPCC 2012). All of the private sector responses and capacities that were enumerated in the previous sections contribute to building resilience. Another aspect of the private sector's role in building resilience that is often overlooked is the development and design of new products or offer of services to help people prepare for, cope with, and adapt to climate change.

A strong case on how companies can design new products to build the resilience of vulnerable populations is the initiative from Sompo Japan Nipponkoa that successfully launched weather index insurance products in Thailand in 2010. This business opportunity responded to the social challenge faced by farmers who depend on agriculture for livelihood. The agriculture and food security is facing the brunt of climate change and extreme weather events, such as drought and flooding that create significant risks to human security. By designing the weather index insurance products, farmers can better cope with and recover from the drought and flood events.

Another case used a partnership model between a corporate (Prudential Corporation Asia [PCA]) and a non-government organization (one of the Prudence foundations) to build the resilience of the elderly in Thailand. The Foundation for Older Persons' Development, which is working with the elderly poor, used its understanding and experience, then provided access to the elderly communities. PCA, on the other hand, designed a community investment program in response to the local resilience challenge and leveraged its network of employees and agents in the country to implement the program. These efforts are aligned with the company's commitment to reduce risk and prepare for uncertainty in the countries where it operates (Avory et al. 2015). This is a mechanism wherein private sector and non-profit groups share resources to overcome a specific resilience challenge and thus build the resilience of local communities.

19.4.3 Challenges

A case study in Nepal found that the main challenge being faced by the private sector in responding to climate change was the lack of understanding on how climate change affects their profitability. Other challenges that were identified include lack of information about technological solutions, lack of clarity on policy provisions, and lack of financial incentives. Only 7% of the private sector respondents claimed to have allocated part of their yearly budget to climate change activities. Lack of information available to the business community, inadequate understanding of climate change, and lack of external support and guidance from the government and other stakeholders were also identified as key challenges faced by the private sector in taking climate action (Jamarkattel, n.d.).

In terms of profitability, corporate activity and investments mainly aim to gain reasonable, fairly quick, and expected returns at tolerable risk. Private investments in adaptation are often targeted to deal with climate risks than to explore marketable business opportunities. With limited potential for short-term profit, the private sector has to mainstream climate risks in their operation just to stay in business. Enabling environments for adaptation investments are not offered particularly in least developed countries as ascertained from their low scores on the "Doing Business" ranking of the World Bank and the "Global Competitiveness Index" of the World Economic Forum (Pauw 2015). Climate change uncertainties also make it more difficult to ensure "financing for adaptation investments with a long return horizon, or based on pre-emptive risk management." This is particularly true for small and medium-sized enterprises (SMEs) with limited credit access (DCED 2016).

Inadequate understanding of the projected climate change impacts on businesses is also a key problem. Companies find it difficult to factor scientific data with long-term trends and generated over large areas into small-scale decision-making. On a more system-based approach in looking at the projected impacts, businesses are not knowledgeable or experienced in quantifying the ecosystems' value to their operations. The natural resources and their ecosystem services often get undervalued, resulting in underinvestment in adaptation measures (DCED 2016).

While there are accessible market options for adaptation and DRM consultancy services, they are mostly based in developed countries and fees are expensive. These types of services are just starting to emerge in many developing countries. SMEs with limited budget and businesses stationed in remote rural areas are unlikely to avail of these kinds of consultancy services (DCED 2016).

Enabling policy and regulatory environments is another challenge as manifested in the lack of clear policy direction from the government, leaving many businesses on a "wait and see" mode. A national adaptation program of action is usually developed by governments with little private sector participation. Another issue is the misallocation of natural resources such as water to ensure adequate access for all users in times of drought or when there is shortage in water supply (DCED 2016).

While the private sector is encouraged by policymakers to actively participate in various climate change programs, the private sector hardly understood the flow and system of engagement and public–private collaboration is usually hampered by a number of barriers. These barriers include a range of market failures that require government intervention. If governments are only clear on how to better engage the private sector, then efforts from the businesses will be widely understood, better appreciated, and vividly seen as good business practice (Biagini and Miller 2013; Villamejor-Mendoza 2014). Incentives would also help encourage more private sector engagement as many adaptation measures require immediate investments, but the benefits may take years to be realized. Common adaptation investment, such as improved flood defenses, has a public good element that may discourage individual companies' investment (DCED 2016).

Despite the abovementioned challenges, the private sector is generally responding to climate change impacts, although on a small scale. Many enterprises, according to Pauw and Pegels (2013), incorporate adaptation measures in their day-to-day activities without realizing it; hence, the question on whether they understand "adaptation" or they just do not bother tagging them as "adaptation" strategies. Few businesses are tackling climate change from a risk management perspective to avoid the worst-case scenario impact of climate or extreme weather events in their operations, rather than from the main causes of vulnerability. These issues are often of a social, systemic, and complex nature that would require more than adjusting business models or building/refurbishing infrastructure (Avory et al. 2015). The required actions are often perceived by the private sector as harder to take by themselves alone. The succeeding section will discuss ways in which the private sector can adapt and build resilience in their operations and in the societies where they operate that will contribute to human security.

19.4.4 Entry Points

International efforts are under way to promote and further encourage private sector climate action engagement in developing countries. Most of these are in terms of larger businesses operating at a global level and linking to global business networks and private sector platforms (Surminski and Eldridge 2015). An important step to explore as an entry point in enhancing private sector involvement is the extent these engagement models could be used. It is essential to document the lessons learned, especially for SMEs, which could be very important for the risk aspects in order to understand impacts and develop response measures and strategies.

Another entry point is through improved "climate proofing" of private sector investments. There are also emerging business opportunities for the private sector that go beyond managing their own climate exposure and start helping others to reduce climate risks. Atteridge (2010) cited a few of these opportunities, including the generation of new financing to help fill the massive deficit in available funds for adaptation; design, manufacture, and distribution of more innovative goods and services that can help reduce the vulnerability of individuals and communities to climate change and the provision of risk management tools. These tools may include subsidies, road maps for developing and disseminating key technologies and services; enhanced communication systems between public and private actors; insurance; equities; loan guarantees; tax incentives; and foreign exchange liquidity facilities (Pauw 2015).

Avory et al. (2015) introduce a framework for potential adoption by companies aimed at building their adaptive capacity to climate change in the form of different "resilience wedges" or "assets." The framework offers an opportunity to adopt multiple assets that companies can pursue together to build greater resilience to climate change, both in their operations and in the communities in which they operate. These are the four key takeaways (Table 19.2) on how business can build resilience to climate change:

Table 19.2 Ways on how business can build resilience to climate change (modified from Avory et al. 2015)

Way	Explanation
Taking resilience building as one dimension of risk management	Incorporate climate change resilience building as one dimension of risk management. For example, climate change and water security can be added as key components of corporate due diligence and risk assessment protocols
Building adaptive capacity through a holistic approach	Take a holistic approach to building adaptive capacity that considers the relationship between the company and society, the natural resources they depend on, the people they employ, the customers or suppliers they work with, and the social license to operate granted by local communities
Collaborating with public-and private-sector players	Identify opportunities to collaborate with public- and private-sector players in order to amplify the impacts of resiliency efforts
Viewing increasing adaptive capacity as a business opportunity	Track corporate costs, investments, and return on investments on CCA projects to help assess and make the business case for investment over time

19.5 Case Stories from the Philippines

This section presents two case stories of private sector response toward human security amidst climate change and disaster risks: (1) the case of Energy Development Corporation (EDC), a private company, and (2) the case of the Philippine Disaster Resilience Foundation (PDRF), an umbrella organization of the private sector for DRRM.

19.5.1 Energy Development Corporation

EDC is a geothermal energy industry pioneer and the recognized world leader in wet-steam field technology. Its missions are to (1) provide the present and future generations a better life with clean and renewable energy and (2) promote customer and investor interests, employee development, community welfare, and environmental stewardship at all times. To achieve its first mission, EDC invests in hydropower, solar power, and wind energy projects in the Philippines aside from its main geothermal energy projects. Its operations are scattered throughout the country.

Three of the greatest threats to its operations are strong winds, landslides, and flooding brought about by strong typhoons or extreme rainfall. Aside from disrupted operations, these events cause covered pipes, impassable roads, warping of transmission lines, and damage to infrastructure, among others. Since EDC power plants and facilities are mostly found in mountainous areas prone to extreme weather events, their major strategies involve measures in ensuring infrastructure integrity and enhancing the design to withstand such events.

EDC is aware of the increasing climate change risks. In the past 3 years, the company has embraced corporate sustainability, giving equal attention to environmental, social, and economic aspects of their operation. They recognize that climate change is their greatest sustainability challenge. Thus, the operations across the value chain are guided by a "sustainability philosophy," which explains the diverse initiatives that are directly or indirectly related to CCA, climate mitigation, and disaster risk. Contributing to the sustainability efforts are the watershed management and CSR groups and the emergency response and disaster preparedness unit. Moreover, whenever there is a typhoon that threatens the operation, each business unit has a Crisis Committee that leads emergency and response activities for the employees and the surrounding communities. The latter is undertaken in coordination with the local government to ensure that efforts are complementary rather than duplicated.

19.5.1.1 Programs and Projects

EDC representatives were interviewed to know more about their Health, Education, Livelihood, and Environment (HELEn) Program and the Leyte Rebuilding Project. Two separate interviews were conducted in June 2017. Company sustainability reports and other useful references were also reviewed.

Health, Education, Livelihood, and Environment Program

EDC has engaged communities as early as 1987, mostly on liaison work related to company's concerns and participation in general activities that promoted good relations. Noting the great need of the communities for socioeconomic development and how it impacts on environmental protection, EDC overhauled its program and launched the HELEn Program in 2005. The revitalized CSR program focused on health promotion, educational assistance, livelihood development, and environmental stewardship, which deepened the relationship of the company with the communities.

A community needs assessment was conducted to help design the program. From the baseline data, several activities were implemented per focus area. Annually, EDC assists an average of 70,000 beneficiaries within 47 partner barangays in 11 host towns where their business units are located. The program attributes its accomplishment to collaborations with relevant stakeholders such as the Department of Health (DOH), Department of Education (DepEd), Commission on Higher Education (CHED), Technical Education and Skills Development Authority (TESDA), Department of Environment and Natural Resources (DENR), Cooperative Development Authority (CDA), Provincial DRRM Council, Philippine Red Cross (PRC), various universities/colleges, partner organizations, local government units (LGUs), some NGOs, and local communities.

Health Since EDC communities are located in remote upland areas where health services are limited, the company helps strengthen the capability of the barangay health centers (BHC) through BHC improvements and equipment upgrade, health worker training and incentives, health education, and focused medical assistance. At the time of writing, 9,470 residents in 45 health centers annually have benefited from these. All of these contribute to achieving functional health systems that play important roles in times of disasters and calamities and are elements of a more secured future. In terms of the health conditions of community members, the fewer the number of residents with chronic illness and disability, the more likely that the community can exhibit greater resilience compared with those communities with more of these characteristics (Fig. 19.1).

Education Education programs ensure that the target student beneficiaries become productive members of society by being gainfully employed or becoming income-generating entrepreneurs and , more importantly, that they have been imparted the correct values. The initiatives start with enhancing the school's capability in terms of classrooms/facilities and teachers' skills, retention and achievement (e.g., scholarships or financial and school supply subsidy, *Gulayan sa Paaralan* [vegetable garden in school]), and career guidance. For the school improvements, EDC implements equity sharing, wherein they provide the materials while the community shoulders the labor. The *Gulayan*

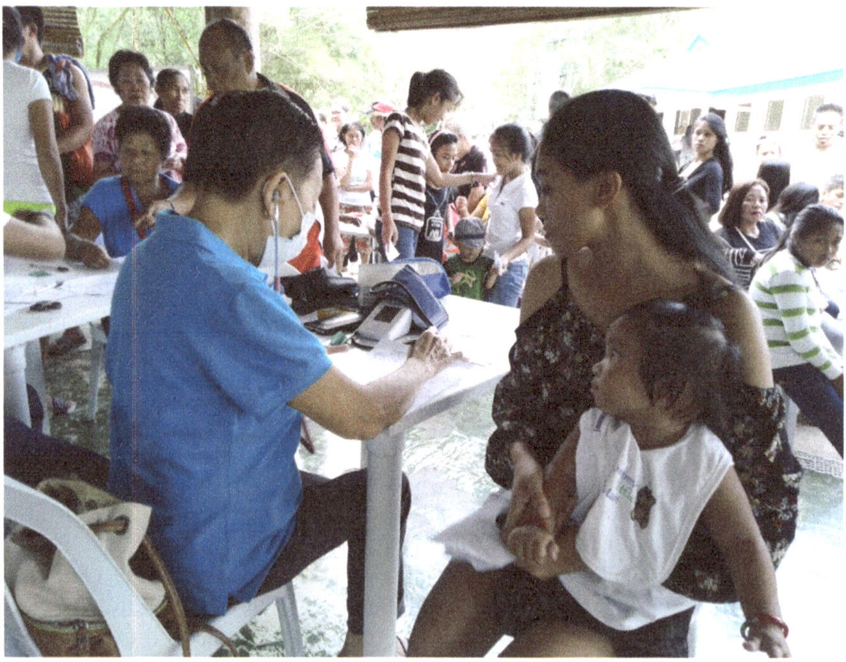

Fig. 19.1 Medical mission in one of the EDC CSR communities (*Source* EDC CSR and watershed management department)

sa Paaralan is done to promote sustainability and self-sufficiency since schools can grow their own vegetables either for feeding program or for fund raising. Upon entering high school, students who are most deserving and indigent are further encouraged to excel through scholarships. Top performers are given the opportunity to advance further to either a technical-vocational skills training or college education. To date, 29,000 students annually (of which 20,254 are elementary students) have benefited from these projects (Fig. 19.2).

There are three more subprograms under Education, namely the Leyte Schools for Excellence, the College Admission Review and Readiness (CAREERS) project, and the Kananga-EDC Institute of Technology (KEITECH).

The *Leyte Schools for Excellence* is a complete elementary school package undertaken in two schools benefiting more than 300 students each. In this partnership, EDC constructed the school buildings and provided school supplies and feeding program ingredients; DepEd handled the teachers' skills training; while parents actively participated in food preparation and parental guidance. The LGU supports EDC by providing a school bus and augmenting school rooms. The project has now gone beyond its original purpose of providing quality education and has become a catalyst toward community development.

The *CAREERS* project bridges top students to avail of quality education in prestigious colleges/universities such as the University of the Philippines (UP) and local state universities and complements this with enrichment workshops and solidarity building sessions that will mold future engineers, scientists, business managers, and

Fig. 19.2 CAREERS Batch 2018 University of the Philippines' graduates (*Source* EDC CSR and watershed management department)

educators. To date, 77 college graduates were produced through CAREERS project, 58 from UP while 19 from different local state universities and colleges (SUCs). Currently, a total of 124 college scholars are enrolled in UP and local SUC's. They are given a monthly allowance of Php 5,500 (~ 106 USD). These scholars are becoming community role models and green ambassadors in their communities.

KEITECH was designed to develop the vocational skills of the students, along with value formation. The program involves 11 months of training of three to four courses offered by TESDA. The courses are on construction, metals and engineering, tourism (hotel and restaurant), and welding and electrical. To date, the following are the achievement highlights of the program:

- 1,861 graduates with 100% consistent passing rate and good work ethics, broken down as 686 multi-skilled graduates, and 1,165 graduates from typhoon Yolanda (international name: Haiyan) areas (3-month training for Leyte rebuilding construction work)
- 22 partner industries from cruise ships, hotels, industrial parks, construction, etc.
- 95% employment rate, with 652 employed locally in six cities and 42 employed abroad in six countries (Fig. 19.3).

The achievements under the education programs can indirectly reduce the vulnerability of the community to climate and disaster risks. According to Lutz et al. (2014), education can have more power in reducing vulnerability than economic development. In a press release posted by the International Institute for Applied

Fig. 19.3 Technical vocational skills training for out-of-school youth (*Source* EDC CSR and watershed management department)

Systems Analysis, Muttarak (2014) said, "Education directly improves knowledge, the ability to understand and process information, and risk perception. It also indirectly enhances socioeconomic status and social capital. These are qualities and skills useful for surviving and coping with disasters."

Livelihood Livelihood support aims to create self-sustaining communities. Its framework revolves around three factors: livelihood for self-sustenance, business enterprise for profitability, and catalyzing local community development. The activities under this initiative are (1) capacity building and (2) livelihood support. To date, EDC has 11,474 cooperative members and six assisted cooperatives, which have attained zero livelihood budget that are now self-sustaining with their own CSR projects. These accomplishments allow communities to improve their economic status through better access to resources that would develop their skills, give more livelihood options, and lessen the dependence on jobs that have unpredictable income. The better the economic status, the higher the chances that the community can better prepare for and cope with the impacts of climate change (Fig. 19.4).

Environment The three areas under the environmental interventions are habitat protection for delivery of ecosystem services, biodiversity preservation, and disaster preparedness and response. The main program under biodiversity preservation is called *Binhi* (seed), a greening program that started in 2008. It consists of four pillars: (1) Tree for Food (e.g., planting agroforestry crops as alternative source of livelihood), (2) Tree for Life (bridging fragmented forests to attract wildlife), (3)

Fig. 19.4 Beneficiaries of the livelihood program (*Source* EDC CSR and watershed management department)

Tree for Leisure (ecotourism), and (4) Tree for the Future (saving endangered tree species).

In line with those pillars, BINHI's Tree for Life and Tree for Food modules push for restoration of denuded lands and conservation of natural forests in the watersheds of EDC's geothermal and hydro project sites. Under these modules, native and indigenous trees, fruit-bearing, and high-value tree crops are planted and grown into forests and agroforests. To date, 6.3 million seedlings have already been planted in 9,196 hectares of degraded lands since the project started in 2009. These were implemented in partnership with 88 organizations consisting of farmer's organizations, indigenous people's organizations, NGOs, federations, and LGUs (Fig. 19.5).

BINHI's Tree for Leisure module focuses on developing areas within its geothermal reservations with ecotourism potential. The main goal is to create alternative livelihood for communities to depart from destructive livelihood practices such as non-traditional *kaingin*, timber poaching, wildlife hunting, and other forest-based illegal activities. In 2017, a feasibility analysis of the different ecotourism models was explored to identify the best model to be implemented in the pilot site–Bacon Manito Geothermal Project (BMGP). The community-based ecotourism model has remained to be one of the most viable options and will be further explored.

The Binhi Tree for the Future project is a CSR-environmental initiative, which includes rescue of priority native tree species from different parts of the country, propagation to preserve the gene pool, planting in safe havens such as schools/parks,

Fig. 19.5 BINHI Tree Planting in Vegetative Material Reproduction Nursery in Antipolo—an annual tree planting activity among Lopez Group of Companies in celebration of National Arbor Day (*Source* EDC CSR and watershed management department)

and advocacy purposely done to increase awareness on the role of each Filipino to rescue these vanishing species. In 2015, EDC identified 96 threatened species (e.g., *yakal, tindalo*, etc.), which the company searched and rescued for 5 years.

From 2008 to 2018, BINHI Tree for the Future already planted 8,327 threatened trees in 16 regions (mostly in schools and parks) and established six arboretum or a collection of threatened trees with an average survival rate of 95%. Native trees were planted because they are already well-adapted to the Philippine climate and are resilient to typhoons; their woods are usually dense with high carbon storage capacity and deeply rooted, which helps stabilize the soil.

Aside from these activities, EDC also established the first automated mist irrigation nursery facility called the Vegetative Material Reproduction (VMR) Nursery to propagate more threatened tree species in the absence of seeds or wildlings. Through VMR, seedlings are planted in hedge gardens and propagated asexually through cuttings (Fig. 19.6).

BINHI extended its coverage in 2016 by entering an agreement with DENR through the Adopt-A-Wildlife Species (AAWS) Program, which aims to update the ecological status of selected threatened tree species through tree inventory, planting, propagation, and conduct of educational awareness in using native tree species.

Forests are sensitive to climatic changes. For example, increased temperature and uncertainty in precipitation can make forest ecosystems more vulnerable (World Bank 2016). Despite the sensitivity, forests have a major role in combating climate change. Through the activities under the BINHI program, the capacity of the forests

Fig. 19.6 BINHI VMR Nursery in Bago City, Negros Occidental (*Source* EDC CSR and watershed management department)

to perform its role is enhanced. Besides serving as a carbon sink to mitigate climate change, the activities also promote food and livelihood security by ensuring that ecosystem services are maintained. Moreover, reforestation efforts contribute to landslide mitigation.

DRRM initiatives recently emerged with realization of the need to be prepared and resilient. While still under the Environment subprogram, these initiatives can benefit other program areas (health, livelihood, education). For instance, EDC has initiatives on promoting disaster preparedness through Barangay Emergency Response Team or village organization and training and school-based orientation and disaster response through rescue, relief, and rebuild. These activities strengthen the disaster preparedness and response of the communities against climate and disaster risks. A prepared community minimizes the chances of post-disaster conflicts and civil unrest, which can further exacerbate the impacts of a disaster.

While all these HELEn focus areas are in place, it is also important for EDC to respect and work with the indigenous peoples in its Mt. Apo geothermal project through cultural preservation, capacity-building, and enterprise development. Indigenous peoples are very vulnerable to climate change and disasters due to their (1) dependency on natural resources, (2) lifestyle being closely related to the environment and its resources, (3) political and economic marginalization, (4) loss of land and resources, (5) human rights violation, and (6) discrimination and unemployment (United Nations nd). This effort of EDC contributes to the resilience of the indigenous community in Mt. Apo by enhancing their personal and cultural/institutional capacities.

Leyte Rebuilding Project (Building of Typhoon-resilient Schools)

The construction of typhoon-resilient school buildings project started in February 2014, wherein EDC and the First Philippine Holdings Corporation, with the help of several donor partners, committed to and acted as the implementing arms of the Office of the Presidential Assistant for Rehabilitation and Recovery whose function has been transferred to the National Economic Development Authority in 2016. The project aims to build ready-to-use typhoon-resilient schools for Typhoon Yolanda-affected communities. As project manager, EDC coordinated with other private groups, the academe, and government on setting building standards and making appropriate designs; facilitated the involvement of beneficiary LGUs and communities; screened and managed the contractors to ensure that they abide by the new specifications; and transparently managed the donated funds.

Currently, the program has four phases. A summary of all the phases can be accessed through this link: https://goo.gl/DXG4zY. As of 2017, EDC has built, on behalf of the donors, a total of 150 classrooms.

The selection of sites was based on the recommendation of DepEd, endorsement of the donors, and evaluation of the project team. The sites for Phases 1 and 2 were mostly selected from the 1,200 fully destroyed schools in Leyte caused by Typhoon Yolanda. The criteria for selection are availability of space, necessity, student population of the school, amount of current donation at the proposed site, hazard and risk assessment, and soil quality, among others. Moreover, the schools should show proof of tenure over the land.

The design of these typhoon-resilient school buildings can withstand up to 250 kph of wind factor and at least 7.2 magnitude of earthquake. These standards were based on DepEd's building standards for schools, with an expected life span of more than two decades. The schools, which can hold 45 students per room, also include ready-to-use facilities such as electric fans, persons-with-disability (PWD) access-concaved designed blackboards, wooden armchairs, teacher's table and chair, gender-based toilet facility inside the rooms and light-emitting diode (LED) light bulbs. After the school building construction, all documents (e.g., notarized deed of donation, building and occupancy permits, building plan) were turned over to the DepEd Regional Office.

To date, there are around 9,000 students who benefit from this program annually. Aside from having new school buildings, it increased the enrolment rate because students were inspired to go to school. Donors also benefited through tax incentives under the Adopt-A-School Program of DepEd. Other benefits that the schools got after the construction were (1) laptops from EDC, (2) educational kits from Knowledge Channel, and (3) school supplies and materials for teachers. EDC also started planting at least 10 seedlings per school in support of the BINHI project.

More than just rebuilding damaged schools, the Leyte Rebuilding Project contributes to the resilience of the communities by providing them better facilities that can withstand disasters and serve as evacuation centers. The project also ensures the continuous learning of the students after the disaster, which, as discussed in the Education section, can increase their ability and capacity to prepare for, survive, and cope with disasters.

Box 19.1 About The Oscar M. Lopez Center for Climate Change Adaptation and Disaster Risk Management Foundation, Inc. (OML Center)

The Oscar M. Lopez Center for Climate Change Adaptation and DisasterRisk Management Foundation, Inc. (OML Center) The OML Center, named after its founder and chairman emeritus, was established in 2012 and envisioned to be a privately funded institution that would support the generation of the science and technology needed for building resilient communities. While initial endowment came from the Lopez Group of Companies (including EDC), currently, majority of the Center's funding is coming from EDC. The practical, evidence-based solutions in the fields of CCA and DRM that will be generated by the Center are expected to provide the knowledge required to make climate-informed decisions (OML Center nd). Since the OML Center was established, it has awarded PhP 25 million (~500,000 USD) worth of research grants that were implemented in at least ten regions across the country, worked on a number of climate change projects, including six internationally funded projects, produced science-based outputs, established networks and partnerships, influenced some policy and business decisions, and affected the lives of individuals and communities that are vulnerable to climate change and climate-related disasters.

19.6 Challenges

HELEn Program

Getting the community to cooperate at the start of the implementation of the program was a challenge, especially for projects that needed more of their engagement (as in the case of livelihood). Meanwhile, a few LGU officials were also hesitant at first as they had other priorities that may not be aligned with the proposed activities. For the communities, their perceived needs were different from what they really need. EDC found out from surveys that the needs they identified are more of what they already have. The company was able to address these issues through constant communication and active engagement with the community. They employed purposive information drives and several informal small group discussions to strengthen their relationship with the communities and LGU partners. Other challenges include the impact of severe weather conditions that affect ongoing projects or shift the focus of attention on rebuilding rather than progressing forward.

Leyte Rebuilding Project

Access to the area and transport of materials, especially during monsoon season, was very challenging (e.g., have to cross rivers; site of Phase 1 is located in a mountain cliff). This caused delays in construction.

It was also a struggle to work in areas outside the scope of EDC's business units. The team had to open several site offices, find new networks, and build new relations. Less security and familiarity in other areas became a challenge as well.

Forming a team with understanding on the project was also a problem. It was hard to find full-time, high-quality engineers and build schools simultaneously, given the size of the team. Another challenge was logistics and purchasing that need swift arrangements to meet deadlines. Getting the contractors follow predetermined materials, standards, and specifications and prevent them from getting subcontractors was difficult and required very close monitoring and evaluation.

The OML Center

In spite of successes in its first 5 years, the Center continues to exert effort in trying to engage broad and lay audiences and create lasting impact from the research and scientific outputs it produces and supports. Building on this experience, as well as on the influences of external socioeconomic political factors, the Center is currently revisiting its organizational purpose and mission to examine how it can best make a difference or add value in the field of CCA and DRRM. The sustained support of EDC is a significant factor in enabling the Center to continue its work in building resilience in the face of changing climate.

19.6.1 Philippine Disaster Resilience Foundation

Following the devastation caused by typhoon Frank (international name: Fengshen), typhoon Pepeng (international name: Parma), and tropical storm Ondoy (international name: Ketsana), the Office of the President issued Executive Order No. 838 creating the Special National Public Reconstruction Commission (Public Commission). Mandated to spearhead effective reconstruction measures addressing the needs of disaster-hit communities, the Public Commission tapped the private sector to channel more support for its reconstruction programs. In 2009, leaders of some of the country's largest private corporations and NGOs established the Philippine Disaster Recovery Foundation (PDRF). It envisions a more coordinated and strategic way to respond to disasters and to maximize the contributions and build on the different strengths of the members of the business community. In the same year, the Public Commission and the PDRF drew a cooperation agreement to solidify the commitment of the private sector to the reconstruction programs of the Philippine government.

After successive large-scale disasters—the civil unrest in Zamboanga, the 7.2-magnitude earthquake in Bohol and Cebu, and the world's strongest recorded super typhoon Yolanda in 2013, PDRF was reorganized and intensified as the umbrella organization of the private sector for disaster preparedness, relief, and recovery. Corresponding programs were created for post-disaster recovery in five key sectors: (a) shelter; (b) livelihood; (c) education; (d) environment; and (e) water, infrastructure, sanitation, and health (W.I.S.H.).

Two years later, its name was formally changed to Philippine Disaster Resilience Foundation to capture the entire DRRM framework. The primary goal of PDRF is to build a resilient Philippines through building the resilience of different stakeholders. They believe that having a resilient business sector, communities, and government will result in a resilient country.

PDRF emphasizes interoperability and cooperation within and among the private sector. Through joint efforts and the strengths of its member companies and partners, it aims to mobilize resources, skills, knowledge, and experience to bolster the disaster resilience of businesses and communities. Its primary functions are (1) protecting from and reducing the impacts of disasters on the assets of member companies; (2) facilitating coordinated relief and response efforts from the business to the community through providing information, training, facilities, and opportunities; and (3) enabling partners to establish sustainable programs in vulnerable communities. It also seeks to give voice to micro, small, and medium enterprises (MSMEs) in the country and to enable them to become disaster-resilient by educating and training them on the concept of business continuity with risk management embedded in the process.

PDRF has many initiatives that help increase the resilience of the communities. Since there are various programs and projects, it uses the "expand and contract" model in implementing them. When one issue "expands," PDRF shifts its focus to that. When the issue "contracts," it refocuses back and goes to the cycle or process from recovery to risk mitigation and back to preparedness.

19.6.1.1 Programs and Projects

PDRF representatives were interviewed and a small focus group discussion was convened last June 2017 to know more about their programs and projects. To date, PDRF has a total of 45,000 beneficiaries.

Initiatives Under the Five Pillars of PDRF Programs
The core activities of PDRF are based on five pillars: (1) education, (2) health, (3) shelter, (4) livelihood, and (5) W.I.S.H. These pillars ensure that the basic needs of a person are attended to. The activities under these pillars also contribute to human security and reduce the vulnerability of the community through education, access to better facilities and equipment, and improved livelihood and mobility, among others. The initiatives under each pillar, the partners for each initiative (mostly from the private sector), and the number of beneficiaries can be accessed through this link: https://goo.gl/kHWH6G.

BREACH Program
The United States Agency for International Development (USAID) also worked with PDRF through the BREACH (Building Resilient and Economically Adept Communities and Households) Program. It aims to develop a transitional community by providing individual shelter units, access to basic services, transport systems to facilitate mobility of goods and people, and economic opportunities, and capability building for disaster preparedness and strengthening risk mitigation for the community.

Through consultation with partners and the Tacloban City government, Barangay Suhi was identified as the suitable location for program implementation. In early 2017, 240 families in the barangay were provided with "butterfly houses." These houses are eco-friendly, containing 100% recycled materials, fireproof, and waterproof. The design can also withstand storms with winds up to 195 kph. Twelve multi-cabs were turned over to the Negros Women for Tomorrow to set up a transportation business, solving mobility and livelihood issues. This could serve at least 14,000 families in the area.

Shelter and access to basic services are important in maintaining personal and health security. Meanwhile, mobility and economic opportunities increase food and livelihood security. The BREACH Program allowed the typhoon survivors to attain these securities that are important in helping them get back to their feet. The more secured they are, the higher the chance that they can help themselves during and after disaster events.

Marikina Watershed Program
PDRF aims to protect 15,000 hectares of forest in the Marikina watershed within 5 years. It was based on their projection that with a business-as-usual scenario, approximately 15,180 hectares of forest cover will be lost from 2016 to 2020. Their four key areas of action are forest protection, forest rehabilitation, advocacy, and sustainable livelihood.

In 2016, they introduced social preparation as another area of action. This will be done through capacity building and value formation among 100 selected families within the watershed.

As mentioned earlier, forests have a major role in addressing the issues associated with climate change. Through the environmental and social activities in this program, the ecosystem services that the Marikina watershed provides are protected and enhanced. These ecosystem services ensure food, livelihood, and environmental security, which are important in adapting to the impacts of climate-related disasters.

Operations Center

The PDRF Operations Center (OC) is the organization's command and coordination center. When activated, it becomes an emergency operation center (EOC) that operates 24 hours a day. It is composed of a dedicated team who promotes disaster preparedness among businesses and partners. It has three goals: (1) ensure coordination and collaboration, (2) empower member companies in building resilience, and (3) strengthen the preparedness of SMEs that work directly with larger organizations. The first goal aims to (1) build an ecosystem of corporations and partner organizations; (2) enable collaborative efforts with government and international humanitarian organizations aimed toward DRR; and (3) ensure swift, strategic, and coordinated response in times of disasters. The second goal specifically involves (1) promoting greater awareness of disaster risks and (2) encouraging disaster preparedness and mitigation measures within individual businesses.

The EOC is divided into eight clusters. These clusters and the companies involved per cluster can be accessed through this link: https://goo.gl/LhxTTF. PDRF has identified government and international humanitarian agencies that they could work with. A summary list of the interoperability of PDRF with the government and international humanitarian agencies is accessible through this link: https://goo.gl/a5LK4R. The companies, organizations, and agencies were grouped into clusters where they are most suitable and most useful.

Currently, the OC is located at Makati City but the long-term plan is to add two sites for backup capacity in Clark, Pampanga (Luzon), and Cebu (Visayas). The aim is to locate it in an area with easy access to a road network, ports, and an airport for airlift capacity, with provisions for stockpiling and parking.

The establishment of various OCs ensures its reliability even if one of the EOCs is affected by a disaster. It also allows PDRF to broaden their scope and reach, making more resources accessible to other communities in the Philippines when needed.

Training and Capacity Building During Peacetime

PDRF has several training programs that they conduct during times when there are no disasters. A list of the trainings/activities that are under their Resilience Training and Education Program can be found here: https://goo.gl/ZsSNrE.

These activities show how PDRF refocused from recovery to resilience. These climate and disaster preparedness training and tools enhance the adaptive capacity of its members, partners, and other relevant stakeholders. This reduces the vulnerability of the stakeholders as they are now more aware of their risks and better prepared to minimize or cope with them.

Public–Private Collaboration During Disaster Response
There were more initiatives that highlight public–private collaboration. One example was during the typhoons Ferdie and Lawin (international names: Meranti and Haima, respectively), which both happened in 2016. Post-disaster needs assessment among businesses was conducted in post-disaster areas. PDRF partnered with Hotel Owners Association (HOA) and FedEx. HOA donated used blankets, bed sheets, towels, and linens, while FedEx provided transportation support. Another example during typhoon Ferdie was the shipping and distribution of relief items in Batanes. However, it was found that rather than distribution, the needed intervention was free supply of construction materials or at least making them affordable. The Philippine Navy provided the transportation of these materials, while the LGU facilitated the transport of these materials from the pier to Basco, Batanes. Meanwhile, Habitat for Humanity provided technical expertise.

19.6.1.2 Challenges

Through Executive Order No. 838 s. 2009 signed by then Philippine President Gloria Arroyo after typhoon Ondoy, the inception of PDRF was made possible. The main objective of this EO was to strengthen public–private partnership (PPP) during disasters. However, during the 2013 back-to-back-to-back disasters, PDRF saw the need for the private sector to not just focus on response and relief but also to strengthen its involvement in other aspects of DRRM.

Stakeholder Relations
Since 2009, PDRF's membership has reached 82 private sector companies and organizations. Engaging each member company remains a challenge, but PDRF emphasizes that it is a win-win situation for them. They can learn from each other and they have access to the organizations' programs and services.

One of PDRF's main strategies to ensure an effective PPP is to work closely with national and various LGUs, regardless of a change in administration. One of the ways that it does this is by working with public sector representatives from all levels.

Organizational Sustainability
It is a challenge to maximize PDRF resources to ensure the continuity of its programs and services. For funding, it relies on private sector network contributions as well as other fund sources such as grants.

Although the organization's operation covers the entire country, at present, it only has 20 staff members. When needed, it is able to add more people through partners, employee volunteers from member companies, and interns.

19.7 Conclusion and Moving Forward

Realizing that extreme weather disturbances and climate change-related disasters have caused economic and non-economic damages and losses to numerous businesses, the private sector is becoming increasingly engaged in helping the government and local communities where they operate to better address climate change and disaster-related risks. Impacts such as disrupted operations leading to dramatic losses have prompted the private sector to start developing solutions in order to avoid or minimize such impacts and vulnerabilities to their businesses. Several private sector responses and initiatives, as previously elaborated, contribute to human security.

The two case stories from the Philippines highlight how the private sector addresses the challenges posed by climate change and disaster risks on human security. Although coming from different principles (EDC programs are grounded on "sustainability approach," PDRF programs on "resilience approach"), the programs of both entities are geared toward ensuring human-secured, progressive, sustainable, and resilient future.

The EDC experience shows the importance of CSR as an entry point in reaching out to their host communities, but their programs later evolved into a more holistic and systemic approach, giving balanced attention to environmental, social, and economic aspects of their business and across the value chain of their operations. Taking advantage of the company's increasing awareness about climate change risks and considering climate change as their greatest sustainability challenge, EDC has launched diverse initiatives that are directly and indirectly related to CCA, climate change mitigation, and DRRM. The company capitalized on the active engagement of the local communities and strong involvement of LGUs, as well as selected national agencies and key partners in trying to achieve their sustainability goals.

The PDRF approach experienced some form of evolution from a disaster response and recovery-centered to a resilience-focused program. Realizing the capacities of the private sector to play an important role in building resilience, projects were launched to encourage interoperability and closer collaboration within and among its member companies and partners. By mobilizing available resources from its corporate network, PDRF has attracted other partners in developing, implementing, and sustaining programs at various scales, supporting a range of beneficiaries from MSMEs to corporate conglomerates as well as local communities where the businesses operate.

The essential requirements common to these two private sector experiences in order to successfully carry out CCA and DRRM-related programs that contribute to human security are awareness and understanding of climate change risks to business operations and value chains; recognition of the capacities to enable solutions and sustain action; and strong commitment to actively cooperate with the government, community, and other actors in collectively sharing resources to achieve mutual interest and goals.

Private sector initiatives should not, however, be taken as substitute to government efforts but as supplement to existing and future plans. The government should provide enabling policy and regulatory frameworks and enhance its basic function of providing weather and climate services. We recommend that the Philippine government (1) continue to develop and improve strategies that will motivate the private sector to leverage resources and create sustainable market-based mechanisms that are not reliant on public sector budgets alone, (2) incentivize good practices of the private sector, (3) manage well the capital and other resources that the private sector has to offer, (4) and provide avenues for open dialogues with the private sector on discussing joint initiatives and monitoring mechanisms to improve CCA and DRRM strategies that would contribute to human security.

Much can be learned on the complexities and dynamic role that the private sector can play in addressing the challenges posed by climate change and disaster risk to human security and continued research is necessary to document their experiences, motivation, good practices, and success. If shared widely, the lessons learned may lead to greater interest and participation by other businesses and stakeholders to work in synergy toward a human-secured, progressive, sustainable, and resilient future.

References

Adger WN, Pulhin JM, Barnett J, Dabelko GD, Hovelsrud GK, Levy M, Oswald Spring Ú, Vogel CH (2014) Human security. In: Field CB, Barros VR, Dokken DJ, Mach KJ, Mastrandea MD, Bilir TE, Chatterjee M, Ebi KL, Estrada YO, Genova RC, Girma B, Kissel ES, Levy AN, MacCracken S, Mastrandrea PR, White LL (eds.) Climate change 2014: Impacts, adaptation, and vulnerability. Part A: Global and sectoral aspects. Contribution of Working Group II to the Fifth Assessment Report of the Intergovernmental Panel on Climate Change. Cambridge, United Kingdom and New York, Cambridge University Press pp 755–791

Agrawala S, Hallegatte S, Shah A, Lempert R, Brown C, Gill S (2011) Private sector engagement in adaptation to climate change: Approaches to managing climate risks (OECD Environment Working Papers, No. 39). Paris, France, OECD Publishing. https://doi.org/10.1787/5kg221jkf1g7-en

Appiah M, Jackson E (2015) Corporate social responsibility and human security in fragile states: private sector engagement in peacebuilding. The Hague Institute for Global Justice, The Netherlands

Atteridge A (2010) Private sector finance and climate change adaptation: how can voluntary private finance support adaptation in developing countries. SEI Policy Brief. Stockholm Environment Institute, Sweden

Avory B, Cameron E, Erickson C, Fresia P (2015) Climate resilience and the role of the private sector in Thailand: case studies on building resilience and adaptive capacity. BSR Report. Business for Social Responsibility, Hong Kong

Berrang-Ford L, Ford JD, Paterson J (2011) Are we adapting to climate change? Global Environ Change 21:25–33

Biagini B, Miller A (2013) Engaging the private sector in adaptation to climate change in developing countries: importance, status, and challenges. Clim Dev 5(3):242–252

Brown J, Stadelmann M, Wang D, Boni L, Jachnik R, Kato T (2015) Estimating mobilized private finance for adaptation: exploring data and methods. Climate Policy Initiative, USA

Buchner B, Trabacchi C, Mazza F, Abramskiehn D, Wang D (2015) Global landscape of climate finance 2015. Climate Policy Initiative, San Francisco

Ceres (2012) Physical risks from climate change: a guide for companies and investors on disclosure and management of climate impacts. Ceres, Boston, MA

Davies C, Klingel F (2015) Adapting to climate-change: the private sector can contribute. Private Sector & Development Proparco's Magazine. Paris, Proparco, p 23
DCED (Donor Committee for Enterprise Development) (2016) Private sector adaptation to climate change and development agency support: DCED Synthesis Note. UK, DCED
Gleditsch NP (2012) Whither the weather? climate change and conflict. J Peace Res 49(1):3–9
Haigh N, Griffiths A (2009) The natural environment as a primary stakeholder: the case of climate change. Bus Strategy Environ 18(6):347–359. https://doi.org/10.1002/bse.602
IPCC (Intergovernmental Panel on Climate Change) (2012) Glossary of terms. In: Field CB, Barros V, Stocker TF, Qin D, Dokken DJ, Ebi KL, Mastrandrea MD, Mach KJ, Plattner G-K, Allen SK, Tignor M, Midgley PM (eds) Managing the risks of extreme events and disasters to advance climate change adaptation. A Special report of working groups I and II of the Intergovernmental Panel on Climate Change. Cambridge University Press, Cambridge, UK, and New York, pp 555–564
Jamarkattel BK (nd) Perceptions and engagement of the private sector in urban climate resilience: A case study from western Terai, Nepal. http://www.carenepal.org/uploads/pdf/publication/_Perceptions%20and%20Engagement%20of%20the%20Private%20Sector%20in%20Urban%20Climate%20Resilience.pdf
Lutz W, Muttarak R, Stiessnig E (2014) Universal education is key to enhanced climate adaptation. Sci 346(6213):1061–1062. https://doi.org/10.1126/science.1257975
Muttarak R (2014) Education is key to climate adaptation. http://www.iiasa.ac.at/web/home/about/news/20141127-Science-Pop.html. Accessed 16 Aug 2017
Oscar M Lopez Center for climate change adaptation and disaster risk management foundation, Inc. (nd) About us. http://www.omlopezcenter.org/. Accessed 11 May 2018
Oswald Spring Ú (2012) Can health be securitized? Hum Evol 27(1–3):21–29
Pauw WP (2015) Not a panacea: private-sector engagement in adaptation and adaptation finance in developing countries. Clim Policy 15(5):583–603. https://doi.org/10.1080/14693062.2014.953906
Pauw P, Pegels A (2013) Private sector engagement in climate change adaptation in least developed countries: an exploration. Clim Dev 5(4):257–267. https://doi.org/10.1080/17565529.2013.826130
PWC (PricewaterhouseCoopers) (2010) Business leadership on climate change adaptation: encouraging engagement and action. PricewaterhouseCoopers LLP, London, UK
Surminski S, Eldridge J (2015) Observations on the role of the private sector in the UNFCCC's loss and damage of climate change work program. Int J Glob Warming 8(2):213–230
UN (United Nations) (nd) Climate change and indigenous peoples. http://www.un.org/en/events/indigenousday/pdf/Backgrounder_ClimateChange_FINAL.pdf. Accessed 16 Aug 2017
UNDP (United Nations Development Programme) (2007) Human development report 2007/2008: fighting climate change: human solidarity in a divided world. Palgrave Macmillan, Houndmills, Basingstoke, Hampshire, UK and New York, p 384
UNEP (United Nations Environmental Programme) (2015) Climate change and human rights. http://web.law.columbia.edu/sites/default/files/microsites/climate-change/climate_change_and_human_rights.pdf. Accessed 24 Jul 2017
UNFCCC (United Nations Framework Convention on Climate Change) (2014) Private sector initiative-database of actions on adaptation. http://unfccc.int/adaptation/workstreams/nairobi_work_programme/items/6547.php. Accessed 24 Jul 2017
Villamejor-Mendoza MF (2014) Private sector engagement in climate change mitigation and adaptation: implications in regional governance. Paper presented at the 3rd international conference on ASEAN Connectivity: current issues and prospects towards ASEAN community, 17 Nov 2014, Mahasarakham University, Thailand
World Bank (2016) Forests combat climate change. http://www.worldbank.org/en/topic/forests/brief/forests-combat-climate-change. Accessed 16 Aug 2017

Perlyn Pulhin-Yoshida is a freelance environmental science researcher-writer and editor. She holds a master's degree in Environmental Science and a bachelor's degree in Development Communication. She has more than 18 years of work experience in the areas of environmental research, science-policy initiatives, knowledge management, and communications and development in international and national research organizations including OML Center, ICRAF, and APN. She has also served as Managing Editor of the Climate, Disaster and Development Journal. Her research and publication interests include climate change, disaster risk management, resilience, and loss and damage. She has co-authored a number of scholarly publications and presented in local and international conferences. She recently moved to Japan and is currently working as a full-time English teacher.

Dr. Rodel D. Lasco has more than 38 years of experience in natural resources and environmental research, conservation, education, and development at the national and international level. His work has focused on issues related to natural resources conservation, climate change, and land degradation. Since 2000, he is an author of several reports by the IPCC, the 2007 co-winner of the Nobel Peace Prize. He is also a member of the National Academy of Science and Technology (NAST) in the Philippines. He is the Philippines Coordinator of the World Agroforestry Centre (ICRAF). Concurrently, he is Executive Director of the OML Center, a private foundation whose mission is to promote research on climate adaptation and disaster risk reduction.

Ana Veronica S. Gabriel started her career in research at the OML Center through projects on climate change and agroforestry, as well as loss and damage linked with climate change adaptation and disaster risk reduction. For the latter, she had the opportunity to showcase the findings at various international and national scientific meetings. She also served as Associate Managing Editor of the Climate, Disaster and Development Journal before becoming Project Officer at the Forest Foundation Philippines, a non-profit organization that provides grants to organizations that empower the people to protect the forests.

Chapter 20
Pathways Toward a Human Secured Asia

Juan M. Pulhin, Makoto Inoue, Rajib Shaw, Ma Louiella Rose O. Catudio, and Millicent Joyce Q. Pangilinan

Abstract In its simplest form, human security is concerned with the provision of "freedom from fear and freedom from want". In the context of climate change and disaster risks, the existing problems that have long been imperiled human lives are greatly enhanced. Looking back, Asia, in its pursuit of resiliency and security, has exhibited various innovative and effective means to ensure human security at all levels which are evident until now. Assessing the present status of security based on the projected threats, pathways toward climate-resilient and sustainable development still seemed to be a long journey. Since Asia accounts for almost 42% of the disasters worldwide over the last century, it should take the lead in defining the pathways toward resiliency, sustainability, and security. With the unceasing pursuit in ensuring human security both in Asia and the world, a resilient and secure world is still attainable for the current and the future generation. This concluding chapter synthesizes the previous chapters as well as looks at the future threats associated with the changing climate to solidify the argument that climate change and disaster risks have the capacity to undermine human security as gleaned from the Asian experience. It also elaborates some of the region's innovative responses to address these threats that can hopefully be the basis for defining the pathways toward a more climate-resilient, human-secured, and sustainable future not only for Asia but with implications for the rest of the world.

Keywords Human security · Climate change · Disaster risks · Resilience · Sustainable development

---

J. M. Pulhin (✉)
Department of Social Forestry and Forest Governance, College of Forestry and Natural Resources (CFNR), University of the Philippines Los Baños, Laguna, Philippines
e-mail: jmpulhin@up.edu.ph

M. Inoue
Faculty of Human Sciences, Waseda University, Saitama, Japan

R. Shaw
Graduate School of Media and Governance, Keio University, Kanagawa, Japan

M. L. R. O. Catudio · M. J. Q. Pangilinan
University of the Philippines Los Baños, Laguna, Philippines

J. M. Pulhin et al. (eds.), *Climate Change, Disaster Risks, and Human Security*, Disaster Risk Reduction, https://doi.org/10.1007/978-981-15-8852-5_20

20.1 Introduction

Perceiving human security in the context of climate change and disaster risk has provided a range of concern from the broad spectrum of the elements that constitute a decent and dignified life. From the definition given by Adger et al. (2014), human security has been elaborated by experts from various fields to encompass both its threats and possible solutions. Incidentally, looking closely at the elements discussed in the previous chapters also magnified the related interacting societal problems such as poverty, increasing population, and weak governance, among others. Despite current efforts to tackle human insecurity, these problems continue to imperil the most vulnerable population in Asia. With this, it can be assumed that human security is not static but a dynamic concept as the threats for the vital core of human lives evolve together with the impacts of climate change and disaster risk.

Climate change, from the viewpoint of human security, reflects its severity by considering human beings as the object greatly affected by its impacts. Observed climate change impacts have significantly compromised food and livelihood security, water quality and quantity, human health, shelter and infrastructure and social services which comprise basic human needs. Manifestations of climate change through the series of extreme and slow-onset hydrometeorological events coupled with geologic hazards have altered different ecosystems, thereby affecting sectors such as agriculture, forestry, and fisheries. These alterations resulted in the decrease in natural capital that directly impaired the livelihoods of the affected population. Further, the instability caused by climate change could ignite conflicts and trigger migration both in the intra and interstate levels. Oftentimes, these lead to changes in the culture of the people, specifically those values and practices that are strongly tied with the natural resources. These impacts are summarized and highlighted in the recent IPCC Special Report on Global Warming of 1.5 °C published in 2018, which argues that even with the 1.5–2 °C increase in the global temperature, human systems, and ecosystems will be significantly affected and will intensify and exacerbate existing threats to human security.

Intensified climate change projections can also translate to more severe natural hazards. Although there is still little evidence for the direct linkage of climate change to the increased intensity and magnitude of natural disasters, another area for consideration is the increasing vulnerability of the affected population due to climate change. The increased vulnerability significantly decreases the coping and adaptive capacities of the population to exhibit resilience after a disaster. Further, if both climate change and natural hazards interact with existing socioeconomic, political, and environmental problems, the intensity and spread of risks may become too overwhelming.

In the wake of evolving and persistent threats to human security, the continuous pursuit to achieve human security in Asia should also be emphasized. Years of confronting human insecurity have provided the Asian region a rich experience on how each country could come up with the most effective and innovative ways to ensure human security at different scales amid escalating threats. Significant improvements have been made such as international collaborations and assistance, mainstreaming of climate change adaptation (CCA), and disaster risk reduction (DRM) into national and local development plans, partnerships with civil society organization (CSOs which includes the non-government organizations (NGOs)) and private sectors and community empowerment.

These innovations can also be credited to the humanitarian assistance provided by various countries and international organizations. The response from the rest of the world has been valuable especially in times of disaster and emergency. Since 1988, the European Union has assisted Asia for the improvement of its disaster response infrastructure. In 2019, it pledged a funding package worth EUR 8.5 million, wherein EUR 1.5 million was allocated for the emergency aid of the monsoon victims in India and Bangladesh, and the rest for disaster risk reduction and improvement of disaster response in the Philippines, Nepal and the rest of the region (European Union 2019). In addition, the Asian Development Bank (ADB) has approved 240 new development projects on disaster resilience, and has established a disaster risk reduction financing mechanism under the 12th Asian Development Fund which will benefit the organization's 18 lower-income members (Asian Development Bank [ADB] 2019). Moreover, data showed that Asia has been receiving attention through the humanitarian aid, which is defined in the Organization for Economic Cooperation and Development (OECD) database as "emergency and distress relief in cash or in kind, including emergency response, relief food aid, short-term reconstruction relief and rehabilitation, disaster prevention and preparedness," The international community has not failed to recognize Asia's vulnerable condition. Figure 20.1 supports the fact that Asia alone receives a big portion of humanitarian aid compared to other regions in the world, and has received 50% of the total world amount since 2013. More importantly, it has played a crucial role in contributing to the region's resilience and disaster preparedness.

This concluding chapter solidifies the argument that climate change and disaster risks have the capacity to undermine human security, while also elaborating the

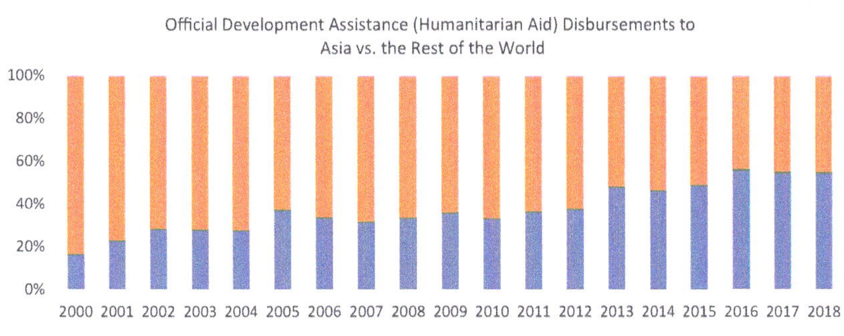

Fig. 20.1 Official development assistance received by Asia vs. the rest of the world (*Source* OECD Database [https://stats.oecd.org/], accessed June 2020)

innovative responses of Asia to address these threats. It then provides the current status of human security facing the future adverse impacts and other interacting stressors that complicate the vulnerability status of the affected population. Lastly, as the region experiences high climate and disaster risks, the chapter lays down the pathways for climate-resilient and sustainable development toward a human-secured future.

20.2 Highlights of the Key Points Discussed in the Book

In the face of persistent threats to human security brought about by climate change and disaster risks, developing policies and valuing experiences that contribute to resilience are of immeasurable importance. Hence, the 19 chapters of this book, divided into four thematic areas, demonstrate the threshing out of key important factors surrounding the discourse on the interplay of climate change, disaster risks, and human security using the Asian lens.

The first theme, which covers the first four chapters, provides the broader theoretical, policy, and institutional contexts of climate change, disaster risk, and human security in Asia. Chapter 1 by Pulhin, Inoue, Shaw, and Pangilinan provided the foundation of the interplay of climate change, disaster risk, and human security. It discussed the need to look at how climate change and disaster risk have impacted people's lives and livelihoods from the viewpoint of Asia, one of the most vulnerable regions in the world. With Asia experiencing the greatest number and the worst impacts of disasters, the region is even more on a disadvantage given the socioeconomic factors that have concurrently contributed to human insecurity. Moreover, the chapter presented an analytical framework employed in the entire book which highlights the importance of policy interventions, international frameworks, and programs to address the interlinked issues on climate change, disaster risks, and human security in pursuit of a more resilient and secure Asia.

With human security needing to be constantly improved, Chapter 2 by Shaw, Pulhin, and Inoue emphasizes ways to reduce exposure and underlying risk factors linked to climate change and disaster risk reduction (DRR). The authors then scrutinized the six core indicators used to determine progress in reducing underlying risk factors, as stated in the Priority for Action Area 4 of the Hyogo Framework for Action 2005–2015, the first global framework for DRR. Through the analysis, the following gaps were identified: (1) little or no consideration of DRR/DRM in national and sectoral public investment in most development sectors and non-recognition of DRR as a cross-cutting issue within different sectoral investments; (2) absence of land use policies and zoning systems, resulting in inappropriate and unsafe development; (3) insufficient institutional capacity in economic and productive sectors; (4) absence of risk financing and risk transfer mechanisms; and (5) disregard for community-based adaptation and DRR. With the determination of the weak areas, the authors proposed guidelines for the formulation of new DRR approaches, considering the new global disaster framework (i.e., Sendai Framework for Disaster Risk Reduction).

For Chapter 3, Prabhakar and Shaw surveyed Asian experts and researchers to elicit response on their perceptions of stakeholders engaged in DRR and CCA, their understanding of the subject, and their recommendations on how these issues should be addressed. Through the results of the survey, the link between development and climate-fragility risks becomes clearer, especially with the presentation of Climate-Fragility Risk Index (CFRI) which offers various climate-fragility indicators.

Prabhakar's Chapter 4 highlighted the importance of institutions and policies in shaping human security in Asia and how these have affected human development in general. His assessment of literature revealed that discussions view human security as both a policy framework and policy outcome. Especially in recent years, human security as a policy outcome has been gaining more support since it has been a recurring concept in civil society programs, and is being mainstreamed into various local and international policies. However, in Asia, human security needs to be given more attention and importance, especially with the high incidence of human insecurity brought about by political instability and other pressing issues including climate change and disaster risks.

The second theme, from Chapters 5 to 10, elaborates on the core elements of human security threatened by climate change and disaster risks. Sabino and Pulhin in Chapter 5 discussed basic needs as the first human security element presented in this book that is threatened by climate change. Basic needs such as food, shelter, clothing, education, clean environment, and health play a central role in the material aspect of human security. When these are not provided and satisfied, people's wellbeing and survival are immensely affected, thereby denying them of the most fundamental human right and security. In Asia, socioeconomic issues such as poverty and lack of access to basic needs have always been a major concern for most governments in the region. Moreover, environmental issues related to climate change and disaster risks have proven to exacerbate these current issues and have allowed them to persist. Hence, the chapter concludes that mainstreaming CCA and DRR strategies into national and local policies and programs is necessary to protect the people in this changing environmental landscape.

The second human security element on the natural resource base was discussed in Chapter 6 by Peras, Pulhin, Gevaña, and Inoue. Various literatures were analyzed in order to draw the relationship among the natural resource base, livelihood and human security in Asia using the sustainable livelihood framework. They found out that climate change and a depleting natural resource base affect human security. To avert these environmental issues, the chapter argued that it is crucial to formulate sustainable livelihood strategies which aim toward building resilience and natural resources base stability. However, this remains a challenge to Asian countries due to inappropriate policies and poor governance, among others.

Agriculture and food security is the third human security element that was discussed in Chapter 7 by Montesclaros and Teng. With agriculture as one of the sectors greatly impacted by climate change, the chapter elaborated how food security in Asia is imperiled. Further, it discussed the relationship between agricultural performance and the four facets of food security, namely, food availability, economic food access, physical food access, and food utilization. As a region that is dependent on the agriculture sector for food and livelihood, the looming impacts of climate change and disaster risks is an urgent concern that needs to be addressed. Hence, the chapter recommended several strategies such as the promotion of urban agriculture, continued assistance to farmers during disasters and mainstreaming of climate change in the local policy level, among others.

Chapter 8 by Ruzol and Pulhin discussed culture as the fourth human security element and the role it plays with underlying factors of vulnerability and adaptation. It reviewed and analyzed studies to look into how knowledge on climate, climate change, human security and other concepts are perceived across cultures in Asia. The authors argue that human security could improve the prevailing science of climate change and disasters by placing the human condition and experience at the core of the discourse. They therefore recommended the use of anthropological approaches to address human security issues such as: (1) diachronic approach, which focuses on evolutionary process of adaption; (2) synchronic approach, which entails understanding how things are related by observing the society and examining case study in relation to others; and (3) critical and feminist approaches, which can be diachronic, synchronic or combination on the role of power in a cultural context.

Population movements as the fifth human security element was discussed by Goméz in Chapter 9, with migration introduced as a means to adapt to climate change and resilience by confronting environmental stresses. He introduced various mechanisms that cause displacement, such as slow-onset disasters, sea-level rise, and catastrophic events. Finally, it underscored important issues triggered by the interrelationship of climate change, disasters, and migration, such as addressing the needs of migrant populations affected by disaster crises and the role of remittances in responding and recovering from them.

Chapter 10 by Howe presented conflicts as the final human security element in the book affected by climate change and disasters. As conflict is considered a major challenge in Asia and with the region's rich experiences brought about by its colonization history and political ideology and beliefs, the chapter drew lessons from country cases and experiences in North Korea, Myanmar, and Lao PDR. The chapter argued that focusing on state-centric security is not only insufficient but also has been proven to worsen human insecurity and environment issues. Moreover, environmental degradation and human insecurity contribute to national insecurity; thus, a holistic governance approach can be used to address such a complex relationship.

The third theme, from Chapters 11 to 15, provided more grounded illustrations of how climate change and disaster risks are undermining human security based on specific selected cases in Asia. Notable disaster events that have undercut human security have been selected for readers to draw concrete and important lessons from. Chapter 11 by Lagmay, Santiago, and Pulhin highlighted the role of science in vulnerability and risk assessments and DRR drawing from the strongest ever recorded typhoon in human history, the Typhoon Haiyan that hit the central part of the Philippines in 2013. The authors underscored the urgent need to mainstream and integrate land use plans and other development plans into policies in the city, provincial, and regional level after the adverse impacts brought about by said highly destructive typhoon. They proposed methods to prepare multi-scenario hazard maps to address pressing climate change issues, and presented actual hazard maps produced with reference to the experiences of areas affected by the typhoon. Further, they proposed for hazard maps to become a major input for local and national governments development planning and decision making in order to have evidence-based and science-based decisions in the future.

Chapter 12 by Okita presented Japan's interesting experience in handling the international assistance it received during the 1995 Kobe Earthquake and the 2011 Great East Japan Earthquake. He argued that receiving an overwhelming amount of aid can be a source of burden for the country and failing to properly manage it may threaten human security. Hence, the inflow of money assistance from donor countries must be considered as an important part of a country's disaster response plan. Protocols must be in place in terms of relaying timely and accurate information as these not only allow securing the urgent needs in the field but also prevent the affected country from receiving unnecessary assistance.

Droughts from the perspective of China were discussed in Chapter 13 by Pink, with emphasis on how it has continued to affect various aspects of human security in the country: community, economy, environment, food, health, personal, and politics. The author argued that droughts will escalate in the country and will continue to affect people and livestock. Hence, the formulation of more aggressive climate adaptation and mitigation policies must be given utmost priority before drought causes more adverse impacts to the country.

Chapter 14 by Fujita, Shaw, and Parvin brings the spotlight to the 2004 Bangladesh flood. Given the number of times that the country experienced catastrophic floods, the government of Bangladesh saw the need to formulate policy, action, strategy, and plan to address the problem that has claimed millions of lives and destroyed property and infrastructure. The chapter thoroughly reviewed Bangladesh's flood history and emphasized the need to have an integrated flood management for DRR and resource management, keeping in mind the four areas in the country, namely, river flood area, flash flood area, urban flood area, and coastal flood area. It concluded that the analysis of different flood phenomena and the assessment of mitigation measures prove to be effective in appreciating the various facets of water management and how social factors influence flood disaster interventions.

The case of the 2004 Indian Ocean earthquake and tsunami and its effects on Indonesia and Sri Lanka was discussed in Chapter 15 by Syamsidik, Nugroho and Fahmi. Considered as one of the deadliest in history, the disaster served as a catalyst to review, revisit and update current disaster management policies. While the intention of resettlement strategies was to address disaster risks, it ironically introduced another risk involving return migration and the increasing coastal population in tsunami-affected zones. The chapter proposed that governments give specific attention to disaster education and disaster-ready infrastructure. Further, it recommended the inclusion of human security aspects in the recovery planning process and the formulation of a more comprehensive disaster management plan to ensure community preparedness in the event of another unfortunate event.

Given the region's vast experiences, it is important to discuss the various initiatives that have been launched and designed to address the challenges and threats posed by climate change and disaster risk to human security. The last thematic area from Chapters 16 to 19 therefore focuses on these initiatives involving regional and national scales as well as from different sectors including the government, civil society, and private sectors. Chapter 16 by Ishiwatari looks into regional policies and the integration of peacebuilding assistance and CCA. The chapter reviewed various approaches in development assistance to respond to climate-fragility risks, with emphasis on peacebuilding efforts toward addressing short-term needs, and CCA efforts that use a long-term perspective. Given these gaps, practical approaches are recommended such as the consideration of climate change impacts to peacebuilding programs and incorporation of fragility risks into CCA programs. More importantly, it emphasized the need to combine the scientific aspect of CC and the socioeconomic aspect of peacebuilding efforts for a more holistic approach in managing climate-fragility risks.

Chapter 17 by Ramirez, Stevenson, and Pulhin underscored the national policies and programs in Asia that address climate change, disaster risks, and human security. It reviewed key national policies and programs in Nepal, Bangladesh, Lao PDR, South Korea, Japan, and the Philippines to underscore innovative solutions and responses, as well as key challenges, which can serve as a guide and reference for countries. Given the increasing risks brought about by climate change and other hazards, the authors emphasized the need to ensure that national policies and programs must be more responsive to changes by incorporating both scientific and indigenous knowledge and experiences. Hence, they strongly urge for a regular monitoring and evaluation of policies and programs in the ground level to ensure that these are still updated and relevant amid a fast-changing landscape of risk and resilience.

Izumi in Chapter 18 focused on the role of CSOs and the challenges they have encountered in the implementation of CCA and DRR projects in Indonesia, Afghanistan and the Philippines. The involvement of CSOs proved to be crucial in the implementation of programs and they vary in terms of the support and assistance that they offer. Given these differences, it is important to explore partnerships between international organizations who have adequate resources and funding, and local organizations who have enough experience and knowledge on the ground. Further, CSOs and its networks can also implement disaster education and assist in capacity development, knowledge sharing, and building partnerships.

Lastly, Pulhin-Yoshida, Lasco, and Gabriel in Chapter 19 shared their experiences with the private sector and their increasing interest in climate change and disaster risks. Citing two cases on private sector initiatives in the Philippines, the chapter highlighted actual practices on the ground and recommended entry points for private sector engagement. Using key informant interviews and small group discussion, results revealed that the private sector is committed to cooperate with the government regarding these environmental issues, and that there is awareness in terms of climate change risks and their capacity to enable solutions. With such a positive response, they strongly urged the Philippine government to continue to motivate the private sector and incentivize their participation to increase collaboration efforts and explore opportunities to improve CCA and DRRM strategies in the country.

20.3 Historical Record of Disasters

Notably, it is indispensable to assume that human security is just an emerging concept since it has been the utmost priority even from the beginning of time. Historical accounts have long foretold various threats that imperiled the vital core of human lives. Accounts on natural phenomena such as famine, drought, flood, and plague have been written and orally passed on to give narratives and warnings for the next generations. Evidently, these narratives have denoted the high dependence of human survival to nature as well as the capacity of these phenomena to impair human lives. With the current advancements, the reality that these natural phenomena, manifested by today's climate change and disaster risk impacts, could greatly undermine human security still persists, hence the creation of this book. Prior to this chapter, experts from different fields featured the interplay of the concepts of climate change, disaster risk and human security in Asia to present the various angles of its operationalization.

Considering the universality of the threats to human security in the context of climate change and disaster risks, the effective and innovative responses of the Asian region has distinguished itself from the other regions. As the most vulnerable region in terms of quantity and magnitude of disaster, Asia has been implementing localized interventions in order to ensure human security at all levels. Interestingly, the uniqueness of each country, shaped by their cultures and socioeconomic and political conditions, has provided an array of concrete solutions that can be distilled to other regions. Revisiting the region's vast experiences from a century of human insecurity unfolds the catastrophic events that molded their current strategies to abate these threats and to ensure human security.

At the onset of the twentieth century, India marked history with one of its greatest and most notorious famines that started in 1899 and continued until 1900. Failure of the monsoon occurrence caused soil moisture drought resulting in major crop failure followed by the widespread famine (Mishra et al. 2018). Decades later, the 1931 Yellow River Flood, one of the deadliest natural scourges of China, inundated the agricultural communities of central and eastern China. The flood claimed almost four million lives and washed away the summer harvest including the substandard houses. Decline of food availability also intensified the existing subsistence crisis, thus, increasing the vulnerability of the impoverished communities. In order to survive, most of the affected population sold their assets at highly inflated prices resulting in a long-term destitution and inequity (Courtney 2018).

Apparently, droughts and flood occurrences were major sources of human insecurity in the early part of the century (1900–1950), particularly in China and India. Yet, it should be noted that the major forces that caused these disaster events are the natural internal processes such as El Niño and La Niña (IPCC 2007). Both processes can influence the global climate pattern by affecting the sea surface temperatures (SSTs) along the equatorial region of the Pacific Basin. These changes disrupt the tropical atmospheric and ocean circulation patterns and alter the mid-latitude jet stream. Thus, these resulted in changes to the temperature and precipitation patterns with extreme weather events particularly in the tropical and extra-tropical regions. These climatic changes are part of the natural climate variability that has been occurring on the planet since then (Hegerl et al. 2018; Ching et al. 2019).

Toward the end of the century, the region of Asia experienced unprecedented natural hazards that took millions of lives and billions of losses. An abrupt increase in the annual disaster occurrences in Asia was observed toward the end of the century (Fig. 20.2). All of the natural disasters significantly increased, primarily, the hydrometeorological disasters. Unlike with the earlier disasters, these phenomena were now beyond the natural climate variability, hence, signaled the start of climate change. According to (Seneviratne et al. 2012), climate change can enhance the frequency, intensity, spatial extent, duration and timing of weather and climate extremes that can further result in unprecedented extremes. Although earth's natural process such as El Niño/Southern Oscillation (ENSO) has been altering the climate even before, the sudden variability was mainly induced by external factors.

Meanwhile, it was also during this time that most of the countries in Asia are building the ruins caused by the recent World War II. Industrialization across Asia spread rapidly to support national developments and emerging economies. Transformations and advancements in different sectors were supported by various anthropogenic activities (Chang and Zach 2018). Unfortunately, these activities resulted in environmental degradation that severely altered the biogeochemical cycle of the earth (Steffen et al. 2007). Particularly, the dramatic rise of carbon dioxide (Fig. 20.3) in the atmosphere contributed to the increase of Greenhouse Gas (GHG) concentration worldwide. Trends of carbon emission both in Asia and the world exhibited the said rise that was attributed to global warming. Figure 20.4 presents the annual land surface anomaly in Asia from 1910–2018 that is a direct manifestation of global warming. Positive anomaly shows that a significant increase in temperature has been recorded as compared with the baseline temperature. This also explains the reason

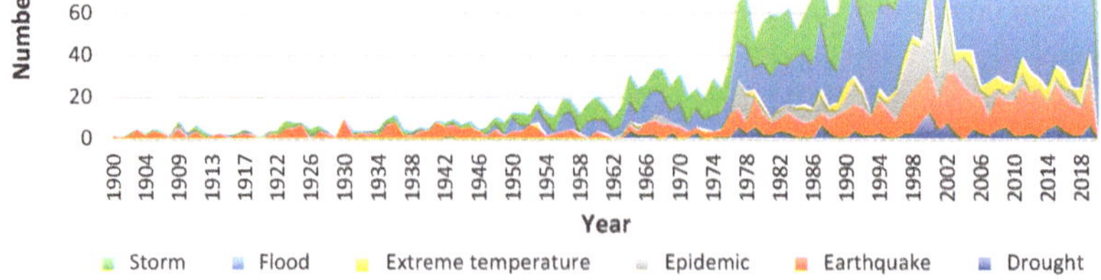

Fig. 20.2 Number of disasters in Asia per year from 1900 to 2019 (*Source* EM-DAT database, accessed June 2020)

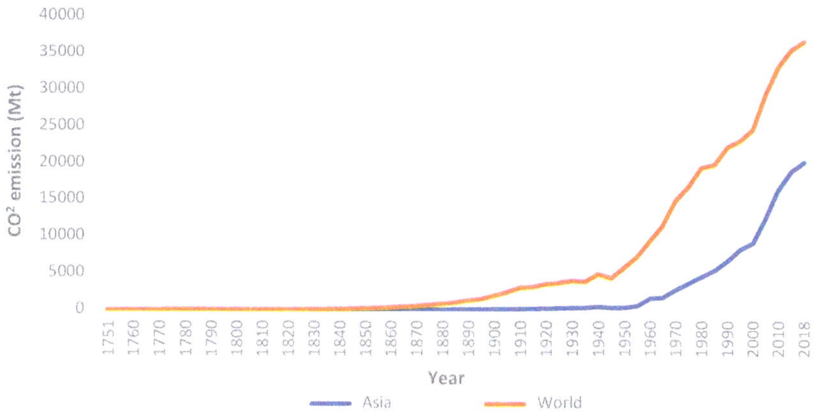

Fig. 20.3 Annual production-based emission of carbon dioxide (CO_2) in Asia from 1751–2018 (*Source* Our World in Data database, accessed August 2020)

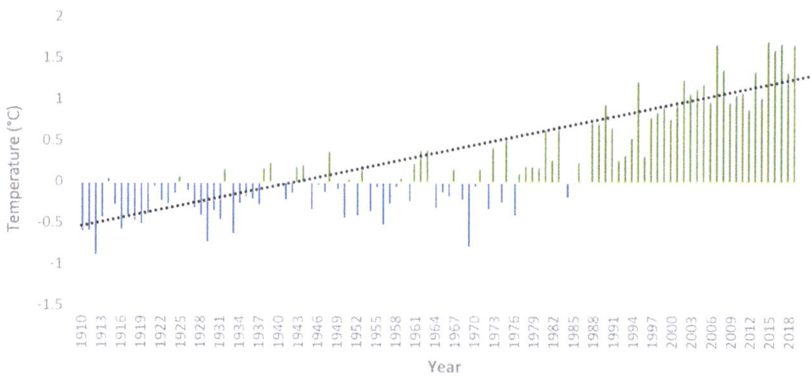

Fig. 20.4 Annual Land Surface Temperature Anomaly in Asia from 1910–2019 (*Source* National Oceanic and Atmospheric Administration [NOAA], accessed August 2020)

for prolonged summer periods and decreased winter in most parts of Asia (Hegerl et al. 2018).

Despite the more frequent and intensified disasters enhanced by climate change, Asia proactively addressed these disasters to ensure human security. For instance, the Great Bhola Cyclone, the occurrence of one of the deadliest cyclones, devastated Bangladesh and claimed the lives of about 300,000 people due to the large storm surge that overwhelmed the island and tidal flats along the Bay of Bengal. However, the aftermath scenario was worse than the initial impact because of the obvious human insecurity faced by the vulnerable population. Food insecurity resulted in the starvation of many people. Prevalence of death due to dehydration and water-borne diseases emphasized the insufficient supply of potable water. Since most of the vulnerable population was deprived of the basic needs of human life, health-related problems amplified the total number of fatalities (Hossain 2017). With this, the government of Bangladesh worked on improving its flood management by shifting to small and medium-scale flood, coastal, irrigation projects. As the country faced more flooding concerns years later, Bangladesh constantly developed its flood management plan as they learn from their previous experiences. In 2004, the country again was struck by a great flood incidence that took thousands of lives and caused massive destruction to more than half of the country. The recent flooding compelled the government to slowly shift from the conventional response and relief practice to a more comprehensive risk reduction system. Indeed, the long flooding history of the country also reflects their long history of protection efforts to ensure the multidimensionality of human security to its most vulnerable population.

In 1995, a magnitude 7.2 earthquake devastated Kobe City, Japan and killed almost 5,378 people and destroyed property amounting to US$100 billion. The earthquake impaired the functionality of the city because of the severe damage to infrastructure in the area which includes the expressways, railways, port facilities, sewer systems and power from water, gas and electricity. Despite the tragic event, it was remarkable that Japan has made use of this circumstance to widen its understanding about earthquakes and its impacts on infrastructure (Holzer 1995). Afterward, necessary revisions on the Japanese seismic design code, reinforcement of infrastructure, and innovative applications of seismic isolation and passive control systems were completed to anticipate a stronger catastrophic event in the future (Nakashima and Chusilp 2003). By 2011, another terrifying earthquake shook the country with the 9.0 magnitude that also triggered a tsunami and destruction of the Fukushima nuclear power plant. Yet, the lessons that they learned from the 1995 Kobe earthquake prepared the country for another more destructive earthquake such as the 2011 Great East Japan Earthquake (GEJE). Similarly, the country improved its management of international assistance that was once overwhelmed during the 1995 earthquake.

Several years later, one of the deadliest disasters in modern history affected a large part of the world with a 9.1 magnitude earthquake strong enough to induce a series of storm surges along the coastlines that border the Indian Ocean Basin. The earthquake resulted in the sudden rise of the ocean floor beyond 40 meters which triggered the first 100-foot wave that hit the shoreline of Banda Aceh in Indonesia within 20 min after the quake. The surge claimed 100,000 lives and left a desolate landscape of the city. Subsequently, the tsunami rolled over to the coastlines of the neighboring nations of Thailand, India, and Sri Lanka, multiplying the deaths and losses associated with the event. Eight hours later, another tsunami was experienced in the coasts of South Africa, which tallied a total of 230,000 deaths across 21 nations (Roos 2018). Ruined coastal communities and the realization of the tsunami impacts compelled the governments of Indonesia and Sri Lanka to construct resettlement and livelihood programs to address the immediate needs of displaced communities including their increasing population rate. Yet, the severe alterations brought by the tsunami took 13 years before the affected population was able to fully grasp these remarkable changes. Despite the long process of recovery, the two countries are constantly improving their disaster management plans to prepare for another catastrophic event in the future.

Tropical cyclone Nargis also left a mark on Myanmar's history due to its severe impact on human lives, environment, and economy. Between the 2nd and 3rd of May 2008, more than 138,000 deaths and 2.4 million affected population were recorded, including an estimated overall economic damage of USD 10 billion (Besset et al. 2017). Next to Bhola Cyclone, tropical cyclone Nargis registered the most number of fatalities because of the lack of cyclone awareness and evacuation plans, absence of high ground or shelters and no comparable knowledge on storm surge of the hardly hit population. Adding up to the vulnerability of the people are the widely deforested, low-lying, and densely populated coastal community with poor housing construct (Fritz et al. 2009). As a response, policy formations and collaborations and partnerships from other nations were organized to increase the resilience of the country. In addition, coastal mangrove forests were replanted to provide a natural barrier against future storm surge scenarios. News reports and early warning systems were also strengthened to give up-to-date reports to the coastal communities (UNEP 2009).

A year after, the recurrent drought scenario in China became the driest event occurrence since 1951. From a humanist approach, the effects of climate change manifested through the alternating drought and flooding episodes and saltwater intrusion aggravated the current crises on water scarcity and pollution, which were enough to undermine human security. For this reason, the government has been reinforcing its socioeconomic development through policy reformation and innovation to beset the severe drought crises and widespread economic failure.

Furthermore, Typhoon Haiyan in 2013, killed almost 6,300 people due to the storm surge and extreme rainfall that caused flooding in most areas in Tacloban City, Philippines. One of the identified reasons for the intensification of this rare event is the anthropogenic climate change manifested by sea-level rise and ocean warming (UNISDR 2008; Santos et al. 2015; Laffoley and Baxter 2016). Since the dreadful event, the country gave much emphasis on the mainstreaming of CCA and disaster risk and vulnerability reduction and management (DRVRM) to the existing local development plan. Science-based information was merged with the local knowledge to provide a profound understanding for the resilience planning of the communities. The country made use of the probabilistic multi-scenario hazard maps to anticipate for unprecedented natural disasters in the future.

Other countries in Asia such as Nepal exhibited their remarkable initiatives to ensure human security in the context of climate change and disaster risk. As one of the countries that are heavily devastated by earthquakes, avalanches, and flooding, it invested in the partnership of the government and non-government institutions through the Nepal Change Support Program (NCCSP). The program aims to implement climate change policies and adaptation, mainstream climate change in key development sectors and implement CCA through public-private partnerships. The country also promotes the integration of community-based adaptation to the management of agriculture, water, forest, and biodiversity sectors through the Local Adaptation Plans for Action (LAPA). Currently, Nepal is preparing for its National Policy and Strategic Action Plan for Disaster Risk Reduction that is in accordance to the Sendai Framework to ensure the increased resilience and enhanced adaptive capacity of the country (Upreti et al. 2013).

Similar to the experiences of the Philippines, Lao PDR is also vulnerable to the impacts of climate change and disaster risks. Admirably, the country focused on the long-term scientific studies to establish the baseline data at the national level that is relevant for the adaptation and mitigation planning. Thereafter, the government implemented the Strategy on Climate Change of the Lao PDR and the National Adaptation Program of Action to Climate Change (NAPA) and the National Disaster Management Plan. These decrees and regulations are still subject to improvement in order to abate the worsening impacts of climate change and disaster risks to the country (RECOFTC 2014; Tong 2009).

Unlike other developing countries, the Republic of Korea holds the notion that contributing to climate change mitigation needs not hamper the country's economic development. With this, the country implemented the concept of green growth development through the enactment of the Framework Act on Low Carbon Green Growth in 2010. This act is in support of their commitment to reduce GHG emissions by 30% in 2020. This initiative mobilized each government ministries to create comprehensive green growth plans at the sectoral, national, and local levels. Since the concept includes climate change mitigation, the country's economy continues to develop as they address climate change and human security. Green growth technology, green industries, and green jobs are promoted to strengthen the implementation of the plan. Similarly, the country has formulated various plans for CCA and disaster risk mitigation (GGGI 2015; Nachmany et al. 2015).

Apart from the government initiatives of the countries in Asia, worth-noting are the significant support from various foreign donors, CSOs and private sectors. For instance, the Japanese International Cooperation Agency (JICA) has been supporting many nations across the world through their CCA and peacebuilding programs. Though some gaps have been found in the operationalization of the two programs, a recommendation to merge the scientific knowledge used for CCA and the socioeconomic approach for peacebuilding was presented in the previous chapter for the improvement of climate-fragility risk management. CSOs, on the other hand, support the CCA and DRR implementation by providing the communities with sufficient knowledge on climate change and risks of disasters (JICA 2016). Some CSOs may also provide financial support depending on their type, knowledge and experience. Private sectors, specifically in the Philippines, have established various programs that directly help the communities in their implementation of CCA and DRRM. Hence, strengthening the public–private collaborations are necessary to provide a synergistic effect on the implementation of CCA and DRRM to ensure human security at all levels (Oscar M. Lopez Center, n.d.).

In reminiscing the past century, it is indeed evident that Asia is the most vulnerable region in terms of the frequency and magnitude of disasters. Regardless of its driving force, natural, or anthropogenic, occurrence of these events severely damaged and impaired the region because of the numerous lives it took and massive damages it incurred. Yet, these catastrophic events proved the resilience of Asia and the solidarity

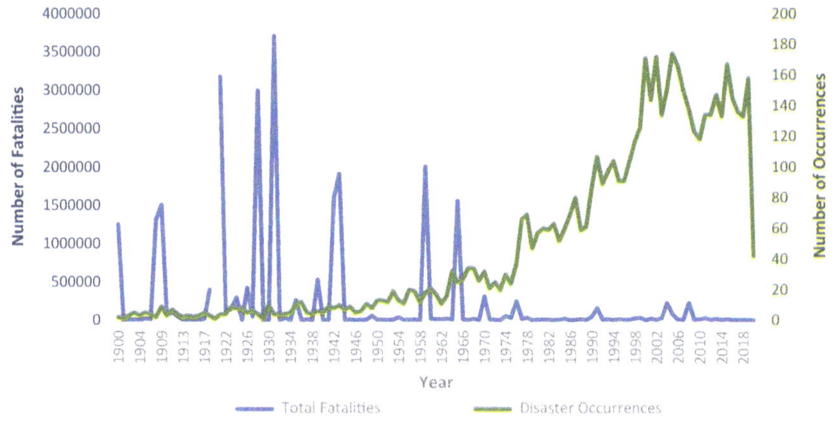

Fig. 20.5 Total Fatalities and Disaster Occurrences in Asia per year from 1900 to 2019 (*Source* EM-DAT database, accessed June 2020)

among the nations. Although there are still gaps and areas for development considering that the countries are still in different phases of implementation, the region's effort to ensure human security should be acknowledged. As reflected in Fig. 20.5, despite the increasing frequency of disasters as influenced by climate change, the number of fatalities is significantly decreased toward the end of the century. This implies that the initiatives of the government, international collaborations, and assistance from various organizations greatly helped in preserving human lives despite the disaster occurrences. However, the concept of human security is not limited to the preservation of life but to its sustenance as well. With the worsening projections of climate change impacts, ensuring the vital core of human life will be more challenging and therefore will require it to be a constant pursuit.

20.4 Future Outlook of Climate Change and Disaster Risk Impacts in Asia

Historically, Asia has experienced extreme events and disasters more than the rest of the world as exhibited in the previous section. While the lessons drawn from these experiences prove to be helpful in both disaster preparation and policymaking, it is equally important to examine projections and scientific evidence that discuss the future adverse impacts of climate change to the region.

Recent IPCC Special Report (IPCC 2018) reveals that global warming is likely to rise to 1.5 °C between 2030 and 2052. Increasing anthropogenic emissions (including GHG, aerosols, and their precursors) coupled with natural climate variability can trigger the projected increase. If the increase will happen, previous economic achievements and improvements of the living standards could be stymied by the worse impacts that the changing climate will pose. Likewise, the advancements in human security could also be impeded because of the overwhelming disasters that the region may be facing. Hence, operationalization of human security should be improved constantly to adapt to the changes that will happen in the next decades.

Table 20.1 summarizes the projected impacts of climate change to Asia and their anticipated effects on various Asian human and natural systems. Major climate-related drivers of impacts have been identified as follows: temperature change, heat extremes, precipitation, sea-level rise, and tropical cyclone. In a report by ADB (2019), impacts have been projected using two scenarios, namely, the Business-As-Usual (BAU) scenario or the Representative Concentration Pathway (RCP) 8.5 and the Paris-consensus scenario or the RCP 2.6. Under the Paris-consensus scenario, maintaining a global temperature increase of 2 °C above pre-industrial levels and, if possible, striving for a 1.5 °C increase is crucial. Various sectors such as agriculture, health, water, economy, and marine ecosystems, will be affected by climate change, and will continue to threaten ecosystem services in the region. These cause a chain reaction leading to unfavorable effects to socioeconomic factors such as livelihood and poverty, and ultimately worsens human insecurity in the region.

20.5 Ways Forward

A retrospection of the previous century in Asia attests to its claim of being the most vulnerable region in terms of the frequency and magnitude of disasters. Various disasters ranging from biological, geophysical, and hydrometeorological constantly devastate the region resulting in massive destruction and loss of lives. Initially, the geological position and topography of Asia are enough reason to put the region in peril. Yet, the apparent manifestations of climate change toward the end of the century enhanced the vulnerability of Asia by increasing the frequency and magnitude of these disasters. Despite these threats, the century of insecurity has also shown the unyielding pursuit of the region to minimize these threats in order to ensure human security at all levels. Evidently, most of the region's initiatives toward CCA and DRRM have shown great improvements. Nonetheless, the projections of worsening climate change and disaster risk impacts place the region in crossroads toward a secure and resilient Asia.

Current initiatives, although considered now as effective, may not be sufficient in the next decades due to the rapid rise of global mean temperature. Asia, as a region of high risk yet advancing toward security, should take the lead in pursuing climate-resilient and sustainable development pathways. These pathways are development trajectories that centralize on the goal of sustainable development realized by the combination of adaptation and mitigation (Denton et al. 2015). In addressing these future threats, Asia should leverage its key resources to provide major solutions in attaining climate security without compromising its development. Since Asia

Table 20.1 Summary of projected climate change impacts and their effects in selected sectors in Asia

Drivers of impacts	Projected impacts in Asia [BAU and Paris-consensus projections from (ADB 2017)]	Examples of sectors affected	Projected sectoral impacts
Temperature change (summer temperature increase)	BAU (RCP 8.5): 4.0 °C by 2070, 6.0 °C by 2100 for the land area of Asia Paris-consensus (RCP 2.6): Increase of 2.0 °C by 2070, below 1.9 °C by 2100 for the land area of Asia Strongest warming in West and Central Asia (IPCC 2018: 190)	Agriculture, health, economy, water	Increased risk of crop failure and lower crop production could lead to food insecurity (Hijioka et al. 2014: 1336) Water shortage in arid areas (Hijioka et al. 2014: 1336) Smaller net reductions in yields of maize, rice, wheat, and potentially other cereal crops in Southeast Asia (IPCC 2018: 179) Increased risk of heat-related mortality (Hijioka et al. 2014: 1336) Increased risk of drought-related water and food shortage causing malnutrition (Hijioka et al. 2014: 1336) Increased risk of water and vector-borne diseases (Hijioka et al. 2014: 1336) Exacerbated poverty, inequalities, and new vulnerabilities (Hijioka et al. 2014: 1337)

(continued)

Table 20.1 (continued)

Drivers of impacts	Projected impacts in Asia [BAU and Paris-consensus projections from (ADB 2017)]	Examples of sectors affected	Projected sectoral impacts
Heat extremes • Unusual heat extreme • Unprecedented heat extreme	BAU (RCP 8.5): 91% of Asian land will be affected Paris-consensus (RCP 2.6): Less than 30% of Asian land will be affected BAU (RCP 8.5): 70% of Asian land will be affected Paris-consensus (RCP 2.6): Around 5% of Asian land will be affected	Agriculture, health, marine ecosystem	Increased risk of crop failure and lower crop production could lead to food insecurity (Hijioka et al. 2014: 1336) Water shortage in arid areas (Hijioka et al. 2014: 1336) Almost complete extinction for coral reefs in Asian tropics at 2 °C warming (ADB 2017) Increased risk of heat-related mortality (Hijioka et al. 2014: 1336) Increased risk of drought-related water and food shortage causing malnutrition (Hijioka et al. 2014: 1336) Increased risk of water and vector-borne diseases (Hijioka et al. 2014: 1336) Exacerbated poverty, inequalities, and new vulnerabilities (Hijioka et al. 2014: 1337)

(continued)

Table 20.1 (continued)

Drivers of impacts	Projected impacts in Asia [BAU and Paris-consensus projections from (ADB 2017)]	Examples of sectors affected	Projected sectoral impacts
Precipitation	BAU (RCP 8.5): Increase in annual mean precipitation by up to 50% in most Asian land areas Paris-consensus (RCP 2.6): Increase in annual mean precipitation by less than 30% in most Asian land; Large increase in heavy precipitation events for North Asia and East Asia (IPCC 2018: 178)	Agriculture, economy, health	Increased risk of crop failure and lower crop production could lead to food insecurity (Hijioka et al. 2014: 1336) Increased riverine, coastal, and urban flooding leading to widespread damage to infrastructure, livelihoods, and settlements (Hijioka et al. 2014: 1336) Increased risk of flood-related deaths, injuries, infectious diseases, and mental disorders (Hijioka et al. 2014: 1336) Increased risk of water and vector-borne diseases (Hijioka et al. 2014: 1336) Exacerbated poverty, inequalities, and new vulnerabilities (Hijioka et al. 2014: 1337)
Sea-level rise	1-meter global sea-level rise: 19 out of 25 cities exposed to a are from Asia (Brecht et al. 2012) BAU (RCP 8.5): 33% of the population of Bangladesh and Vietnam will permanently fall below the high tide line (Kulp and Strauss 2019) Paris-consensus (RCP 2.6): High tide lines higher than land now home to 19% of the population of Bangladesh, 26% of the population of Vietnam, and 17% of the population of Thailand (Kulp and Strauss 2019)	Agriculture, water, economy	Increased risk of crop failure and lower crop production could lead to food insecurity; Water shortage in arid areas (Hijioka et al. 2014: 1336) Increased riverine, coastal, and urban flooding leading to widespread damage to infrastructure, livelihoods, and settlements (Hijioka et al. 2014: 1336)

(continued)

Table 20.1 (continued)

Drivers of impacts	Projected impacts in Asia [BAU and Paris-consensus projections from (ADB 2017)]	Examples of sectors affected	Projected sectoral impacts
Tropical cyclones	1.6 billion affected by tropical cyclones by 2050 in Asia and the Pacific	Agriculture, health, economy	Increased risk of crop failure and lower crop production could lead to food insecurity; Water shortage in arid areas (Hijioka et al. 2014: 1336) Increased risk of flood-related deaths, injuries, infectious diseases, and mental disorders (Hijioka et al. 2014: 1336) Increased riverine, coastal, and urban flooding leading to widespread damage to infrastructure, livelihoods, and settlements (Hijioka et al. 2014: 1336)

accounts for almost 42% of the disasters worldwide over the last century, bridging the identified gaps and strengthening the implementation of its mitigation and adaptation strategies from the community up to the international level could render a great impact toward global resiliency and sustainability.

Recent advancements in science and technology have played an important part in the current status of CCA and DRR implementation. Significant transformations have resulted in increased human security, particularly for the developed country. For instance, the green growth development strategy of the Republic of Korea is a prototype of a growing economy without compromising its goal on climate change mitigation. Developed countries, with the abundance of their resources, should invest in this type of development for the actualization of their commitment in reducing their carbon emission. Conversely, most of the developing and underdeveloped countries do not have the same access and capacity like the developed countries for science and technology. The evident inequality among nations is one of the stumbling blocks that impede the developing and underdeveloped countries. With this, international and regional collaboration and support should be amplified to bridge the gap among the nations.

Even so, building the scientific capacity from regional, national down to the local level manifested through the generation of policies and programs that are scientifically sound and effective could still be possible. For instance, Asia Pacific Network for Global Change Research (APN) is one of the distinguished intergovernmental network systems that exemplifies on innovating global change at the regional and local scales. With 22 member countries, APN majors on collaborating the research community for the provision of grassroots results and the policymakers for the implementation of scientifically sound and evidence-based decision making. In the last five years, the organization has supported 115 distinct and relevant projects on climate change mainly from Southeast Asia, South Asia, and Temperate Asia. Researches focused on asserting community-based adaptation and mitigation surrounding the issues on ecosystems and biodiversity, extreme weather events, water-food-energy nexus, sustainable waste management, and climate education (Uchiyama et al. 2020).

One of the projects supported by APN is the research conducted in the Philippines that assessed the institutional capacity of Local Government Units (LGUs) in implementing the localized CCA and mitigation and disaster risk. It was revealed that capacity development for LGUs through the increase in their technical capacity and know-how should be at par with the national actors to actualize the perceived resiliency and security at the local level (Grefalda et al. 2020). Another project funded by APN revealed that environmental deterioration, exacerbated by climate change and socioeconomic changes, could result in resource insecurity, thereby jeopardizing human wellbeing. Hence, leading to increased poverty and unsustainable land use practices, overexploitation and resource degradation. As a response, policy-relevant solution in the regions of India, Nepal, and Bangladesh should be done to achieve a climate, and integrated landscape based development (APN, n.d.).

Investments in the Information and Communication Technology (ICT) are also encouraged for effective communication, specifically the early warning system and response time during the disaster at all levels. Information sharing through knowledge portals such as the EM-DAT, NOOA, and Our World in Data is essential in giving up-to-date information for a more comprehensive understanding of climate change and disaster risk at all levels. Similarly, information from these portals can be used for knowledge sharing and science and evidence-based policymaking.

Moreover, existing financial support from humanitarian assistance from various countries and international organizations are essential in providing emergency response in times of disaster. Foreign donors should also make sure that these aids should be realized in the ground as a support for the coping and adaptive mechanism of the vulnerable population. On the other hand, discernment on the provision of humanitarian assistance during disaster should also be considered to avoid putting a burden because of too much assistance provided. The capacity of the receiving country to effectively manage relief assistance and operations should likewise be enhanced to maximize the use of valuable resources.

On the national level, the government, as one of the main drivers toward the secured future, should seek to improve its governance. Science-based decision making informed by local knowledge and experience, employment of adaptive collaborative governance by harnessing the participation of the different sectors—government, private, and civil society. Similarly, the government should seek to increase the awareness and understanding of the communities on climate change and disaster risks. Increased awareness and understanding are imperative in the empowerment of the communities since this is the first step in increasing their adaptive and coping capacity.

As emphasized in the previous chapters, enabling policies and responsive institutions create a significant impact on climate resiliency and sustainable development. With the comprehensive framework of human security, the policy as the channel, and institutions as the influencer of change, providing tangible solutions on the ground could be possible. Capacity building for the communities and all national and local government units is imperative to aid in the effective management of multiple stresses that interact with climate change impacts. Through science-based policy reforms and interventions, institutions should be able to simultaneously address climatic and non-climatic factors that exacerbate vulnerability risk. Integration of CCA and DRR strategies to the comprehensive national and local development plans is essential in the implementation of sustainable development in the context of the changing climate. Specifically, it is crucial to utilize science and technology and innovations to effectuate an integrated response to climate change that combined mitigation and adaptation with effective disaster disaster preparedness, response, and recovery strategies.

Providing tangible solutions to these concerns through the integration of asset building (five capital assets—natural, human, social, physical, and financial) strategies is necessary to address the multidimensionality of human security. This can also be used in monitoring the progress toward climate resiliency and sustainability at the community level.

Academes, research institutions, and similar organizations are encouraged to proactively take part in this initiative by integrating scientific and local knowledge as a tool for decision making. Particularly, they can provide support through the establishment of baseline data at the national and local level for the creation of localized tools for vulnerability and risk assessment and resilience building. Since the impacts of climate change and disaster risks vary based on the spatiotemporal components, having a localized understanding of climate change and disaster risk could aid in the creation of appropriate frameworks to support CCA and DRR without compromising the central goal of sustainable development. Integration of science and local knowledge is also necessary for providing suitable CCA and DRR strategies. A good example of this is the probabilistic multi-scenario hazard maps from the Philippines that depicts the future unprecedented hazards.

Underlying factors of human security should be emphasized since they are the most urgent concern that should be addressed. Widespread poverty, inequality and increasing populations are the facets of human insecurity that these vulnerable populations are facing on a daily basis. Increasing the knowledge and understanding on climate change and disaster risk is a preliminary step toward community-level resiliency and sustainability. Thus, capacity building for the vulnerable communities should be done to also increase their coping mechanism and adaptive capacity.

Likewise, some of the Asian countries have greatly advanced in this area by incorporating topics on climate change and disaster in educational curricula. In Vietnam, a 12-month program entitled "Introducing Disaster Preparedness in Primary Schools" has been implemented to integrate the disaster preparedness education into school curricula throughout Vietnam. Thereafter, the government mandated the inclusion of disaster preparedness education in school curricula. The Philippines has also integrated climate change and disaster risk in its curriculum that includes training on DRR measures and emergency preparedness skills. Information dissemination through multimedia educational packages on disaster preparedness has been carried out that resulted in the active engagement of children and community stakeholders on DRR initiatives. Sri Lanka on the other hand, has focused on developing the capacity of both school personnel and children through trainings that covered topics on disasters, effective disaster risk management, disaster education in Japan, simulation exercises/emergency drills in schools and demonstrations on lessons on natural disasters (De Juan et al. 2020).

For higher education, Indonesia, Sri Lanka, India, and Maldives, after the 2004 Indian Ocean Tsunami, implemented the Selamat project that aimed to bridge the academic research education and field practices. The project invested on high quality knowledge and information on disaster risk reduction. An extensive network on the tsunami-affected areas was established to ensure largest circulation in future. To guarantee the sustainability of disaster preparedness in the targeted countries and communities, the project developed certificate courses and customized courses that will provide young professional development (Shaw et al. 2015). Innovative educational programs on climate change and disaster risk management at different levels from elementary to postgraduate level should be scaled up and sustained in the different countries to educate both the present and future generation and hence build resilience in these areas.

Indeed, there are a number of concrete ways that can be considered, replicated and multiplied whether locally, nationally, and regionally to level up the efforts toward a more secure and resilient Asia. Some CCA and DRRM strategies may be applicable across the counties but it is likely that modifications are necessary to match the local context and needs. The immense potential of scientific and technological advancements to carry Asia forward to a desired state of security and sustainability should be matched with innovative institutional arrangements, enabling policy, appropriate social interventions, a culture of resilience, and a strong sense of concern for the next generation. There should be one and only pathway which is to transform Asia from the most vulnerable to the most climate-resilient and secure region. The road to transformation has just begun.

References

Adger WN, Pulhin JM, Barnett J, Dabelko G, Hovelsrud G, Levy ML, Oswald Spring U, Vogel C (2014). Human security. In Field CB, Barros VR, Dokken DJ, Mach KJ, Mastrandrea MD, Bilir TE, Chatterjee M, Ebi KL, Estrada YO, Genova RC, Girma B, Kissel ES, Levy AN, MacCracken S, Mastrandrea PR, White LL (eds) Climate change 2014—impacts, adaptation and vulnerability. Part A: global and sectoral aspects: working group II contribution to the IPCC fifth assessment report: volume 1: global and sectoral aspects. Cambridge University Press, pp 755–791

APN E-Lib (nd) Interlinkage of ecosystem services and human wellbeing to enhance climate smart landscapes in small watersheds: analysis for policy-relevant solutions in South Asian context. https://www.apn-gcr.org/resources/items/show/2115. Accessed 29 August 2020

Asian Development Bank (2017) A region at risk: the human dimensions of climate change in Asia and the Pacific. Mandaluyong City, Philippines. http://dx.doi.org/10.22617/TCS178839-2

Asian Development Bank (2019) Strengthening disaster resilience. Mandaluyong City, Philippines. http://dx.doi.org/10.22617/FLS190070-3

Besset M, Anthony EJ, Dussouillez P, Goichot M (2017) The impact of Cyclone Nargis on the Ayeyarwady (Irrawaddy) River delta shoreline and nearshore zone (Myanmar): towards degraded delta resilience? Comptes Rendus—Geoscience 349. Académie des sciences, pp 238–247. https://doi.org/10.1016/j.crte.2017.09.002

Brecht H, Dasgupta S, Laplante B, Murray S, Wheeler D (2012) Sea-level rise and storm surges: high stakes for a small number of developing countries. J Environ Dev 21:120–138. https://doi.org/10.1177/1070496511433601

Chang HJ, Zach K (2018) Industrial development in Asia-trends in industrialization and industrial policy experiences of developing Asia. In: WIDER working paper 120

Ching H, Lam Y, Haines A, Mcgregor G, Ying E, Chan Y, Hajat S (2019) Time-series study of associations between rates of people affected by disasters and the El Niño Southern Oscillation (ENSO) cycle. Int J Environ Res Public Health 16:3146

Courtney C (2018) The nature of disaster in China: the 1931 Yangzi River Flood. Cambridge University Press. https://doi.org/10.1017/9781108278362

De Juan A, Pierskalla J, Schwarz E (2020) Natural disasters, aid distribution, and social conflict—micro-level evidence from the 2015 earthquake in Nepal. In: World development 126. Elsevier Ltd., p 104715. https://doi.org/10.1016/j.worlddev.2019.104715

Denton F, Wilbanks TJ, Abeysinghe AC, Burton I, Gao Q, Lemos MC, Masui T et al. (2015) Climate-resilient pathways: adaptation, mitigation, and sustainable development. Climate change 2014 impacts, adaptation and vulnerability. Part A: global and sectoral aspects, pp 1101–1131. https://doi.org/10.1017/CBO9781107415379.025

European Union (2019) Humanitarian aid: € 8.5 million for South and South East Asia. https://ec.europa.eu/commission/presscorner/detail/en/IP_19_4451
Fritz HM, Swe T, Blount C, Thu MK (2009) Cyclone Nargis storm surge in Myanmar. Nat Geosci 2:448–449. https://doi.org/10.1038/ngeo558
GGGI (Global Green Growth Institute) (2015) Korea's greengrowth experience: process, outcomes and lessons learned. GGGI, South Korea
Grefalda LB, Pulhin JM, Tapia MA, Anacio DB, De Luna CC, Sabino LL, Garcia JE et al. (2020) Building institutional resilience in the context of climate change in Aurora, Philippines. Environ Res 186:109584. Elsevier Inc. https://doi.org/10.1016/j.envres.2020.109584
Hegerl GC, Brönnimann S, Schurer A, Cowan T (2018) The early 20th century warming: anomalies, causes, and consequences:1–20. https://doi.org/10.1002/wcc.522
Hijioka Y, Lin E, Pereira JJ, Corlett R, Cui X, Insavrov G, Lasco R, Lindgren E, Surjan A (2014) Asia. In: Field CB, Barros VR, Dokken DJ, Mach KJ, Mastrandrea MD, Bilir TE, Chatterjee M et al (eds) Climate change 2014: impacts, adaptation, and vulnerability. Part B: regional aspects. Contribution of working group II to the fifth assessment report of the intergovernmental panel on climate change
Holzer T (1995) The 1995 Hanshin-Awaji (Kobe), Japan, Earthquake. GSA Today, pp 154–167
Hossain N (2017) The 1970 Bhola Cyclone, nationalist politics, and the subsistence crisis contract in Bangladesh. Disasters. http://doi.org/10.1111
IPCC (2007) Climate change 2007: synthesis report. In: Core Writing Team, Pachauri RK, Reisinger A (eds) Contribution of working groups I, II and III to the fourth assessment report of the intergovernmental panel on climate change. IPCC, Geneva, Switzerland, 104 pp
IPCC (2018) Global warming of 1.5°C. In: Masson-Delmotte V, Zhai P, Pörtner HO, Roberts D, Skea J, Shukla PR, Pirani A, Moufouma-Okia W, Péan C, Pidcock R, Connors S, Matthews JBR, Chen Y, Zhou X, Gomis MI, Lonnoy E, Maycock T, Tignor M, Waterfield T (eds) An IPCC special report on the impacts of global warming of 1.5 °C above pre-industrial levels and related global greenhouse gas emission pathways, in the context of strengthening the global response to the threat of climate change
JICA (2016) Annual report. Retrieved from https://www.jica.go.jp
Kulp SA, Strauss BH (2019) New elevation data triple estimates of global vulnerability to sea-level rise and coastal flooding. Nat Commun 10:4844. https://doi.org/10.1038/s41467-019-12808-z
Laffoley D, Baxter JM (2016) Explaining ocean warming: causes, scale, effects and consequences. In: Full report. IUCN, Gland, Switzerland, 456 pp
Mishra V, Tiwari AD, Sha R, Xiao M, Pai DS, Lettenmaier D (2018) Droughts and Famines in India, 1870–2016. Geophys Res Lett 46(4):1–14
Nachmany M, Fankhauser S, Davidová J, Kingsmill N, Landesman T, Roppongi H, Schleifer P, Setzer J, Sharman A, Singleton CS, Sundaresan J, Townshend T (2015) Climate change legislation in South Korea: an excerpt from the 2015 global climate legislation study "A review of climate change legislation in 99 countries." Grantham Research Institute on Climate Change and the Environment, The Global Legislator Organization and Inter-Parliamentary Union, United Kingdom
Nakashima M, Chuslip P (2003) A partial view of Japanese post-Kobe seismic design and construction practices. Earthq Eng Seismol J 4:15–25
Oscar M. Lopez Center for Climate Change Adaptation and Disaster Risk Management Foundation, Inc (nd) About us. http://www.omlopezcenter.org/. Accessed 28 August 2020
Roos D (2018) The 2004 Tsunami wiped away towns with 'mind-boggling' destruction. History. Retrieved from https://history.com
Santos CT, Toda L, Orduña JR, Santos FD, Ferrão J (2015) The impacts of typhoon Haiyan in the Philippines: implications to land use planning. Climate Disaster Dev J 1:57–66. https://doi.org/10.18783/cddj.v001.i01.a06

Seneviratne SI, Nicholls N, Easterling D, Goodess CM, Kanae S, Kossin J, Luo Y, Marengo J, McInnes K, Rahimi M, Reichstein M, Sorteberg A, Vera C, Zhang X (2012) Changes in climate extremes and their impacts on the natural physical environment. In: Field, CB, Barros V, Stocker TF, Qin D, Dokken DJ, Ebi KL, Mastrandrea MD, Mach KJ, Plattner GK, Allen SK, Tignor M, Midgley PM (eds) Managing the risks of extreme events and disasters to advance climate change adaptation. A special report of working groups I and II of the intergovernmental panel on climate change (IPCC). Cambridge University Press, Cambridge, UK, and New York, NY, USA, pp 109–230

Shaw R, Takeuchi Y, Krishnamurthy R, Pereira JJ (2015) Community-based disaster risk reduction article information. Oxf Res Encycl. https://doi.org/10.1093/acrefore/9780199389407.013.47

Steffen W, Crutzen PJ, McNeill JR (2007) The anthropocene: are humans now overwhelming the great forces of nature? Ambio 36:614–621. https://doi.org/10.1579/0044-7447(2007)36%5b614:TAAHNO%5d2.0.CO;2

The Center for People and Forest (RECOFTC) (2014) Community forestry adaptation roadmap to 2020 for Lao PDR. RECOFTC, Bangkok (Thailand)

Tong PS (2009) Lao People's democratic republic forestry outlook study: Asia-Pacific forestry sector outlook study II working paper series. Food and Agriculture Organization, Bangkok (Thailand)

Uchiyama C, Stevenson LA, Tandoko E (2020) Climate change research in Asia: a knowledge synthesis of Asia-Pacific network for global change research (2013–2018). Environ Res 188(January):109635. https://doi.org/10.1016/j.envres.2020.109635

UNISDR (2008). Climate change and disaster risk reduction, September, 1–12. http://www.unisdr.org/files/4146_ClimateChangeDRR.pdf

United Nations Environment Programme (UNEP) (2009) Learning from Cyclone Nargis. https://wedocs.unep.org/bitstream/handle/20.500.11822/14116/myanmar_cyclonenargis_case_study.pdf

Upreti BR, Bhattarai R, Wagle GS (2013) Human security in Nepal: concepts, issues and challenges. Nepal Institute for Policy Studies (NIPS) and South Asia Regional Coordination Office of NCCR (North-South). NIPS, Kathmandu (Nepal)

Dr. Juan M. Pulhin is full Professor and former Dean of the College of Forestry and Natural Resources, University of the Philippines Los Baños, Philippines. He was Lead Author of the 2007 Fourth Assessment Report of the Intergovernmental Panel on Climate Change (IPCC), Coordinating Lead Author of the Human Security Chapter of the 2014 IPCC Fifth Assessment Report and currently Lead Author of the Asia chapter of the IPCC Sixth Assessment Report. He has been a four-time a visiting professor at The University of Tokyo in 2007, 2010, 2016 and 2017 and a visiting scholar at Waseda University, Tokyo in 2018. He has written and edited 11 books and published more than 100 technical papers and book chapters on natural resources policy and governance, and on the human dimension of climate change and disaster risks. He is a fellow, invited expert, and council member of several international organizations and a member of the National Academy of Science and Technology (NAST), Philippines.

Dr. Makoto Inoue is Professor at the Faculty of Human Sciences, Waseda University, Japan and former Professor at The University of Tokyo. He serves as Dean of Graduate School of Human Sciences, Waseda University since September 2020. His specialty is environmental sociology, forest governance, and Southeast Asian studies. He has written and edited/co-edited 34 books, and published more than 140 original articles in peer-reviewed international and national journals on the social aspects of forestry and the environment. He serves as President of The Japanese Association for Environmental Sociology (June 2019–present), Vice President of Japanese Society for Current Anthropology (January 2020–present), Vice President of Society for Environmental Economics and Policy Studies (April 2018–March 2020), Councilor of The International Association for the Study of the Commons (January 2019-present), a board member of various academic societies in Japan, and an editorial board member of several international journals.

Dr. Rajib Shaw is Professor at the Keio University in Japan and former Executive Director of the Integrated Research on Disaster Risk (IRDR) programme. He is also Senior Fellow of Institute of Global Environmental Strategies (IGES) Japan, and Chairperson of SEEDS Asia, and CWS Japan, two Japanese NGOs. Previously, he was Professor at the Graduate School of Global Environmental Studies, Kyoto University, Japan. His expertise includes community-based disaster risk management, climate change adaptation, urban risk management, and disaster and environmental education. He is the editor of a book series on disaster risk reduction, published by Springer, Editor-in-Chief of the journal Progress in Disaster Science. He severs as Chair of United Nations Science Technology Advisory Group for disaster risk reduction, and is Co-chair of the UNDRR's Asia-Pacific Science Technology Academic Advisory Group (ASTAAG). Prof. Shaw has published more than 45 books and more than 300 academic papers and book chapters.

Ma Louiella Rose O. Catudio obtained her bachelor's degree in Forestry from the University of the Philippines Los Baños (UPLB). She is currently a graduate student of MS Forestry specializing in silviculture and forest influences in the same university, and currently serves as a graduate assistant for a forestry project on the assessment of the Community-based Forest Management (CBFM) program in the Philippines. Prior to this, she was involved in the studies of water resource management and community and program development of the Department of Environment and Natural Resources (DENR). Her researches delve into the study of social forestry, forest resource management, agroforestry, and climate change.

Millicent Joyce Q. Pangilinan is currently a researcher for a forestry project at the University of the Philippines Los Baños. She has 10 years of experience in research and project management in both the private and development sectors. She started her career in Japan doing economic data and market analysis intended for mergers and acquisitions projects in Asia and in the USA. After 4 years, she transitioned into the development field, where she was engaged in research projects on agriculture, climate change, and forestry in Southeast Asia and the Philippines. She earned her master's degree in International Public Policy (Economics track) at Osaka University, Japan and bachelor's double degree in Economics and Finance at De La Salle University Manila, Philippines.

Index

J. M. Pulhin et al. (eds.), *Climate Change, Disaster Risks, and Human Security*, Disaster Risk Reduction, https://doi.org/10.1007/978-981-15-8852-5

The manufacturer's authorised representative in the EU is Springer Nature Customer Service Centre GmbH, Europaplatz 3, 69115 Heidelberg, Germany. If you have any concerns regarding our products, please contact ProductSafety@springernature.com

Printed and bound by CPI Group (UK) Ltd, Croydon, CR0 4YY
30/09/2026
02225254-0017